FUNDAMENTALS OF
PLANT SCIENCE

Join us on the web at

agriculture.delmar.com

FUNDAMENTALS OF
PLANT SCIENCE

Marihelen Glass

Rick Parker

DELMAR
CENGAGE Learning™

Australia • Brazil • Japan • Korea • Mexico • Singapore • Spain • United Kingdom • United States

DELMAR
CENGAGE Learning™

**Fundamentals of
Plant Science,**
Marihelen Glass and Rick Parker

Vice President, Career and
Professional Editorial:
Dave Garza

Director of Learning Solutions:
Matthew Kane

Managing Editor:
Marah Bellegarde

Product Manager:
Christina Gifford

Editorial Assistant:
Scott Royael

Vice President, Career and
Professional Marketing:
Jennifer McAvey

Marketing Director:
Debbie Yarnell

Marketing Coordinator:
Jonathan Sheehan

Production Director:
Carolyn Miller

Production Manager:
Andrew Crouth

Content Project Manager:
Katie Wachtl

Senior Art Director:
David Arsenault

Technology Project Manager:
Tom Smith

Production Technology Analyst:
Thomas Stover

For product information and technology assistance, contact us at
Professional & Career Group Customer Support, 1-800-648-7450

For permission to use material from this text or product, submit all
requests online at **cengage.com/permissions.**
Further permissions questions can be e-mailed to
permissionrequest@cengage.com.

Library of Congress Control Number: 2008934876
ISBN-13: 978-1-4180-0081-3
ISBN-10: 1-4180-0081-7

Delmar
5 Maxwell Drive
Clifton Park, NY 12065-2919
USA

Cengage Learning products are represented in Canada by Nelson Education, Ltd.
For your lifelong learning solutions, visit **delmar.cengage.com**
Visit our corporate website at **cengage.com**.

Notice to the Reader

Publisher does not warrant or guarantee any of the products described herein or perform any independent analysis in connection with any of the product information contained herein. Publisher does not assume, and expressly disclaims, any obligation to obtain and include information other than that provided to it by the manufacturer. The reader is expressly warned to consider and adopt all safety precautions that might be indicated by the activities described herein and to avoid all potential hazards. By following the instructions contained herein, the reader willingly assumes all risks in connection with such instructions. The publisher makes no representations or warranties of any kind, including but not limited to, the warranties of fitness for particular purpose or merchantability, nor are any such representations implied with respect to the material set forth herein, and the publisher takes no responsibility with respect to such material. The publisher shall not be liable for any special, consequential, or exemplary damages resulting, in whole or part, from the readers' use of, or reliance upon, this material.

Printed in the United States of America
1 2 3 4 5 6 7 12 11 10 09 08

Dedication

To my family, husband Ron and our daughters, Holly, Heidi, and Alyssa for their encouragement and support. Especially Ron for his patience and understanding during these years and to my father for showing me the wonderful world of horticulture when I was a child.

Marihelen Glass

To Sam, our special son, who taught me the most important lessons in life.

Rick Parker

CONTENTS

Preface xi

Author's Acknowledgments xiii

About the Authors xiv

PART I: 🌱 Plants and Nature

Chapter 1   Why Plant Science? 2

*Objectives • Key Terms • Plants and Nature •
Economic Importance of Plants • Aesthetic and
Recreational Significance • Plants in Science and
Technology • The Scientific Method • More on the
Scientific Method • Plants and Society • Summary •
Something to Think About • Suggested Readings •
Internet*

Chapter 2   Plants and Ecology 18

*Objectives • Key Terms • How Plants Affect Ecology •
Climatology • Biosphere • Oxygen-Carbon Dioxide
Balance • Cycling in the Ecosystem • Trophic Levels •
Ecological Succession • Recolonization • Summary •
Something to Think About • Suggested Readings •
Internet*

Chapter 3   Biomes 48

*Objectives • Key Terms • Terrestrial Biomes • Aquatic
Biomes • Summary • Something to Think About •
Suggested Readings • Internet*

PART II: 🌱 Form and Structure

Chapter 4   The Basic Design I 78

*Objectives • Key Terms • Vegetative Morphology and
Adaptations • Root Morphology/Function • Water
and Nutrient Absorption and Conduction • Stem
Morphology/Function • Leaf Morphology/Function •
Vegetative Reproduction • Summary • Something to
Think About • Suggested Readings • Internet*

**Chapter 5   Design Basic II: Morphology
and Adaptations of Reproductive
Structures** 109

*Objectives • Key Terms • Flower Morphology • Fruit
Morphology • Seed Morphology • Development:
Seedling to Adult • Monocots and Dicots • Botanical
Names • Supermarket Botany • Summary •
Something to Think About • Suggested Readings •
Internet*

**Chapter 6   The Inside Story: Molecules
to Cells** 139

*Objectives • Key Terms • Molecules of Life • Basic
Cell Structure • Prokaryotic and Eukaryotic Cells •
Organelles and Other Inclusions • Summary •
Something to Think About • Suggested Readings •
Internet*

Chapter 7   Growth: Cells to Tissue 153

*Objectives • Key Terms • DNA Replication • Protein
Synthesis • The Gene • Cell Division: Mitosis and
Cytokinesis • Tissues • Tissues and Development •
Anatomy of the Leaf • Summary • Something to
Think About • Suggested Readings • Internet*

Chapter 8   Wood 193

*Objectives • Key Terms • Wood Structure and
Secondary Growth • Wood Uses • Forests and
Forestry • Summary • Something to Think About •
Suggested Readings • Internet*

PART III: 🌱 Function and Control

Chapter 9   Plant-Soil-Water Relationships 212

*Objectives • Key Terms • Water and Nutrition •
Soil • Water Movement • The Soil-Plant-Air
Continuum • Applications • Hydroponics • Summary
• Something to Think About • Suggested Readings •
Internet*

Chapter 10   Energy Conversions 250

Objectives • Key Terms • Laws of Thermodynamics • Oxidation Reduction • Photosynthesis • Respiration • Implications of Metabolism • Summary • Something to Think About • Suggested Readings • Internet

Chapter 11   The Control of Growth and Development 282

Objectives • Key Terms • Principles of Growth and Development • Limitations of Growth and Development (Stresses) • Plant Hormones • Plant Movements • Summary • Something to Think About • Suggested Readings • Internet

PART IV: Evolution and Diversity

Chapter 12   Sexual Reproduction and Inheritance 312

Objectives • Key Terms • Meiosis • The Angiosperm Life Cycle • Inheritance • Molecular Genetics • Summary • Something to Think About • Suggested Readings • Internet

Chapter 13   Genetic Engineering and Biotechnology 341

Objectives • Key Terms • Biotechnology Defined • Genetic Engineering • Targeting Agriculture • Transgenetic Plants • The Future • "Omics" Revolution • Biotechnology Policy, Public Perception, and the Law • Summary • Something to Think About • Suggested Readings • Internet

Chapter 14   Diversity: Vascular Plants 367

Objectives • Key Terms • Movement to Land • Evolution and Distribution of Vascular Plants • The Seed Plants • Summary • Something to Think About • Suggested Readings • Internet

PART V: Plants and Society

Chapter 15   Putting Down Our Roots 402

Objectives • Key Terms • Beginnings of Agriculture • Earliest Crop Plants • U.S. Agricultural Crops • Summary • Something to Think About • Suggested Readings • Internet

Chapter 16   Vegetables 451

Objectives • Key Terms • Scope of Vegetable Production • Categorizing Vegetables • Steps in Vegetable Production • Site Preparation • Planting Vegetables • Cultural Practices • Harvesting • Storing Herbs • Summary • Something to Think About • Suggested Readings • Internet

Chapter 17   Small Fruits 474

Objectives • Key Terms • Grapes • Strawberry • Blueberry • Raspberry • Blackberry • Currant and Gooseberry • National Grower/Horticultural Organizations • Summary • Something to Think About • Suggested Readings • Internet

Chapter 18   Fruit and Nut Production 498

Objectives • Key Terms • Fruit and Nut Production Business • Types of Fruits and Nuts • Selecting Fruit or Nut Trees • Site and Soil Preparation • Planting • Cultural Practices • Harvesting and Storage • National Grower/Horticultural Organizations • Summary • Something to Think About • Suggested Readings • Internet

Chapter 19   Flowers and Foliage 513

Objectives • Key Terms • Flower Business • Value of Floriculture • Flowering Herbaceous Perennials • Flowering Annuals • Bulbs • Flowering Houseplants • Commercial Greenhouse Environment • Summary • Something to Think About • Suggested Readings • Internet

Chapter 20   Forage Grasses and Sod 538

Objectives • Key Terms • Forage Grasses • Managing Forages • No-Till and Minimum-Till Seeding • The Sod Farm • Summary • Something to Think About • Suggested Readings • Internet

Chapter 21   Plants of Medicine, Culture, and Industry 574

Objectives • Key Terms • History of Herbalism • Medicinal Plants • Herbal Remedies • Psychoactive Plants • Poisonous Plants • Industrial Plants • Summary • Something to Think About • Suggested Readings • Internet

Chapter 22 Modern Agriculture and World
Food: Why Plant Science? 605

*Objectives • Key Terms • The Mechanization and
World Food Supply • Food Commodities • Historical
Perspective • The New Agriculture • The Farmer's
Bargaining Power • The Limits of Production • The
Future of Agriculture • Gardening: Number One
Leisure Activity • Social and Political Considerations •
Summary • Something to Think About • Suggested
Readings • Internet*

Appendix 637

Glossary 645

Index 671

PREFACE

A quick glance at a plant scientist's bookshelf might give the impression that there are enough books about plants and plant science available to satisfy all student competency levels and teaching styles. However, many of the words and most of the content in those texts seem to be directed toward those teaching the course, not the students and their learning styles. Furthermore, these same texts fail to project excitement and the importance of plants in daily life. After 35 years of teaching, we believe that the subject can be taught in an interesting way and an applied format. Awakening the student to the importance of plants in day-to-day living is an achievable goal.

This text was written for two different types of courses: (1) A course for students from many diverse backgrounds who are grouped together (such courses are typically beginning biology, agriculture, and horticulture majors as well as students from other fields of study) and (2) A course intended for students who are not plant science majors and who use this course to fulfill a science requirement or to satisfy their curiosity about plants.

A horticulture major would be intimidated if marched into a voice class and asked to sing something on the first day, and many students are apprehensive speaking about or even listening to lectures on anything to do with the sciences, introductory or not. Students do appreciate a challenging but a manageable menu in a course. Although chapters may be rearranged to tailor a course to the objectives of an instructor, there is much to be said for a front-to-back approach in an introductory course. Such a text is more likely to provide a logical, building sequence for the students as concepts gradually unfold. The *Fundamentals of Plant Science* is written in such a sequence.

The development of this text is based on the approach that has been found most successful in teaching introductory courses. This approach considers three basic elements: organization, balance of information, and depth of coverage.

ORGANIZATION: Approach the subject from a most familiar to least familiar sequence of subject matter. From experience, students (especially those already apprehensive) can be drawn into the subject best by first covering the material that is most understandable and real to them. This text proceeds from general to specific concepts, introduces ecological and community-related material first, and then discusses the whole organism. Finally, it considers cellular level structure and function and diversity, returning to matters of ecological significance.

BALANCE OF INFORMATION: Applied information is integrated throughout the basic plant science material. Students can understand the most basic and detailed information, and this is more palatable when some measure of applied information and examples are included. Material is also easier to retain when it relates to one's own experiences and to the everyday world.

DEPTH OF COVERAGE: Present the basic plant science information in appropriate depth and coverage. Even though all pertinent information is included and makes a special point to be accurate, this text is not intended to be a compendium or an encyclopedic reference in all concepts. What to include and what to delete is always a matter of choice, and this text was written with the student in mind. The finished work has resulted in a text with more than enough traditional material for majors but without the overabundance of detail found in some books. An introductory text need not be a reference source for the professor, but rather an educational tool for the student.

The philosophy followed in the organization of topics, the depth and breadth of coverage, and the incorporation of applied material throughout the book are all aimed at helping the student learn. This approach has been more successful in introductory courses than the traditional hierarchical organization (i.e. molecules, cells, tissues, organisms, ecology).

No longer does the philosophy that states that the capable students will learn the information and our only job is to cover everything thoroughly and carefully. Rather, it is the professor's obligation to help every student develop an understanding of how plants interact, grow, reproduce, function, and relate to the student's daily life. That understanding will promote information and decision making concerning the role of plants in a functional world.

The goal of this book is to present the basic plant science information in a depth appropriate for beginning majors, while providing enough applications to keep the nonmajor interested. The last several chapters at the end of the book show how plants influence society stressing the applied aspects of plant science while possibly deleting some of the classic material.

Finally, the design is to develop a text that can be covered in a single semester; instead of providing far more information than could possibly be taught and leaving it to the professor to select what to delete. This textbook provides a challenging one-semester course that includes extra material from which supplemental information may be drawn.

The overall theme of this book encompasses ecological and applied components. The need for the enlightened management of plant resources is a large part of the message. Another is that humans need to function within the natural framework of all biological species.

Also Available for the Instructor:

Instructor Resource to Accompany Fundamental of Plant Science
ISBN: 1-4180-0081-7
ISBN 13: 978-1-4180-0081-3

This powerful electronic resource is an essential classroom tool. This thorough instructor resource includes everything you need to prepare for, teach, present and test the concepts of plant science.

Instructor's Guide includes Overview and Summary, Objectives, Key Terms, Chapter Outlines, and Something to Think About answers.

Computerized Testbank containing more than 900 questions with answers and internet testing capability.

PowerPoint Presentations containing over 350 slides that focus on each chapter's key points to facilitate classroom presentations.

ACKNOWLEDGEMENTS

First, I appreciate the valuable input and patience of my husband Ron, and our three daughters, friends, colleagues, and students, especially William Harrison and Alexandria Wofford who helped with the digitizing of the photos, plus Plant Materials II Class, 2007, who were willing to help in any way, while I was writing this book. I would also like to acknowledge Julia Blizen and Nancy Cavanaugh of Cove Creek Gardens, Greensboro, North Carolina. They have shared photographs, information, and support through this entire work.

I want to particularly thank Rick Parker, who without his help this would still be just a dream, and Christina Gifford with Delmar Cengage Learning, Jolynn Kilburg, and Tiffany Timmerman with S4Carlisle Publishing Services. Thanks Rick, Chris, Jolynn and Tiffany.

In addition, Delmar Cengage Learning and the authors would like to thank the following individuals for their invaluable review of the manuscript throughout its development process:

Richard Harkess
Mississippi State University

Harrison Hughes
Colorado State University

W. Dennis Clark
Arizona State University

Gail A. Baker
Lane Community College

ABOUT THE AUTHORS

Marihelen Glass

MARIHELEN GLASS completed her undergraduate degree at Texas Tech University, Lubbock, Texas, and completed her MS and PhD in Floriculture and Horticulture, respectively, at Texas A&M University, College Station, Texas. She taught plant materials and urban horticulture classes at Texas A&M University. At Texas Tech University, she taught in the Department of Biological Sciences, teaching botany, plant taxonomy, and plants and man. She also taught classes in the Department of Plant Science; they were plant materials, greenhouse production, and landscape horticulture design. A move to North Carolina gave Dr. Glass the opportunity to teach horticulture classes at North Carolina A&T State University, where she was program coordinator for the Landscape Architecture and Horticulture Programs. Dr. Glass was instrumental in introducing biotechnology in the curriculum of seven university departments with funding provided by the North Carolina Biotechnology Center. She had the first Plant Biotechnology Laboratory on the NCA&T campus. Dr. Glass has received 10 teaching awards, including one from the North Carolina University System, one from the American Horticulture Society, one from The American Association of Nurseryman's and one from the American Society for Horticultural Sciences and ASHA Southern Region to name a few.

She and her husband (Ron) are the parents of three daughters and one son.

R. O. Parker

R. O. (RICK) PARKER completed his undergraduate degree at Brigham Young University–Utah and he completed is PhD in Reproductive Physiology at Iowa State University. After two postdoctorate positions and a stint as a coauthor with M. E. Ensminger, he served as a division director and instructor at the College of Southern Idaho (CSI) in Twin Falls for 19 years. As director, he worked with faculty in agriculture, information technology, drafting, marketing and management, and electronics. Dr. Parker also taught computer, biology, and agriculture classes at CSI. Currently, he serves as the director for AgrowKnowledge, the National Center for Agriscience and Technology Education, a project funded by the National Science Foundation. Additionally, he is the editor for the peer-reviewed *NACTA Journal*, which focuses on the scholarship of teaching and learning. He is the 2008 recipient of the NACTA (North American Colleges and Teachers of Agriculture) Distinguished Educator Award.

Dr. Parker is also the author of these other Delmar texts: *Aquaculture Science* (second edition), *Introduction to Plant Science, Introduction to Food Science, Plant and Soil Science: Fundamentals and Applications,* and *Equine Science* (third edition).

He and his wife (Marilyn) of 38 years are the parents of eight children and grandparents to nineteen.

Plants and Nature

PART **1**

Why Plant Science?

The study of plant science is one discipline of science that can help people understand the various changes in the environment in which they live. Plants are so essential in our daily life, and we need to understand them better.

Objectives

After completing this chapter, you should be able to:

- Express an awareness of how plants survive in nature
- Relate the importance of plants in daily life
- Realize the esthetic and recreational significance of plants
- Describe the scientific method
- Discuss what the future holds for plant and man's relationship

Key Terms

ecosystem	oxygen	scientific method
photosynthesis	respiration	natural law
communities	sociobiologist	
carbon dioxide	fact	

"PRAIRIE BIRTHDAY" BY ALDO LEOPOLD

Every July, I watch eagerly a certain graveyard that I pass in driving to and from my farm. It is time for a prairie birthday and in one corner of this graveyard lives a surviving celebrant of that once important event.

It is an ordinary graveyard, bordered by the usual spruces, and studded with the usual pink granite or white marble headstones, each with the usual Sunday bouquet of red or pink geraniums. It is extraordinary only in beginning triangular instead of square, and in harboring, within the sharp angle of its fence, a pinpoint remnant of the native prairie on which the graveyard was established in the 1840's. Heretofore unreachable by scythe or mower, this yard-square relic of original Wisconsin gives birth, each July, to a man-high stalk of compass plant or cut leaf Cutleaf Silphium, spangled with saucer-sized yellow blooms resembling sunflowers. It is the sole remnant of this plant along this highway, perhaps the sole remnant in the western half of our county. What a thousand acres of Silphium looked like when they tickled the bellies of the buffalo is a question never again to be answered, and perhaps not even asked.

Figure 1-1

Leopold's beloved sunflower.

This year I found Cutleaf Silphium in their first bloom on 24 July, a week later than usual; during the last six years the average date was 15 July. (See Figure 1-1.)

When I passed the graveyard again on 3rd of August the fence had been removed by a road crew, and the Cutleaf Silphium cut. It is easy now to predict the future; for a few years my Silphium will try in vain to rise above the mowing machine, and then it will die. With it will die the prairie epoch.

The Highway Department says that 100,000 cars pass yearly over this route during the three summer months when the *Silphium* is in bloom. In them must ride at least 100,000 people who have "taken" what is called history, and perhaps 25,000 people who have "taken" what is called Horticulture. Yet I doubt whether a dozen have seen the Silphium, and of these hardly one will notice its demise. If I were to tell a preacher of the adjoining church that the road crew has been burning history books in his cemetery under the guise of mowing weeds, he would be amazed and uncomprehending. How could a weed be a book?

This is one little episode in the funeral of the native flora, which in turn is one episode in the funeral of the flora of the world. Mechanized man,

Figure 1-2

As seen from outer space, the planet Earth may look like a blue world, but in truth it is a green world powered by energy from the sun.

oblivious of floras, is proud of his progress in clearing up the landscape on which, willy-nilly; he must live out his days. It might be wise to prohibit at once all teaching of real horticulture and real history, lest some future citizen suffer qualms about the floristic price of his good life (Leopold, 1949).

Aldo Leopold, one of history's great conservationists, and a cofounder of The Wilderness Society, wrote the preceding essay out of concern not only for *Silphium* but for all plants and their importance to our world. Likewise, this book is concerned with helping you to develop an appreciation for plant science—the study of plants—and the impact of plants on your life. Leopold's statement that the teaching of plant science should perhaps be prohibited was made tongue-in-cheek. He meant to convey that studying plant science fosters an awareness of the crucial role of plants to the functioning of the world.

We appreciate plants for their contributions both to natural beauty and to our recreational pursuits in homes, gardens, parks, and wilderness areas. They enhance our enjoyment of life. However, other roles of plants come to mind; for example, plants are important for shelter, clothing, and diet. Wood is the world's most common building material, cotton is one of our basic fabrics, and fruits and vegetables are part of a balanced diet. The utility of plants in everyday life should not be underrated.

The most fundamental value of plants, however, lies in yet another dimension. They are vital members of our **ecosystem** (see Figure 1-2). Plants produce the oxygen that we breathe approximately every 5 seconds. Also, by the process of **photosynthesis**, they convert light energy to food energy. This benefits not only human consumption and sustenance, but also other members of the animal kingdom, who are in turn food sources for each other and for us. Just how crucial is the role of photosynthesis in the function of the world? *No other organisms, including humans, would exist on earth were it not for plants and other photosynthetic organisms.*

A Sand County Almanac, with other essays on conservation for Round River by Aldo Leopold.
Oxford University Press.

🌐 Plants and Nature

The role of a single species, let alone a single plant, in the balanced functioning of any natural habitat can be overlooked easily. Leopold's awareness that this lone *Silphium* plant was a remnant of the native prairie and a relic of original Wisconsin demonstrates his understanding of natural **communities**. Before clearing and plowing for cultivation were undertaken, the prairie species provided food for the wild animals of the region. From the buffalo to the smallest field mouse, diverse and numerous plants and animals found there coexisted in balance. Even ranching activities placed an unnatural burden on part of the native flora, resulting in permanent changes in the number of species and their densities. There are precious few native prairies, or even small plots, still intact in North America. Their use to human society as valuable pasture or farmland has irretrievably altered their floristic composition. Unfortunately, the same circumstance is true in all the major prairies of the world. Forests and even deserts are now being irreversibly changed according to the short-term interests of human society. The importance of plants in the balanced functioning of the biological world cannot be overemphasized. Even though plants may have applied uses and cultural importance in the development and maintenance of the human population, they act most significantly in the natural balance of all biological energy and of atmospheric oxygen and carbon dioxide.

The source for all energy available to living organisms is the sun. The warmth provided by the sun's rays make the earth habitable, and in plant tissues sunlight energy is converted to food in the form of carbohydrates. Green plants, then, are the primary producers of food for the rest of the biological world, food that is subsequently converted to growth energy nutrients from the soil and **carbon dioxide** from the atmosphere in a process called photosynthesis. One of the end products of this process is plant tissue, which can be used as food by animals. Another end product of this energy conversion is the production of **oxygen**.

The two important atmospheric components for life on earth are oxygen and carbon dioxide. Animals must have a constant supply of oxygen to carry out the metabolic process that converts food tissues into a form of energy that allows muscle contraction, brain functioning, new growth, and ultimately the perpetuation of the species. Plants, too, need oxygen to carry on their life processes. The source for all atmospheric oxygen is the same plant process that converts light energy to carbohydrates—the process of photosynthesis is the most important chemical process known to humankind.

Since **respiration**, the utilization of food for producing a form of energy to do biological work, results in the release of carbon dioxide into the atmosphere, the balance of these essential gases is maintained.

Plants carry out both processes, releasing both oxygen and carbon dioxide into the atmosphere, whereas animals are unable to produce oxygen. Water is an ingredient in both processes, cycling through plant, animals, the atmosphere, and soil.

Thus, if the natural functioning of plants is disrupted, the flow of energy, the exchange of oxygen and carbon dioxide, and the availability of fresh water are all affected. These balances are essential; disturbing them ultimately affects the very organisms that have brought about changes in the natural balances—humans.

In his forward to *Sand County Almanac,* Leopold summarized the cause for these changes, "We abuse land because we regard it as a commodity belonging to us. When we see land as a community to which we belong, we may begin to use it with love and respect," which is a true appreciation of what our responsibilities are within the natural standing of the plant world.

Economic Importance of Plants

Although Leopold made no direct allusion to any potential economic value of *Silphium,* his description of "saucer-sized yellow blooms resembling sunflowers" brings to mind an economically valuable relative. Prior to the commercial development of the giant Russian sunflower for seed and oil, it too was only a common prairie weed (*Helianthus annuus*). This, of course, is true for all economically valuable plants; at one time the native plants that gave rise to today's crops were of no greater value than the *Silphium.*

The development of agricultural crops is the foundation on which civilization was built and one of the key factors in the continuation of successful human population. Modern agriculture produces an incredible amount of food annually, but with continued growth of the world population further increases are necessary, as shown in Figure 1-3. From approximately 700,000 plant species known to exist, over 95% of all food consumed by humans comes from less than 20 species, and over 80% of the food consumed is from only six plant species. How many potentially valuable plant species are yet to be discovered and developed, and how many are no longer available to fulfill such potential?

Figure 1-3

As population levels increase, food supply becomes more precarious. Food must be produced on a constantly larger scale, or localized food supplies will fail to feed the multitudes. Courtesy of USDA Photo Gallery.

The natural world still contains a vast array of different kinds of organisms, yet each day sees the extinction of additional species. No one can adequately predict the consequence of that loss to the world. It is quite possible that some other mowing machine in some other graveyard has just destroyed a plant with the genetic potential for becoming a major world food crop. It is equally possible that there are many plants still plentiful that have undiscovered potential. These plants need to be found and studied.

Plants influence both conscious and unconscious decisions, even though most of us are unaware of their presence until we really need them, as at mealtime. The entire economic stability of many nations depends heavily on importing and exporting plant products. As the world's human population continues to grow, the economic importance of plants and plant products becomes even more significant.

Every time we pay for groceries we buy, the economic importance of plants for food is clear—sometimes painfully so. From the produce aisle with its array of fresh fruits and vegetables through the canned goods to the bakery items, essentially everything we buy is directly or indirectly produced from plants or plant extracts (see Figure 1-4). The meats and dairy products would be unavailable without adequate feed for the domesticated animals from which they come. Even the cash register tape, the boxes, or sacks in which our groceries are placed, and the cash used to pay for these purchases are plant products.

Figure 1-4

Shopping at the supermarket is a constant reminder of the economic and ecological importance of plant science.

Figure 1-5

The fossil fuels so important to the technological world come from plants that lived several hundred million years ago. Courtesy of Bureau of Land Management.

In winter, many tons of firewood provides heat throughout the world. Additional energy comes from organic materials that lived many millions of years ago, as shown in Figure 1-5. Essentially all of these organisms were plant materials at one time. Thus, products from plants are available to us in the form of synthetic fibers, plastics, and other materials manufactured from petroleum by-products.

The opium poppy shown is Figure 1-6 is one of the plants that provides two types of drugs, a widely used pain reliever and the hard core drug heroin. Synthetic pharmaceuticals are based on natural compounds first found in plants, bacteria, or fungi. The two most widely used pain relievers, morphine and aspirin (salicylic acid), are such compounds. Socrates was put to death with juice from the hemlock plant, and the Roman emperor Nero ruled after his father, Claudius, was assassinated with poison mushrooms and monkshood juice. Curare-tipped arrows and blowgun darts have aided many native tribes in the Amazon jungles in centuries of successful hunting. *Cannabis,* the genus that produces hemp fibers for rope and the hallucinogen marijuana, is a well-known member of the plant world.

Figure 1-6

Some plants produce drugs that have been used and misused by humans throughout recorded history. The opium poppy, Papaver somniferum, *contains a mixture of opiates from which one of the world's most important pain relievers, morphine, and the world's most seriously abused illegal drug heroin are produced.*

Many plants have industrial applications as well. Their impact on the historical development of modern civilization is vast. Natural rubber is from a tropical tree, and from another tropical species comes the only source for carnauba wax, the hardest of the natural waxes for polishes and industrial application uses. Many of the finest lubricating oils for industrial machinery are found in plants, and various other plant extracts are important to a broad spectrum of industrial applications. Seeds from the jojoba, a desert shrub, yield a substitute for the sperm whale oil with which it shares almost identical chemical properties. Fibers for clothing, string, twine, and canvas come from plant sources, the most widely used being cotton. Dyes, essential oils for linoleum, plastics, soaps, and many other products are also derivatives of plants and plant parts. Thus for food, forage, fiber, fuels, medicine, and varied industrial uses, plants have by far the greatest economic impact of any group of living organisms.

🔆 Aesthetic and Recreational Significance

Leopold's aesthetic sense must have been acute indeed. In the introduction essay, his finely tuned appreciation for such a simple scene as the decorative yellow *Silphium* blossom against the backdrop of a country graveyard rings out. Leopold imagined that scene in another era, when millions of *Silphium* plants must have created the illusion of large, yellow inland seas. In our parks and botanical gardens, plants provide panoramic beauty. On a somewhat smaller scale, they provide shade and allow us to bring nature into our homes, businesses, and schools.

Sociobiologists contend that humans are genetically adapted to be comfortable in "natural" settings, but their immediate environment has evolved to include ever-increasing surroundings of metal, concrete, and plastic. Many people have learned to spend most of their time in such "unnatural" surroundings without any outward signs of physical or psychological discomfort. Others have not. In any case, a great many find enjoyment and relaxation when they are able to "return to nature" by camping, picnicking, sitting under a tree, walking barefoot in the grass, or backpacking as you can see in Figure 1-7. But many urban dwellers enjoy such opportunities infrequently, if at all. Thus, state and national parks and wilderness areas understandably are chosen more and more often as vacation spots (see Figure 1-8). As you can see in Figure 1-9, the natural beauty and integrity of such areas must be preserved if they are to continue to provide the relaxation and recreation people seek there, in other words, we are loving our parks and natural areas to death.

Among the multitude of recreational activities humans have devised, and one of the most consistently popular, is golf. The proximity of the trees, hills, water, and grass of a golf course gives the feeling of bringing in a natural setting without having to invest hours of travel to reach it. Walking and jogging through city parks are also popular forms of recreation that can bring the participant in closer contact with plants and nature. Even gardening and weekend yard work is a popular recreational activity that provides a much welcome involvement in the plant world.

We humans also attempt to bring nature into our plastic worlds by "greening ourselves in" with houseplants and landscaping our living spaces. The popularity of houseplants in working and living

Figure 1-7

Crowded urban living leads people to seek a temporary escape. The population density of recreational areas is sometimes greater than in the city itself, but the appeal of communing with nature continues to attract greater numbers of people.

Figure 1-8

Scenic view of Eldorado National Forest, California. The wilderness defies a unique natural beauty.
Courtesy of National Park Service.

Figure 1-9

Spring in bloom at an arboretum in your area. This is a quality of life that is important to us all. Landscaping provides relief from the hurried and disorderly world.

environments supports the theory that humans are genetically adapted to an environment in which we coexist with other living components of nature. There is even considerable evidence that mixing soil, pruning, watering, fertilizing, and tending to living plants is an excellent form of therapy, a subtle means of communing with nature by substituting houseplants for the larger and less accessible natural world (see Figure 1-10).

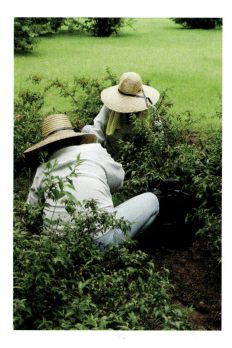

Figure 1-10

Gardening can be therapeutic. Horticultural therapy is recognized as an important tool in treating mental illness and emotional problems of people. Courtesy of Guilford County, Cooperative Extension.

An understanding of how biological, geological, and environmental factors combine to produce the beauty of the natural world maximizes our appreciation and enjoyment of them, but it also helps us to understand how such beauty can best be maintained and preserved.

Plants in Science and Technology

The term *science* permeates every aspect of modern society. Food, consumer products, and even ideologies are scientifically designed, produced, and developed. There are natural, physical, political, and even social sciences. The scientific label has become a symbol of credibility, and yet a larger segment of the human population still perceives science as a secretive and/or a mystical activity. Scientists are often depicted as rather strange white-collared people with exceptional IQs who converse only in five-syllable words and mathematical formulas, often with beeping, blinking computers. Scientists are a diverse group of people observing, experimenting, and looking for new technologies and solutions to world problems.

Heredity and environment combine to shape the unique course of each of our lives. Similarly, our unique circumstances determines the type of associations we have with other parts of the biological world. Whatever our life's pattern, each of us will always have one or more relationships with the plant kingdom, be it as a consumer of shelter, clothing, and food, a guardian of houseplants and gardens, a wilderness lover, an agriculturist, or perhaps a biologist. In any event, an understanding of individual plants, and plant groups and their relationship to the rest of the biosphere, affords us a viewpoint that will enhance our associations with plants.

But how does one understand plants? How does one know about any aspect of the physical and biological world? Most of our understanding is based on previous observations and discoveries of others. This text, for example, contains information based on a systemic assemblage of information from prehistoric to present times, organized into facts and theories. A **fact** is an idea that is known and specific. A theory is an idea expanded from facts via reasoning but is not currently known to be entirely or universally true. The investigation process that incorporates reasoning and observation, with the objective of reaching theory or fact, is called the scientific method.

Leopold's careful observation of the flowering dates for *Silphium* would have been missed by most of us. Who cares if flowering occurred on July 24, a week later than usual? Why did he make such an observation? Leopold was practicing science, that is, he was increasing his understanding of an event by using the scientific method. To this purpose he probably kept a diary describing personal observations in nature.

The Scientific Method

The scientific method always starts with a certain base of facts—what is known. In observing the flowering dates of *Silphium,* Leopold may have been collecting an information base from which he then could have generated a theory about the general flowering time for *Silphium.* Or, perhaps already having a hypothesis or testable assumption about why the flowering time for *Silphium* was later than usual, he might have been seeking out a cause and effect relationship for flowering time. The goal of the scientist is to observe with accuracy so both reasoning and observation come into play. Science is the systematic accumulation of knowledge through the use of logic based on factual evidence. It is both an organized body of facts and a method of problem solving. The scientific method is the process of acquiring factual knowledge and for solving problems.

Briefly, the **scientific method** is a repeated sequence of events that includes the following steps:

- Recognition of a problem.
- Establishment of a hypothesis to explain or solve the problem.
- Careful observation to gather factual information relative to the problem.
- Reconsideration of the hypothesis in light of the accumulated facts.

- Testing of the hypothesis to establish repeatability and thus validity of the facts.

The scientific method is not to be regarded as inflexible or infallible. The important parts of an attempt to approach anything scientifically are accurate observations and objective interpretation. The number of observations made, the number of different observers involved, the analysis of the accumulated facts, the questions asked, and the sequence in which all of this transpires are not the same from one situation to the next. The basic goals, however, are the same; observation, hypothesis formation, testing, and repeatability to confirm validity all must be present for science to function properly.

When a theory or hypothesis has been repeatedly tested many thousands of times with the same results, it can be said to be a **natural law**. Few theories are considered laws, since there is a lack of sufficient experimental evidence to prove that there are no exceptions to the predicted results; one of the laws of nature is that of gravity. It can safely be predicated that, on the earth, every time an object is dropped it will move toward the center of the earth. To date there are no exceptions to this law; even through it has been tested in countless different ways.

Science does not attempt to determine what is "good" or "evil." Nor does it have a moral purpose, or does it attempt to convince anyone of anything. Science is simply an approach to establish what is and what is not factual. Science is not a world only for the scientist; everybody can and does practice science.

Approaching an everyday situation scientifically is a matter of being objective and following a logical sequence through to the end. Comparison shopping for the best value for your dollar is essentially a scientific process of deciding among many possibilities that are closest to satisfying one's needs. Science in this context is basically an attempt to remove rumor and hearsay from the process, replacing them with facts and objective evaluation of reliability, performance, and durability. Actually, purchasing an item is even a continuation of the process; using the item in question can be viewed as an experiment. The owner makes observations and comparisons of actual and claimed performance levels, which establish scientifically the validity of the manufacturer's claims.

So science is not mystical or limited to unique intellect or personality. Science is basic and understandable in design, although many of the facts and hypotheses are specific to each area of specialization. It is the framework in which knowledge is accumulated. Biology is the study of all living organisms; plant science is the study of plants. More and more specialized subdisciplines delve into greater understanding of organisms, furthering the accumulation of knowledge.

More on the Scientific Method

Nonscientists seem to believe that the scientist is a person with secret means of obtaining knowledge to benefit humankind. Explanations put forth by research scientists may be wrong as often as they are right. And not all discoveries directly benefit humans. Indeed, some seem completely useless or detrimental. Some individuals take another view of the scientific method. They infer that science is simply doing one's best with one's mind with no limits to creative approaches. This view indicates that the means used by scientists in solving problems are not unique to science. In fact, the scientific method can be used by anyone to solve a variety of problems.

There are thought processes that have to be used, when working with the scientific method. Experimental results are used, when necessary, to modify a hypothesis and to test predictions, but the cycle of testing really never ends. One could say it is a process of continuous improvement.

HOW TO USE THE SCIENTIFIC METHOD

The scientific method begins with a hypothesis, and then moves forward by testing the hypothesis. A hypothesis is simply a tentative explanation—a guess—put forth to account for a set of observations. Sometimes hypotheses have been called "educated guesses."

A hypothesis is a tentative statement or assumption that is made in order to be tested. To formulate a hypothesis is to make a testable prediction about the relationship between variables. A hypothesis is usually stated before any sensible investigation or experiment is performed because the hypothesis provides guidance to an investigator about the data to collect. A hypothesis is an expression of what the investigator thinks will be the effect of the manipulated (independent) variable on the responding (dependent) variable. Hypotheses can usually be formed as an "if...then" statement.

The concept of hypothesis testing is basic to all science, and it is also the most misunderstood by the public. The word *prove* should not exist in the scientist's vocabulary, since an infinite number of examples are needed to prove a hypothesis.

No matter how much evidence we gather to support a specific hypothesis, we can never be certain that the same data would not equally support any number of unknown alternative hypotheses.

After forming a hypothesis, a scientist proceeds by designing and performing experiments. The primary purpose of scientific experimentation is to test hypotheses. So, any hypothesis selected by a scientist to explain a natural phenomenon must meet a very important requirement: It must be testable.

Plant scientists rarely deal with cases in which every prediction made by a hypothesis turns out to be correct. The question then becomes: How many or what proportion of a given number of predictions must be verified in order to make the hypothesis a useful one? For this reason, experimental data are subjected to a statistical analysis—mathematics used to determine whether deviations from the pattern that is predicted by the hypothesis are significant.

Plants and Society

The twenty-first century has suddenly thrust on both developed and underdeveloped countries complications not ever dreamed of by our forebears. Events that now affect our daily lives would have sounded like scientific fiction in the 1940s or 1950s. Technological advancement has developed at an incredible rate causing some sociologists to wonder if it would not have been better had certain advancements never been made. As developed countries take advantage of the luxury of goods and services, all at the expense of tremendous energy consumption, underdeveloped countries, seeing the splendor, suddenly rush to share it. The world's transportation and communication systems have decreased at a phenomenal rate; we can know "what's going on" virtually anywhere on the globe. What does all this have to do with plant science? The realization is that conventional food, fiber, forage, and energy resources are not inexhaustible. This has forced world leaders to consider where all the above-mentioned resources came from in the first place. Fortunately, the sun continues to supply sufficient radiation to keep the earth warm, but that is not enough to guarantee future human success. As long as the sun does provide sufficient heat and light, the role of plants to the very existence of humans on this planet remains critical. The interdependence of plant and animal life, especially human life, is and always will be of primary ecological importance. It is now crucial for average citizens to understand botanical implications so they can make intelligent decisions concerning the future. These are no easy decisions, either for the politician or the voter. Some of today's social problems are overpopulation, inadequate food supply, and a reduced quality of life as you can see in Figure 1-11.

Figure 1-11

Overcrowding in human society detracts from the quality of life. Social problems can increase when population densities reach high levels. Courtesy of U.S. Fish and Wildlife Service.

Does one nation with food have the right to dictate philosophy to another nation and use that food as a political weapon? Is it worth the environmental risk necessary to develop a supply of nuclear energy? Does one state have the right to take water from another state (nation) that has an abundant supply but might need it in the future? Are you willing to pay a great deal more for environmental monitoring to ensure that the water you drink, the food you eat, and the air you breathe are safe?

In a botanical sense, there exists a serious set of questions: How can we grow enough food to feed a world population two, three, or four times as large as our current one? What do we do if the quality of both drinking water and irrigation water becomes so poor that the animals and plants being grown no longer produce effectively? What do we do if the air becomes so poor that plants will no longer grow?

These are not hypothetical questions but rather real, current concerns faced by today's leaders. In a democracy, only a qualified electorate can make those decisions. As part of that electorate, voters need to understand that knowledge plays a serious part in the botanical world and everything that depends on it.

Summary

1. Plants are critical to continued human existence for aesthetic, economic, and ecological reasons.

2. Plants are the producer organisms for the world, converting sunshine into chemical energy for all organisms while releasing oxygen into the atmosphere. Ecosystems function properly only as long as nature is in balance.

3. Plants are the producers for all living organisms. Those, which grew millions of years ago, provide the world with fossil fuels, and modern plants provide humans and animals with food, shelter, medicine, fibers for clothing, and thousands of industrial products.

4. Plants permeate our conscious and subconscious lives in the form of nature, from which we have never totally separated ourselves. Parks, wilderness areas, and both interior and exterior landscaping remain at the center of human attempts to bring tranquility and beauty into a hurried, competitive, synthetic world.

5. Scientific inquiry begins with recognition of a problem, followed by establishment of a hypothesis, observation, reevaluation of the hypothesis, and finally testing of the hypothesis for repeatability and validity.

6. National and worldwide food, energy, and water shortages are problems that must be faced by both governmental leaders and citizens. As informed voters, those of us living in a democracy have the opportunity to help protect our botanical world.

Something to Think About

1. How can we grow enough food to feed a world population two, three, or four times as large as it is today?

2. What do we do if the quality of both drinking water and irrigation water becomes to poor that the animals and plants being grown no longer produce effectively?

3. What do we do if the air becomes so polluted that plants will no longer grow?

4. List how plants are used in everyday life.

5. Why are our national parks being loved to death?

6. Explain the scientific method and how it is used.

Suggested Readings

Biello, D. 2007. Many plants can adapt when climate goes against the grain. *Scientific American.*

Holland, A., and K. Rawles. 1993. Values in Conservation, *ECOS, 14*(1).

Leopold, A. 1949. *A sand county almanac.* New York: Oxford University Press.

Leu, A. 2004. Organic agriculture can feed the world. *Acres USA. Plants in daily life.* 2005. London: Royal Horticulture Society.

Internet

Internet sites represent a vast resource of information. The URLs for Web sites can change. Using one of the search engines on the Internet, such as Google, Yahoo! Ask.com, or MSN Live Search, find more information by searching for these words or phrases: **scientific method, hypothesis, testing, repeatability, and aesthetic and recreational significance of plants: Aldo Leopold, aesthetic significance of plants, recreational significance of plants, scientific method, hypothesis formation, testing, validity, and urban living.**

CHAPTER 2

Plants and Ecology

Why do specific plants grow where they do? Worldwide climate patterns, the water and nutrients cycle important to plant growth, and balances responsible for plant distribution are some considerations. All these topics are essential to understanding life and living organisms. The earth is the correct size and distance from the sun to allow life, as we know it, to exist. The size of the earth is important because it provides a gravitational force that holds an atmosphere sheath comprising a unique mixture of gases, including water vapor. Perhaps the most critical factor is the 150 million-kilometer distance from the sun, which allows an optimum amount of heat and light energy to reach the earth's surface. The possibility that these conditions exist on other planets has been investigated within the field of exobiology.

After completing this chapter, you should be able to:

■ Present the relationship of worldwide climate to plant growth

■ Explain the term *ecology*

■ Describe the limiting factors for the growth of plants

■ Discuss the cycling of the ecosystem

■ Illustrate the food web and food pyramid

■ Describe ecological succession and recolonization

Key Terms

ecology	evolution	methane
abiotic	parasitic	nitrogen fixation
climatology	commensalism	cyanobacteria
condensation	mutualistic	nitrite
humid	symbiotic	autotrophic
circulation cells	biomes	heterotrophic
evaporation	community	herbivore
transpiration	population	carnivores
windward	trophic	omnivores
leeward	carrying capacity	biomass
rain-shadow desert	compensation point	primary consumers
diurnal	food chain	scavengers
intensity	percolate	phytoplankton
daylength	turgidity	zooplankton
visible spectrum	organic matter	ecological succession
biotic	greenhouse effect	seral
biosphere	nitrogen gas	climax community
ecosystem	nitrate	succession
niche	ammonia	recolonization
habitat	decomposers	
competition	inorganic nitrogen	

How Plants Affect Ecology

In recent years, conflicts between resource exploiters and conservationists have led to a public awareness of **ecology**, even though many of the vocal advocates of both points of view are poorly informed of what ecology really means. The term is derived from the Greek *oikos*, meaning "home." Thus, in a broad sense the home is simply the habitat for all living organisms, and ecology is a study of the factors that allow organisms to grow, compete, reproduce, and perpetuate the species. It is a study of the total environment and interrelates with many traditional fields of plant science.

A more specific question can be asked: Why do the producer organisms—green plants—live there? Here the answer is much more complicated. This question can be answered by analyzing the **abiotic** (nonliving) environment. One plant may live where it does because the light, water, temperature, nutrients, soils, relative humidity, wind, and other factors are satisfactory for growth and reproduction of the species. But mere survival is not enough; the plant must have sufficient quality of growth to allow *reproduction and perpetuation of the species*. Some time humans settle for less. For example, the cotton plant grows from year to year in the warm environment of the tropics, but agriculturists have seen fit to move the cotton plant to a more temperate climate where it cannot survive the winter. The farmer's only concern is with an economic yield (fibers and seeds from the fruit). It makes no difference that frost in late fall kills the plant, providing that the fruits (cotton bolls as in Figure 2-1) have matured. The farmer will simply plant cottonseed again next spring.

Figure 2-1

Mature cotton boils. Cotton is a major crop plant in temperate climates throughout the world.
Courtesy of USDA-NRCS Photo Gallery.

Buying a tropical foliage plant for your home or office, on the other hand, is a completely different situation. With varying degrees of success, you will attempt to care properly for your plant year round. Those of us who have had our houseplants die anyway may feel that we are not blessed with the innate ability to grow plants; we lack a 'green thumb.' In fact, we may actually feel that we are afflicted with the brown thumb syndrome; somehow, mysteriously, the ultimate death is not incurable. In fact, the trick to successfully growing plants is no trick at all; rather, it is an understanding of the combined effects of water, nutrition, light, humidity, temperature, and pest control. Houseplants are not specifically bred for inside growth but are really outdoor plants that are well adapted to their native environment, usually a tropical one. When plants are introduced; into areas that are genetically adapted, their needs go unsatisfied. Therefore, we must compensate artificially for the requirements of houseplants—the better we can modify their immediate environment, the more successful we will be in ensuring their healthy growth. The more we know, the greener our thumbs will become.

WHAT IS AGROECOLOGY?

Our language produces many new words. For example we understand *agronomics, agronomy,* and *ecology* but what about *agroecology?* K. H. W. Klages, an agronomist at the University of Idaho, is credited as one of the first to discuss ecology and agriculture. The term *agroecology* appeared in the late 1970s. It started from the recognition that green revolution-era agroecosystems were highly dependent on inputs such as pesticides, capital-intensive machinery, and specific seed varieties.

According to various sources, agroecology is the application of ecological concepts and principles to the design and management of sustainable food systems. Agroecology infers the linking ecology, socioeconomics, and culture to sustain agricultural production, farming communities, and environmental health. It is holistic in its approach meaning it is comprehensive and integrated when considering all elements of the many systems associated with agriculture.

The methods of agroecology have as their goal achieving sustainability of agricultural systems balanced in all spheres. This includes the socio-economic and the ecological or environmental.

While farming methods vary, traditional manipulated "agroecosystems" generally differ from natural ecosystems in six ways: maintenance at an early successional state; monoculture, crops generally planted in rows; simplification of biodiversity; intensive tillage, which exposes soil to erosion; use of genetically modified organisms; and artificially selected crops. Agroecology tends to minimize the human impact.

An agroecosystem is a key idea in agroecology. Agroecosystems are defined as semidomesticated ecosystems that fall on a gradient between ecosystems that have experienced minimal human impact and those under maximum human control, for example cities. They are generally defined as novel ecosystems that produce food via farming under human guidance.

Practitioners of agroecology are called agroecologists, and they take a critical view of modern industrial agricultural techniques, seeing the

industrial model as fundamentally or radically (at its roots) unsustainable. The agroecologist views any farming system primarily with an ecologist's eye; that is, it is not firstly economic (created for a commodity and profit), nor industrial (modeled after a factory). In fact, agroecosystems are both understood and designed following ecological principles. For example, integrated pest management aims to control problematic pests through the introduction of other species, not application of pesticides to kill that pest. A common example of this would be intercropping to attract beneficial insects within rows of a given plagued crop. The insects would balance the disturbed ecology represented by the pest, thus, eliminating unsustainable practices such as increasingly intensified pesticide use.

Climatology

The factors of precipitation, temperature, and light combine to provide most of the abiotic environment that controls worldwide plant distributions. Collectively these represent the climate to which a given green plant species must be adapted to survive. The study of climate, **climatology**, is essential to a basic understanding of where plants do and do not grow.

Precipitation

Of the climatological factors, water is undoubtedly the most important. Precipitation, which includes all forms of moisture deposited on the earth's surface, is the source of essentially all fresh water.

Water vapor, although always present to some extent in adequate amounts in the lower layer of the atmosphere, must exist in adequate amounts if **condensation** into water droplets or ice crystals is to occur. The droplets or crystals then must attain a sufficient weight to fall to the earth as precipitation. For condensation to occur, the air containing the water vapor must cool to a temperature below its condensation point, the temperature at which water vapor condenses into a liquid state. The condensation point varies with the amount of water vapor present and the air temperature at which the vapor is being held.

Air is generally warmed by radiant heat from the sun, particularly near the equator and at lower elevations. Warm air is capable of holding more water vapor than is cold air; thus, at and near the equator and at lower elevations, the air is warm and **humid** (full of water vapor). Warm air rises, cooling as it does so, and condensation occurs.

A complex set of events serves to produce high elevation air currents moving north and south, from the equator toward either pole. These **circulation cells** do not reach the poles, however, but descend at about 30° north and south latitude. This descending air has already lost most of its moisture as it nears the earth's surface. As it warms, its capacity to hold moisture increases, enabling it to absorb water vapor near the ground and produce a drying effect on the earth. This cell completes its circulation by moving back across the earth's surface to the equator and outward toward either pole. As the air travels, it continues to warm (especially air moving from 30° to the equator), and absorbs water vapor from the soil, ponds and streams (**evaporation**), and plants (**transpiration**).

As a result of the movement of these circulation cells and their effects on precipitation, most of the earth's natural deserts occur between 20° and 30° north and south latitudes (see Figure 2-2). Not all land between these two latitudinal belts is desert, and there are desert areas outside these belts. These anomalies occur because the precipitation affects the air circulation patterns in these cells, which are modified by topography, elevation, proximity to the coast, and ocean current temperatures.

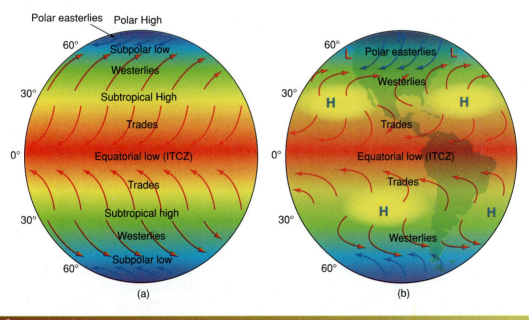

(a)   (b)

Figure 2-2

The circulation of air cells toward and away from the earth's surface produces bands of precipitation and dryness between the equator and approximately 60° north and south of latitude. Cold has more effect on precipitation patterns than do air cells.
Courtesy of National Weather Service.

A coastal mountain range has a significant effect on precipitation patterns. Warm, moist sea air moves inland and is forced upward in elevation by a mountain range; it cools as it rises, causing condensation of water vapor to turn into rainfall. The ocean (**windward**) side of the mountains therefore received considerable precipitation as a result of the same physical phenomenon described for air cell patterns. As this now cold, dry air moves over the range and descends the inland (**leeward**) side of the mountains, it begins warming as it descends, and its capacity to hold moisture increases. As it moves farther inland, the dry air takes moisture from the land and the plants, producing a **rain-shadow desert**—like the leeward side of the Andes Mountains in Argentina.

Generally, large continental landmasses are dry in their interior because cold air masses collide with the warm, moist air from the sea. This forces the warm air to rise, and the condensation causes precipitation to occur before the moisture can reach very far inland. Even though no mountains may block the flow, the effect is similar to a mountain-induced rain shadow. Also, cold temperatures greatly modify precipitation patterns, so higher elevations tend to receive less rainfall.

Although the effects of air cell circulation patterns do produce increased precipitation near 60° north and south latitudes, where air again rises away from the earth's surface (see Figure 2-2) as one moves farther north and south from 60°, the effect of temperature, especially cold, becomes increasingly important. North and south of approximately 60° is in fact the most significant factor influencing precipitation.

Temperature

Maximum and minimum annual ranges and **diurnal** (daylight) fluctuations in temperature produce another important set of climate conditions to which plants must be adapted if they are to survive in a given area. Because of the relative position of the earth to the sun, the angle of incidence for sunlight and the daylength results in the equator being warmer year-round than any other zone. As one moves north and south away from the equator, the annual temperature fluctuations become progressively more extreme, as shown in Figure 2-3a. The sun's rays that strike the earth's surface most perpendicularly do not necessarily result in higher temperatures. This is because of another unique property of water. Water (including water vapor) gains and loses heat very slowly. An area of high humidity, therefore, has less extreme temperature fluctuations than do areas with low humidity.

Light

Two considerations are important when discussing light as a component of the climate affecting plant distribution and activity: both light **intensity** and **daylength** can have strong regulatory effects on plants.

Of the total radiation produced by the sun, light is only that portion known as the **visible spectrum**. The intensity of this light varies with latitude and season, and is greatest at the point the earth is most perpendicular to the rays of the sun at any given moment. Light intensity is progressively reduced from the point, as the rays become more oblique. In addition, not all the light emitted by the sun reaches the earth (see Figure 2-3b). Of the light energy that arrives at the outer limits of our atmosphere, only about half manages to get past the reflection and absorption by dust, clouds, water, and other gases, including the important ozone layer, which absorbs most of the ultraviolet radiation. Some of the light reaching the earth's surface is reflected from rocks.

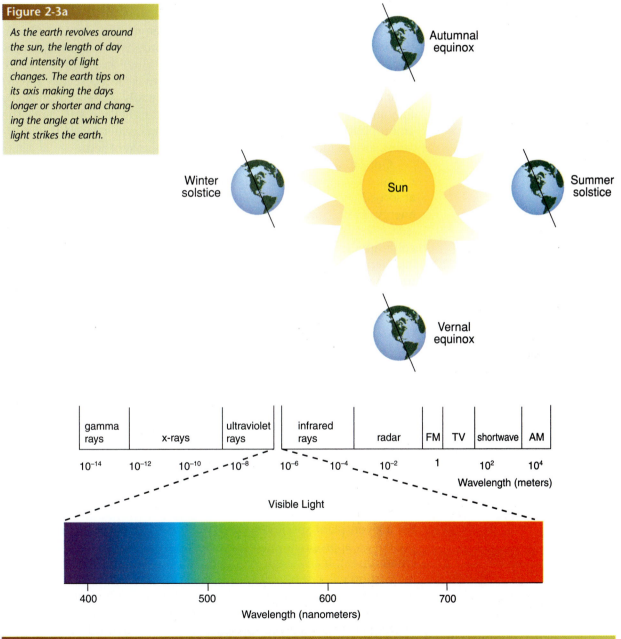

Figure 2-3a

As the earth revolves around the sun, the length of day and intensity of light changes. The earth tips on its axis making the days longer or shorter and changing the angle at which the light strikes the earth.

Figure 2-3b

The portion of the total spectrum of light visible to humans is only a small range of wavelengths. Courtesy of National Weather Service.

Light

~~Water,~~ is absorbed as heat or, in the case of green plants, converted to the chemical energy of carbohydrates (the process of photosynthesis). More than half is in the infrared region of the spectrum and warms the earth. About 7% of the ultraviolet radiation manages to get past the ozone. The human eye sees light beginning at the short wavelength of violet and then in successively longer wavelengths as blue, green, yellow, orange, red, and finally fading at the beginning of the infrared region. The visible spectrum as shown in Figure 2-3b, represents only a tiny fraction of all radiation, which includes the very short waves (gamma rays and X-rays) and very long waves (radio waves).

Visible light contains the entire wavelength absorbed by photosynthetic pigments. In particular, the chlorophyll molecule absorbs light at wavelengths of about 4,330 nm (blue) and 660 nm (red). Wavelengths capable of exciting a chlorophyll molecule may have a great deal of energy (short, blue light) and easily cause excitation, or they may have less energy (long waves, red light) and still cause the chlorophyll molecule to be excited. The consequences of that excitation are included in the discussion of photosynthesis.

Our eyes perceive sunlight as white light, simply because it is a mixture of all colors of the rainbow. Yet specific wavelengths are absorbed and reflected differently; shorter wavelengths have more energy. Leaves are green because they have chlorophyll molecules that reflect green light while absorbing red and blue light; thus, any color appears to the human eye as that color because light of that wavelength is being reflected. If all wavelengths are being absorbed, the object appears black; if all wavelengths are being reflected, the object appears white (see Figure 2-4).

Daylength varies by season and according to latitude. As discussed in the preceding section on temperature, during the winter the sun's rays hit the earth at a more oblique angle. This produces both a shorter day and less light intensity (and less heat). There are actual effects of changing daylength in plants relative to hormonal control of plant activity. In general, however, plants become less active during the shorter days of winter. With the gradually longer days of early spring, increased light intensity and warmer temperatures combined to initiate new activities in most plants. Reproduction and other development phenomena are affected.

Limiting Factors

For a given plant species to successfully inhabit an area, it must have present in appropriate quantities all the necessary components of its environment. If any single requirement is insufficient, then the species in question will not be able to survive. Whichever environmental component causes the organism to fail is said to be the limiting factor.

Even though any of the requirements could be theoretically the limiting factor for a given species, only the one that actually prevents the species from surviving is so designated. Usually one of the abiotic factors, especially water or temperature, is the most likely candidate.

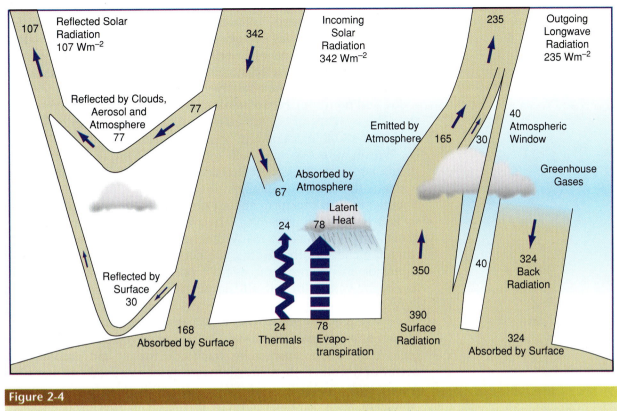

| 107 | Reflected Solar Radiation 107 Wm^{-2} | 342 | Incoming Solar Radiation 342 Wm^{-2} | 235 | Outgoing Longwave Radiation 235 Wm^{-2} |

Reflected by Clouds, Aerosol and Atmosphere 77

77

Emitted by Atmosphere 165

40 Atmospheric Window

30

Greenhouse Gases

Absorbed by Atmosphere 67

Latent Heat 78

24

78

350

40

324 Back Radiation

Reflected by Surface 30

168 Absorbed by Surface

24 Thermals

78 Evapo-transpiration

390 Surface Radiation

324 Absorbed by Surface

Figure 2-4

Only part of all the sun's rays reaches the earth's surface, and some of that is reflected back into space. Courtesy of National Weather Service.

For example, many plant species are limited from growing in a desert because there is not enough water. In fact, if irrigated, many of these plants could survive well in a desert. Others, however, would not be able to survive even with supplemental water because of the high summer temperatures. For them, excessive heat would be the limiting factor. Other plants are limited from growing in more northern latitudes because of freezing winter temperatures that they cannot endure.

Many other abiotic and **biotic** (biological) components can be limiting factors. Nutrients, space, soil characteristics, and damaging interaction with other living organisms sharing the area are among the probabilities controlling plant success or failure. Other components of the total environment contribute to the success or failure of given organisms in various settings; they are considered in following sections.

Biosphere

Our planet has often been referred to as "Spaceship Earth," traveling through space with a critical atmosphere, critical energy input, and finite resources. Ecologists think of our entire earth as a giant system—the **biosphere**. Any portion of the earth that represented a relatively closed system in terms of nutrient cycling, energy input, and therefore a definable set of plants, animals, soils, and climate may be defined as an **ecosystem**. It may be very small, as in a backyard pond with a few organisms, or it may be very large, as in a deciduous forest or grassland. The size of the ecosystem is in the eye of the beholder, and its limits must always be a relatively homogeneous area made up of living and nonliving components that have something in common; then the system can be studied as a unit. The ecosystem concept is useful for all biological study.

Within an ecosystem, each species has a unique set of requirements, a spot in which it is better adapted than competing organisms. **Niche** is the term that denotes this position within the ecosystem. The niche of an organism should not be confused with the habitat in which the organism lives. The physical **habitat** is much more general and can include many different species, each with its own unique niche. In other words, the concept of niche includes every aspect of an organism successfully positioned within the habitat. The niche would include all the physical habitat characteristics of temperature, water, soil, and light plus all possible interactions with other organisms in the ecosystem. When two individuals of different species have some of the same specific requirements, they are sharing a part of the same niche. Such coexistence can theoretically occur for a while if there are abundant resources available, but ultimately they will be in **competition** with each other for this specific need. The more similar the niche of these two species, the sooner they will find themselves in competition with one another.

It is easier to visualize animals than plants competing for the same resource, but the term is still appropriate. Plants compete for water, light, nutrients, space, and other components in their habitat by being better designed to acquire those needs. The depth or total volume of the respective root system, the vertical growth rate and leaf production, and the abilities to withstand climatic and herbivore stress all enable one species to compete more successfully than another for common needs. The results of direct competition depend on the situation. When their respective niches are almost identical, one species could succeed while causing the other to become extinct. When their respective niches are less similar, one species may succeed in one area while the other species is superior in an environmentally different area. Such a situation might allow both species to survive, but with reduced total ranges. A third possibility is rapid **evolution** in divergent directions. Each of the two competing species would have individuals selected that were less directly competitive; their respective niches would become less similar. Whatever the end results, no two species can share exactly the same niche in the same place.

Other interactions between individual of different species are not competitive. In such situations, each organism functions in some way. If organism A benefits from its interaction with the organism but in the process affects B negatively, it is a **parasitic** relationship. For example, dodder (*Cuscuta*) is a parasitic vine that derives its nutrition from whatever host plant it lives on. The host plant ultimately dies in such a relationship.

Commensalism is an interrelationship in which one organism benefits and the other is unaffected. Some plants require partial shade to survive. Growing under a tall tree does not provide the necessary shade and the tree benefit nor does this association harm it. When both individuals benefit from an interrelationship, they are said to have a **mutualistic** or **symbiotic** relationship. A classic example of mutualism is the lichen, an organism composed of fungus and an alga (although some evidence indicates that the fungus benefits more than the alga).

In addition to plant-plant interaction, there are many plant-fungus, plant-insect, and plant-animal interrelationships that exemplify these systems. Some plant-herbivore (plant-eating animals) interactions significantly affect the plant's total environment. Certainly the most obvious is the plant being a food source for the herbivore. This is not considered parasitic, however, because a parasite usually lives off its host for a long time before irreversible damage is done. There are also plant-animal interactions in which the plant causes a negative effect on the animal: there are toxic plants, plants with spines or thorns, and plants that provide shelter to one animal that in turn attacks herbivores that could damage its home. Some plants have more complex interrelationships and interdependencies with a wide variety of other organisms.

The likelihood of understanding all the possible niche interactions of any single species is improbable, so true comprehension of total ecological balance within even a single habitat therefore is impossible. Studying selected components shared by the many species within a given area, however, is worthwhile and provides ecologists with some basis for comparison of the general relationship. Therefore, even though the earth is extremely diverse in terms of numbers of species and numbers of individuals within a species, there are good reasons to consider the planet as a whole. Sometime broadscale conclusions about changes in gas concentrations in the atmosphere increase or decrease in temperature, changes in precipitation patterns, and other serious problems can be viewed only on a global basis. Ecologists have chosen to divide the world into a few major vegetation types based primarily on effective precipitation, temperature, and soil. These specific groups of ecosystems, representing recognizable types of vegetation that are remarkably stable with time (sometimes over hundreds or thousands of years) are called **biomes.**

Within a particular area in an ecosystem, all the living components are collectively referred to as a **community**. The term may encompass a bit more than what we refer to as a human community, since from an ecological standpoint the community represents all the plants, animals, and microorganisms living together in an area. Within that community, there may be groups of organisms, all of the same species, which constitute a **population**. A community might have a population of bluegrass, a population of rabbits, a population of foxes, and a population of grasshoppers. Each population could have only the number of individuals that its **trophic**, or feeding, level and food supply would support.

This limit on the number of organisms that a given area can support without causing degeneration of the area is termed the **carrying capacity**. For example, the carrying capacity of a specific area of prairie may be 1,000 field mice, 20 deer, and 1 coyote. In practical use, ranchers need to know how many sheep or cattle given sections can support without overgrazing the land, which could cause a smaller carrying capacity in subsequent years.

Many of the specific components of an ecosystem cycle go through set cycles within the normal functioning of the system. These cycles are all in balance at a worldwide level, but it is possible to throw them out of balance if the natural habitat is significantly altered. Overgrazing is only one of many such negative alterations possible.

Oxygen-Carbon Dioxide Balance

The earth's atmosphere is made up of a mixture of nitrogen (N_2), oxygen (O_2), carbon dioxide (CO_2), water vapor (H_2O), and a number of other gases of lesser importance. Practically all these substrates are critical for life processes, and the recycling of both oxygen and carbon dioxide is absolutely essential to almost all living organisms. The air around us contains only about 21% oxygen and 0.33% carbon dioxide. Essentially all the remainder is nitrogen (about 78%). The opposing process of photosynthesis, which produces sugars, and respiration, which allows those sugars to be used for energy, also involve oxygen and carbon dioxide exchange.

As the photosynthesis process starts with water, carbon dioxide and sunlight is producing the sugars, oxygen is given off into the atmosphere. At the same time, the respiration process consumes oxygen, "burns" sugars, and releases carbon dioxide and water. The relative rates of photosynthesis and respiration, then, determine the amount of oxygen and carbon dioxide in the atmosphere. All organisms carry on respiration and give off carbon dioxide, but only green plants carry on photosynthesis and produce oxygen. Therefore, green plants must reach an acceptable balance between the two processes. If the amount of carbon dioxide released is exactly equal to the amount of carbon dioxide consumed, the plant is said to be at the **compensation point**. Such plants cannot accumulate materials and thus do not grow. In agriculture, for example, it is important that photosynthesis far exceeds respiration if crops are to be productive. The balance of atmospheric oxygen and carbon dioxide is a critical interrelationship in which plants (which photosynthesize and respire) play the central role. The **food chain**, photosynthesis, respiration, and other organismal interactions are not isolated events. Their interrelationships and their responses to environmental factors are complex, so a broad understanding of the entire system is necessary.

By 1772, Joseph Priestly had discovered that if an animal, such as a mouse, were placed in a closed container, it would die after a period of time. If a green plant were placed in the same container and even if the container were glass so that sunlight could enter, the plant would also die. But if both the mouse and plant were placed in the container at the same time, they would coexist. If the gas balance (O_2 and CO_2) were correct, they could theoretically live in this condition indefinitely. A closed terrarium represents a similar system. Green plants within it consume carbon dioxide and synthesize sugars, giving off oxygen: The microorganisms, worms, and other animals living in the soil (or perhaps on the plants) carry on respiration along with the green plants, consuming oxygen and releasing carbon dioxide back to the atmosphere of the terrarium. In effect, a balanced terrarium is an ecosystem.

Cycling in the Ecosystem

The ecological success of an ecosystem depends on its efficiency and stability. A great deal of that efficiency is related to nutrient and water cycling, essential components for all living organisms. If a single factor is missing, the entire system loses efficiency and slows down, and in a small system the balance may be altered permanently. The broadscale ecosystems the biomes comprise have evolved and increased in complexity over millions of years.

Organisms survive, grow, and produce because they have the energy, water, and nutrients to do so. The energy for our entire biosphere is derived from the sun (see Figure 2-5). This energy does not cycle, but it is converted from one to another, eventually ending up as heat, which has little value in the overall functioning of the system. Because the energy is lost as heat, there must be new input every day.

Figure 2-5

Sunlight penetrating the forest floor. Courtesy of National Forest Service.

On the other hand, our supply of water and nutrients is finite. At any given time, a portion of the water and nutrients is tied up in various parts of the system—in the air, the soil, the oceans, or the living or dead organic matter. As water and nutrients are transferred from one part of the system to another, a cycle is eventually completed and begins anew, depicted in Figure 2-6.

Figure 2-6

The water cycle. Water falls as precipitation and either evaporates directly, runs off, or percolates into the soil for plant use. About 99% of all water taken in by plant root systems is released into the atmosphere by transpiration (evaporation from plants). Some of the percolated water penetrates past root systems to replenish underground water supplies.

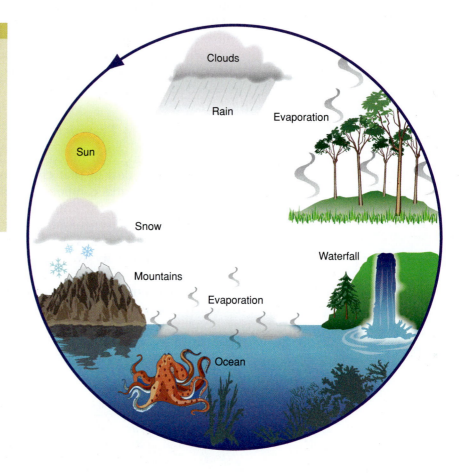

We derive part of our understanding of ecosystem from knowing that water and nutrients cycle within certain physical boundaries. It makes no difference whether we are discussing a desert or a tropical rain forest; the manner in which nutrients cycle is exactly the same, although the rates may differ.

The Water Cycle

On the earth the total amount of water is enormous and essentially constant, but 99.4% of it is composed of salt water and ice found in oceans, inland seas, glaciers, and polar icecaps.

The salt water provides a saline habitat for organisms adapted to those conditions, and it provides a reservoir from which pure water molecules can evaporate. The energy for this process is derived from the sun.

As the surface waters of the oceans are warmed, evaporation occurs and water is moved into the atmosphere. When physical conditions are proper for condensation, clouds occur, and eventually precipitation is produced: rain, snow, hail, or sleet. As described earlier, air currents cause these cells of moist air to move around the earth, and precipitation may fall thousands of kilometers away from where the water was released by the ocean or land mass.

Some precipitation falls on land, a portion of it is evaporated back into the atmosphere, some goes to surface runoff (see Figure 2-7), some **percolates** through the soil to recharge underground water supplies, and a portion is stored in the soil as a water source for plants and ultimately animals.

Figure 2-7

Runoff of water can produce scenic phenomena such as Bold River Falls in the Cherokee National Forest, Tennessee.
Courtesy of National Forest Service.

The plants absorb the water from the soil, transport it through their cells, and eventually release it back into the atmosphere through transpiration. About 99% of all water absorbed by the root system is given up this way. Only a small portion is stored in cells to be used for metabolism and maintaining water pressure, or **turgidity**. At any given time, a very small fraction of the total precipitation is tied up in the living tissues of plants and animals.

Both surface and underground runoff water eventually returns to the oceans. Thus, the problems of water management are not a matter of total supply but simply of having enough fresh water at the right place at the right time. Water is fast becoming our major nonrenewable resource.

The Carbon Cycle

All **organic matter** includes carbon and hydrogen. That carbon is also cycled through the living/nonliving systems in an orderly flow that allows for organic molecules to be constructed, as you can see in Figure 2-8. Carbon comes directly from the atmosphere as carbon dioxide. Earth's atmosphere contains about 0.033% or 330 parts per million (ppm) CO_2. In the comparison with the amounts of nitrogen and oxygen, which dominates the air we breathe, this is a small percentage. It is remarkable that a gas so important should be present in such small concentrations. We shall see in a later discussion of photosynthesis that CO_2 concentrations in these ranges are absolutely critical to the photosynthesis process, and slight reductions in concentration can drastically reduce the fixation of carbon into sugars.

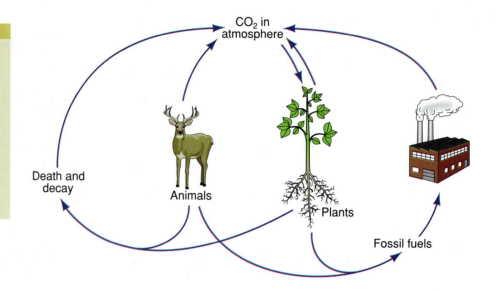

Figure 2-8

The carbon cycle. Carbon in the form of atmospheric carbon dioxide is produced as a by-product of respiration in plants and animals, the burning of fossil fuels and firewood, and even volcanic action. It is used by plants, both natural and cultivated for photosynthesis.
Courtesy of Rick Parker.

In the process of photosynthesis, CO_2 is taken into the green plant and incorporated first into sugars and then later metabolized into all the organic molecules important to life. In the normal process or respiration, some CO_2 is cycled directly back into the atmosphere. Most of it is stored in living tissue, animals in normal food chains consume portions of it, and some occur in dead plant parts such as leaves, flowers, fruits, and seeds.

The level of CO_2 in the atmosphere is relatively constant, although there is considerable concern about its fluctuation. For example, there is good evidence that the level in the atmosphere has risen considerably during the past 100 years, primarily from the additional input from the burning of fossil fuels. In that period of time, the level has increased from 300 ppm to the present 330 ppm. At the current rate of increase, a recent National Academy of Sciences study predicts that the concentration of CO_2 in our atmosphere will double by the year 2020. Thus far the oceans have acted as an effective buffer by absorbing excess CO_2 as carbonates, but there is considerable controversy about when this buffering capacity might be overridden. If this were to happen, the CO_2 in the atmosphere would rise dramatically, cause the sun's rays to be trapped at the earth's surface (the so-called **greenhouse effect**), and cause the temperature of the earth to rise. Studies indicate that only a 2° or 3°C increase in the earth's temperature would cause the polar icecaps to melt and result in serious ecological and economic damage. Such melting would raise the level of the oceans by as much as 5 m, according to some estimates, and significantly change the outline of our continents by submerging many millions of hectares of coastal cities.

Other sources of carbon released into the atmosphere include volcanic eruptions and the weathering of rocks, processes that are relatively stable over geologic time and about which we can do little. From the standpoint of the photosynthetic process, an increase of CO_2 concentration in the atmosphere should not be a problem. In fact, CO_2 enrichment of the environment in greenhouses and other confined spaces is now practiced commercially to improve yields. The ecological consequences of such an event worldwide, however, would far override any benefits derived from additional photosynthesis in nature.

The Nitrogen Cycle

The air we breathe is about 78% **nitrogen gas** (N_2). As we shall see later, nitrogen is an extremely important element in the building of organic molecules, but most plants and animals have absolutely no way of incorporating atmospheric nitrogen gas. Instead, plants obtain nitrogen as **nitrate** (NO_3) or **ammonia** (NH_3) from the soil. Once nitrate or ammonia nitrogen has been taken up by the plant, it is converted onto organic matter and becomes a part of living cells. If the plant is eaten by an animal, the organic nitrogen is partially converted into new organic matter in the animal. Whenever the animal dies, the **decomposers** break down the organic matter into **inorganic nitrogen** in the soil. Most of this nitrogen will be recycled in the form of nitrate ammonia; some will be volatilized into the atmosphere in a form that produces the pungent odor of decay. Bacteria present in the soil are capable of converting nitrate to ammonia, and vice versa. Environmental conditions such as moisture, temperature, pH, and oxygen in the soil determine the balance of the two compounds. As Figure 2-9 shows, the cycle begins as plants take up nitrogen from the soil.

One way to provide these nutrients is to make them synthetically, usually involving **methane**, a fossil fuel. There was a time, not too many years ago, when fossil fuels were so inexpensive we felt that essentially all of our nitrogen fertilizer needs could be met this way. Now the energy crisis has forced us to look more seriously at natural **nitrogen fixation**. Many bacteria and **cyanobacteria** are capable of converting atmospheric nitrogen into inorganic nutrients in the soil that plants can use directly or that can be converted by other microorganisms into usable forms. They may do so as part of their independent, natural metabolism as free-living, nitrogen-fixing

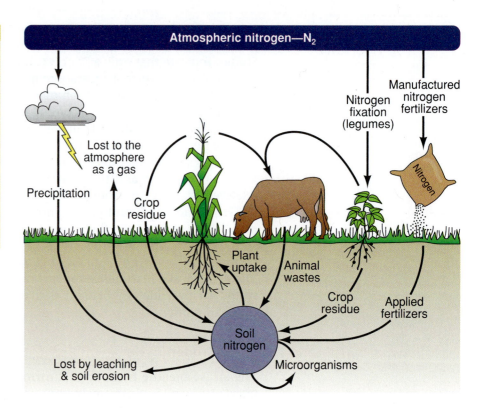

Figure 2-9

A. *The nitrogen cycle. Nitrogen occurs as atmosphere gas (N), in the bodies of plants and animals, and in the soil as organic molecules from the decomposing action of soil microorganisms.* **B.** *The cycling of the various compounds is shown schematically in relation to where each compound is located.*

Courtesy of Rick Parker.

microorganisms, or in a mutually beneficial relationship in the roots of a higher plant as symbiotic, nitrogen-fixing bacteria. It has been known for many years that many legume (bean family) plants become infected with certain kinds of bacteria that cause a nodule (tumor) on the roots of those plants (see Figure 2-10).

The nodules become infected with certain kinds of bacteria that have the ability to fix nitrogen from the air into NH_3. In return for this "free lunch," the plant provides the bacteria with certain organic

Figure 2-10

When members of the legume family are symbiotically infected by bacteria, root nodules form, nitrogen fixation in these nodules result. The bacteria convert atmosphere nitrogen to a form that can be taken up by plants. The plant on the right was treated with rhizobium, the left without.

molecules necessary for the bacterial growth. These tumors do not harm the plant and, in fact, are considered highly desirable. Farmers routinely pull up alfalfa or soybean plants to see how many nodules occur on the root system and thereby judge the nitrogen nutrition of the plant.

In recent years it has been discovered that different types of bacteria that do exactly the same thing infect many nonlegume plants. Such plants require essentially no nitrogen fertilization. They are usually found on soils very poor in nitrogen, and they play an important role in the fertility of such soils. Free-living bacteria and cyanobacteria appear to be far more important in nitrogen cycling than was originally believed. For example, the cyanobacteria that inhibit the water and soil of rice paddies in Asia are responsible for much of the nitrogen fertility of land farmed continuously for centuries. Most of these organisms have the ability to become dormant during periods of drought; then they grow vigorously in a matter of hours after receiving moisture, fixing nitrogen at a remarkable rate until water again becomes the limiting factor.

One other source of nitrogen from the atmosphere is important in certain regions. Electrical discharges during thunderstorms are capable of putting **nitrite** (NO_2) into the air; rainfall carries it to the ground, and bacteria oxidize it to nitrate (see Figure 2-11). The burning of fossil fuels may put some ammonia and nitrogen oxides into the air. Ammonia is also soluble in water and may be brought to the ground with precipitation.

Figure 2-11

Lightning can put nitrite (NO_2) into the air, and rainfall can then carry it to the ground. Courtesy of USDA Photo Gallery.

The nitrogen cycles form the atmosphere by several fixation schemes to be captured in the soil or directly by the plant. It is taken up by the plant, and fixed into organic matter, and later broken down by decomposers to the inorganic from where the cycle begins again.

These cycling nutritional materials combine with the abiotic compound of the environment and the energy of the sun to control what kinds of plants grow where. Understanding these interrelationships increases our ability to grow them in other regions.

Trophic Levels

Organisms that make their own food directly from sunshine, carbon dioxide, and water are called **autotrophic** (*auto*, "self"; *tropic*, "feedings"). All green plants are autotrophic. Other organisms, including humans, are said to be **heterotrophic** (*hetero,* "others"). All heterotrophics lack the ability to make their own organic molecules from inorganic substances and light energy. Not only do plants produce their own food, but they also produce tissues that can be used as food by animals. Therefore, autotrophs provide food for the heterotrophs; plants are the producers.

Food Chain

A food chain is a hierarchy of organisms in which the producer organisms, the green plants, are the base. Figure 2-12 gives an example of a food chain. The green plants are consumed by plant-eating animals (**herbivores**), which are in turn consumed by meat-eating animals (**carnivores**). An example of such a simple food chain is a grass plant (producer) that is eaten by a cow (herbivore) that is then eaten by a person (**omnivore**). Consider that many grass plants are required to feed one cow, and the number of grams of grass required for conversion to 1 gm of beef is fairly large; in fact, the

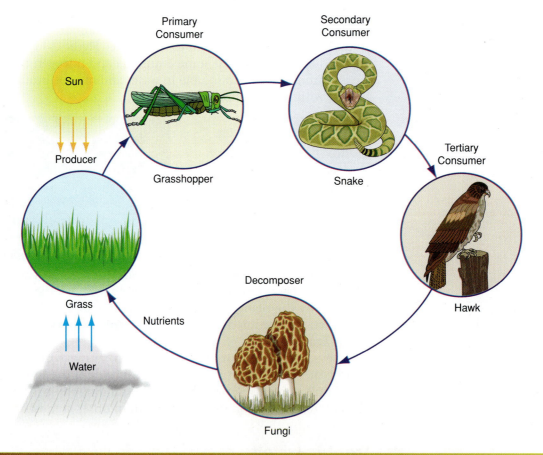

Figure 2-12

A food chain. Plants are the base of the chain. Herbivores (mouse) eat plants, and carnivores eat animals. The snake is a list-level carnivore, the hawk a second-level, or top, carnivore.

ratio is about 9:1 at best. Similarly, it takes at least 9 gm of beef to provide 1 gm of weight gain for an actively growing human child (the relationship becomes much more complicated for organisms that are fully grown).

What is apparent in this food chain is that it is indeed oversimplified. Cows may choose many different kinds of plants for forage, and humans eat many foods other than beef. A great deal depends on availability (many humans in the world have few choices).

We refer to the organic matter present in organisms as **biomass**, an ecological description of the weight of the matter itself. We measure our biomass by simply weighing ourselves, and the biomass of a plant can be determined by weighing both the above ground part and the below ground part. Such information is important in determining how efficiently the energy from the sun is being converted into the chemical energy of organic molecules.

Food Web

For most organisms there is a choice of food, and therefore it makes more sense to think of a food web, which as shown in Figure 2-13 is a hierarchy of consumption in which alternate food sources are represented. The number of organisms available becomes very important because the total energy loss at each trophic level (actually a feeding level) is approximately 90%. Thus only about 10% of the total energy is incorporated (as organic matter) at each trophic level.

Since the green plants are at the first trophic level and are called producers, the herbivores that feed on them at the second level are called **primary consumers**. Animals that eat the primary consumers are the third trophic level and are called secondary consumers. The hierarchy continues until one reaches the top trophic level, occupied by the top carnivore. Nothing feeds on the top carnivore except **scavengers**, which consume the flesh after the top carnivore dies. Finally, decomposers (mostly bacteria and fungi) break down the organic matter into simple inorganic nutrients, which are cycled back to the soil, later to be absorbed by new plants, which start the cycle all over again.

A complex food web is one in which the organisms at each level eat many different species available to them. Such flexibility of food sources develops in nature out of necessity.

In unstable or unpredictable environments, where a given plant species may not be sufficiently available (or even at all) in a particular year, animals that would normally feed on that plant must feed on another species or fail to survive. These interrelationships are true at the secondary and tertiary consumer levels of the food chain. A desert, such as the one in Figure 2-14, is an example of such an environment because the precipitation is so unpredictable, and thus plant growth is not always ensured.

A simple food web, on the other hand, is one in which a high degree of specificity exists in food sources. In many cases, given

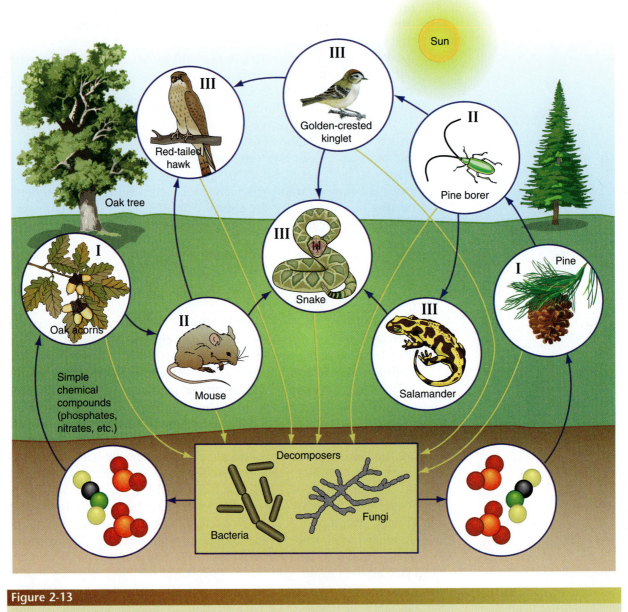

Sun

III Red-tailed hawk

III Golden-crested kinglet

II Pine borer

Oak tree

I Oak acorns

III Snake

III Salamander

I Pine

II Mouse

Simple chemical compounds (phosphates, nitrates, etc.)

Decomposers

Fungi

Bacteria

Figure 2-13

Food webs depict more accurately the feeding relationship between the producer plants (P).

animals feed on only a single plant species (for example, koalas feed only on *Eucalyptus* leaves). In turn, given carnivores may depend exclusively on a single species of animal as their sole food source. Such extreme specialization could have evolved only in a very stable environment, where each plant and animal species is always present in essentially the same abundance. Because these animals have never needed to search for and successfully compete for other food sources,

Figure 2-14

Deserts have unpredictable environments; thus complex food webs develop. Sonoran Desert, Saguaro National Monument near Tucson, Arizona. Courtesy of National Park Service.

they have lost the ability to do so. Their continued existence, therefore, is linked to the future stable supply of their food source.

Probably the most predictable and stable environments exist in the tropical rain forest. In these regions many highly specific ecological interrelationships exist—a simple food web. The elimination of any single species in such environments could well cause a chain reaction affecting several other species. That we are unaware of many of these specific interdependencies is one of the main reasons ecologists and conservationists concern themselves with protecting known endangered species.

Interestingly, another of the most predictable (stable) of the world's environments is the ocean, and yet a very complex food web exists there. The base of the aquatic food chain consists of single-celled algae generally referred to as **phytoplankton** (*phyto*, "plant"; *plankton,* "small, free-floating aquatic organisms"). These are in turn eaten by **zooplankton** (*zoo*, "animal"), single-celled animals that are then eaten by larger aquatic animals. Size and agility in avoiding capture govern the food source here almost entirely. Flavor, texture, and other factors that we find important in foods apparently have little to do with fish diet. The little fish must be large enough to be an enticement to the larger fish, but it must not be so large that it cannot be swallowed. Without benefit of knife or fork, size becomes very crucial in marine diets. Even on other food chains, size is important. Remember the difficulty in chasing a single pea around on a slick plate? It hardly seems worth the effort.

Food Pyramid

The best way to depict these energy relationships is three-dimensionally in the form of a pyramid. The broad base of the pyramid is made up of the first trophic level, the producer organisms with a very large biomass. Generally, that means a very large number of individual plants. The second trophic level is represented by the primary consumers, the herbivores. Fewer individuals and less biomass occur here because the energy conversion efficiency is only about 10%. In other words, 90% of the plant tissue eaten by a herbivore does not go to make animal tissue. Much of the energy lost is in the form of heat, some as indigestible waste that goes to the decomposers. At the third trophic level, decomposers take their toll as some organisms die and are recycled to the soil and air (see Figure 2-15). Even the green plants lose leaves, flowers, and fruit during growth, and the decomposers break them down. A compost pile is an example of this principle.

The pyramid concept is used to demonstrate that very little energy and biomass is left after only three or four trophic levels. Thus, food chains are never very long; the energy losses are simply too great. The lesson to be learned from this generalization is that there is more food for organisms that feed on more than just meat.

The food supply for omnivores such as humans is greatly enhanced if we eat plants rather than animals. If we are willing to shift our diet

Figure 2-15

Decomposition of the forest is a natural phenomena that occurs in all forests over time. Courtesy of National Forest Service.

so that plants constitute a greater portion of our total intake, we have access to more food and the energy costs are much lower. High-priced beef is rapidly bringing this lesson home. This is not to say that all humans should become vegetarians, but modification in diet toward more plant foods would mean additional food for a hungry world.

In summary, the food pyramid clearly shows the impossibility of having a larger top than can be supported by the base. The plant material on earth can support only a finite amount of animal life, including humans. It would be foolish, therefore, to presume that human population can increase beyond a given size without proportionately increasing the available food base. This, of course, is what modern agriculture attempts to do. Continued pressure for ever-increasing productivity requires modifications in the habitats of the crop plants to allow production beyond what natural conditions would yield. Such modifications are not always in the best interests of the surrounding environment. In later chapters, we will look more closely at energy flow and changes in energy from plants as the all-important first step in the sequence.

Ecological Succession

Ecological succession is the sequential replacement of one community type by another through a series of development stages. These are known as **seral** stages until the final community structure is reached. This last stage in succession is called the **climax community** because it is the optimum assemblage of species that the environment can support in the area. The measure to determine whether a climax community has been reached is stabilization of the dominant species; when those species begin replacing themselves rather than being replaced by a new species, the climax community has been achieved.

The environment determines the community structure as the process of change occurs in spite of the fact that the climate patterns remain the same. The change from one community to the next is actually brought on by the modifications produced by each temporary community. As the dominant species alter the area in which they are growing, they actually produce conditions less favorable for themselves and more favorable for a new assemblage of species.

When **succession** occurs in a new or pristine habitat or in one that has not previously had a similar community occurring there, it is called primary succession. The classic example of primary succession is the normal transition of a pond or a bog and then to woodland as it slowly fills with silt and organic material. The south shores of Lake Michigan have been undergoing primary succession for many years as the lake slowly retreats to the north.

Secondary succession occurs when land that has been cleared for pasture or for farmland is no longer maintained. Such abandoned fields slowly revegetate with plants native to the area. The actual species vary with the region, but the first year finds annual weeds dominating the site. In the second year, perennial grasses and some herbaceous perennial broadleaf species join the annual weeds. For the next several years, the grasses dominate, but an ever-increasing number of shrubs begin to appear, as do some tree seedlings. While the trees are slowly reaching maturity, the shrubby species and grasses codominate. The fast-growing tree species shade out the shrubs and grasses as they become large enough to form a dense forest, but they might be replaced in time by slower-growing shade-tolerate deciduous tree species. If a deciduous forest is the dominant climax vegetation, a new group of shrubby plants take their place in the understory, maintained by the fact that deciduous trees do not form a shade canopy over them all year long.

Primary succession in ponds results from the slow filling of the body of water with silt and organic debris until plants can gradually invade from the banks toward the middle. As soon as submerged plants are able to root in the mud near the edge of the pond, the buildup of silt occurs much more rapidly, trapped and held by these plants. As the bottom continues to fill, floating surface plants like water lilies become common near the banks, and the submerged plants move farther toward the middle of the pond. Next, reeds, cattails, and similar rooted emergent plants become established in the deeper accumulation of sediment in the shallow water near the banks of the pond. The filling of the pond occurs much more rapidly now with plants occupying the entire area. Eventually true terrestrial plants establish along the original shallow zones of the pond, now dry land. The center of the rapidly filling pond becomes smaller and smaller, with an accompanying succession of seral vegetative zones, with it is a bog and finally dry land.

Generally, successional events share several common characteristics regardless of the wide variety of localities and plant species, and the rate of species replacement is higher in the initial stages, slowing and stabilizing in the older stages. In addition, the size of the plants and the total biomass increase through the seral stages until the climax community is reached. Finally, the food webs become more complex as the seral stages move toward climax, and thus a higher percentage of organic materials synthesized by the producer organisms are used in the older stages.

Recolonization

Occasionally, a natural phenomenon totally destroys life in an area. The eruption of a volcano in a vegetated area, a devastating forest fire, and the building of a shopping mall parking lot are all examples of such areas. Given time, all of these will again become revegetated through the process of **recolonization** and succession.

An extreme example of a recolonization occurred on the Pacific Island of Krakatau. In 1883, a volcanic eruption destroyed almost half the island and covered the remainder in a thick layer of lava pumice and ash. All forms of plant and animal life were eradicated, but as soon as the substrate cooled, the process of recolonization began. Single-celled algae quickly established in pockets of rainwater caught by the lava folds. Slowly, organic materials combined with ash in these pockets to provide a shallow layer of soil sufficient for hardy species of vascular plants. Gradually, more and more plant species reestablished, and they helped provide more humus for an ever-improving habitat, which contained a greater diversity of life forms. By 1934, only 51 years later, over 270 species of plants occupied the island with a commensurate number of animal species also in full residence.

The eruption of Mount St. Helens on May 18, 1980, in southern Washington, produced large areas denuded by lava flow and ash. The recolonization of this area is being monitored carefully by ecologists. Undoubtedly, succession to a climax forest community will take a long time.

The formation of a new volcanic island would undergo the same sequence of biological habitation, but technically this would be an example of colonization rather than recolonization. The island of Surtsey was formed in the 1960s off the coast of Iceland by volcanic eruption, and within only a few months living organisms had colonized the still warm lava.

Summary

1. The term *ecology* implies a thorough understanding of the biotic and abiotic components of the environment. Organisms exist where they do in nature because the combination of these factors is appropriate for their existence. Crop plants and houseplants need to be grown with the total needs of the plant in mind.

2. Precipitation is the single most important factor in determining plant distribution. The annual distribution of precipitation is dictated by a complex pattern of air circulation cells, which distribute moist air to certain parts of the world. These cells are modified by elevational changes and collisions with other air masses.

3. Temperatures are modified by the amount of water in the atmosphere. Moisture-laden air heats and cools more slowly than dry air. The equator is warm year-round because of the angle of incidence for the sun's rays striking the earth.

4. Only humans perceive a small portion of the entire radiation spectrum emanating from the sun as visible light. Fortunately, most of the harmful ultraviolet radiation is absorbed by the ozone layer in the outer layer of the earth's atmosphere and does not reach the surface. The ozone layers are being depleted by the abundance of pollutants, thus, causing a hole in the atmosphere ozone. Light in the red and blue portions of the spectrum excites the chlorophyll molecule to begin the energy capture process of photosynthesis.

5. Any of the abiotic factors critical to continued existence of an organism could also be the limiting factor for the same organism.

6. Within the earth's total ecological system, the biosphere, balanced ecosystems include many different physical habitats. Plants occupy a specific niche within a given habitat, and competition for resources occurs when different species have similar requirements. Plants and animals also develop specific interrelationships with each other: parasitism and commensalism are very common associations.

7. The earth's atmosphere is primarily nitrogen, oxygen, and a small amount of carbon dioxide. Water vapor also occurs in varying amounts. Photosynthesis puts oxygen into the atmosphere and uses carbon dioxide, whereas respiration uses oxygen and produces carbon dioxide. As long as worldwide photosynthesis and respiration rates remain equal, the relative concentration of oxygen and carbon dioxide are stable.

8. Water and nutrients are important to plant and animal growth cycles. At any given time they may occur in living or dead plants and animal tissue as organic matter, or they exist in the soil atmosphere as inorganic chemicals. Carbon and nitrogen both cycle through such stages. Carbon dioxide levels are critical to the carbon cycle; decomposition plus nitrogen fixation provide the necessary nitrogen for plant growth. The energy from the sun does not cycle but must be constantly replenished.

9. Organisms that obtain energy directly from the sun are called autotrophs or producers. These green plants are eaten by herbivores, which in turn are eaten by carnivores. A group of

organisms that represent the producer-herbivore-carnivore sequence is called a food chain or more commonly a food web. The biomass relationships are best represented in the form of a pyramid.

10. The natural change in community structure is termed *succession.* This process results in the climax community. Primary succession and secondary succession have different points of origin but proceed in a parallel manner. Recolonization is the successional replacement of organisms in a habitat denuded by a natural disaster, such as a volcanic eruption.

Something to Think About

1. Explain how biotic and abiotic components of the environment relate to ecology.

2. What is the single most important factor in determining plant distribution?

3. Identify limiting environmental factors to the growth of an organism.

4. Explain a simple food web and food pyramid.

5. What is ecological succession?

6. Explain the oxygen–carbon dioxide balance.

7. What is recolonization?

Suggested Readings

Billings, W. B., *Plants, man, and the ecosystem* (2nd ed). Fundamental of Botany Series, Belmont, CA: Wadsworth Publishing.

Dale, V. H., L. A. Joyce, S. McNulty, and R. P. Neilson. 2000. The interplay between climate change, forests, and disturbances, *Science of the Total Environment 2*, 201–204.

Dale, V. H., P. Mulholland, L. M. Olsen, J. Feminella, K. Maloney, D. C. White, A. Peacock, and T. Foster. 2004. *Selecting a suite of ecological indicators for resource management.* West Conshohocken, PA: ASTM International.

Kapustka, L. A., H. Gilbraith, M. Luxon, and G. R. Biddinger, Eds. 2004. *Landscape ecology and wildlife habitat evaluation: critical information for ecological risk assessment, land-use management activities and biodiversity enhancement practices.* West Conshohocken, PA: ASTM International.

Kratsch, H. A., W. R. Graves, and R. J. Gladon. 2006. Aeroponic system for control of root-zone atmosphere. *Environmental and Experimental Botany, 55*, 70–76.

Internet

Internet sites represent a vast resource of information. The URLs for Web sites can change. Using one of the search engines on the Internet, such as Google, Yahoo!, Ask.com, or MSN Live Search, find more information by searching for these words or phrases: **ecology, exobiology, abiotic, biotic, prevailing winds, rain shadow desert, diurnal, visible spectrum, daylength, biosphere, commensalism, and ecosystem.**

CHAPTER 3

Biomes

Ecosystems occur in a different geographical setting where latitudinal and elevation differences combine with proximity to mountain ranges and coastline to produce four unique sets of indigenous plant and animal species. Although different, these organisms are all typical of deserts in general. The Chihuahuan, Sonoran, Mojave, and Great Basin deserts of North America, the Sahara, Gobi, Namib, Patagonian, and all the other deserts of the world make up the desert biome. The plant types and the abiotic environmental factors common to all these otherwise unique ecosystems are the basis for describing this biome.

After completing this chapter, you should be able to:

- Understand the biomes in the worldwide ecosystems
- Identify the plants and their relationship to tropical rain forests, savannas, deserts, grasslands, temperate deciduous forests, coniferous forests, tundras, or aquatic biomes

Key Terms

biomes	spines	tundra
terrestrial biomes	ephemeral	permafrost
tropical rain forest	desertification	marine biome
canopy	mesic	littoral zone
epiphytes	grasslands	limnetic zone
holdfast	xeric	profundal zone
lianas	temperate deciduous forest	thermal stratification
thorn forest	chaparral	eutrophic
savannas	taiga (snow forests)	algal blooms
succulents		eutrophication

Biomes are worldwide groups of similar ecosystems. An ecosystem is a balanced and self-perpetuating assemblage of all the living organisms and the nonliving environmental factors in a given area. Ecosystems differ in size according to the geographic limits and climate ranges that control the types of living organisms within them (see Figure 3-1). The two broad categories of biomes include the terrestrial biomes and the aquatic biomes.

Figure 3-1

The vegetative biomes, world view. Courtesy of National Geological Survey.

🌱 Terrestrial Biomes

Only one-fourth of the earth's surface is land, and yet the vast majority of all plant and animal species are found there. Continents range from equatorial to polar latitudes and from sea level to more than 9,000 m in elevation. Climatic and soil differences, combined with ranges in latitude and elevation, result in a phenomenal number of ecological settings for plants and animals to inhabit. Even though this variation results in many different plant types, reasonably accurate and useful assemblages or categories according to vegetation can still be recognized. The number of distinct **terrestrial biomes** varies according to the authority, but most commonly seven are described.

BIOSPHERE EXPERIMENT

Biosphere 2 is a three-acre structure built to be an artificial closed ecological system in Oracle, Arizona, near Tucson. Constructed between 1987 and 1989, it was used to test if and how people could live and work in a closed biosphere, while carrying out scientific experiments. It explored the possible use of closed biospheres in space colonization, and also allowed the study and manipulation of a biosphere without harming the earth's biosphere. The name comes from the idea that it is modeled on the first biosphere—the earth. Funding for the $200 million project came from Edward Bass, a Texas oil and investments businessman.

The project conducted two sealed missions: the first from September 26, 1991, to September 26, 1993, and the second for six months in 1994. In 1995, the Biosphere 2 owners transferred management to Columbia University. Since 1996, hundreds of college students have spent at least one semester at the Biosphere 2 Center campus. The site has its own hotel and conference center. Columbia has since divested itself of all Biosphere-related responsibilities. While the structure is no longer maintained in an airtight state, Biosphere 2 is still open for tours and plans to be open in the future.

Biosphere 2 was an achievement of engineering more than science. The above-ground physical structure of Biosphere 2 was made of steel tubing and high-performance glass and steel frames. The window seals and structures had to be designed to be almost perfectly airtight, such that the air exchange would be extremely slow, to avoid damage to the experimental results. To deal with atmospheric expansion during the daytime and contraction during the nighttime, diaphragms permitted the building to grow larger during the day and shrink at night, thus, keeping the atmospheric pressure inside constant.

Whether considered a success or a failure, an interesting consequence of the experiment is that it showed the difficulty of copying the functions of the earth's biosphere with human knowledge and technology. Despite an expenditure of over $200 million, this attempt at a new biosphere did not sustain eight humans for even a relatively short period of time. Biosphere 2 developers and experimenters learned that small, closed ecosystems are complex and susceptible to unplanned events, and that such events occur unexpectedly. Perhaps a lesson for future use!

The property, which is outside of Tucson, Arizona, is scheduled to be redeveloped for a planned community.

Tropical Rain Forest

Found predominantly at or near the equator, a **tropical rain forest**, as shown in Figure 3-2, is characterized by having 200 to 500 cm of precipitation per year with some areas occasionally having over 1,000 cm in a year. Because of the equatorial location, there is no seasonal but rather a continual growing season with cold periods. The daily temperature is minimal because of the insulating effects of the water, which gains and loses heat very slowly. Maximum daytime temperatures of only 30°C are typical, but the heat is oppressive due to the high humidity. Diurnal temperature fluctuations are on the order of only 5°C.

Figure 3-2

A tropical forest, with vines growing on the trunks and branches of the tall forest trees to collect sun needed for the vine to grow.

This consistent temperature warms the atmosphere, therefore creating a moist climate with no extreme temperature fluctuation and no cold season provides a stable and favorable environment for plant growth. The vegetation is dominated by tall (50 to 60 m), broadleaf, evergreen trees that branch near the crown to form a solid layer, or **canopy**, of leaves. Because of the density of the trees in the forest, light availability is the primary limited factor to plant growth below the canopy. Competition for light has resulted in tall, fast-growing trees that always have leaves on them. Certainly, leaves do become old, diseased, or shaded so that they die and fall to the forest floor, but this happens throughout the year rather in accordance with seasonal variations that affect most deciduous trees.

The tree canopy shades the forest floor so that smaller trees and herbaceous plants cannot survive; therefore, the floor is open, dark, and damp. Where the canopy is broken and light penetrates, dense undergrowth of plants results. This may occur at the banks of a river, where an old and diseased tree falls, or in areas where the frost has been cleared.

The warm, moist conditions of the tropics are ideal for the bacterial and fungal processes that decompose dead plants and animals and return their rich organic nutrients to the soil. Because of the density of living plants needing these nutrients, deep topsoil rich in these decomposition products is not able to accumulate. The trees therefore have shallow root systems that quickly take water and nutrients out of the soil to be used for new growth. Support roots aid the stabilization of the shallow-rooted trees.

In addition to having nutrients confined to shallow topsoil, many tropical soils are lateritic—they have certain metals in combination with large amounts of clay that pack very tightly if not kept loose with organic materials. Annual plowing for cultivation and crop harvesting removes the source of organic matter. Without plants to take up the larger amounts of water, rainfall causes leaching of existing nutrients out of the topsoil. The lateritic soils become compact and harden, so that after only a few years of cultivation such areas are as hard as concrete and further cultivation or revegetation is impossible. Therefore, clearing vast tropical forests for cultivation does not solve world food shortages, and it is an ecological disaster, resulting in permanent loss of these areas. The loss of this vegetation affects not only ecological balance but broad climate patterns as well. In addition, the extinction of species eliminates their unique genetic potential.

The tropical rain forests are the oldest vegetative biome because their equatorial position has shielded them from the effects of past periods of glaciation. This great age (some 200 million years long), combined with climate stability and abundance of all resources, has produced phenomenal diversity in plants and animals native to these areas. Although tropical rain forests are relatively unexplored and poorly understood biologically, there are still more species of plants and animals described from this biome than in all the other biomes combined (see Figure 3-3). It would be sad indeed to see this unparalleled natural resource destroyed. Currently, these forests are being destroyed faster than they can be studied and understood. This situation has resulted in a race against time to classify and catalogue tropical plants before they become extinct. Soon a herbarium (plant museum) may be the only place to find samples of the diverse tropical vegetation.

Tropical diversity is exemplified in plant groups that have developed unusual methods of survival. **Epiphytes** are plants that grow with roots attached to another plant by a **holdfast**. These plants do not harm their benefactors in any way; they are not parasitic, but rather coexist with them. Bromeliads and orchids include many common epiphytes. Long **lianas** (hanging vines) use the trunks and branches of the tall forest trees to climb into the canopy, where they produce leaves for photosynthesis. Since these vines are rooted, they obtain water and nutrients from the soil. They often grow from one tree to the next, trailing long looping sections between trees. These would be ideal for Tarzan to swing from tree to tree without having to touch the ground. Many native primates do in fact travel throughout the forest in such a manner. Entire communities of animals inhabit the forest canopy; exotic and beautiful plants, birds, insects, and other organisms are all part of this spectacular biome.

Figure 3-3

Tropical rain forest biome world view. Courtesy of National Geological Survey.

Although the given species change from one tropical rain forest to the next, the general picture is very similar.

Nearly half the forested areas of the earth are tropical rain forest. The Amazon River Basin of South America, the Congo Basin in Africa, and certain areas of Southeast Asia are the three largest areas; but, tropical forests also exist in Australia, Central America, New Guinea, the Philippines, Malaysia, the East Indies, and many of the Pacific Islands. Since many of these forests are found in countries with overpopulation and food shortage, they are in danger of being cleared and cultivated.

Occasionally rain forests occur in regions where one would not expect to find them, such as the Olympic Peninsula in the state of Washington. Located on the Pacific in a cool temperate region and warmed by the Japan Current, the land mass receives constant precipitation. Many of the species found there are of genuine rain forest origin.

Savannas such as the one in Figure 3-4 are usually found between tropical rain forests and deserts. Their proximity to either of these two areas greatly affects the annual rainfall. The "normal" range often falls between 80 and 160 cm per year. Because of their latitudinal distance from the equator, savannas have seasonal temperature fluctuations even though they have no true cold period. Precipitation is also scattered; long dry periods, which are often very hot, are followed by heavy warm-season thunderstorms. There is little rain during the cool season. This lack of rainfall for a prolonged period apparently excludes many species that otherwise might occur there and gives the savanna its characteristic vegetative composition.

Figure 3-4

Savanna biome, world view. Courtesy of National Geological Survey.

Savannas are primarily grassland with scattered deciduous trees. Since there is no true cold season, these trees generally lose their leaves during the long dry season each year, leafing out again when the rains come, and generally flowering while leafless. There are few annual plants in the savannas because of the density of the perennial grasses. A number of perennial herbs do thrive here, emerging from underground heat-resistant bulbs after the rains begin. The trees have smaller leaves than do those in the tropical forests. Because it is necessary to reduce water loss in the dry season and because light is not a limiting factor, the increased surface areas characteristic of tropical forest species are not found. In savanna areas that border the drier desert regions, the trees are smaller, denser, and often thorny. These areas are called **thorn forest**.

Some of the most extensive **savannas** are in the central and eastern African veldt (pronounced "felt"), as seen in Figure 3-5. A large variety of animal life, including many species of antelope, zebras, giraffes, elephants, and their predators depend on the grass species found there. Other savannas are found in Brazil, India, Southeast Asia, northern Australia, and North America.

Figure 3-5

A lush savanna in Africa.

The maintenance of this grassland component requires periodic burning. Burns clear the dead dry grass so that the new lush grass can grow when the rains come. Fire also keeps the trees thinned and scattered by removing young seedlings. The mature trees often have a thick bark, and since they are leafless during the dry season, minor trunk scorching is usually the only damage done. If young tree seedlings were allowed to grow to a size that would protect them from these grass fires, their density would form a canopy that would shade out the undergrowth of grasses and perennial herbs. For the proper balance of grasslands, animal forage, and scattered trees for shade and nesting sites, fires are essential. Lightning is a natural source of fire in the savannas as well as in other biomes.

For centuries, inhabitants have set fire to grasslands, recognizing its role in the balance of plant and animal populations. In some parts of the world, including North America, controlled burns are used to bring about the same kind of ecological manipulation that occurs in nature. If kept under control, these burns are not a misuse of fire. However, this practice has come under scrutiny because of the devastating forest fires that started as controlled burns. These fires have accidentally destroyed millions of acres in the forests of

North America. The study of fire ecology is an important part of overall management in several biomes.

Most of the desert areas of the world are found in a belt from 20° to 30° north and south latitude with rain-shadow deserts at other latitudes (see Figure 3-6). A few deserts have no vegetation and lots of shifting sand dunes, such as parts of the Sahara, but most have scattered low-growing vegetation.

Figure 3-6

Desert biome, world view. Courtesy of National Geological Survey.

Deserts receive 25 m or less average annual precipitation. The driest deserts, including the Sahara, have less than 2 cm average rainfall, and all desert areas can have extended droughts with no rainfall for several years. In the Atacama Desert in northern Chile, for example, the total rainfall over a period of 17 years was 0.05 cm, with only three showers in that 17-year span that were heavy enough to measure. It is important, therefore, to note that the rainfall figures mentioned previously are average over many years, with considerable fluctuation possible. When the rains do come, they can be heavy enough to produce flash floods or can be light showers.

Because of the dryness, diurnal temperature fluctuations are great. It is not unusual to have a 25°C or greater drop in temperature at night, since there is little moisture to hold the heat produced during the day. Once the sun is down, the heat source is gone. Conversely, it warms up rapidly after sunrise. It is common for vacationers, camping in the North America desert in the summer months, to bring no blankets or sleeping bags. A thin sheet provides little warmth, and these campers usually spend a sleepless night in the car!

Desert temperatures also follow the seasons, with occasionally harsh winters, freezing temperatures, and snow. Elevation, proximity to the coast, and latitude also combine to produce some "warm" deserts, which seldom have freezing temperatures. Such is the case of the Sonoran Desert of North America, the only North American desert having the giant saguaro cactus, a plant that cannot tolerate the freezing temperatures found in the Chihuahuan, Mojave, and Great Basin deserts.

Desert vegetation is amazingly diverse and uniquely beautiful, falling into three major categories based on the physical adaptation that allow their survival in times of drought. **Succulents** are plants that store water in thickened leaves or stems and protect that supply with thorns and spines. Most of the cactus family are stem succulents with either jointed pads, as in the prickly pears and chollas, or barrels with a single thickened stem. The **spines** are modified leaves that protect the cacti from herbivores in search of moisture. Other plants, called leaf succulents, store water in modified leaves. The century plant (*Agave*), Spanish dagger (*Yucca*), and the low-growing *Sedum* and *Portulaca* are common examples, although many different plant groups have succulent members adapted to dry zones, see Figure 3-7.

Figure 3-7

The small leaved shrubby plants are two or three vegetation types found in deserts. Courtesy of National Parks Service.

Low-growing, small-leafed shrubs with spines or sharp branches survive in the desert by conserving water in their woody stems and reducing water loss through reduced leaf surface area. In addition to or instead of spines, these shrubs often have foul-tasting compounds in their leaves to discourage herbivores from browsing. Creosote bush (*Larrea*) and tar bush (*Flourensia*) are such examples. Both spines and thorns protect cat claw (*Acacia*) and mesquite (*Prosopis*).

A third group of desert plants survives in a totally different manner. Instead of storing water, they wait until an adequate supply is available. They exist as heat- and drought-resistant seeds that will not germinate unless a heavy rain washes off a self-produced chemical inhibitor to germinate. The rain also provides ample soil water for these plants to germinate, grow to maturity, flower, reproduce, and set seeds before the supply of water is gone. This all happens in a short period of time—possibly a week or two, at most a few weeks, but certainly in one season. These desert annuals are sometimes called **ephemeral** because of their short life cycle. Their seed may have to wait as long as 20 or 30 years before conditions are right for germination and completion of their life cycle. Consider the fortitude of a seed able to survive so long in desert soils where surface temperatures are regularly over 60°C in the summer. If a year is wet, the flowering desert is as beautiful as any area.

The largest and driest desert in the world is the Sahara, followed in size by the Australian desert. Figure 3-6 shows where many of the other major deserts are located, including the four North American deserts, the Gobi of Mongolia, deserts in India, the Middle East, other parts of Africa, and in South America. Approximately one-third of the earth's surface is arid or semi-arid. Deserts are recent in origin, some no older than 12,000 to 15,000 years and possibly none older than 5 to 6 million years. The relative youth of deserts is even more striking when compared with the age of the tropics, which have existed for some 200 million years.

Deserts have expanded due to worldwide drying trends since the last glaciation period, which ended some 15,000 years ago. Human activities are accelerating this process of **desertification** (conversion to deserts) in many areas through overgrazing, cultivation of marginal areas, and general removal of plant life and water for our own uses. It is estimated that the Sahara is advancing to the southwest at approximately 17 km per year. With wise use and controlled developmental studies being conducted by government agencies, universities, and private groups in essentially every country containing desert lands, some of the most interesting and promising research is being done in the area of desert agriculture. There have been attempts to develop strains of already existing crop plants that require less water, can withstand higher temperatures, and can tolerate a great level of soil salinity.

Some studies are also aimed at finding commercial uses for plants that naturally occur in desert regions and are therefore already well suited to desert climates. It is probable that this latter approach has the greatest potential, and plants such as saltbush (*Atriplex*), prickly pear Cactus (*Opuntia*), and ironweed (*Kochia*), show great promise as forage plants. Guayule (*Parthenium argentatum*) for rubber, jojoba (*Simmondsia chinensis*) for oil, and several plant species for fuel biomass also have significant potential for commercial use.

Below ground desert dwellings using solar energy have been designed for experimental habitation, as have large, plastic dome greenhouses for single-family food production. Industries not requiring water can use desert land, which is plentiful and inexpensive for building new factories. Almost one-third of the earth's surface may some day be used more extensively than even before.

Grasslands

This biome is dominated by areas of perennial grasses (see Figure 3-8). Predominately in temperate latitudes, grasslands receive from 30 to 150 cm annual precipitation and have distinct seasons. Temperatures are often above 40°C in summer and far below freezing in winter, with occasional extremes. The annual precipitation is usually distributed throughout the year with occasional summer peaks. On the more **mesic** (wetter) end of the range, **grasslands** grade into savannas or temperate deciduous frost, whereas on the **xeric** (dry) end, they grade toward deserts.

Figure 3-8

Grassland biome, world view. Courtesy of National Geological Survey.

Because of the matted turf of fibrous grass roots, there are very few annual plants in the grasslands. The herbaceous plants found there are mostly perennial with underground storage structures such as tubers, bulbs, and rhizomes. Occasional prairie fires help maintain the integrity of the grasslands (much as in savannas), but in the dried grasslands the invasion by desert species is common.

Overgrazing of such areas has increased their desertification, allowing mesquite, cacti, and weedy annuals to become well established in the thinned turf. There are few native types of grassland (prairies) intact in the United States, most of these areas having been overgrazed by domestic animals or cleared for cultivation, as seen in Figure 3-9. Other major grasslands are found in Eastern Europe and parts of Russia, Central Asia, Argentina, and New Zealand.

Figure 3-9

Overgrazing of grasslands.

The dust bowl of the 1930s in the Oklahoma and Texas panhandles was caused by an extreme drought that drastically reduced productivity, and the few water wells were not adequate for irrigation. Without the cultivated plants to hold down the soil against the ever-present spring winds, the topsoil was literally blown away, as Figure 3-10 shows.

This biome is found in all the major continental areas of the Northern Hemisphere but is almost absent from the southern hemisphere (see Figure 3-11). The average precipitation of 75 to 225 cm per year is usually scattered throughout the year. Warm summers and cool to cold winters are typical. These areas rarely have droughts and only limited periods of snow and subfreezing weather. The trees drop their leaves each fall, hence the name **temperate deciduous forest** (see Figure 3-12a).

Figure 3-10

Dust storm approaching from the west on June 4, 1937. Four minutes after this picture was taken, total darkness occurred from the dust blocking the sun. Courtesy of USDA Photo Gallery.

Figure 3-11

Temperate deciduous forest biome, world view. Courtesy of National Geological Survey.

Because the forest leaf canopy is not intact year-round, the ecology of these regions is unique. Low growing, understory vegetation dominates in the early spring before the trees completely leaf out to form the shade-producing canopy. Once the canopy is formed, the forest trees are the dominant vegetation, followed again in the fall by a second-story understory assemblage that develops in response to the available light resulting from the leaf drop (see Figure 3-12b). Thus three-district vegetational assemblage exists during the growing season, resulting in a complex ecological situation.

Figure 3-12a

In temperate deciduous forests, in fall, the leaves turn colors prior to dropping from the trees.

Figure 3-12b

Dogwoods are a typical understory plant of North America.

Courtesy of National Forest Service.

Amazingly, however, similar types of trees are found in temperate deciduous forests throughout the world. Many of these trees are hardwoods, highly valued in the furniture-making industry (Chapter 8).

In drier portions of the biome, where the winters are cool and moist but the summers are hot and dry, a unique vegetational association called **chaparral** exists. Characterized by smaller, often thorny or roughly branched evergreen trees and shrubs and deciduous trees, the chaparral has a short spring growing season interrupted by the heat and drought of the summer. Often referred to as a Mediterranean climate because of the winter rainfall and extensive chaparral areas along the shores of the Mediterranean Sea, such localized areas are also found in southern California, southern Africa, coastal Chile, and coastal western and southwestern Australia. Although they are isolated from one another and contain different species of plants, their climatic conditions give these areas a similar appearance.

Coniferous Forest

Coniferous trees are the cone-bearing members of the gymnosperms. Conifers, valuable for their use in the lumber industry, are evergreen, except for larch, bald cypress, and tamarack. Their needle-shaped leaves have a thick covering (cuticle), which helps prevent water loss. Many conifers have shallow soils found commonly in mountainous areas (see Figure 3-13). Mountains in Europe and Asia, as well as the Rocky Mountains and Appalachian Mountains of North America, have coniferous forests with similar climates and rainfall patterns.

The ability to thrive in thin, rocky or sandy soil that often contains little moisture explains the significant strands of coniferous forests (see Figure 3-14) in the southeastern United States and in the western coastal areas of California, where giant redwoods (*Sequoia*) grow.

The far northern coniferous forest, found almost exclusively in the Northern Hemisphere north of the 50° latitude, is referred to as a **taiga** (**snow forests**). The average precipitation ranges from about 35 to 100 cm per year, most of it falling in the summer. Winters are very long and cold and have a persistent snow cover. Although the winter air is dry, the ground remains moist because of the low evaporation rate resulting from the cold temperatures. These forests grade into either grasslands or temperate deciduous forest to the south, depending on precipitation levels.

At their northern limits, the coniferous forest gradually gives way to the **tundra**, as shown in Figure 3-15. Found predominately in the Northern Hemisphere north of the Arctic Circle, the tundra comprises approximately one-fifth of the earth's total land surface. With less than 25 cm of annual precipitation and strong, dry winds, the subzero temperatures and long periods of winter darkness create an exceptionally harsh environment for plant growth. The ground is frozen solid for most of the year, thawing only to a depth of about

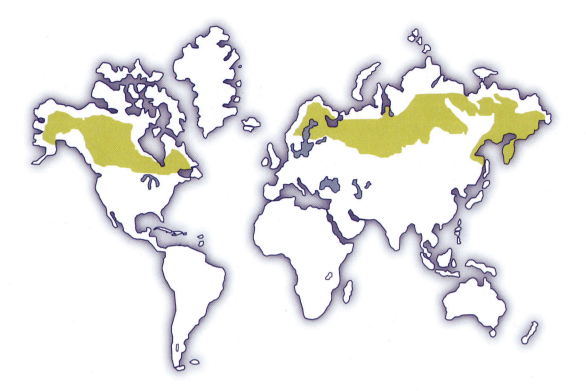

Figure 3-14

Rocky Mountains, Colorado.

1 m during the short summer. This soil **permafrost** causes plant root systems to be relatively shallow yet extensive. In the moist sedge-dominated communities, underground plant parts may be as much as 10 times greater than above-ground biomass.

Figure 3-15

Tundra biome, world view. Courtesy of National Geological Survey.

Tundra vegetation is typically composed of scattered, low-growing woody perennials that are well adapted to the drying winds and extreme cold. Certainly, tundra plants must survive long periods of moisture unavailability, and their morphological adaptations commonly include thick waxy cuticles and dense leaf pubescence. Additionally, their prostate to shrubby woody trunks are often covered by a protective coat of lichens or moss that helps prevent desiccation as well as provide protection from the cold. Although no tundra vegetation grows taller than 1 m, and the harsh climate has limited the number of well-adapted species, the successful plants are abundant. Essentially every native plant community is dominated by only two or three species, but there are huge expanses of such areas, as shown in Figure 3-16a.

The mean daily temperature is above freezing for only about one month of the year, providing tundra plants with a very short growing season. During this period, soil moisture is available because the ground thaws above the permafrost, and photosynthesis is possible for all 24 hours of summer daylight. Growth is generally minimal, however, because the replenishment of stored food reserves in the roots and woody stems is the primary plant function during this short period of favorable conditions. Some plants add only a few new leaves to each twig before the cold temperatures again become restrictive.

Sexual reproduction is an even more tenuous function than vegetative growth, since the plants only have 4 to 6 weeks to complete the entire cycle of flower development, pollination, fertilization, and fruit and seed maturation. Many species have developed dormant buds at the end of the previous summer, thus, when the snows melt; the mature flowers emerge within a few days (see Figure 3-16b). Many of these flowers are modified to concentrate the heat of the sun, thus increasing the maturation process by speeding up metabolic activity. The arctic poppy has a white cup-shaped flower that tracks the sun, focusing the sun's rays on its reproductive parts. In full

sunlight, the temperature inside the flower can be as much as 28°C higher than the air temperature around it. Insects attracted to this warmth affect pollination, and seeds are mature about 3 weeks after the flower opens. Not all the arctic plants have such sophisticated flowering adaptations: some depend on vegetative reproduction, only managing to set mature seeds in unusually long and mild growing seasons, which may not occur for 50 years or more.

Once mature, seeds of arctic plants must be able to remain dormant for long periods. It is a rare year that provides a growing season sufficient in length to allow for seed germination and development to mature protective woody growth. The deep-freeze conditions of the arctic enable seeds to retain their viability for an unusually long time, however. As an extreme example, a seed of an arctic lupine (genus *Lupines*, family Laminaceae), found in a Yukon deposit dated at 10,000 years old, germinated after dry storage for 12 years at room temperature. Long dormancy enables seeds to remain viable until the conditions are optimum for success.

Some mountainous areas have climatic conditions and vegetation similar to those of the arctic tundra. These mountain tundra zones are at elevations above tree line in essentially the entire world's larger mountain ranges. The farther from the poles these ranges occur, the higher the elevation before tundra conditions can be found.

At extreme elevations and polar latitudes, no vascular plants occur. The icecaps of both polar regions support a few algae and fungus species but no higher or vascular plants. The tundra is therefore the farthest limit of plant growth.

Aquatic Biomes

The aquatic biome is comprised of marine (saltwater) and freshwater.

Marine

Covering nearly 75% of the earth's surface with an average depth of approximately 5 km, the **marine biome** is phenomenally large (see Figure 3-17). The plant life there exists under unique circumstances. The most important is the light penetrates to an average effective depth of only a few meters. Below this shallow zone of adequate light, only the short wavelength blue and green portions of the spectrum penetrate effectively, and even then it is essentially dark below 660 to 750 m. Thus the vast majority of plant life is limited to the lighted surface, with a few organisms capable of using shortwave light for photosynthesis at greater depths. Bacterial, fungal, and animal forms inhabit the oceans even at their greatest depth, but as on land, the base of their food chain is still plants.

Water itself is obviously not in short supply; however, only organisms that can grow in salt water exist in the oceans and seas. These organisms are, for the most part, single-celled algae having no need for the complex vascular system, supportive tissues, and reproduction organs of most terrestrial plants. Only in the shallow waters

Figure 3-17

Marine biome. Oahu, Hawaii. The oceans of the world cover about 75% of the earth's surface. The greatest vegetative diversity occurs near the shoreline, where habitats vary. Courtesy of National Sea Grant Program.

along the coastal shelves do more complex algal types exist. Simple transport systems, anchoring devices, and other multicellular modifications have been developed by these organisms in response to the constant wave action along the shores.

The climate for marine plants is to a great extent a function of the ocean currents. Plant distribution and temperature are especially current dependent. The most important factor in producing ocean currents is patterns of air circulation. In combination with temperature, which affects water densities, and the deflection of currents off the continental landmasses, these predictable wind patterns create massive water movements around the world.

Ocean currents affect not only the distributions of plants and animals in the oceans, but some climate patterns on land as well, as shown in Figure 3-18.

Figure 3-18

Worldwide ocean currents affect vegetative assemblages close to the shorelines. Gray arrows show cold currents, and colored arrows show warm currents. Courtesy of National Sea Grant Program.

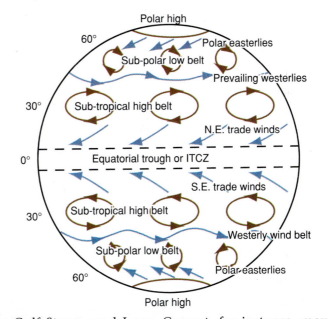

The Gulf Stream and Japan Current, for instance, move water warmed in the tropical latitude northward across the Atlantic and Pacific oceans, respectively, to northern latitudes. The British Isles and Alaska are affected by these currents, which change not only the climate, but also the terrestrial vegetation along the coasts and for some distance inland. The coniferous forests of North America, therefore, do not extend as close to the coasts in the west as they do in the east because of the Japan Current. These currents also affect the entire assemblage of marine organisms along these coasts. For example, seals can be seen off the coast of southern California when the cold water moves from northern latitudes south with the California Current. Another ocean phenomenon that occurs is the El Niño—Spanish for "the Christ child"—which refers to an interval of especially warm ocean temperatures that intermittently appear around Christmas in the equatorial Pacific. The phenomenon is associated with weather changes around the globe, including in the

Pacific Northwest, where it causes the winters to be especially mild. The effect occurs with a frequency that varies from two years up to a decade. Unusually cold ocean temperatures in the equatorial Pacific characterize la Niña—Spanish for "little girl." This other phenomenon occurs after an El Niño and has the opposite effect on the weather, which in turn makes winters colder with more precipitation. These are but a few examples of how ocean currents modify climate and vegetation worldwide in both aquatic and terrestrial biomes.

Marine organisms are sometimes classified as pelagic or free floating and benthic, or bottom dwelling. The free-floating organisms are primarily phytoplankton (single-celled plants) and zooplankton (single-celled animals). Phytoplankton is composed of single-celled algae, primarily diatoms. The open waters nurture millions of these organisms, plus eggs and larval forms of fish and invertebrates, to provide the early stages of the marine food chain. The larger pelagic animals are thus provided with a reliable and abundant food source. Animals of the benthic zone are usually sedentary or slow-moving clams, starfish, snails, worms, and sea anemones, sponges, and the larger fish. Bacteria and fungi also inhabit this zone, thriving on organic debris that settles from the pelagic zone.

Not all ocean zones are considered productive; some regions are almost devoid of essential nutrients, and therefore few phytoplanktons can survive. Since there is no food source for the zooplankton and larger marine animals, such regions have been referred to as the ocean deserts. The commercial fishing industry is well aware of these unproductive zones.

One interesting and important part of the marine biome is the coral reef. Reefs have been formed only in warm, well-lighted waters of the world (see Figure 3-19). The largest one is the Great Barrier Reef off the coast of northeastern Australia. Extending for a distance

Figure 3-19

Coral reefs only form in warm, well-lighted waters of the world.
Courtesy of National Sea Grant Program.

of about 200 km, it, like other coral reefs, is composed of colonial coelenterates and encrusting algae. These organisms secrete calcium and become quite hard. Primary production (photosynthesis) is provided by symbiotic algae living among the coral. Coral exists in a variety of colors, and the reflection of light rays in tropical waters shows off a spectacular display of natural beauty.

Freshwater

Essentially all terrestrial organisms require a supply of fresh water. Plants obtain their moisture from the soil. Some animals are capable of securing all the water they need directly from eating plant tissue. The vast majority of animal life, however, needs additional freshwater supplies, and they find it in lakes, ponds, rivers, and streams (see Figure 3-20). Of the basic natural resources, humans require oxygen, food, fuel, fresh water, and raw materials for shelter, clothing, medicine and industry. Loss of any one could theoretically limit future human growth. The one that first becomes inadequately available, however, is the limiting factor. Since only slightly more than 2% of the world's total land surface area is covered by standing or running fresh water, this resource is a prime candidate as the limiting factor for human population.

Figure 3-20

Devils Watering Hole at Inks Lake State Park, Burnet, Texas. Freshwater biome ecology is complex and occurs near the surface of the fresh water and provides an array of different habitats.
Courtesy of Texas Fish and Wildlife Service.

The ecology of freshwater areas is complex. From the banks, where many vascular plants such as trees, shrubs, and herbaceous plants grow, to shallow water, where some specially adapted vascular plants such as cypress trees can exist, to progressively deeper water, where nonvascular plants, primarily algae, thrive, each slight change produces a new zone of plant species. Whether the water is still, as in ponds and lakes, or flowing, as in streams and rivers, it plays an essential ecological role.

Scientists who study freshwater biology divide this biome into standing water and running water. In a sense, it is difficult to classify running water as an ecosystem because the water and nutrients are not recycling within given boundaries. Standing water lakes may be large or small, and the life zones are classified as the **littoral zone**, at the edge of the lake and quite productive; the **limnetic zone**, a region of open water where phytoplankton are abundant in the upper layers, and the **profundal zone**, the region below the limnetic zone where there is no plant life. Principal occupants of this third zone are the scavenging fish, fungi, and bacteria.

The climatic zone of lakes is similar in composition to the open water in the ocean (although the species may be different), but the littoral zone of the lake is unique. The shoreline may contain bottom-rooting aquatic angiosperms (flowering plants) such as cattails and rushes; water lilies and other rooted plants may extend further out for a considerable distance.

Running water arises from melting ice or snow, from artisan water below the soil surface, or as an outlet from lakes. The flux or quantity of water transported per unit time determines to a large extent the kinds of organisms that prevail. Slow-moving streams may be rich in phytoplankton, whereas rapids and fast-moving water may have very few. Rapid water movements also preclude the attachment of angiosperms to the bottom of the river or stream. Under such conditions most productivity is confined to the quiet shallow areas, where algae and mosses can attach to rocks.

Even latitude, which affects temperature, has a major effect on biological productivity, especially in **thermal stratification** (which causes turning over) of lakes and ponds. In winter, as the air temperature drops, so then does the water temperature, gradually cooling from the surface downward. Since water is an excellent insulator, gaining and losing heat slowly, several days of subfreezing air temperature may be required before the surface water finally begins to form ice at 0°C. During this gradual cooling toward the freezing point, the top layer of water reaches 4°C, which is the temperature at which water is most dense (the water molecules are actually packed most closely together). Thus, at 4°C the top layer is more dense than water below, which is not in direct contact with the air and does not cool as rapidly at the lower levels. The denser upper layer of water sinks to the bottom; since this top layer is more saturated with oxygen than are lower levels, as it sinks it oxygenates the pond or lake and stirs up organic matter. This results in a period of growth for freshwater organisms.

When ice forms, which is less dense than liquid water, it floats on the surface. This ice layer will slowly thicken if the air temperature remains cold enough, and the ice layer helps insulate the water below from the air temperature. Thus, the entire body of water does not become an ice block. In the spring, as the surface ice warms from 0°C to 4°C, the lake turns over again, with results comparable to those at the onset of cold weather. This may happen repeatedly until all surface water is warmed above 4°C.

Water is truly a unique liquid. If, as with other liquids, the freezing point produced a solid state denser than the liquid, ice would sink to the bottom. The pond or lake would be filled with ice, which would thaw slowly from the top down in the summer and probably never completely melt. This would eventually result in a solid block of ice in winter and only the upper portions even being in the liquid state for part of the year. A vast majority of our deep freshwater bodies then would not support aquatic life, nor would it be available for human's needs.

Because standing bodies of water have living organisms that reproduce, grow, and ultimately die, these bodies of water slowly fill with organic debris. During periods of **eutrophic** (nutrient-rich) conditions, bursts of growth and reproduction occur. As in the oceans, a majority of plant activity is at the surface where light is available. This often produces **algal blooms**, which cover the surface and initiate oxygen-poor conditions below. Excess algal growth increases the animal populations that feed on them. The ultimate effect is a population crash in the pond or lake. This process of **eutrophication** is often artificially accelerated by human activities such as the dumping of raw sewage; the runoff of agriculture fertilizers and hog waste provides rich nutrient supplies for the plant life to grow.

Other human activities that not only disturb the ecological balances but also render the water unsafe for use include various forms of pollution. Chemicals, trash, and sewage are among the most serious water pollutants. Whatever the source, any negative use of the limited supplies of fresh water, could have very serious consequences. We need to understand as best we can freshwater ecology and use this knowledge to wisely manage this important resource (see Figure 3-21).

Figure 3-21

Human use of fresh water includes recreational activities such as the fishing done by these fishermen. Courtesy of Texas Fish and Wildlife Service.

Summary

Biomes are worldwide groups of similar ecosystems. The terrestrial and aquatic biomes are summarized next.

1. **Tropical rain forests:** Equatorial, large evergreen trees with a dense leaf canopy, little diurnal or annual temperature fluctuation, 200 to 500 cm of rainfall annually, shallow topsoil that is often lateritic, and many species of plants and animals.

2. **Savannas:** At subtropical latitudes, there is some seasonality, 80 to 160 cm annual precipitation, perennial grasslands with scattered deciduous trees, maintained by periodic burning; the driest savanna areas are called thorn forest.

3. **Deserts:** Primarily found in a latitude belt between 20° and 30° in both northern and southern hemispheres, only 25 cm or less annual precipitation, extreme daily temperature fluctuations, and annual season; low growing, scattered vegetation is either small leafed shrubs, succulents, or herbaceous annuals.

4. **Grasslands:** Receiving 30 to 150 cm of annual precipitation, distinct seasonality; constant coverage of perennial grasses is normal but cultivation and overgrazing are permanently changing much of the world's native grasslands; in the more xeric grasslands, desertification is a common direction of change.

5. **Temperate deciduous forest:** Found primarily in the Northern Hemisphere, 75 to 225 cm of annual precipitation is seasonal; deciduous trees (hardwoods) dominate the vegetation in the fall, an autumn and spring understory of shrubbery and herbaceous plants becomes evident; the xeric end of the precipitation range produces what is termed chaparral or Mediterranean vegetation.

6. **Coniferous forest:** Known as taiga in the northern latitudes and high elevation mountain areas; predominately Northern Hemisphere; annual precipitation of 35 to 100 cm limits vegetation to coniferous gymnosperms (softwoods), which are evergreen; at more southern latitudes this biome is produced by sandy soils.

7. **Tundra:** Mostly north of the arctic circle, about one-tenth of the earth's surface area; less than 25 cm of precipitation annually; tundra plants are scattered and low-growing perennials; permafrost soil conditions exist year-round and the growing season is very short; mountain tundra occurs above the tree line in higher elevation mountainous areas; vegetation and climatic conditions are very similar to arctic tundra but at more southern latitudes.

8. **Aquatic biomes:** The aquatic biomes include marine and freshwater systems; the marine biome includes nearly 75% of the earth's total surface area and, although the oceans average 3 km in depth, most plant life is limited to the top few meters; most plant life is single celled, with the shallow water along the shores the only locality for the more complex kelps and other multicellular algae; ocean currents such as the Gulf Stream and Japan Current affect terrestrial vegetation by modifying the temperatures of the area and thus the climate; the freshwater biome includes flowing water and standing water ecosystem types; vegetation in both is more complex with more species than in the marine biome.

Something to Think About

1. What is a biome?
2. Identify the vegetation in a tropical rain forest.
3. Describe the vegetation in a savanna.
4. List the vegetation in a desert.
5. Identify the vegetation in a grassland.
6. Describe the vegetation in a temperate deciduous forest.
7. Identify the vegetation in the coniferous forest.
8. Discuss the vegetation in the tundra.
9. Name the vegetation in the aquatic biome.
10. Where are the following biomes found: tropical rain forest, savanna, desert, grassland, temperate deciduous forest, coniferous forest, tundra, and aquatic biome?

Suggested Readings

Dale, V. H., C. M. Crisafulli, and F. J. Sawanson. 2005. *Ecological responses to the 1980 eruptions of Mount St. Helens.* New York: Springer-Verlag.

Dashini, K. M. M., and C. L. Foy. 1999. *Principles and practices in plant ecology: Allelochemical interaction.* New York: CRC Press.

Kalman, B., and J. Langille. 1998. *What are food chains and webs?* New York: Crabtree Publishing.

Sprent, J. 1987. *The ecology of the nitrogen cycle.* Cambridge: Cambridge University Press.

Wigley, T. M. L., and D. S. Schimel. 2000. *The carbon cycle.* Cambridge: Cambridge University Press.

Internet

Internet sites represent a vast resource of information. The URLs for Web sites can change. Using one of the search engines on the Internet, such as Google, Yahoo!, Ask.com, or MSN Live Search, find more information by searching for these words or phrases: **biomes, grassland, tundra, tropical rainforest, temperate deciduous forest, coniferous forest, savanna, desert, aquatic, carbon cycle, and nitrogen cycle.**

PART 2

Form and Structure

CHAPTER 4

Basic Design I

Adaptation basically means adjusting to or fitting a specific environment. Adaptive changes can occur in any part of the plant, and the outcome of this adjustment may be the development of unique or unusual morphological modifications. Theses natural morphological adaptations allow plants to exist in different environmental settings. Plants have become adapted to an amazing number of environmental settings and variations. Although plants are placed under stress by cold, drought, heat, and monsoons, plants found in these extreme areas are remarkably well designed, unable to move away from the inhospitable conditions. While the variability is great, the external morphology (overall form) of higher plants is composed of the same basic parts: roots, stems, and leaves.

After completing this chapter, you should be able to:

- Use the terms that describe vegetative parts of the plant
- Name three types of roots
- List three functions for roots and for stems
- Describe a stem
- Name four types of stems
- List all of the parts of typical leaf
- Give three types of venation in leaves
- Explain the difference between an incomplete and a complete flower
- Identify all of the parts on a complete flower
- Describe the functions of the essential reproductive organs of the flower
- Name three types of flowering characteristics found in plants
- Discuss four ways flowers vary
- Name the two large categories of fruit
- List four terms used to describe fleshy fruits and four terms used to describe dry fruits
- Discuss how plant structure is used to classify plants
- Identify all of the parts of a typical seed

Key Terms

nonvascular plants	conduction	adventitious roots
vascular systems	taproots	prop roots
vascular plants	storage taproot	bromeliads
gymnosperms	fibrous roots	haustorial
angiosperms	branch roots	rhizobium
anchorage	epidermal cells	mycorrhiza
absorption	root hairs	symbiotic

nodes
internode
axil
lateral buds
terminal buds
bud scales
annual tree ring
alternate leaf
   arrangement
opposite leaf
   arrangement
whorled
basal rosette
rhizome
stolons
tuber
corms
cladophylls

convergent evolution
thorns
prickles
spines
tendrils
blades
petiole
pinnately compound
palmately compound
venation
netted
parallel
palmate
midrib
margins
entire
dentate
lobed

apex
peltate
sessile
sheath
pubescence
mesic
insectivorous plants
vegetative reproduction
budding
sexual reproduction
grafting
rootstock
scion
vascular cambium
stem cutting
layering

Vegetative Morphology and Adaptations

One of the consistent features common to plants is a lack of motility. Except for a rare example of motility in some aquatic plants, plants are stationary—fixed in their location and usually anchored by a root system. This immobility subjects plants to the dictates of nature. Like animals, the typical plant body is over 90% water. Thus, in addition to being able to withstand locations of light, temperature, space, and nutrient availability, plants do not have the ability to move to a place with more water.

Lower plants (those which are less complex) get water and dissolved materials to each cell of the plant body by being in contact with moisture. These single-cell or small multicellular organisms are known as **nonvascular plants**. Higher plants (those which are more highly evolved and complex) do have **vascular systems** in which water and solutes are transported to the different parts of the plant body, and these plants are known as **vascular plants** (see Figure 4-1).

The two most visible, largest, and complex of vascular plants are the **gymnosperms**, notably represented by the cone-bearing trees such as pine, fir, spruce, and juniper, and the angiosperms, or flowering plants. Since **angiosperms** are very common and extremely important to human existence, our discussion centers on them more than the gymnosperms.

Root Morphology/Function

The four main functions of the plant root systems are: (1) **anchorage**, (2) storage of food, (3) **absorption** (uptake) of water and nutrients, and (4) **conduction** (movement) of water and nutrients dissolved in it to the aboveground parts of the plant. Anchorage holds the plant in position. Winds, which can buffer a tree at high velocities, exert an enormous force. The flexibility of the tree allows it to bend rather than blow over. But it is the root system that ultimately anchors the tree. Although grasses have a relatively shallow root system, it is so extensively intertwined among the soil particles that it forms a dense mat of roots, which provides an effective anchorage against grazing animals.

Some plants need extensive storage capabilities within their root systems. Sugar beets and carrots are examples of root adaptations for storage of water and carbohydrates. These materials are used for new shoot growth during the second growing season. Many such highly modified storage roots produce important food products.

The amount of water (and dissolved minerals) absorbed by roots can be extensive. Water absorption and conduction in a typical corn plant, for example, can surpass 2 liters per day. This requires an enormous root surface area through which water can be absorbed.

Root Structure

Figure 4-2 shows the three basic types of root systems: **taproots**, which have a main central root, and a **storage taproot**, which acts as a food reservoir to retain surplus food during the winter or adverse periods, **fibrous roots**, which have many roots of equal size. Each of these has extensive lateral branching, with the fibrous root system fashioning an interwoven mass of roots.

The depth that taproots penetrate the soil varies from only a few centimeters to a reported 35 m (for the mesquite). The taproot grows straight down, developing secondary **branch roots** (lateral roots) as it grows. These in turn have tertiary roots, which have quaternary roots, which successively branch and lengthen to provide increased surface area for water absorption. As the root system develops, the aerial portion of the plant body root and the increased volume of tissues demand a constant supply of water. Some trees are known to have taproots that extend far into the soil to ensure a supply of water even in periods of drought. Young pecan trees often have a taproot extending to a depth exceeding their height.

Fibrous root systems are generally much more diffused and closer to the surface than the taproot system. Each of the equal-sized roots

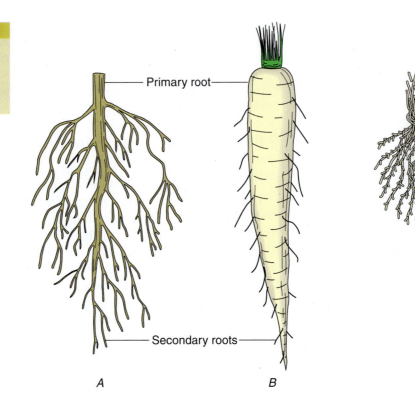

Figure 4-2

Types of plant root system:
(A) Taproot (B) Storage
(C) Fibrous. Courtesy of Rick Parker.

Primary root

Secondary roots

A   B   C

develops secondary and tertiary branch roots, as in the taproot system. This root network can effectively prevent any other plant from becoming established. The grassland biome can be solid grasses without many other plants except where the grass turf has been thinned or removed.

Having such a large fibrous root system close to the soil surface is very important for plants in environments with relatively little rainfall. What little moisture there is can be more efficiently used. Conversely, in such an environment, plants with taproot systems must develop the taproot very quickly to reach the soil moisture present below the fibrous root zone. Early development for the desert plants having a taproot system is much greater below the ground than above, and lateral roots do not develop as soon nor as close to the surface as they do in other plants.

Root Hairs

In most plants, many of the roots' **epidermal cells** elongate, forming long slender **root hairs** that increase the total root surface area many times, as shown in Figure 4-3. This increase allows much greater water absorption because these root hairs extend out among the soil particles not in direct contact with the root itself. On plants with actively growing root systems, new root hairs are continually being produced at the rate of many million each day (see Figure 4-4). Without root hairs, most plants would not be able to absorb even a fraction of the water they need. A newly germinated seeling is an excellent subject for observation of root hairs.

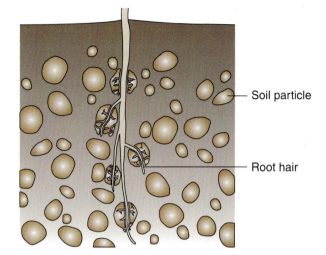

Soil particle

Root hair

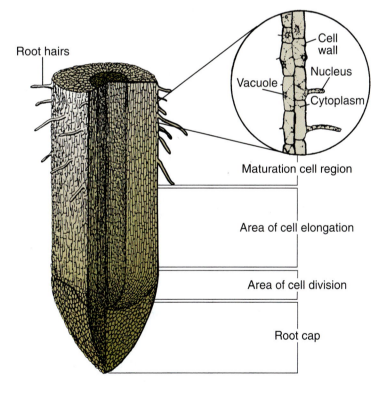

Root hairs

Cell wall

Nucleus

Vacuole

Cytoplasm

Maturation cell region

Area of cell elongation

Area of cell division

Root cap

Adventitious Roots

Some plants form roots on other parts of the plant body. These are **adventitious roots**, and they most commonly develop on a stem. The ability is propagated from leaf cuttings. Many new plants can be propagated from one adult *Dieffenbachia* (dumb cane) if its stem is cut into sections and rooted in moist soil or vermiculite.

Adventitious roots often augment the normal root function of anchorage, such as the mangrove plant in Figure 4-5. Developing from the lowest nodes of their stem, they extend from the stem into the soil and help brace the plant. These aptly named **prop roots** are found most commonly on plants growing in marshes and mud flats.

Figure 4-5

Adventitious prop root of a mangrove plant.

Modified Roots

Structural modification of roots can affect any of their five primary functions.

Anchorage

Prop roots are modified adventitious roots proving anchorage and stability. Short adventitious roots of English ivy (*Hedera helix*) develop at the nodes of the stem. In Virginia creeper (*Parthenocissus quinquefolia*), pad-like appendages arising at the nodes secrete a sticky substance for anchoring the stem to surfaces. The substance cements the ivy to a surface so tightly that, when the ivy is removed, mortar and even the building's facing can break away. Although appearing root like, these are actually modified stems.

Plants that grow on other plants usually have modified roots. Epiphytes such as **bromeliads**, as shown in Figure 4-6, and more temperate ball moss (*Tillandsia recurvata*) and Spanish Moss (*Tillandsia usneoides*) (also bromeliads, not mosses), are anchored by their roots to the trunks and branches of trees.

Water and Nutrient Absorption and Conduction

True parasite plants have a detrimental effect on their host. Mistletoe (*Phoradendron*) and dodder (*Cuscuta*), for example, produce **haustorial** roots, which not only anchor the parasite to its host but also penetrate the host's vascular system. In this way the parasite avails itself of water and nutrients from the host plant.

Since roots require an oxygen supply to carry on respiration, some plants that grow in poorly drained soil or in soil covered by stagnant water develop root modifications that grow out of the water to aerate the plant. Roots of the black mangrove (*Avicennia nitida*) develop pneumatophores, which stick straight out of the swampy water. The bald cypress (*Taxodium distichum*) has characteristic "knees," which may provide oxygen to the root system although there is some doubt about this function.

Storage

Many plants have extensive root storage capacities. Starch and other molecules are stored for growth or flowering and are a reserve against periods of harsh environmental conditions. Beets and carrots are well-known examples. The sweet potato (*Ipomoea batatas*) also has enlarged storage roots, as does the tropical cassava dish, poi, which is a staple in Polynesia and Southeast Asia made from crushed and fermented taro corm (*Colocasia esculenta*).

Nodulation

Some plants, notably members of the pea family (*Fabaceae* formally *Leguminosae*) are unique in their ability to increase the levels of soil nitrogen in a useful form. Nitrogen fixation involves the formation of a root nodule in response to an infection initiated by bacteria from the genus. **Rhizobium** bacteria convert, or fix, atmosphere nitrogen (N_2 that is found in the air spaces of the soil). Plants cannot use atmospheric nitrogen, but ammonia can be used. In this way the modified roots and their bacterial guest effectively fertilize the soil with nitrogen.

ADAPTING TO A BIOME

In order to survive and thrive in a desert biome, organisms must solve several problems that their relatives living in wetter environments do not face. Chief among these are obtaining and preventing water loss. For example, the grizzly bear cactus (*Opuntia polyacantha*), is a member of the rose family (*Rosaceae*), and is native to the American Southwest. Hoodia (*Hoodia gordonii*) is a member of the milkweed family (*Apocynaceae*) and is native to the Kalahari Desert of Africa. The hoodia is a stem succulent, described as "cactiform" because of its remarkable similarity to the unrelated cactus family.

These unrelated plants resemble each other in many ways. Both are adapted to a desert biome. Through convergent evolution, both plants developed characteristics that allowed them to survive a desert biome; for example, both plants are essentially unbranched cylinders. Their leaves are highly reduced compared to the wetter-biome relatives. They are also succulents with thick, waxy epidermis. All these adaptations prevent water loss.

Mycorrhiza

Mycorrhiza usually has short roots that form **symbiotic**, or mutually beneficial, associations with certain fungi found in the soil, as shown in Figure 4-7. The fungi actually enter the tissue of these modified roots, benefiting nutritionally from the roots. The roots, in turn, become more efficient in the absorption of certain minerals needed by the plant.

Stem Morphology/Function

Plant stems come in a great array of shapes and sizes, but they all share the same basic functions: (1) the attachment and support of leaves, flowers, and fruits, (2) the conductions of materials, (3) storage, and (4) growth.

Stems are the place of attachment for leaves, flowers, and fruits. Erect stems provide structural support for leaves, raising them to allow for adequate exposure to light. Similarly, flower stems allow

Mycorrhiza fungi on plant roots. They grow in a symbiotic relationship with the roots of the Ericaceae family and other plant families, which have no root hairs, to collect the needed nutrients and water.

more visibility to pollinators, and the resulting fruit and seeds are better dispersed from their elevated positions of attachment on the stem.

A second function of the stem, in combination with roots, is conduction. The stem conducts water and minerals from the roots in all the aboveground parts, carries food produced by photosynthesis in the leaves to the roots, stem, flowers, and fruits and transports hormones from the tissue in which they are synthesized to those areas where the effects are produced.

Third, most stems store nutrients, organic molecules, water, and by-products; a large amount of storage occurs in certain modified stems, but even unspecialized stems contain some storage tissues.

Finally, the stem contains meristematic tissue. In general, meristematic tissues are areas of rapid cell proliferation; this results in growth elongation of stem tips, increase in stem diameter, and production of tissue and organs such as leaves and flowers.

Stem Structure

Regardless of whether stems are erect, prostate, or in some other position, all stems have areas of leaf and bud attachment called **nodes**. The portion of stem between one node and the next is termed the **internode**. Normally the leaf is attached to the stem at the node, with a bud (embryonic shoot) arising in the leaf **axil** (where the leaf connects to the stem). Because woody plants also have terminal buds, they are called axillary or **lateral buds**. See Figure 4-8 for a diagram of the parts of a leaf. All herbaceous and woody plants also have **terminal buds** at the end of each shoot. Since a bud is an undeveloped shoot, a new leaf, branch, or flower may arise from it when and if it becomes active. Not all axillary buds develop, and the hormonally controlled pattern of bud growth and inhibition, as well as the length of the internodes, is responsible for the vast array of plant shapes.

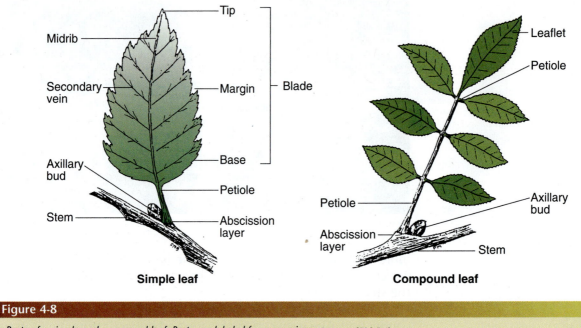

Simple leaf   **Compound leaf**

Figure 4-8

Parts of a simple and compound leaf. Parts are labeled for comparison. Courtesy of Rick Parker.

Herbaceous stems are so called because they do not develop wood from year to year and in fact, normally live only one season. Some herbaceous stems have short life cycles, dying before winter; others are killed by the freezing temperatures of winter. Woody stems, on the other hand, do produce new growth each season, over wintering in a dormant state. Dormant lateral buds on most woody plants are protected by a number of overlapping **bud scales**. As the bud begins new growth, these scales fall off, leaving bud scale scars on the stem.

The dormant terminal bud is also protected by a series of scales that leave a complete ring of scars around the stem when they fall off at the onset of new growth in the spring. The length of a year's growth can be determined by measuring the distance between any two rings of terminal bud scale scars (see Figure 4-9). The total annual increase in length (new growth) of a stem varies from one species to the next, but it can also vary from year to year on the same plant in response to environmental conditions such as water availability and length of frost-free weather.

Growth patterns in plants are entirely different from growth patterns in animals. In plants elongation is restricted to the tips of stems and roots, whereas in animals growth is not limited to specific tissues. Once a woody stem has completed a year's growth, new elongation will take place in the section. This can be illustrated by driving a nail into a stem at a measured distance above ground. The nail will remain at that height for the life of the tree, although the tree will certainly become taller. Annual increases in stem girth (thickness) also occur. Yearly growth is calculated by the analysis of the internal wood anatomy, measuring the thickness of each **annual tree ring** produced.

The most common pattern of leaf and axillary bud attachment is an alternating sequence of one leaf and bud per node. This is termed an **alternate leaf arrangement**. Two leaves (and their axillary buds) attached at the same node directly across the stem from each other produces an **opposite leaf arrangement**. Three or more leaves attached at each node produce a **whorled** arrangement. Plants such as the dandelion (*Taraxacum officinale*) have all leaves grouped in a **basal rosette** at ground level, which obscures whether they are alternate, opposite, or whorled. For other examples, see Figure 4-10.

Although most plants have only one bud in each leaf axil, some have several. In such case, the central bud is the true axillary bud that will develop into a new lateral branch; the others are called accessory buds. Buds that develop on the plant at positions other than the leaf axil are termed adventitious buds. These buds may develop on stems, roots, or leaves, producing new shoots from any of these positions. Adventitious buds are often produced on stems as

Figure 4-10

Examples of various leaf arrangements and leaf types.
Courtesy of Rick Parker.

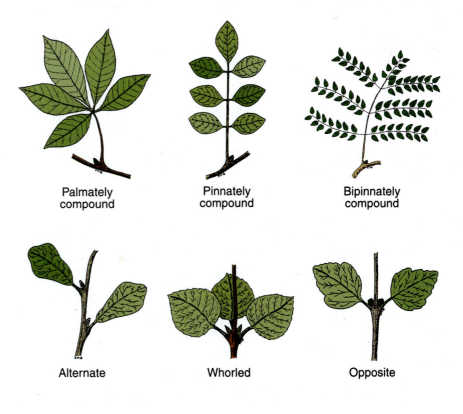

Palmately compound   Pinnately compound   Bipinnately compound

Alternate   Whorled   Opposite

a response to pruning or injury. As with axillary buds, they initiate new growth in response to hormonal controls, which are in turn linked to growth at the apex of each stem.

Modified Stems

Stem modifications include a variety of changes in both external morphology and internal anatomy. Minor adaptations include production of thicker bark, greater height, more flexible stems, basal buttressing, and increased photosynthesis. All enable the plant to adjust to some environmental stress. Major adaptations significantly alter the form of the stem. These specialized stems are still recognizable, however, in that they retain some or all of the typical stem structures, such as nodes, internodes, buds, and leaves.

Bulbs

Bulbs are cone-shaped stems surrounded by many scale-like leaves that are modified for food storage (see Figure 4-11). The conical stem produces a single aboveground shoot from the terminal bud and a new bulb from a lateral bud. Unlike corms, the food reserves of the bulb are in the modified leaves. These reserves are exhausted in the production of the leafy aboveground shoot, and food for storage in new bulbs is produced in the leaves by photosynthesis. Both onions and daffodils have bulbs. The food storage leaves of the onion bulb constitute the edible part.

Figure 4-11

Bulb showing leaf modification. The white layers are modified leaves. Courtesy of North American Bulb Exports.

Rhizomes

These are horizontal stems usually located beneath the soil surface. Superficially, they resemble roots, but like typical aerial stems they have nodes, internodes, buds, and often leaves. Adventitious roots develop at the bottom side of the **rhizome** in the area of the node, and shoots emerge above ground from the same location. The chief functions of rhizomes are food storage and vegetative reproduction. Most are perennial, increasing the length each year and sending up new plants at their nodes. Some plants, such as irises, produce new leaves and flowers only at the actively growing stem tip. If the rhizomes are divided between the nodes, new plants may be produced.

Stolons

Stolons, as shown in Figure 4-12a, are also horizontal stems that produce roots and shoots at the nodes, but they form above ground.

Figure 4-12a

Stem modifications: Stolon is a horizontal fleshy stem that can produce new shoots when it touches the soil. Courtesy of *www.TurfFiles.edu.*

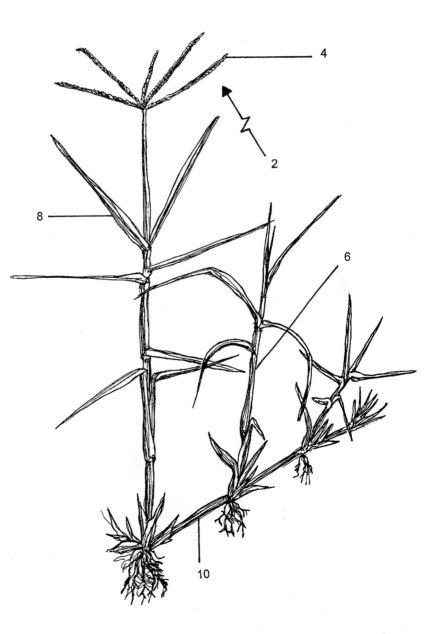

Bermuda grass (*Cynodon dactylon*) and a number of other grasses spread vegetatively—they produce new growth at every node.

Tubers

These are short enlarged organs that develop at the end of slender rhizomes. The potato (*Solanum tuberosum*) is a well-known **tuber** (see Figure 4-12b). The "eyes" of the potato are actually groups of buds with underdeveloped internodes. The eyes are used for vegetative propagation.

Corms

Corms, as shown in Figure 4-12c, are short, thickened underground stems that act as food storage structures. Gladiolus is an example of a corm from which a single shoot develops by using the food stored there. Once leaves are produced above ground, photosynthesis provides the food necessary to continue growth, produce a flower, and develop a new corm with the next year's food store.

Figure 4-12b

Tuber Irish potato shows (eye pieces) buds. Courtesy of Rick Parker.

Figure 4-12c

Corm is another modified stem. Courtesy of North American Bulb Exporter.

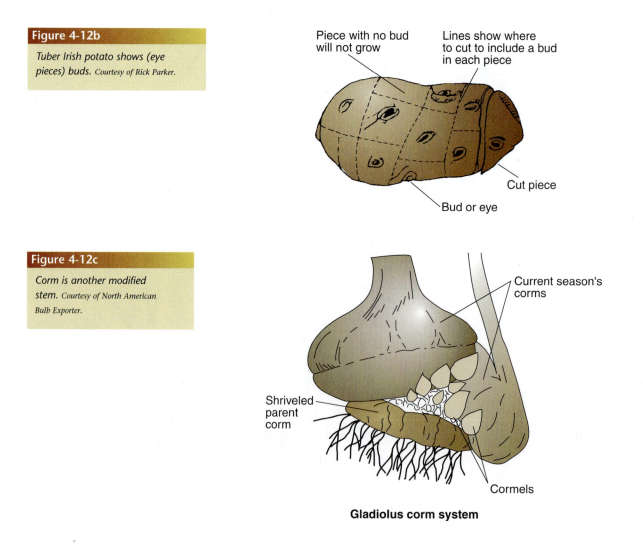

Piece with no bud will not grow

Lines show where to cut to include a bud in each piece

Cut piece

Bud or eye

Current season's corms

Shriveled parent corm

Cormels

Gladiolus corm system

Cladophylls

Cladophylls are green leaflike stems that not only look like leaves but also are actually photosynthetic. As a stem, however, they may also develop flowers, fruits, and even true leaves. The "spear" of the asparagus plant (*Asparagus officinalis*) is the edible shoot produced from underground rhizomes. Small scales on these spears are the true leaves. If allowed to continue growing; the shoots develop branches in the axils of their scales—the green, featherlike cladophylls. A few species of cactus (*Opuntia*) and some orchids (*Epidendrum*) have cladophylls.

Succulents

The succulent stems have tissue modified for water storage. Cacti are easily recognizable examples. There are two main types: those with jointed stems, such as the prickly pear and cholla, both of the genus *Opuntia*; and those with a single unjointed stem, the barrel cacti. Succulent stems are found not only in the cactus family (*Cactaceae*) of North and South American deserts but also in members of the *Euphoribiaceae* native to the African deserts. Members of these two families that are often remarkably similar in external morphology because of the analogous habitats is termed **convergent evolution**.

Thorns

As modified stem branches, **thorns** arise from the axils of leaves, as do regular branches. The honey locust (*Gleditsia triacanthos*), hawthorn (*Crataegus*) and fire-thorn (*Pyracantha*) are thorn-bearing plants (see Figure 4-13a). Rose (*Rosa*) "thorns" are not true thorns but are stem surface outgrowths called **prickles**, which Figure 4-13b

Figure 4-13a

Thorns are a part of the vascular system of the plant.

Figure 4-13b

Prickles are attached to the epidermis of the stem. Roses have prickles.

shows. Spines and thorns are terms often used synonymously; however, most spines are modified leaves arising from below the epidermis. Both are stiff, sharp-pointed, woody structures that can be equally painful if encountered inadvertently, and thus the technical difference is understandably unimportant to the victim. Both are equally effective in reducing predation by herbivores (plant-eating animals).

Spines

As modified leaves, **spines**, as shown in Figure 4-13c, are technically different from thorns, which are modified stem tissue. The most familiar example of plants bearing spines are the members of the *Cactaceae*; whose stems are modified for water storage. The ocotillo (*Fouquieria splendens*) is another desert plant with true spines that are a modified petiole and midrib of the first season of growth for that stem.

In subsequent years, true leaves are produced in the axil of the spines after rain.

Figure 4-13c

Spines are modified leaves.

Tendrils

Tendrils can be modified leaves or stems. Grape plants (*Vitis*) and Virginia creeper (*Parthenocissus*) are two well-known examples of plants with stem tendrils. Grape tendrils twine around a support structure; Virginia creeper tendrils anchor the plant with a sticky substance. Once anchored, these tendrils coil or contract to pull the plant closer to its anchoring surface.

Leaf Morphology/Function

Green plants are green because of the light-trapping pigment chlorophyll. In the leaf, light energy is converted into food energy in the form of carbohydrates. Although many herbaceous stems are green and photosynthetic, the greatest amounts of photosynthesis carried on by terrestrial plants occur in their leaves. Leaves, then, can be thought of as energy factories providing food energy necessary to sustain all life on earth.

Gas Exchange

The process of photosynthesis involves the exchange of carbon dioxide (CO_2) and oxygen with the atmosphere. Carbon dioxide is taken into the leaf, and oxygen is released into the atmosphere. During this gas exchange at the surface of the leaf, water is evaporated from the leaves. This water loss, the process of transpiration, is the end of the water movement through the plant, which began with the roots absorbing water from the soil. Thus, the interrelated processes of photosynthesis, gas exchange, and transpiration take place in the leaves.

Leaf Structure

Leaves are well designed to carry on their functions, usually having broad, flat, and thin **blades** that are connected to the plant stem by a stalk called a **petiole**. Leaf blades also come in an incredible array of sizes and shapes, and although sometimes consistent within a given plant group, the amount of variability displayed is often so great as to prevent use of leaf morphology as a diagnostic character for the identification of plant groups.

The leaf blade can be either simple, having only one unit attached to a petiole or stem node, or compound, having two or more separate leaflike subunits (leaflets) making up the blade. Even though some compound leaves are large enough that the individual leaflets might appear to be separate leaves, confusion can be avoided by remembering that buds always occur in the axil of leaves; leaflets do not have buds in their axils. The arrangement of the leaflets of a compound leaf can be **pinnately compound** (with leaflets arranged along the length of a central stalk) or **palmately compound** (with the leaflets attached to the end of the petiole like the fingers radiating from the palm of the hand). See the diagram in Figure 4-10.

Leaves also have a specific pattern of vascular arrangement in the blades. Leaf **venation** can be either **netted** (pinnate), **parallel**, or **palmate**, as shown in Figure 4-14. Netted, or pinnatifid, vein arrangement has a main central **midrib** with secondary veins branching from it. Tertiary and then quaternary lateral leaf venation arises from the primary branches, with each becoming progressively smaller. The midrib, then, is the most visible of all these levels of vein branching.

Parallel venation has no central midrib; rather, it has several equal-size veins running the length of the leaf blade parallel to each other. Between these major veins is a network of venation. Palmate

Figure 4-14

Different leaf venation.
Courtesy of Rick Parker.

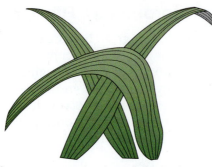

Grass, an example of parallel venation

Ginkgo, an example of dichotomous venation

Poplar, an example of pinnate venation

Sweergum, an example of palmate venation

venation is similar to parallel venation in that several equal-size veins exist, one going to each palmately arranged lobe of the blade. As in pinnate venation, each palmate vein is equivalent to a midrib before the leaf lobe, or leaflet, to which it runs.

Innumerable combinations of overall leaf shape (see Figure 4-15a), apex, margin, and base modifications are possible. Leaf **margins** are generally said to be **entire** (smooth), **dentate** (with sharp teeth), or **lobed** (with rounded intrusions). The **apex** (tip) and base, where the petiole attaches, can be variously rounded, angled, pointed, or indented. In addition, leaves can be smooth, rough, thin, thick, leathery, or succulent (fleshy and juicy with stored water), as shown in Figure 4-15b.

The petiole can also add to the overall leaf variability by being short or long, attached to the middle of the blade (**peltate**) or to the edge as is normal or even absent. When the petiole is absent, the leaf is said to be **sessile** on the stem and can further modify to partially wrap around (clasping) the stem or even form a **sheath** (very common in grasses). Additionally, the presence or absence of **pubescence** (hairs), which can be simple, branched, granular, barbed, long, short, fine, coarse, dense, sparse, and specifically localized or generally distributed, makes it evident how complex and variable a thorough discussion of leaf morphology would be.

Modified Leaves

Leaf surface-to-volume ratio is a factor in water loss; the greater the surface area of the leaf from which water can evaporate (transpiration),

Needle-like   Awl-like   Scale-like   Oblong   Linear

Lanceolate   Oblanceolate   Spatulate   Ovate   Obovate

Elliptic   Oval   Cordate   Peltate

Reniform   Pinnately lobed   Palmately lobed

Figure 4-15a

Various leaf shapes. Courtesy of Rick Parker.

Margins

Entire   Serrate   Double serrate   Serrulate   Crenate

Dentate   Incised   Undulate   Lobed   Spinase

Tips

Acute   Obtuse   Acuminate   Mucronate   Cuspidate

Truncate   Emarginate   Round   Retuse

Bases

Acute   Cuneate   Cordate   Oblique   Truncate

Rounded   Hastate   Sagittate   Obtuse

Figure 4-15b

Leaf margins, tips, and bases help in identification. Courtesy of Rick Parker.

the greater the amount that is lost. Since the total leaf volume controls the amount of water available, a large surface area with a small volume can rapidly result in wilting. A large volume and a small surface area conserve water against loss. Desert plants usually have small, often thicker leaves with a small surface-to-volume ratio. Plants in more **mesic** (wetter) areas often have much larger and thinner leaves, resulting in a larger surface area for the leaf volume. Some desert plants have large thick leaves with heavy wax cuticles. These succulents have greater water storage capacities and are well designed to allow only minimal water loss.

The century plant (*Agave*) and yucca have stiff, sharply pointed apical leaf spines, and *Agave* has sharp teeth on the edge of the leaf margin. Other desert species have marginal "ornament"; some exhibit extreme indentation, and others display sharp, stiff teeth, which are neither thorns nor spines. These effectively discourage predators (see Figure 4-16).

Figure 4-16

Agave are succulent, and well protected by a sharp spine.

Insectivorous Plants

Insectivorous plants display some of the most unusual adaptations of leaves to specialized functions. In general, the leaves are modified to entrap insects but occasionally trap small frogs and rodents. Enzymes secreted by the plant then digest the insects, providing necessary nutrients, especially nitrogen. Insectivorous plants grow in nitrogen poor areas such as bogs and swamps. Inadequate soil oxygen in these habitats results in little decomposition of organic material, the normal source of nitrogen for plants.

The manner of entrapment varies among the species. The Venus flytrap (*Dionaea muscipula*) has a folding leaf, the sundew (*Drosera*) a sticky surface, and the pitcher plant (*Darlingtonia*) columnar tubes. Pitcher plants differ from one species to the next in their insect-trapping modifications, as shown in Figure 4-17. Some have only water at the bottom of the column in which the insect drowns, others have stiff hairs on the inside of the tube pointing down to prevent the insect from crawling back out, and in others, sticky or gummy surfaces hold the insect.

Figure 4-17

Insectivorous plants entrap their prey with various leaf modifications. Courtesy of Julia Blizen, Cove Creek Gardens.

The misleading term *carnivorous plant* implies a general meat-eating capability. This occasionally has produced an astounding science fiction vision of giant man-eating plants with tendrils or vine-like stems that can reach out, entangle their hapless victims, and draw them into the plant's dark digestive interior. Such imaginary activities are, of course, just that. Insectivorous plants are hardly to be feared (unless you are a fly) but are remarkable examples of modified plant parts that result in increased adaptability to an otherwise inhospitable environment.

Vegetative Reproduction

Although most **vegetative reproduction** occurs by **budding** from the parent stem or from rhizomes or stolons, some plants are capable of vegetatively (asexually) reproducing via modified leaves. Bryophyllum leaves develop small plants at the notches along the dentate margin, which has foliar embryos. These small plants are complete with leaves and short stems; these baby plants fall from the still healthy and active parent leaf, root, and begin growing. Rex begonias produce new plants from the upper leaf surface once it is in contact with the soil. The process of propagating plants asexually is important in ornamental plant commercial operations, fruit tree and grapevine production, and other horticultural production.

Reproduction in plants is, for the most part, a completion of a sexual life cycle. The advantage of **sexual reproduction** is that it produces variations within the species (Chapter 12). Should the environment change, at least some members of the population might be adapted to the new condition. These individuals would survive, allowing for the successful continuation of the species.

On the other hand, it is sometimes advantageous to have an asexual reproductive system that produces genetically identical individuals. Individuals produced from a single plant are referred to as members of a clone. Although there might be short-term advantages, even in nature, to producing individuals that are identical and perfectly suited to the environment, it is difficult to see how long-term survival could be served by a species with no means of sexual reproduction. The advantages of genetically identical and uniform plants in agriculture are obvious. Mechanical harvesting, the timing of crops for market, and reducing perishability all depend on reliable and constant factors. Much the same results are achieved through traditional breeding however; genetic engineering is being used in agriculture to achieve the same and better goals as traditional breeders.

Although many plants reproduce asexually in nature, few of them have totally lost the ability for sexual reproduction. One plant that reproduces exclusively asexually is the commercially grown banana. As far as is known, there is no record of this plant having reproduced by seeds. Small, vestigial seeds are found in bananas, but they are not fertile. Thus all commercial banana propagation is accomplished by offshoots from the mother plant (see Figure 4-18). As the central stalk flowers and produces bananas, the stalk dies, but offshoots at the base create new plants, which can simply be pulled off and planted. Commercial banana plantations have generated their crops in this manner; however, banana plants are being micropropagated for commercial growers.

Figure 4-18

There is no proof that bananas produce seed. There are small vestigial seed seen in the fruit, but they are not fertile.

Budding/Grafting

Specific adaptations of organs allow asexual propagation from essentially all vegetative parts of the plant: leaves, stems, buds, roots, and even single cells. **Grafting** (see Figure 4-19a) and **budding** (see Figure 4-19b) are vegetative methods used to propagate plants of a clone whose cuttings are difficult to root or to make use of a **rootstock** rather than having the plant on its own roots. The **scion** is that part of the graft combination that is to become the upper or top portion of the plant. Just as transplants in animals, genetic compatibility is essential, although it is not quite as specific. Often two species within the same genus can be grafted, and they are perfectly compatible. The critical factor in ensuring that the graft or bud "takes" is the matching of the **vascular cambium** of both the rootstock and the scion. If this tissue fails to join, the donor scion will die. For a graft, an entire stem piece is placed on a stock, according to the size of its stem, time of year, and various other factors. The art (as well as the science) of grafting ensures a high degree of success. In budding, a single bud is placed on a stem of the stock. Essentially all commercial roses are propagated in the manner.

(A)
The scion before any cuts are made.

(B)
The first cut is made in the scion.

(C)
The second cut is made in the scion.

(D)
The root, before any cuts are made.

(E)
The first cut is made in the root.

(F)
The second cut is made in the root.

(G)
The scion and root are positioned for joining.

(H)
The scion and root are pushed together. (Cambrium must match on at least one side.)

(I)
The two pieces are tied together.

(J)
A covering of grafting wax may be necessary to prevent drying. This is especially important if the scion and the rootstock are not the same.

Figure 4-19a

There are various ways to graft plants, however, one must have rootstock and the scion that is compatible for this procedure to be effective. The vascular cambium from both plants must meet and the new cells arise in the lateral meristematic tissue. Courtesy of Rick Parker.

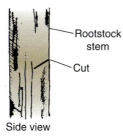

Side view

(a) A 45° cut is made in the rootstock about one-quarter of the way through the stem.

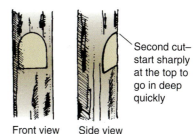

Second cut– start sharply at the top to go in deep quickly

Front view   Side view

(b) A second cut is made starting about 1 1/2 inches above the first cut and extended down to meet the first cut. *Cut deeply enough at the top so that an upside-down "U" form is made. A cut that is made too shallow results in an "A" shape.*

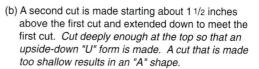

Chip removed

Side view

(c) The chip produced by the two cuts is removed.

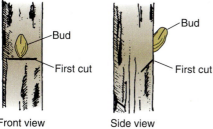

Bud — First cut
Bud — First cut

Front view   Side view

(d) The bud to be inserted is cut from the bud stick exactly as the chip was removed from the rootstock.

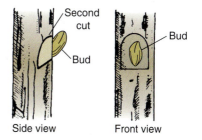

Second cut
Bud
Bud

Side view   Front view

(e) Second cut on the bud.

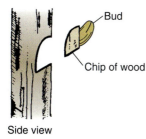

Bud
Chip of wood

Side view

(f) The bud is removed from the bud stick.

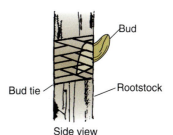

Bud
Bud tie — Rootstock

Side view

Note: Be sure to align the cambium of the bud and the rootstock on at least one side

(g) The bud is inserted in the rootstock and tied with rubber bud tie.

Figure 4-19b

As in grafting, there are different methods for budding plants. This illustrates how the chip bud and T-bud methods are performed.
Courtesy of Rick Parker.

Stem Cuttings

Many ornamental and horticultural important species are propagated by **stem cutting**, which is shown in Figure 4-20. When stem cuttings are placed in a suitable propagating medium (sand, peat moss, vermiculite), adventitious roots form at the cut end of the cutting, after which the cutting can be potted and grown in a normal manner. Certain hormones promote the rooting process in most cases, and their use is standard practice in commercial operations.

Stem cutting with unwanted portions
(leaves, seed heads, and flowers) removed

Figure 4-20

Stem tip cuttings are made by cutting a 2- to 6-inch piece of stem. Remove leaves that will touch the media and remove other unwanted portions (seed heads, flowers, and other leaves).

Courtesy of Rick Parker.

A few ornamental species are propagated by single leaves (African violets and begonias are often propagated this way). The base of the leaf or the base of the petiole will root in a suitable medium and produce a new shoot from the leaf.

On rare occasions, roots may be cut into pieces, and adventitious buds will form on them. Sweet potatoes are propagated in this manner. The common perennial morning glory (*Ipomoea purpurea*) is spread during cultivation because it propagates from root cuttings; plowing increases the number of root sections, which then produce a new plant.

Layering

This is a method of propagating plants with long flexible stems, such as grapes or berries. The stem is pinned to the ground at a node and covered with soil, and adventitious roots form at the node. Once the roots have appeared, the stem may be cut from the parent plant, and the new plant is on its own. **Layering**, as shown in Figures 4-21a and 4-21b, is one of the surest methods of asexual propagation because the new part is still receiving nutrition from the parent plant; there is tip layering and simple layering.

Figure 4-21a

Tip layering is when a stem that remains attached to the root-stock is placed in the ground and the tip of the stem is allowed to remain above the ground and new roots appear at the bend in the stem. A new plant forms and can be removed. Courtesy of *Rick Parker.*

Figure 4-21b

Simple layering is similar to tip layering, but you make a notch in the stem before you put it under ground. Leave 6 to 12 inches of the top exposed, stake that portion, and when it roots you have another plant. Courtesy of *Rick Parker.*

Air Layering

Air layering is the process of packing a stem node with moist peat moss and wrapping it in plastic. Once adventitious roots form at the node, the top of the plant can be cut off just below the rooted node and planted as a new individual. The base of air layered plant can initiate new terminal growth (see Figure 4-22).

Figure 4-22

Some plants can be propagated by air layering: A section of bark is removed, the area is packed in damp sphagnum moss, and wrapped in plastic. The roots form at the cut surface. Courtesy of *Rick Parker.*

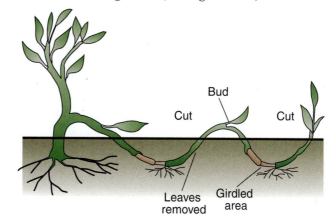

Summary

1. Higher plants, especially angiosperms, are adapted through their external morphology and survive the stresses of the environment, since plants do not have the advantage of motility.

2. Roots anchor plants, absorb and conduct water and minerals to the rest of the plant, and store starch as a food resource. Both taproot and fibrous root systems increase their surface areas many times by the production of root hairs. Adventitious roots include roots developing at stem nodes as well as prop roots, which help support the stem.

3. Root modification includes attaching to surfaces (as in ivy) for support and to other plants for nourishment. Parasitic plants drain their host so much that they may ultimately kill them. Roots also enlarge for storage. Nutrient uptake is improved through root nodulation for nitrogen fixation and by fungal mycorrhiza associations.

4. Stems conduct organic solutes, water and nutrients throughout the plant body, produce new growth; support leaves, flowers, and fruits; and store materials. Stems can be prostate, various upright shapes, or even twining on other structures. Stems are either herbaceous or woody and have leaves and buds attached in a variety of arrangements.

5. Stem modification includes rhizomes and stolons; both producing roots and shoots at the nodes for vegetative propagation. Enlarged storage tubers and reproductive corms and bulbs are all underground stems. Cladophylls are leaflike, succulent stems modified for water storage. Thorns protect the plant from herbivores.

6. Primary leaf functions are photosynthesis and gas exchange; huge quantities of water can be lost during the latter activity. Leaf structure is incredibly variable, with blade shape ranging from needlelike to flat and broad. Leaves can be simple or variously compound, and their margins can be smooth to deeply lobed or divided. Leaf surfaces are smooth, pubescent, prickly, or many other textures.

7. Leaf modifications include succulence for water storage, bulb leaves for storage, spines, and climbing tendrils. Insectivorous plants have leaves modified for insect entrapment and digestion to supplement nitrogen uptake.

8. Vegetative (asexual) reproduction by rhizomes, stolons, underground-modified stems, or leaves provide mechanisms for commercial plant propagation. There are a number of advantages to cloning plants although the variability resulting from sexual reproduction is lost. Budding and grafting are very common techniques for vegetative plant propagation, as are stem cuttings and layering.

Something to Think About

1. From where do the primary roots arise?

2. Where are the nodes found?

3. Which plants have stems?

4. Name three types of roots.

5. Identify four types of underground stems.

6. Describe four types of venation found in leaves.

7. Why is asexual reproduction included in this chapter?

8. List what leaf modifications are and how they function.

9. What are the vegetative reproduction structures?

10. Name root modifications and how they work.

Suggested Readings

Barden, J. A., R. G. Halfacre, and D. J. Parrish. 1987. *Plant science*. New York: McGraw-Hill.

Collinson, A. S. 1988. *Introduction to world vegetation*. London: Unwin Hyman Ltd.

Hartmann, H. T., A. M. Kofranek, V. E. Rubatzky, and W. J. Flocker. 1988. *Plant science: Growth, development, and utilization of cultivated plants* (2nd ed.). Englewood Cliffs, NJ: Prentice Hall.

Janick, J., R. W. Schery, F. W. Woods, and V. W. Ruttan. 1974. *Plant science: An introduction to world crops* (2nd ed.). San Francisco: W.H. Freeman and Company.

Janick, J. Ed. 1989. *Classic papers in horticultural science*. Englewood Cliffs, NJ: Prentice Hall.

Reiley, H. E., and C. L. Shry, Jr. 1997. *Introductory horticulture*. Albany, NY: Delmar Publishers.

Shroeder, C. S., E. D. Seagle, L. M. Felton, J. M. Ruter, W. T. Kelley, and G. Krewer. 1995. *Introduction to horticulture: Science and technology*. Danville, IL: Interstate Printers and Publishers, Inc.

United States Department of Agriculture. 1961. *Seeds: The yearbook of agriculture*. Washington, DC: United States Department of Agriculture.

Internet

Internet sites represent a vast resource of information. The URLs for Web sites can change. Using one of the search engines on the Internet, such as Google, Yahoo!, Ask.com, and MSN Live Search, find more information by searching for these words or phrases: **primary roots, asexual propagation, sexual propagation, inflorescence, seed germination, leaf shapes, leaf modifications, imbibition, plumule, hypocotyl hook, dormancy factors, stem cuttings, budding, grafting, layering, bulbs, rhizomes, tubers, corm, herbaceous, perennials, annuals, terminal bud, compound leaves, and simple leaves.**

CHAPTER 5

Design Basic II: Morphology and Adaptations of Reproductive Structures

Sexual reproduction in higher plants always takes place in specialized plant parts that develop during specific periods of the plant's life. In angiosperm, sexual reproduction occurs in the flower, resulting in the production of fruit containing seeds, which house the new plant embryo. Flowers come in a vast array of sizes, shapes, colors, and arrangements, as do the resulting fruits and seeds. As with root, stem, and leaf morphology, however, the basic components are the same from plant to plant.

After completing this chapter, you should be able to:

- Identify the different floral parts
- Distinguish the male and female parts of the flower
- Know flower classifications
- Recognize different fruit types
- Relate flower production to fruit production
- Learn morphology of different seeds
- Understand seed germination
- Discover the difference between monocots and dicots
- Link plant science to everyday life

Key Terms

receptacle	placental	dioecious
sepals	carpel	regular
petals	simple pistil	irregular
stamens	compound pistil	fused
pistil	fertilization	vectors
calyx	superior ovary	nectar
corolla	inferior ovary	self-pollinated
perianth	inflorescence	cross-pollinated
sex cells	complete flower	outcrossing
gametes	incomplete flowers	self-incompatibility
stamens	bisexual	pollinators
ovary	perfect	co-evolution
pollen	unisexual	parthenocarpy
anthers	imperfect	pericarp
filaments	staminate	exocarp
ovules	pistillate	mesocarp
style	carpellate	endocarp
stigma	monoecious	simple fruits

dry fruits
dehiscent
indehiscent
compound fruit
aggregate fruits
achenes
multiple fruits
dispersal
scarification
double fertilization
endosperm
zygote
embryo
viviparous

viability
germinate
imbibition
cotyledons
digestion
epicotyl
hypocotyl
plumule
lignified
embryonic axis
senescence
annuals
biennials
rosettes

perennials
herbaceous perennials
monocarpic
polycarpic
monocotyledonae
  (monocot)
dicotyledonea (dicot)
scientific names
common names
genus
species
binomial

Flower Morphology

Although the flower has deservedly been made the object of poetry and often symbolizes beauty, love, peace, and happiness, the basic biological functions of the flower is sexual reproduction. Exotic colors and shapes of flowers are actually devices to attract specific pollinators. This ensures their reproductive success, since pollen will be carried among flowers of the same species. Not all flowers are large or beautifully showy, but even the plainest of flowers functions successfully. Some flowers are not conspicuous at first glance, but close observation reveals remarkable, complex, colorful, and beautiful design.

Floral Parts

A typical flower possesses four different floral parts that are attached to the **receptacle** in the following order, from outside to inside: **sepals**, **petals**, **stamens**, and **pistil** (see Figure 5-1). The sepals are collectively referred to as a **calyx**, which encloses and protects the bud as the flower develops within. The **corolla** (crown), the collective name for the petals, is frequently colorful and exhibits diverse sizes and shapes. Together, the calyx and corolla are termed the **perianth**.

111

The actual production of **sex cells**, and the **gametes**, takes place in the **stamens** and **ovary** of the pistil. Stamen produces **pollen** in its **anthers**, which are saclike structures attached to the end of the slender **filaments**. These filaments raise the anthers to a position where the pollen is more accessible to visiting pollinators.

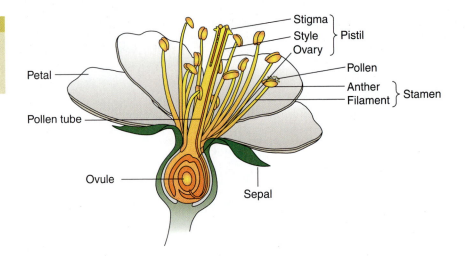

The filament of a stamen is sometimes fused to the corolla tissue, especially in corollas with petals joined to form a tuber section. Some flowers have nonfunctional, sterile stamens called staminodia, which often have flattened petaloid filaments. In fact, petals originally evolved from stamen through such modifications.

At the center of the flower is the pistil, which often resembles a pharmacist's pestle. The base of the pistil is the ovary, which contains one or more **ovules**. The slender neck-like portion of the pistil is the **style**, which elevates the **stigma** to a favorable position relative to contact with pollen. Whether air born or brought by a visiting pollinator, the process of pollen landing on the stigma is pollination. Many stigma surfaces are sticky, pubescent, or otherwise modified to help ensure successful pollen attachment. In addition, pollen grains of many species are "ornamented" with spines, ridges, and barbs to further aid pollination.

A pistil can have a single **placental** surface to which one or more ovules are attached, or it can have two or more placentas. When an ovary has more than one, a wall usually separates them. Each separate reproductive unit composed of a placental surface and ovules is called a **carpel**. Each carpel also has its own style and stigma; however, these are usually fused together and appear as a unit of the total pistil. When a pistil is composed of a single carpel, as in the flowers of peas and beans, it is called a **simple pistil**. When there are multiple carpels, such as those found in tulips, lilies, grapefruits, and poppies, it is termed a **compound pistil**. The sections of a cut grapefruit are the carpels of the compound ovary. The ovules of each of these carpels are where sexual **fertilization** (gamete fusion) takes place in flowering plants. The resulting plant embryo is housed in the seed (mature ovule) within the fruit (mature ovary).

In most flowers, the pistil is attached to an enlarged apical portion of the stem called the receptacle. Usually it is located above the attachment points of the other floral parts (stamen, petals, and sepals). Where so attached, it is termed a **superior ovary**. In some flower groups, however, the ovary portion of the pistil has a modified receptacle or perianth tissue surrounding it and fused to it with the other floral parts attached above the ovary at the top of this tissue. Such flowers are said to have an **inferior ovary** because of the lower position of ovary attachment relative to the other floral parts. The position of the ovary is an important taxonomic characteristic.

Flowers can occur singly or in an **inflorescence** of several to many flowers, as you can see in Figure 5-2. Inflorescence can be as simple as only a few flowers attached near one another on the flowering stem or as complex as the head of a sunflower, which is composed of hundreds of tightly grouped flowers with the head actually

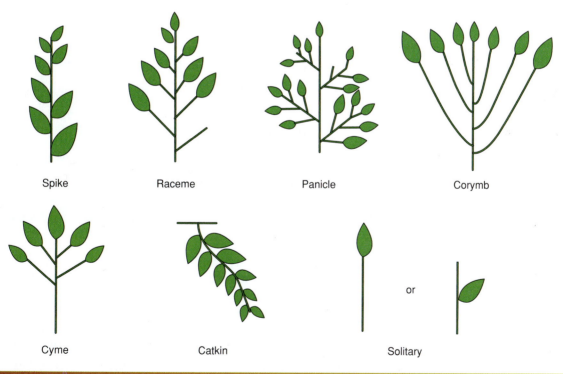

Spike   Raceme   Panicle   Corymb

Cyme   Catkin   Solitary   or

Figure 5-2

Inflorescence types. The peduncle is the stem just below a solitary flower or an inflorescence. Pedicels are the stalks of each flower in the inflorescence. Spike inflorescence type, raceme, panicle, corymb, umbel, catkin, and head.

being in the shape of a single larger flower. The range of inflorescence complexity and shape and the modifications of the component flowers within some inflorescence are great.

All these arrangements and modifications in some way ensure pollination and fertilization and thus the reproduction success of the species. Unlike leaf morphology, the reproductive parts of a given plant species are much less affected by fluctuations in the environment such as temperature, wind, and availability of water. Flower appearance is usually consistent because pollinator-flower specificity often exists. If a pollinator fails to recognize the flowers of some individuals of that species, reproduction cannot occur. Failure to successfully reproduce sexually could ultimately result in few offsprings, reduced genetic variability, less adaptability, and finally extinction.

A **complete flower**, such as the one in Figure 5-3, has sepals, petals, stamens, and pistil all present. Many successful species, however, lack one or more of these four basic floral parts. Such flowers are termed **incomplete flowers**. Flowers having both stamens and pistil (that is, both sexes) are called **bisexual**, or **perfect**, flowers whereas **unisexual**, or **imperfect**, flowers contain either stamens or pistil but not both. Those containing only stamens are logically called **staminate** flowers, and those lacking stamens but having carpels are termed either **pistillate** or **carpellate**.

Figure 5-3

Complete flower.

Plants having both staminate and pistillate flowers on the same plant are **monoecious**, literally meaning "one house" (see Figure 5-4). **Dioecious** (two houses) plants, such as the holly shown in Figure 5-5, have the staminate flowers on one plant and the pistillate flowers on another. Corn (*Zea mays*) and squash (*Cucurbita*) are monoecious; willow trees (*Salix*) are dioecious. Historically, the designation "male" and "female" parts have often been applied to the stamens and carpels, respectively, and thus with unisexual flowers one might see reference to the male or female flower. But such designations should be made only with a full understanding of what is meant, and that is best accomplished with the more descriptive terms *staminate* (male) and *pistillate* (female).

Figure 5-4

Female flowers and male flowers are separate but located on the same plant—a monoecious plant.

Figure 5-5a

Staminate flowers: holly (Ilex). A holly plant is dioecious and will only produce pollen. Courtesy of USDA Photo Gallery.

Figure 5-5b

Pistillate flowers on the holly plant will produce fruit. Courtesy of USDA Photo Gallery.

Flower Shape

Flower shapes are either **regular** (radially symmetrical), as a sunflower (see Figure 5-6) or irregular (bilaterally symmetrical), as the snapdragon in Figure 5-7. A regular flower can be cut through the center in more than one plane and results in identical halves. An **irregular** flower is bilateral, having only one plane through which it may be cut to result in mirror-image halves.

Additionally, the petals can be either district (not fused to one another) or partially to completely **fused** into a single structure. Sepals, stamens, and carpels also may be distinct or variously fused. Nor is fusion limited to members of the same floral part. The filaments of many species fuse to the corolla, and there are isolated examples of other floral parts fused to each other.

Flower Modification

The array of flower shapes, sizes, color, and arrangement seems endless. The variability that exists, however, is not haphazard. Flower modification heightens the likelihood that pollination agents, or **vectors**, may be general visitors, or they may be attached to specific flowers. They are enticed by a unique floral presentation that result from alternation of the flower parts. The changes include different sizes, shapes, fusion, colors, patterns, odors, and edible materials, as well as relocation of stamens and stigma. Thus, flower modifications promote reproductive success.

Petal Modifications

Petal color, size, shape, and fusion are the most common variables in flower morphology. Each such modification exists as a result of increased reproductive success through pollinator visitation. Some flowers are successful exclusively through visual attraction (see Figure 5-8). Others must additionally produce **nectar** as a food reward for a visiting pollinator or have a strong, detectable odor. Odors can be sweet and fragrant or foul. An aroma similar to rotting meat attracts flies as the pollinator vector in certain plants. Some petal surfaces are marked to signal to appropriate insects which way to approach, where to land, and which way to enter the corolla tube. These markings often change after successful pollination. This signals to other potential visitors that the flower has already been visited; significant efficiency results.

The human eye cannot appreciate all these petal markings, but certain insects can. Reflections of ultraviolet wavelengths of light from a given portion of the petal produce a specific pattern for bees and wasps, which can be seen in the ultraviolet range (see Figure 5-9). Such insects, however, do not discriminate colors at the opposite end of the spectrum, making them essentially red color blind.

Other floral and nonfloral parts display modifications involved in pollinator attraction. In some flowers, the sepals appear petaloid; in others, highly modified staminodia are petal-like. The stamens are

visually attractive through elongated and colorful filaments or large, colorful, and showy anther sacs (*Caesalpinia* and *Tradescantia*). The bright red poinsettia "petals," shown in Figure 5-10, are in fact colorful leafy bracts located below the cluster of small relatively inconspicuous flowers. Bougainvillea also has petaloid bracts and less conspicuous flowers.

Figure 5-10

The poinsettia bracts are much showier than the small tubular flowers located in the center of the bracts.

There are only a few examples of modified flowers and flower parts, whose evolution increased pollinator visitation and reproductive success. The complex mosaic of natural beauty that has resulted from these modifications is intellectually intriguing but also enjoyable to simply observe and appreciate.

Pollinating Systems

There are two general categories of pollination strategy: plants that are **self-pollinated** have their own pollen land on their own stigma; those that are **cross-pollinated** receive pollen from the flowers of other individuals of the same species.

Self-Pollination

Most self-pollinating plants have bisexual flowers; their flower design allows their own pollen to be shed onto their stigma (or a visiting insect can initiate this). Since all the flowers of a single plant have the same genetic origin, pollen transfer to any flower on the same plant is still considered to be a self-pollinating event. It does not increase genetic variability.

Cross-Pollination

When genetic information from two different plants is mixed, the seeds have a larger pool of genetic information. The resulting progeny will display variations in form and function. This cross-pollination is also called **outcrossing**. Because greater genetic variability is desirable, outcrossing has been ensured in many plants by the development of genetic **self-incompatibility**.

Self-Incompatibility

This can be physical or chemical in nature. Some flowers are structurally designed so that the pollen of that flower cannot be shed onto its stigma. In some the style is elongated more than the stamen filaments, positioning the stigma well above the anther sacs. In others, the pollen is not released from the anthers when the stigma is receptive; it is shed either before the stigma is ready or after the stigma has already been pollinated from a different plant.

Chemical Incompatibility

This type of incompatibility is less easily determined. Certain proteins in the outer layer of the pollen grain are involved in a recognition reaction with the surface of the stigma, which ensures that pollen from the same plant will be rejected.

Wind Pollination

For a plant to be successfully cross-pollinated or outcrossed, there must be a dependable source of pollen from other plants of the species. Wind-pollinated plants generally have many small, inconspicuous flowers that produce large quantities of lightweight pollen grains. Unlike much of the pollen of insect-pollinated plants, the pollen grains are usually smooth and do not stick together. Since wind carries the pollen, colorful petals, edible tissues, attractive odors, and nectar production are generally lacking; these adaptations are necessary only for plants that attract insect visitors. Wind-pollinated flowers are usually modified, having long styles with well-exposed stigmas that are often feathery or branched and fairly large. Their stamens are also well exposed to the wind, sometimes hanging down away from the flower on long, slender, flexible filaments to better shed their pollen when the wind blows across them. Since they are usually very small flowers, they most often occur in inflorescence; their great numbers and density further increases the chances of pollination. Most of these flowers have only a single ovule and produce a single-seeded fruit, such as a grass grain, the winged fruit of the elm, or the acorn of an oak.

Because wind pollination is chancy, copious amounts of pollen are shed from each flower (see Figure 5-11). The vast majority never lands on a stigma but end up on the ground not far from its source. Many wind-pollinated species are unisexual (monoecious) with small staminate flowers grouped together on one part of the same plant, and pistillate flowers together on a different part of the same plant. Other wind-pollinated species, such as cottonwood and honey locust, are dioecious and have staminate and pistillate flowers on different plants.

Since wind pollinate is inefficient, it is most successful where many plants of the same species grow close together in open areas. The tropics, therefore, have few wind-pollinated species. They are found mostly in temperate zones. All grasses and deciduous hardwood trees are wind-pollinated, the latter flowering primarily in the early spring before the leaves develop.

Gymnosperms are also wind pollinated. Most of the cone-bearing trees (conifers) have small pollen-producing male cones and much larger female cones on the same tree. It is probable that angiosperms evolved from gymnosperms, so for a long time plant scientists theorized

Figure 5-11

Wheat flowers with the anthers exerted so that their pollen is picked up by the wind.

that, since gymnosperms are wind pollinated, wind-pollinated angiosperms must be relatively primitive. Now it is generally accepted that they are not primitive but in many cases fairly advanced, having evolved from insect-pollinated groups.

During the flower season, many areas have so much pollen shed by wind-pollinated trees (both angiosperms and gymnosperms) and grasses that the air appears hazy. Ponds and shorelines of lakes and still streams have a layer of billions of yellow pollen grains floating on the surface. Hay fever and many other allergic reactions are common ailments during these periods.

Insect Pollination

By far the most common **pollinators** are insects. The only group of organisms that displays as great a diversity in numbers of species, distributional range, and overall complexity as the flowering plants are the insects. The origins and developments of such variability and numbers in these two groups have been processes of **co-evolution**.

In most plants, after fertilization, the ovary of the flower matures into fruit. Fertilization actually takes place in the ovules, which in turn develop into seeds. Fruit development, therefore, is normally triggered by pollination-fertilization-seed development, and if there is no fertilization or if the seeds fail to develop, the ovary will not mature into a fruit (see Figure 5-12).

As with most other "normal" events, there are exceptions; some fruits, bananas for example, develop without fertilization. This process is called **parthenocarpy**, and occurs in fruits that are seedless. Crop scientists specializing in fruit production have studied seedless grapes and oranges found in natural stands. In some unusual cases, the ovary does not enlarge, but the receptacle grows and surrounds the ovary. Such is the case in the apple.

Figure 5-12

Simple fruits develop from a single ovary.

Fruit play an important role in the reproductive cycle of flowering plants, providing continued protection for the enclosed seed and siding in their dissemination. The step from naked seed (in gymnosperms) to enclosed seed (in angiosperms) was a major one in the evolution of plants.

Fruit Morphology

As the ovules begin development into seeds, the ovary wall matures into the fruit wall, which is then called the pericarp. The **pericarp** usually has an outer **exocarp**, a middle **mesocarp** layer, and an inner **endocarp**. The distinctiveness of these three subunits of the fruit wall varies from plant to plant. On the outside of the developing ovary are shriveled stamens, corolla, and calyx. The calyx, in fact, is often not only still visible, but in some species remains healthy and even enlarges as the fruit matures.

Kinds of Fruit

Simple fruits are fruits that develop from a single ovary, whether it is composed of only one or several fused carpels, such as the peach in Figure 5-13.

Figure 5-13

The peach is a fleshy simple fruit.

Fruits can be further subgrouped as either fleshy or dry, depending on whether the pericarp is soft and juicy. Additionally, **dry fruits** can be either **dehiscent**, where the dry pericarp splits open at maturity, releasing the seeds, or **indehiscent**, where the seeds remain with the fruit and the entire fruit falls from the parent plant at maturity (see Table 5-1).

Table 5-1

<table>
<tr><th colspan="3">Types of Simple Fruits</th></tr>
<tr><th>Type</th><th>Development</th><th>Examples</th></tr>
<tr><td colspan="3" align="center">Simple Fruits: Dry and Dehiscent</td></tr>
<tr><td>Follicle</td><td>Develops from single-carpel ovary, splits open down one side</td><td>Columbines, magnolia, milkweeds</td></tr>
<tr><td>Legume</td><td>Develops from single-carpel ovary, splits open along both sides</td><td>Pea family; all peas and beans (*Fabaceae*)</td></tr>
<tr><td>Silique</td><td>Develops from two-carpel ovary, halves fall</td><td>Mustard family (*Cruciferae*)</td></tr>
<tr><td>Capsule</td><td>Develops from compound ovary with two or more carpels; capsules dehisce in many different ways</td><td>Cotton, poppy, primrose, pinks</td></tr>
<tr><td colspan="3" align="center">Simple Fruits: Dry and Indehiscent</td></tr>
<tr><td>Achene</td><td>Small, one-seeded fruit; pericarp easily separable from seed coat, although closely encasing it</td><td>"Dry" fruits of strawberry, buckwheat, and sunflower family (*Asteraceae*)</td></tr>
<tr><td>Samara</td><td>Winged one- or two-seeded achenelike fruit; wings form from outgrowth of ovary wall</td><td>Elms, ash, on seeded maples</td></tr>
<tr><td>Schizocarp</td><td>Two- (or more) carpel ovary that, on maturity, splits into separate, one-seeded sections that fall</td><td>Maples considered samaras or winged schizocarps</td></tr>
<tr><td>Caryopsis</td><td>One-seeded, usually small fruit with pericarp completely united to seed coat</td><td>"Grain" of all grass family (*Graminacea*); includes wheat oats, rice, corn, barley, oats, and other important grasses</td></tr>
<tr><td>Nut</td><td>One-seeded fruit with hard pericarp (shell)</td><td>Walnut, hazelnut, chestnut, acorns</td></tr>
<tr><td>Fleshy berry</td><td>Two- (or more) carpel ovary, each usually having many seeds; inner layer of pericarp (mesocarp and endocarp) is fleshy</td><td>Tomatoes, grapes, dates</td></tr>
<tr><td>Hesperidium</td><td>Berry with thick leathery "peel" (exocarp and mesocarp) and juice; pulpy endocarp arranged in sections; rind has oil glands</td><td>Oranges, grapefruit, lemons, limes; all citrus fruits</td></tr>
<tr><td>Pepo</td><td>Berry with outer wall or rind formed from receptacle tissue fused to excocarp; fleshy interior is mesocarp and endocarp</td><td>Gourd family (*Cucurbitaceous*) cucumbers, watermelons, squash, pumpkin</td></tr>
<tr><td>Drupe</td><td>Usually only one-carpel ovary and with only one seed developing; endocarp is hard and stony, fitting closely around seed: mesocarp is fleshy, and fruit is thin skinned (thin, soft exocarp)</td><td>Many members of the rose family (Rosaceae), including cherry, peach, plum, almond, apricot; not in the Rosaceae, olive and coconut are also drupes (coconut has a fibrous outer coat rather than fleshy)</td></tr>
<tr><td>Pome</td><td>From compound, inferior ovary in surrounding receptacle or perianth tissue (one embedded); fleshy edible part is ripened tissue surrounding ovary, which matures into "core" and contains seeds</td><td>Apples and pears, both members of the Rosaceae family</td></tr>
<tr><td colspan="3" align="center">Aggregate and Multiple Fruits</td></tr>
<tr><td>Aggregated fruits</td><td>Development of numerous simple carpels from a single flower; some are dry fruits, attached to fleshy receptacle, others an aggregation of simple fleshy fruit (drupe)</td><td>Strawberry, blackberry, raspberry</td></tr>
<tr><td>Multiple fruits</td><td>Individual ovaries; nutlets on enlarged fleshy receptacle or group of berries</td><td>Mulberry, pineapple</td></tr>
</table>

Courtesy of Rick Parker.

Compound Fruits

These are formed by the development of a group of simple fruits. There are two general kinds of **compound fruit**.

Aggregate fruits

Aggregate fruits are formed from many separate carpels (ovaries) of a single flower, as in the strawberry, raspberry and blackberry. Strawberries differ from the latter two in having small, dry, individual fruits or **achenes** attached to an enlarged juicy receptacle. Blackberries and raspberries are an aggregation of separate fleshy fruits (drupes) attached to a common receptacle.

Multiple fruits

Multiple fruits are the result of the development of the ovaries of several separate flowers that have fused on the axis of the inflorescence. Pineapples and figs are both multiple fruits (also called accessory fruits).

Fruit Modification

As among flowers, the difference among fruits has evolved directly as a result of increased reproductive success. The role that fruits play is directly a result of increased reproductive success. The role of fruits in the reproductive cycle of plants is seed **dispersal**. The modifications therefore reflect the methods by which the fruit are transported—by wind, water, or animal. If the fruit (and the seed within) dispersed to areas away from the parent plant, each new plant seedling has a better chance for adequate space, water, and nutrients, which it needs to survive. Additionally, the species thereby has the opportunity to increase its total distributional range. The genetic variability of new generations contributes to their success in new environments.

Wind Dispersal

Very small, lightweight fruit can be carried some distance from the parent plant by the wind. Larger fruit must be modified in some way for effective wind dispersal. Dandelion fruit, with their parachute tuft of soft fine bristles and the winged fruit of maple trees are wind-borne (see Figure 5-14).

Water Dispersal

Some fruits have thick, fibrous outer coverings that provide buoyancy and protection from salt water. The coconut's husk has ensured it dispersal to virtually every tropical sandy shoreline in the world (see Figure 5-15). Freshwater streams and lakes also are dispersal agents for buoyant fruit of many of the plants found only in these habitats. Some fruits float because of low-density buoyant tissue, whereas others have air inside them. Other fruits are not particularly buoyant or water-resistant but can float long enough to be carried at least a short distance from their source.

Figure 5-14

Fruit modifications play a large part in the dispersal of seeds.

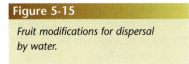

Figure 5-15

Fruit modifications for dispersal by water.

Rainwater can act as a dispersal agent. A reasonable heavy downpour will help knock some fruits from the parent plant and wash them away if there is any incline available to produce a runoff.

Animal Dispersal

When ripe, many fruits are bright colored, thin walled, and juicy, especially the red fruits. Although not attractive or even visible to most insects, red fruits are highly visible to birds and other animals. Many such fruits are also sweet, making them even more attractive to animals. When the fruit is ripe, the enclosed seeds are dry and hard enough to pass through the digestive tract of the animals. The animals distribute the seed effectively throughout their range, as shown in Figure 5-16. Some fruits even taste bitter before complete ripening discouraging animals from eating them before they are ripe.

Figure 5-16

Bright colored fruit attracts birds and animals that eat the fruit and distribute the seeds effectively in their range.

Although the seeds pass through the digestive tract intact, the digestive acid acting on them alters the seed coats. Many actually could not germinate without this **scarification**. Immature seeds cannot survive the action of these juices; thus, most fruits are green when unripe, camouflaged against the green leaves of the plant. The level of sugar in these fruits remains low until maturity, further reducing their attractiveness to animals before the enclosed seeds are mature.

Not all fruit that depends on animals for dispersal are juicy and sweet; many hard-shelled nuts and dry fruits, such as grass grains, are gathered and stored away for the winter by squirrels and other small mammals such as packrats and field mice. Since not all of these seeds are eaten, some may germinate when conditions are appropriate.

Still others are dispersed externally by hitching a ride on a person or an animal. Such fruits are externally modified with barbs, hooks, bristles, or sticky surfaces by which they adhere to the fur, hair, feathers, or skin of animals. These fruits can also be carried by humans on their clothing, cars, camping equipment, and tires of their vehicles.

Seed Morphology

Sexual reproduction in flowering plants culminates when ovules develop into seeds. Some fruits have only one seed and others thousands; some seeds are very large (palm) and others microscopically small (orchids); some must germinate immediately, and others can remain dormant for many years. All seeds, however, have certain common characteristics.

In flowering plants there are two-gamete fusions in the ovule. Thus **double fertilization** is unique to angiosperms and results in the formation of a fertilized egg, and a nutritive tissue, the **endosperm**. The endosperm tissue developed first and is allowed by divisions of the **zygote**, which produces the **embryo**, as shown in Figure 5-17. The embryo and endosperm are enclosed within the seed coat, and when both reach maturity the seed often becomes dormant or inactive until the germination process begins.

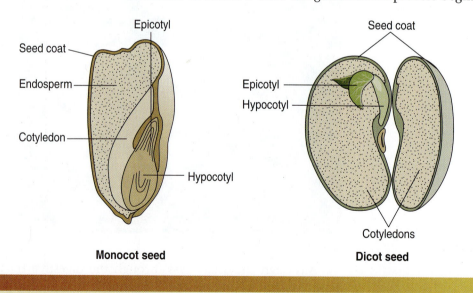

Monocot seed   **Dicot seed**

Figure 5-17

Steps in seed germination. Courtesy of Rick Parker.

Seed Modifications

Seeds that display adaptive features do so for improved dispersal, and most of their modifications are similar to those found in fruit. Seeds release from their surroundings by dehiscent fruit fall either to the ground or into water. Those, which land in water, are often as buoyant as fruit and are dispersed as readily. Small lightweight seeds can be carried away by the wind, whereas larger seeds develop wings to allow them to spin or flutter through the air. It is possible for the seeds to land far away from the parent plant. Some seeds are actually edible. Some like that of *Erythrina* seeds possess brightly colored coats that are displayed openly to entice passing birds. Seeds also develop a full array of surface modifications to promote adherence to animals. Some others have been found economically useful, such as the long surface fibers on the cottonseed.

One of the most interesting dispersal mechanisms is the physical ejection of seeds by the fruit. Several species are explosively dehiscent, propelling seeds up to several meters away from the parent plant. Some split open and eject violently when a certain degree of dryness is reached. Other seeds actually germinate while still attached to the parent plant. These **viviparous** seeds later fall to root in the soft soil or mud. Once dispersed, seeds next must successfully germinate and grow to maturity. Although not normally considered modifications, the different strategies that have evolved controlling seed germination and seedling growth are responses to environmental factors.

ECONOMIC IMPORTANCE OF SEEDS

Many seeds are edible. In fact, the majority of food calories for humans come from seeds, especially from cereals, legumes, and nuts. Seeds also are the source of most cooking oils, for example soybean oil, canola oil, corn oil, and safflower oil. Seeds or extracts are used in many beverages and used as spices and as food additives.

Some seeds are also poisonous. One of the deadliest poisons, ricin, comes from seeds of the castor bean (*Ricinus communis*). Another seed poison is strychnine from the seeds of the *Strychnos nux-vomica* tree, an evergreen tree native to Southeast Asia. Other poisonous seeds are those of the yew, wisteria, apple, horse-chestnut, and peach.

The world's most important clothing fiber grows attached to cottonseed (*Gossypium* spp.). Other seed fibers are from kapok (*Ceiba pentandra*), used as an alternative to down as filling and for insulation, and milkweed (*Asclepias* spp.), used as a substitute for kapok during World War II.

Important nonfood oils are extracted from seeds, for example flax or linseed (*Linum usitatissimum*) oil is used in paints. Two plant seeds produce an oil

with characteristics similar to whale oil: jojoba (*Simmondsia chinensis*), a shrub native to the Sonoran and Mojave deserts of Arizona, California, and Mexico, and crambe (*Crambe abyssinica*), native to southwest and central Asia and eastern Africa. Castor oil from the castor bean is a nonfood oil.

Other seed uses include: toys and beads; resin from *Clusia rosea;* nematicide from milkweed seeds; animal feed from cottonseed meal, soybean meal, canola meal, and others; and birdseed.

Seed Germination

The germination of the seed begins a new plant life. Even though the embryo is formed and has its primary tissue well developed, the mature seed may be stored for varying periods of time and still retains its **viability**—the ability to **germinate**. Only when the proper environmental conditions are provided does the seed revitalize and produce a seedling. For seeds such as the maple (*Acer*), such longevity may be only 1 week; but other seeds, such as *Lotus*, may retain viability for hundreds of years under proper storage conditions. Most common cereal plants retain viability for about 10 years.

The environmental requisites for seed germination include suitable oxygen concentrations, temperatures, moisture, and in some cases light. Mature seeds are dry for germination to begin; these dry tissues must be hydrated. It may be difficult to tell that a dry and dormant seed is really living, but respiration and metabolism continues throughout dormancy at a much-reduced level.

Imbibition

If moisture enters the seed coat, a strictly physical process called **imbibition** causes the tissue to swell with enormous expansion forces. This process will occur even in dead seeds, or in sticks of wood that become wet. Dry seeds can be placed in plaster of paris and hardened into a block. When the block is wetted, the imbibing seeds will swell so dramatically that they will shatter the plaster. It is said that the Egyptian pyramid laborers drove pegs of wood into holes bored in rock, poured water around the pegs, and thereby split huge boulders with ease.

The amount of moisture required for seed germination varies greatly among species. Some seeds germinate in what might be considered a very dry soil. Modifications of the seed coats allow some surfaces to act as a wick and absorb more water. Some seeds have mucilaginous coatings that attract eaters and enhance the imbibition process.

Good drainage and aeration are just as essential for seed germination as they are for obtaining enough water to begin imbibition. Since imbibition triggers the onset of high rates of metabolism and respiration, adequate amounts of oxygen are absolutely critical. If the soil is flooded, oxygen levels may be so low that respiration is impossible, and the seed will rot.

Once water has penetrated the seed, the tissues of the endosperm consisting of stored foods (macromolecules of starch, protein or lipids) and **cotyledons** (embryonic leaves) start the metabolic process called **digestion**. The breakdown of these macromolecules into their simpler components. For starch, the breakdown product is glucose; for protein, the breakdown product is amino acid; and for lipids, it is fatty acids.

These simpler molecules become the "fuel" or substrate for respiration. Respiration goes on in a dormant seed at a reduced rate; the rate accelerates greatly when new substrate becomes available, and the enzymes of metabolism begin to function. This extra energy results in the synthesis of new compounds necessary for growth and development, and the embryo begins its enlargement.

To the naked eye, the imbibition process produces a swollen seed, with a large increase in volume and weight simply from the uptake of water. The next observable step is usually the protrusion of the radicle, the embryonic root. It may protrude hours, or even days, before the first sign of the shoot.

The growing point of the shoot above the point of cotyledon attachment is called the **epicotyl**, and the section of stem below the cotyledons is called the **hypocotyl**. At the base of the hypocotyl, the transition zone separates the shoot from the root (see Figure 5-18). The growing tip (apex) of the epicotyl is usually referred to as the **plumule**.

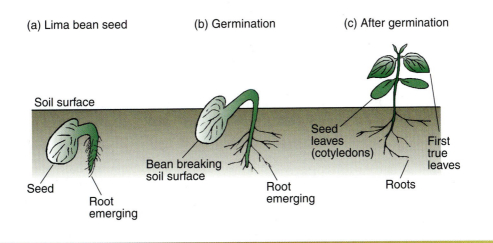

(a) Lima bean seed   (b) Germination   (c) After germination

Soil surface

Seed leaves (cotyledons)

First true leaves

Seed

Root emerging

Bean breaking soil surface

Root emerging

Roots

Figure 5-18

Cross section of different types of plant stems. Courtesy of Rick Parker.

Some seedlings emerge with the cotyledons rising above the ground and participating for a short while in photosynthesis before they shrivel and fall off. Others, such as the pea, emerge with the cotyledons still underground, in approximately the same place where the dry seed was placed. The difference is strictly a genetic one and is determined by whether the hypocotyl elongates sufficiently to elevate the cotyledons above the soil. In such seeds, like the bean, the hypocotyl is U shaped at the time the seedling emerges from the soil, and the bent shoot is literally pulled up through the soil. This mechanism is an adaptation for exerting force at the soil surface to penetrate the soil crust while protecting the delicate shoot. This apparatus is referred to as the hypocotyl hook.

Oxygen

As the ovule matures, the cell layers that make up the seed coat become rather impermeable to water, and this ensures that during the long period of dormancy and storage the seed will not lose too much water and that respiration will remain at a low rate. On the other hand, it is essential for the seed coat to retain sufficient permeability to begin the absorption of oxygen wherever renewed growth begins. Some seeds, such as many beans, become so impermeable to both water and oxygen that germination cannot proceed. These seeds must go through an abrasion of the seed coat (scarification) to allow penetration by water and gases.

Temperature

Even though the physical process of imbibition may occur; internal factors that control the metabolic processes are responsive only at suitable temperatures. Again, the critical temperature varies. Some species germinate in the winter and others only during very hot weather. The adaptive strategy of each species determines the appropriate timing for the greatest chance of survival. Within certain limits, the rate of chemical reactions in living systems doubles for each 10-degree (Celsius) increase in temperature. Therefore, germination tends to be more rapid at higher temperatures, although temperatures that exceed certain limits may be inhibitory.

Light

Light can influence germination in some species. Since most seeds germinate underground, the general trend is for germination to occur in darkness. However, some seeds, particularly those that are very small and germinate right at the soil surface, may be light sensitive. Certain types of lettuce seeds, for example, will germinate only if they are exposed to light. Researchers studying this phenomenon discovered that red light at a wavelength of about 660 nm stimulated germination. Interestingly, if the same seeds were exposed to light of a slightly longer wavelength, about 730 nm (far red), germination was effectively inhibited.

Environmental Interactions

Most adaptive strategies, including those for germination, do not depend on a single environmental factor. More likely, several factors may interact to trigger the germination. In the desert, for example, ephemerals complete their life cycle in a matter of a few weeks. Their chance of success is dictated almost entirely by an interaction of temperature and rainfall. Moisture is needed for imbibition to initiate germination, but a sudden summer thundershower could spell doom for a seed that began to develop at that time of year. Fortunately, summer temperatures are too high for germination to begin; the seedling could never survive the intense summer heat. Germination must wait until early spring or fall. At that time there is adequate moisture and cooler temperatures.

The same situation holds true for species in alpine meadows, where the growing season is very short because of temperature restrictions.

Moisture

Water may be always plentiful, but germination too early might cause a young seedling to die because of frost. Even though a few warm days may occur in early spring, the moisture-temperature combination signals the delay of germination until later in the season, when all danger of frost is past. Once the seedling has been established, the life cycle must be completed rapidly before the onset of cold weather in the late summer or early fall.

Dormancy Factors

Many times during ovule development the cells of the seed coat become so **lignified** (impregnated with lignin) or otherwise become so hard that uptake of water and oxygen is essentially impossible. As the seed ages, various chemicals and physical forces gradually break down the seed coat so that it finally does become penetrable. This might come about by freezing and thawing, by the seed passing through the digestive tract of animals, or by wind abrasion and water erosion. Not all seeds from the same plant, even if produced in the same year, will have the same coat thickness.

Some of them may germinate the first year, those of medium thickness may germinate the second year, and the very hard ones may germinate several years later. Such variability increases the chances for survival, even if no seeds at all are produced in certain years.

Physiological Factors

Occasionally, seeds appear to be mature, but in fact the development of the embryo is not yet sufficiently advanced to allow germination to proceed. Whenever this happens, it is necessary to delay germination until embryo maturity is completed. Even under the microscope, the embryo may appear to be mature, but certain chemical adjustments are necessary before germination will proceed.

Chemical Factors

Sometimes germination is inhibited by the accumulation of chemicals. This appears to be an adaptive strategy to spread out the germination process over time. As the seed ages, certain chemicals may be broken down until the concentration is so low that germination can proceed. More often, the chemical inhibitors are water soluble, and with rainfall they are simply leached from the seed. Sometimes the concentration is such that very little leaching is needed; other seeds may have very high concentrations, which requires years of leaching before the seed will germinate. Again, the adaptive strategy is clear: if all seeds were to germinate at one time, and it happened to be the wrong time, the species could become extinct.

Development: Seedling to Adult

The mature seed contains a new embryo complete in every detail—an **embryonic axis** (the stem and root) and one or more leaves. In the germination process, the new plant simply expands its existing tissues and adds new ones. The digestion of storage molecules (carbohydrates, proteins, or lipids) found in either the endosperm or cotyledon(s), provides the energy necessary for building these new parts.

Environmental factors are critical in the germination process and in the establishment of the seedling. It is not strictly by chance that the radicle begins to emerge and grow faster than the shoot. Although proper temperature and oxygen are obviously important in the process, proper moisture is probably the most critical factor in the success or failure of the new plant. An emerging root system must remain in contact with moist soil, or the embryo will die. It is absolutely essential that the root system grow rapidly enough to penetrate soil to depths that provide additional moisture. In many situations, seeds may germinate on a moisture layer even if the surface layers are dry. This is one of the critical factors in dryland farming (no supplemental irrigation). The same principle also holds in natural ecosystems. Not only is rainfall critical to germination per se, but stored soil moisture also is essential to seedling success. Some genetic stains are particularly adapted for seedling vigor and are thus better able to survive under stress conditions.

If the embryo does become established, it proceeds through its life cycle as a rapidly growing seedling, reaching vegetative maturity, passing into reproduction maturity (the ability to flower and set seed), and finally into **senescence** (literally old age). Plants that complete this process in one growing season are called **annuals**. Most crops like corn are annuals. In some cases the entire growing season (early spring to fall) may be required to accomplish all these processes. Other annuals do so in a very short period. These are adapted to environmental conditions that force them to reproduce quickly if they are going to survive.

A second group of plants, the **biennials**, require two growing seasons to complete their life cycle. Many members of the cabbage family (*Cruciferae*), among others, require one season of vegetative growth followed by a cold period to induce flowering the following year. Apparently some chemical stimulus required for flowering, perhaps a flowering hormone, is synthesized only under low temperature conditions. During the first year, growth is strictly vegetative, and many leaves are produced on short internodes of stem; thus the leaves are "telescoped" close together and near the ground. Such plants are referred to as **rosettes** because the leaf placement resembles the petals of a rose. In the spring of the second year, the stem begins to elongate rapidly and produces flowers.

Interestingly, humans handle biennials as "annual" crops. When cabbage is grown as a vegetable crop, it is necessary to use only the first part of the life cycle. The rosette in this case is a group of overlapping leaves that make up the head, and the plant is strictly vegetative. On the other hand, when cabbage is grown from seed, it is absolutely essential to allow the plant to proceed through normal development into the second year of flowering and seed maturity.

The third group of plants is **perennials**. They complete their life cycle in more than 2 years and often continue to live for many years, producing flowers, fruits, and seeds each year. All woody plants are perennials and many are specially adapted to survive harsh winters and revive in the spring. Many perennials go through a long period of vegetative growth before they begin to flower. It is not known why these plants require such lengthy leaf production. Citrus trees, for example, may need 8 to 10 years of vegetative growth prior to the onset of flowering. If farmers could hasten the beginning of reproduction activity, they could bring orchards into production sooner. Some perennials, such as chrysanthemums, die back to the ground level each winter but produce new growth each spring from an underground crown of stem tissue. Such plants are called **herbaceous perennials**.

Botanists categorize plants as annuals, biennials, or perennials according to life cycle. However, there are complications where certain plants are concerned. Consider the century plant (*Agave americana*) so named because it remains vegetative for many years, flowers once, and dies. Is this plant truly perennial, or should it be considered an annual or some modified form of biennial? One way to simplify the problem is to classify plants only according to whether they flower once or more than once. Since botanists often refer to fruits as ripened ovaries or carpels, we can call all plants that flower once **monocarpic** plants. Those plants, which flower over and over again, are called **polycarpic** plants. The later scheme is definitely simpler, but the terms *annuals*, *biennial*, and *perennial* will probably continue to be used, since they are readily understood.

The vegetative phase of plant growth consists of the addition of more roots, stems, and leaves. Certain hormonal relationships, to be discussed later, determine whether the axillary buds at the nodes initiate active growth. If they do begin growth, then each bud develops into its own new shoot with stems and leaves. The overall effect of such development is to produce a bushy plant with a rounded head. If, on the other hand, the lateral (axillary) buds fail to develop, the plant continues growing from the terminal bud and becomes much taller with very little lateral spread. Such plants have strong apical dominance, and the terminal bud exerts control over all growth. The bushy, round-head plant has weak apical dominance.

The plant continues its vegetative activities—adding new photosynthesis surface, and storing organic materials—until some internal or external mechanism initiates a changeover to reproductive activities. When this happens, internal biochemical forces cause the terminal bud, and perhaps the lateral buds, to stop producing leaves and begin producing flowers. Exactly what triggers the meristematic cells (those capable of division) at the shoot apex to stop producing leaves and begin producing flowers is one of the mysteries of botanical research.

Monocots and Dicots

One of the impressive development phenomena of angiosperms is that the two major subgroups, **monocotyledonae (monocots)** and **dicotyledonae (dicots)** as shown in Figure 5-19, are so easily distinguished throughout their life cycles. As seeds, the monocots (*mono*, "one"; *cot*, "cotyledon") have only a single cotyledon, or embryonic leaf; the dicots (*di*, "two") have two. As they develop into seedlings and adult flowering plants, their morphological distinctiveness continues to be evident. Table 5-2 summarizes the major differences between these two groups of flowering plants.

Additionally, within each of these two groups there are several categories of gradually less inclusive groupings of plants. The next useful level of inclusiveness is the plant family.

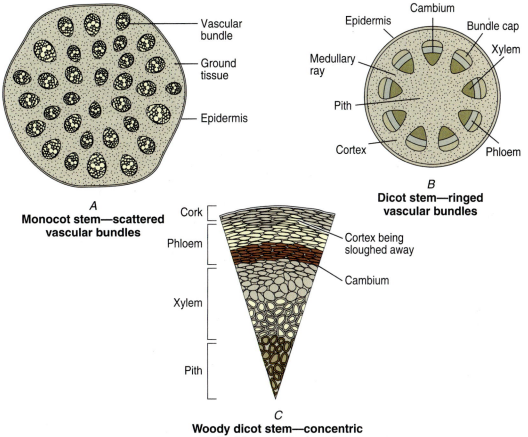

A
**Monocot stem—scattered
vascular bundles**

B
**Dicot stem—ringed
vascular bundles**

C
**Woody dicot stem—concentric
cylindric vascular bundles**

Figure 5-19

The structure of a monocot seed and a dicot seed. Courtesy of Rick Parker.

Table 5-2

Monocot and Dicot Characteristics	
Monocots	**Dicots**
One cotyledon	Two cotyledon
Flower parts in threes or multiples of three	Flower parts in fours or fives or multiples of these numbers
Herbaceous almost never woody	Can be woody or herbaceous
Usually linear leaves with parallel venation	Leaves with netted (reticulate) venation
Scattered vascular bundles in the stem	Vascular bundles in a ring

Botanical Names

Two of the main plant groups discussed so far are the angiosperms and gymnosperms. As mentioned, angiosperms are either monocots or dicots, and within each of these two categories plants are grouped by relationship into families. The family level is usually a recognizable and coherent group that also has subgroups within it.

There are both **scientific names** and **common names** for most plant families. For example, the rose family is *Rosaceae* and the nightshade family is *Solonaceae*. Although the use of common names predominates among nonscientists, the need for scientific names is very real, especially at more specific levels, such as **genus** and **species**. Every organism has a unique scientific **binomial** consisting of the genus and species; but not every one of these organisms has a unique common name. For accuracy and ease of classification, therefore, the scientific name is essential.

Supermarket Botany

Early in human history, small populations of predominately nomadic people gathered plants for food or medicine. In time humans evolved from food gatherers to food growers. Members of the primitive farming communities recognized the value of some of the plant modifications and were able to capitalize on them. Wild strains that processed beneficial variations could be selectively propagated. These adaptations might have included plants with larger edible roots, increased seed yields, and the capacity to withstand unusually adverse environmental conditions.

Once humans began domesticating plants, they were able to select individual plant species systematically according to desirable traits, and develop them through traditional hybridization and plant breeding. These techniques are employed today in the development of many commercially valuable consumer products. Although these products have a wide variety of applications, certainly the most important is as a food source.

The vast majority of food for human consumption comes from the flowering plants. Most of us are familiar with the broad selection in the canned fruit and vegetable aisles of the supermarket. The commercial designation of these plant products does not always reflect their botanical function. Squash, cucumber, pepper, tomatoes, and corn are commonly called vegetables ("Eat your vegetables, dear"), even though they are reproductive not vegetative, parts of the plant. Many grocery store "vegetables" are misnamed. Beans and peas, for instance, are actually fruits (green beans) or seeds (lima, kidney, pinto, navy, soybeans, and green and black peas). In addition, many of these consumer items belong to the same plant family.

Table 5-3 lists some of the more common edible plant materials by family and identifies the actual part of the plant that is eaten. Scientific binomials (genus and species) are also included. Note how frequently several species of a single genus (for example, *Brassica*) are commercially developed for the consumer.

Table 5-3

Common Food Plants Found in the United States			
Family Name	**Common Name**	**Scientific Name**	**Fruit Part**
Monocots			
Poaceae (grass family); all cereal grains belong to this family of plants, as does sugarcane, a great source of granulated sugar	Wheat	*Triticum aestivum*	Fruit (grain)
	Rice	*Oryza sative*	Fruit (grain)
	Corn	*Zea mays*	Fruit (kernel)
	Barley	*Hordeum vulgaris*	Fruit (grain)
	Oats	*Avena sativa*	Fruit (grain)
	Rye	*Secale cereale*	Fruit (grain)
	Sorghum	*Sorghum bicolor*	Fruit (grain)
	Sugarcane	*Saccharum officinarum*	Stem
Lilliaceae (lily family)	Onion	*Allium cepa*	Bulb
	Garlic	*Allium sativa*	Bulb
	Asparagus	*Asparagus officinalis*	Young stem
Bromeliaceae (pineapple family); a family of mostly epiphytic plants; pineapples rooted in soil	Pineapple	*Ananas comous*	Fruit
Dicots			
Brassicaceae (mustard family); most food plants in this family have a tangy, sharp taste	Mustard	*Brassica alba*	Seeds and leaf
	Broccoli	*Brassica oleraceae*	Inflorescence
	Cabbage	*Brassica oleraceae*	Leaves
	Cauliflower	*Brassica oleraceae*	Young inflorescence
	Brussels sprouts	*Brassica oleraceae*	Lateral buds
	Turnips	*Brassica rapa*	Root
	Radish	*Raphanus sativus*	Root
	Watercress	*Nasturtium officinale*	Leaves
Fabaceae (bean family); a very large family having many members with high protein levels and nitrogen fixation in roots nodulated with the bacterium Rhizobium	Broad bean	*Vica faba*	Seed
	Green bean	*Phaseolus vulgaris*	Seed (pod)
	Pinto bean	*Phaseolus vulgaris*	Seed
	Navy bean	*Phaseolus vulgaris*	Seed
	Kidney bean	*Phaseolus vulgaris*	Seed
	Lima bean	*Phaseolus lunatus*	Seed
	Black bean	*Phaseolus mungo*	Seed
	Soybean	*Glycine max*	Seed
	Green pea	*Pisum sativum*	Seed and pod
	Black-eye pea	*Vigna sinensis*	Seed
	Peanut	*Arachis hypogaea*	Seed
Rosaceae (Rose family); a large family with many different kinds of fruits	Cherry	*Prunus avium*	Fruit (Drupe)
	Apple	*Malus sylvestris*	Fruit (Pome)
	Pear	*Pyrus communis*	Fruit (Pome)
	Peach	*Prunus persica*	Fruit (Drupe)
	Plum	*Prunus domestica*	Fruit (Drupe)
	Apricot	*Prunus armeniaca*	Fruit (Drupe)
	Blackberry	*Rubus canadensis*	Fruit (Berry)
	Strawberry	*Fragaria virginiana*	Fruit (Achene)

(Continues)

Table 5-3 (*Continued*)

Family Name	Common Name	Scientific Name	Fruit Part
Dicots			
Solanaceae (nightshade family); also contains other economically important species, including tobacco and a number of poisonous members	Tomato	*Lycopersicon esculentum*	Fruit (Berry)
	Potato	*Solanum tuberosum*	Tuber
	Pepper	*Capsicum* (jalapeno, bell, cayenne,etc)	Fruit
	Eggplant	*Solanum melongena*	Fruit
Cucurbitaceae (gourd family)	Squash	*Cucurbita*	Fruit
	Pumpkin	*Cucurbita pepo*	Fruit (Melon)
	Cucumber	*Cucumis sativus*	Fruit
	Gherkin	*Cucumis anguria*	Fruit
	Watermelon	*Citrullus vulgaris*	Fruit (Melon)
	Honeydew melon	*Cucumis melo*	Fruit (Melon)
Chenopodiaceae (goosefoot family)	Beet	*Beta vulgaris*	Root
	Spinach	*Spinacia oleracea*	Leaves
Asteraceae (sunflower family); although one of the largest families, it contains very few economically important plants	Sunflower	*Helianthus annus*	Seeds
	Artichoke	*Cynara scolymus*	Inflorescence
	Lettuce	*Lactuca sativa*	Leaves
Apiaceae (carrot family)	Carrot	*Daucus carota*	Root
	Celery	*Apium graveolens*	Petiole
	Parsley	*Petroselinum crispum*	Leaves and stem
Convolvulaceae (morning glory family)	Sweet Potato	*Ipomoea batatas*	Tuberous root

 # Summary

1. The flower, the site of sexual reproduction in angiosperms, has four male parts: sepals, petals, stamen, and pistil. The stamens produce pollen; the ovary of the pistil contains the ovules. A simple pistil contains only a single carpel; a compound pistil is made up of two or more carpels. The ovary, which can have a superior or inferior position, matures into the fruit, and the ovules mature into the seeds. Flowers occur singly or in an inflorescence and can be bisexual or unisexual. Unisexual flowers are either monoecious or dioecious.

2. Flower shape is highly variable, but is either regular (radially symmetrical) or irregular (bilaterally symmetrical). Floral parts can be fused or distinct and have a wide variety of colors, patterns, sizes, and other modifications. All flower modifications exist for a pollinator's attraction. Most pollination is by animals, especially insects, but wind pollination is also common in nonshowy flowers.

3. Flowering plants are either self-pollinated or cross-pollinated. Self-incompatibility ensures outcrossing. Flower-insect specificity can be very highly developed and results from long periods of co-evolution.

4. Fruits develop only if fertilization occurs in the ovules. Parthenocarpic fruits develop without fertilization and are normally seedless. A mature fruit usually has three layers to its pericarp or fruit wall. The pericarp can be fleshy or dry; when dry, it can be either dehiscent or indehiscent.

5. There are simple fruits and compound fruits, which include aggregate and multiple fruits, according to how they develop. Most fruit modifications are dispersal mechanisms. Small light fruit are wind dispersed, whereas other fruits can be water dispersed or animal dispersed.

6. Seeds contain the embryo and nutritive tissue either in the endosperm or cotyledon(s). The embryo and endosperm are formed because of double fertilization of the embryo and the polar nuclei. Seed modifications are also a result of developing successful dispersal mechanisms.

7. Seed germination can occur immediately following seed dispersal or up to many years later. The first step is imbibition, the absorption of water. The emergence of the radicle precedes the emergence of the cotyledons, epicotyl, hypocotyl, and plumule.

8. Several mechanical and environmental components affect seed germination. The thickness of the seed coat or the presence of chemical inhibitors produced dormancy until scarification or washing breaks the dormancy. Oxygen, temperature, and light all can affect the germination of seeds, either individually or in combination.

9. The development of a seedling until it is an adult plant involves many physiological and anatomical changes over time. Annuals develop much more quickly than biennials and perennials, the latter being woody or herbaceous. The terms *monocarpic* and *polycarpic* refer to the number of times a given plant flowers in its lifetime.

10. Angiosperms contain two natural subgroups, the monocots and dicots. Each has several sequentially less inclusive groups, including families, genera, and species. The supermarket name for a plant part often differs from the botanical name for the part. Many commercially useful food plants reflect vegetative and reproductive modifications.

Something to Think About

1. Why is the flower so important in people's lives?

2. What is sexual propagation?

3. What are considered female flower parts and male flower parts?

4. What role does pollination have on the production of fruit?

5. What role does fertilization have in the production of fruit?

6. What environmental components affect seed germination?

7. What are some of the physiological and anatomical changes that occur within the seed?

8. What are the differences in a monocot and dicot?

9. How are angiosperms and gymnosperms different?

10. What are scientific binomials?

Suggested Readings

Bewley, J. D., and M. Black. 1994. *Seeds: Physiology of development and germination.* New York: Plenum Press.

Bird, R., et. al. 1998. *The complete book of plant propagation.* Newtown, CT: Taunton Press.

Bowes B., and B. Bowes. 1999. *A colour atlas of plant propagation and conservation.* Australia: Blackwell Publishing.

Hartmann, H. T., et. al. 2002. *Plant propagation principles and practices.* Upper Saddle River, NJ: Prentice Hall.

Internet

Internet sites represent a vast resource of information. The URLs for Web sites can change. Using one of the search engines on the Internet, such as Goggle, Yahoo!, Ask.com, and MSN Live Search, find more information by searching for these words or phrases: **incomplete flower, complete flower, pistil, vascular system, root structure, modified roots, meristems, nitrogen fixation, mycorrhiza, apical dominance, botanical names, morphology, dry fruits, fleshy fruit, angiosperms, and gymnosperms.**

CHAPTER 6

The Inside Story: Molecules to Cells

Anton van Leeuwenhoek (1632–1723)

was among the first to examine tissue through a primitive optical system, the forerunner of the light microscope. About the same time, Robert Hooke saw tiny boxlike structures, compartmentation, on a microscope scale, which he described as cells. Scientists later discovered that all living organisms except viruses are made up of remarkably similar cells. Cells come in all sizes and shapes, but the nature of cells is surprisingly similar in all organisms

After completing this chapter, you should be able to:

- Compare eukaryotes to prokaryotes
- Describe the basic chemical composition of cells
- Describe the functions of the parts of a cell
- Draw and label all parts of a plant cell
- Know the function of the mitochondrion
- Explain the function of the chloroplast
- Understand how Golgi apparatus function
- List other organelles in the plant cell

Key Terms

molecules	organelle	mitochondrion
proton	compartmentalization	aerobic respiration
neutron	of enzymes	cristae
electron	membrane-bound	anaerobic
glycoproteins	nucleus	stroma
tumors	plastids	chromoplasts
turgor pressure	mitochondria	leucoplasts
glucose molecules	ribosomes	starch
microfibril	golgi apparatus	chloroplast
middle lamella	lysosomes	thylakoids
primary cell wall	glyoxysomes	granum
secondary cell wall	peroxisomes	endoplasmic reticulum
plasmodesma	microtubules	(ER)
prokaryotic	nucleus	rough endoplasmic
eukaryotic	nuclear envelope	reticulum
bacteria	nuclear pores	smooth endoplasmic
lumen	nucleolus	reticulum

Molecules of Life

Cells all have a plasma membrane, which encloses an aqueous solution called the cytoplasm. In the watery world enclosed by this membrane, the mechanics of life occur: building large molecules from small molecules, making smaller molecules by tearing down larger ones, and always moving substances around. Constantly in motion, the cellular "factory" knows exactly where each "nut" and "bolt" goes.

Within this aqueous medium are even smaller structures, organelles (or, little organs) of various shapes and sizes. Each type has a specific job in the overall function of the cell.

Under a microscope, cyclosis (movement of the cytoplasm and many of the organelles) can be observed in the living cell; some go in one direction and others go in another. It resembles a network of freeways, with cars going in many directions. These movements are apparently not random, but very highly organized.

Chemistry is the center of cell structure and function. Certain chemical and physical laws govern how molecules are assembled from simple atoms, how the bonds are broken, and how they are reformed in new molecules.

All matter is made from various combinations of the chemical elements, substances which cannot be broken down by ordinary chemical means. The earth's crust contains 92 naturally occurring elements, and several others have been made artificially. The basic particle of an element is the atom, and atoms are combined in various ways to form **molecules**, the building blocks of life.

Atoms are composed of three primary subparticles: **proton**, **neutron**, and **electron**. The nucleus of the atom contains the protons and neutrons. Electrons move in specific orbits around the nucleus. The proton has a specific mass and positive charge; neutrons also have mass but no charge. Electrons have essentially no mass but have a net negative charge.

Various combinations of protons, neutrons, and electrons produce the chemical elements, and the structure of the elements is determined by number of protons, which is the atomic number. The atomic weight is the sum of protons and neutrons. The simplest chemical element is hydrogen, with only one proton. Helium has two protons, carbon has six, and oxygen has eight. Each chemical element has its own symbol, often but not always indicated by the first letter. Some chemicals' symbols are derived from Greek or Latin terms. C is the symbol for carbon; H is the symbol for hydrogen, and He is the symbol for helium. But K, the symbol for potassium, is derived from the word *kalium*. The symbol P is reserved for phosphorus, Na (*natirum*) stands for sodium, and Fe (*ferrum*) for iron (see Table 6-1).

Table 6-1

Selected Chemical Elements of Protons and Neutrons		
Chemical element	**Number of protons**	**Number of neutrons**
Hydrogen (H)	1	0
Helium (He)	2	2
Carbon (C)	6	6
Nitrogen (N)	7	7
Oxygen (O)	8	8
Sodium (Na)	11	12
Chlorine (Cl)	17	18
Calcium (Ca)	20	20

Note: The number of protons does not always equal the number of neutrons. On the other hand, the number of electrons always equals the number of protons—one negative charge for each positive charge.

Basic Cell Structure

Although plant cells and animal cells are very much alike, plant cells have a wall that becomes more or less rigid. Animal's cells lack this cell wall. This is not to say that the plasma membrane surrounding an animal cell is "naked." Large molecules called **glycoproteins** (carbohydrate protein) occur on the outside of the membrane and serve as a recognition surface. These glycoproteins form characteristic three-dimensional surfaces that allow cell sliding past each other to recognize similar surface features and then cling together. Such recognition may be less important in plant cells because the cellulose wall fixes the cells in place; there is little movement. Even in some plant cells, however, protein layers on the exterior surface of the cell wall are important in recognition. Plant cells are grouped together in tissues by virtue of their proximity at the time of division. The plane of division determines whether a group of plant cells will divide in only one plane to form a chain, in two planes to form a sheet, or in all three planes to form a cube.

If cell divisions were strictly random in respect to plane of division, a sphere of cells would result. Such is the case in **tumors**, which can develop in both plants and animals.

The **turgor pressure** is the water pressure within each cell. Cells are literally blown up, and all push against each other. These cells are inflated by water. A cellulose molecule may be as many as 2000 **glucose molecules**, a 6-carbon sugar (hexose). The glucose units are connected end to end, forming a macromolecule. Cellulose molecules may be as many as 2000 glucose molecules in length, and these long molecules are bundled together in packages to form a **microfibril**. Groups of microfibrils are wound together much like a steel cable to form a macrofibril. The macrofibrils are a major component of the cell wall and are held together with other kinds of macromolecules, including hemicelluloses and pectic compounds. These substances glue the entire structure together in a sheet of fibers. The first microfibrils of the primary wall form a network with a predominately transverse pattern. When turgor causes the cell to expand and the wall increases in surface area, the **microfibrils** become more parallel to the longitudinal axis of the cell. The overall effect is a cross-hatched appearance of the various layers. Two adjacent plant cells are held together by the **middle lamella**, composed primarily of cementing pectic substances, and the **primary cell wall** on each side of the middle lamella. In many plants a **secondary cell wall** may be laid down at a later date, adding strength and rigidity to the tissue, particularly if lignin is present. Tree trunks, for example, have cells with very thick secondary cell walls.

Elongation in some plant cells reaches many centimeters. The cotton fiber, for example, often reaches lengths of 4 to 6 cm. The size of certain cells is determined by the species in question, and the genetics will ultimately determine how large a cell can become. Generally, both plant and animal cells are microscopic.

Multicellular plants communicate by tiny passages that connect the cells through one plasma membrane, the primary cell wall, the middle lamella, the adjacent primary cell wall, and finally the adjacent plasma membrane. These connections occur in spots in the wall called primary pit fields, and each strand is called a **plasmodesma**. Although too small for the interchange of larger organelles, cytoplasmic connections apparently allow for the transfer of chemicals from one cell to another.

The wall itself is extremely permeable to all kinds of substances. Cross-linking of the molecules still allows water and various kinds of solutes to penetrate the wall; the barrier, determining what gets in and out of the cell, is the plasma membrane itself, just as in animal cells.

Prokaryotic and Eukaryotic Cells

Within the living world there are basically two types of cells: Those that are **prokaryotic**, and those that are **eukaryotic**. Prokaryotic means "before the nucleus"; therefore, prokaryotic cells are those that do not have a well-defined nucleus. All other cells in the world, both plant and animal, do contain a nucleus and are therefore referred to as eukaryotic. Prokaryotic is the more primitive cell. Probably the prokaryotic cells that exist today are similar to the very first cells on earth. Prokaryotic cells first made their appearance in the oceans some 3.5 billion years ago and today are represented by only two types of organisms, **bacteria** and cyanobacteria. Now, as then, all prokaryotic organisms were single celled. Some chains, groups, and colonies exist, but no truly multicellular organisms with cellular differential occur. Prokaryotic cells are comparatively simple and small, rarely more than 1 or 2 μm in diameter. They consist of a rigid wall surrounding a plasma membrane that holds the components of the cytoplasm. They do contain ribosomes, but for the most part organic molecules are simply free in the cytoplasm. Packaging and partitioning in these cells is not nearly so dramatic as it is in the eukaryotic cells. The hereditary material within the cells, DNA, consists of a single long molecular thread. It is circular, but it is not organized into any sort of regular chromosome as occurs in the nucleus of eukaryotic cells.

Primary Cell Wall

In newly forming cells, the wall that first surrounded the plasma membrane is referred to as the primary cell wall. Initially, the primary cell wall is relatively plastic, gradually becoming more rigid as the cell ages and enlarges. It finally stretches more and more, synthesizing new cell wall material, until the cell has reached its ultimate size. Certain cells grow more in certain directions than in others, leading to the elongation process. Elongation in some plant cells reaches many centimeters. The cotton fiber, for example, often reaches lengths of 4 to 6 cm. The size of certain cells is determined by the species in question, and genetics will ultimately determine how large a cell can become. Generally, both plant and animal cells are microscopic.

Secondary Cell Wall

Many plant cells develop a secondary cell wall, which is laid down between the primary cell wall and the plasma membrane and may become much thicker than the primary wall. The secondary cell wall adds strength and rigidity to the cell. In some cases, for example the flax fiber, the secondary cell wall gradually fills the space inside the cell by the cytoplasm, and the **lumen** or living space inside the cell becomes very small indeed. The same is true for many other types of fibers. By the time the secondary cell wall is complete, some plant cells die, and the cytoplasm is simply absorbed.

Cell Membranes

Cell membranes are approximately half phospholipid and half protein. The three-layered structure of cell membranes consists of a double layer of phospholipids, with the insoluble phosphate portion in the center and the water-soluble phosphate portion oriented toward the outside in each direction. Proteins are inserted on each side of the lipid layer; some of the protein molecules extend across the lipid layer and protrude out the other side; others are a component of only one side of the membrane. The proteins are laterally mobile in the double layer. In a biological membrane seen through an electron microscope, the fixation process causes the protein layers to appear as dense lines, but the phospholipid layers are transparent. Thus, one sees two black lines with a space in between. This typical structure occurs in all biological membranes and is referred to as the unit membrane. This does not mean that all membranes are exactly the same. They have different permeability characteristics, and just because a substance can get across a chloroplast membrane does not mean that it can get across a mitochondrial membrane. Membrane selectivity suggests that each kind of membrane has subtle molecular characteristics that allow it to function in its own conditions. Scientist currently picture the biological membrane as a fluid mosaic in which large protein molecules float in a sea of lipids.

Substances must pass across biological membranes to get into or out of a cell. These membranes tend to be very permeable to water and certain gases, including oxygen and carbon dioxide. Other kinds of molecules may have difficulty traversing the membrane because of size or polarity. Ions and polar molecules tend to move through the protein portion of the membrane. Many of these proteins are involved in the process called active transport (from ATP), which is used to move substances across membranes against a concentration gradient.

Organelles and Other Inclusions

There is some disagreement among cell biologists concerning the definition of an organelle. For the purpose of this book, an **organelle** is a distinct entity within the cell that performs a particular function as a **compartmentalization of enzymes**. Thus, we include ribosomes and vacuoles as organelles; even though some scientists would classify only **membrane-bound** structures as organelles.

Typical organelles of a photosynthetic plant cell include the **nucleus**, vacuole, **plastids**, **mitochondria**, **ribosomes**, **Golgi apparatus**, **lysosomes**, **glyoxysomes**, and **peroxisomes**. A number of other cell inclusions, including **microtubules**, are important.

Nucleus

The **nucleus** is a fairly conspicuous organelle within the plant cells. During cyclosis it can be observed to remain in a relatively static position, attached by strands of membranes that form a network to suspend it in space; other organelles seem to slide by. In the process of cell division, the nucleus undergoes dramatic changes as the hereditary material, DNA, is replicated and partitioned to daughter cells.

A typical young plant cell is approximately 30 to 40 μm in diameter, whereas the nucleus itself is about 10 μm in diameter. It is enclosed by a double membrane system that makes up the so-called **nuclear envelope**. Viewed through the electron microscope, the nuclear envelope is seen to contain relatively large holes, referred to as **nuclear pores**, through which certain kinds of small and large molecules may pass. Many dramatic changes take place in the nucleus. A dense region in the nucleus is called the **nucleolus**. Some cells contain only one; others two or three, and others have literally hundreds of nucleoli. They appear to function as the synthesis of rRNA. During interphase, the so-called resting stage of cell division, the nucleoli can be observed in great detail, but as the cell begins the division process, they usually disappear at about the same time the nuclear envelope disappears. Since the nucleus contains the genetic information, it directs the framework of activity for the entire cell—when and how to divide.

Mitochondrion

The **mitochondrion** is the organelle responsible for the process of **aerobic** (uses oxygen) **respiration.** It is capable of converting sugars into CO_2 and H_2O and releasing energy in the process. Energy from these molecules is produced as ATP, which is the main energy source for the cell. Mitochondria tend to be far more numerous than chloroplast, with perhaps as many as a thousand per cell. They may be oblong, oval, or round and approximately 1 μm in diameter, about the same size as a bacterial cell (see Figure 6-1). Their structure consists of an outer membrane and an inner membrane that is involuted to form the **cristae.**

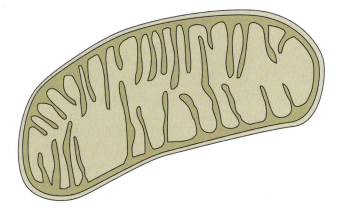

Figure 6-1

The mitochondrion is the site of aerobic respiration consisting of an outer membrane and an inner membrane that invaginates and folds to form the cristae. The mitochondrial matrix is the site of enzymes showing folding of the Krebs cycle, and the surface of the cristae provides a surface for the enzymes of electron transport. The enzymes responsible for glycolysis occur outside the mitochondrion, free within the cytoplasm.

Courtesy of Rick Parker.

The involutions give a tremendous increase to the surface area of the cristae and provide a surface on which the enzymes of respiration occur. Plant cells that carry out a great deal of respiration and are required for producing a tremendous amount of ATP energy tend to have many mitochondria. Other cells that function primarily to provide some service other than respiration may have very few. It is important to remember that all cells carry on respiration, although some may do so under **anaerobic** (no free oxygen) conditions.

Plastids

Other than the nucleus and vacuole, the plastids constitute the most conspicuous organelles of a plant cell. A double membrane, just as the nucleus and mitochondria, bound all plastids, and the internal structure is a system of membranes separated by a fairly homogeneous ground substance of membranes separated by a fairly homogeneous ground substance called the **stroma**.

There are three types of plastids. **Chromoplasts** are pigment organelles, as the name implies, but are specialized to synthesize and store carotenoid pigments (red, orange, and yellow) instead of chlorophyll. In the process of fruit ripening and in other pigmented tissue, they accumulate large quantities of carotenoids to give the characteristic color to the tissue.

Leucoplasts are nonpigmented plastids but contain enzymes responsible for the synthesis of **starch**. Large starch gains may accumulate in plastids, as in a potato tuber.

Chloroplast is the green plastids associated with the entire photosynthetic process, and they represent the functional unit in the transfer of light energy into the chemical energy of sugar production (see Figure 6-2). All plastids begin as nonpigmented protoplastids and then differentiate into one of the three basic types. Chromoplasts may begin as chloroplasts but lose chlorophyll and accumulate carotenoids to become chromoplasts during fruit ripening and other processes. Leucoplasts may be transformed into chloroplasts when exposed to light. They still retain the ability to store starch, as do all chloroplasts when exposed to light. Sometimes chloroplast, following several hours of sunshine, will accumulate several large starch grains as products of photosynthesis, which distort the internal membrane structure. As opposed to prokaryotic cells in which pigment molecules are attached to peripheral membranes of the cell, the chloroplast represents highly organized arrangements of the chlorophyll, and the other pigment molecules are arranged in specific double membrane layers called **thylakoids**. Stacks of thylakoids constitute a **granum** (plural, *grana*). The matrix inbetween the grana is called the stroma, and the grana and stroma together make up the body of the chloroplast.

Figure 6-2

The chloroplast consists of a double membrane bound structure enclosing dense stacks of membranes called grana (GR) surrounded by a matrix called stroma (ST). The chlorophyll molecules occur in layers embedded in protein to make up the grana. Enzymes associated with the light reactions of photosynthesis occur in the grana; those responsible for the dark reactors of photosynthesis occur in the stroma. Courtesy of Rick Parker.

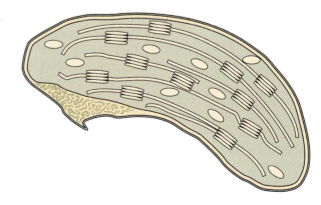

Chloroplasts tend to be elliptical and 5 to 10 μm in diameter. In a green plant cell there might be 20 to 100 chloroplasts. During cyclosis they move freely throughout the cytoplasm. In carrying on the process of photosynthesis, they respond directly to the energy from the sun by orienting themselves perpendicular to the rays of the light. In case the light energy becomes too great, they have the capability of moving away from the sun and orienting themselves at an oblique angle so that less light hits them.

Endoplasmic Reticulum and Ribosomes

The process of synthesis can occur only in the presence of ribosomes found on a series of interlacing membranes that traverse the cytoplasm and form the framework, on which certain important functions are performed, including protein synthesis. This membrane system, the endoplasmic reticulum (ER), provides the scaffolding to which the ribosomes are attached to the ER. ER may have a group of ribosomes, much like buttons attached to a piece of cloth, in what is called rough endoplasmic reticulum; or it may consist of a membrane with ribosomes, in which case it is referred to as smooth endoplasmic reticulum. The ribosomes themselves appear as dark round dots on the endoplasmic reticulum at low magnifications, but as the magnification increases, it becomes apparent that they consist of two parts—a small spherical body and a large concave body. This organelle is about 15 nm in diameter and therefore of much smaller dimensions than the other organelles already described. The ribosome is made of rRNA and protein. In this structure, the amino acids are aligned in proper order for incorporation into the protein. In any given cell, there might be many thousands of ribosomes. Thus, even though the process of protein synthesis may at first seem relatively slow, it is possible to make many molecules in a short period of time because each ribosome may be involved in the synthesis process.

The **vacuole** begins as a very small organelle that eventually increases in size until, in a mature plant, it dominates the entire cell, as shown in Figure 6-3. As a matter of fact, the cytoplasm may be stretched to the outer limits adjacent to the cell wall. In some cases the nucleus is displaced into or adjacent to the cell wall. In some cases the nucleus is displaced into a "corner" of the cell, and the vacuole actually occupies most of the space within the cell.

Vacuolar sap is mostly water and much less viscous than is the cytoplasm proper. It is probably best to think in terms of the vacuolar as the storage area of the cell, a place where nutrients and various solutes are maintained until they are needed in general metabolism or stored as water material. For the most part, macromolecules are not part of the vacuolar system but are maintained within other organelles or directly in the cytoplasm. The membrane that surrounds the vacuole is the **tonoplast**. The tonoplast selectively acts to determine what gets in and out of the vacuole. There are many

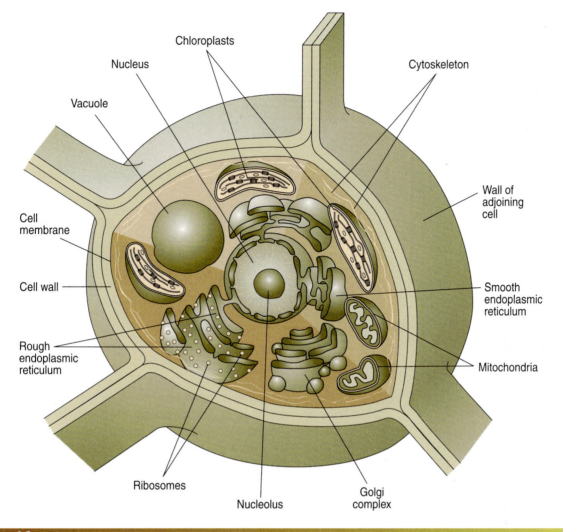

Figure 6-3

The typical green plant cell consists of a relatively thick cell wall (CW), a plasma membrane (PM), and a cytoplasmic matrix (C), filled with various organelles, including the nucleus (N), chloroplast (CH), mitochondrion (M), rough endoplasmic reticulum (RER), golgi apparatus (G), and vacuole (V). This cell is not yet mature. The vacuole is not yet mature. The vacuole will continue to expand and dominate the interior of the cell. Courtesy of Rick Parker.

different kinds of relatively small molecules within the vacuolar sap, including the ions and small molecules such as sugar and amino acids.

Golgi Apparatus

Located throughout the cytoplasm is a group of organelles collectively called the Golgi apparatus, as shown in Figure 6-4. They appear as flattened membranes, much like a stack of pancakes. At the edges of these flattened membranes one can observe small pieces of membranes called **vesicles** being pinched off from the periphery of the "pancake." The vesicles contain the macromolecules used in construction of both the membranes and primary cell wall. As the cell grows under the influence of turgor pressure against the plasma membrane, the membrane must enlarge and be strengthened by the deposition of new material. This packing function is performed by the Golgi apparatus, ensuring that as the interior expands the expanded membrane and wall will be able to take the additional stress imposed.

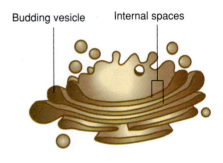

Figure 6-4

The Golgi apparatus consists of stacks of membrane, like a layer of pancakes. At the margins, vesicles are pinched off and carry macromolecules to new plasma membranes or cell walls. Golgi bodies may often be found near the exterior of the cytoplasm, close to their site of transport.

Budding vesicle   Internal spaces

Other Organelles

Sometimes other small organelles are found in specific plant tissues. Lysosomes, glyoxysomes, and peroxisomes are organelles bounded by a single membrane and containing a package of enzymes for a specific task. Lysosomes contain acid hydrolytic enzymes capable of breaking down proteins and certain other macromolecules. Glyoxysomes are found primarily in fatty seeds such as cotton and peanut and they provide enzymes for the conversion of fats to carbohydrates during the germination process. Peroxisomes provide a compartment of enzymes important in the glycolic acid metabolism associated with photosynthesis.

Microtubules

Microtubules are relatively small structures found in all eukaryotic cells and characterized by a tubular or spaghettilike appearance. For the most part, they are found in the cytoplasm, but they may also be a part of cilia and flagella, whiplike projections on the surface of motile cells.

Cytoplasmic microtubules are rather uniform in size and remarkably straight. They are about 23 nm in outside diameter and several micrometers in length. The wall of the microtubules consists of individual linear or spiraling filamentous structures, about 13 subunits. There is a lumen (open area in the center), but occasionally dots or rods are observed in the center portion. Microtubules are apparently composed of a special type of protein called **tubulin**.

Although the function of microtubules is still not perfectly clear, their orientation and distribution suggest that they form a framework, which somehow shapes the cell and redistributes its contents. We will soon see how cell specialization brings about different shapes and functions, and cell differentiation begins to occur at the same time that numerous microtubules begin to appear. Microtubules may also transport macromolecules, possibly forming channels in the cytoplasm. In plant cells microtubules occur inside the plasma lemma and are oriented tangentially to the cell. It has been suggested that they function in cell wall deposition, and microtubules have been noted to underlie the points where the secondary cell wall is being deposited in spiral or reticulate patterns. **Spindle fibers**, to be discussed with the cell division process, are proteins composed of microtubules.

Summary

1. The basic unit of life is the cell, and essentially all organisms are composed of remarkably similar cells.

2. Water is basic to all life as we know it, and its unique chemical and physical properties are related to hydrogen bonding. All the features of water density, cohesion, adhesion, and heat gain and loss can be attributed to this compound's special composition.

3. Energy is moved through living systems by a number of transfer molecules. These molecules allow for an orderly passage of energy from one chemical compound to another, ensuring that the efficiency of conversion is maintained.

4. The macromolecules of life are primarily carbohydrates, proteins, lipids, and nucleic acids. These large molecules, constructed from simpler molecules, provide the chemical framework of life.

5. In the process of plant metabolism, many plants synthesize so-called secondary compounds important to human needs. Some of these, such as alkaloids, volatile oils, anthocyanins, and tannins, are probably parts of an adaptive strategy in plant-herbivore interaction. Others, such as terpenes, resins, and sterols, are used in the manufacture of various industrial products.

6. Prokaryotic cells have no nucleus, whereas eukaryotic cells do have a distinct nucleus that can undergo division to produce new cells. Plant cells have a distinct cellulose wall; animal cells do not.

7. Membranes are made of lipids and proteins, and substances move through biological membranes according to properties of size and solubility. Membranes provide a large surface area to allow for collisions of molecules responsible for biochemical reactions.

8. Within the green plant cell are found the nucleus, mitochondria, chloroplasts, and endoplasmic reticulum with ribosomes, the Golgi apparatus, and microtubules. Many other smaller organelles are sometimes included. Mature plant cells usually have a single, large vacuole dominating the water relationships within the cell.

Something to Think About

1. What are the differences between eukaryotes and prokaryotes?

2. Name and tell the function of six plant cell parts.

3. What is the function of microtubules in a plant cell?

4. Name some small organelles found in plant tissue.

5. What is the special type of protein found in microtubules?

6. What are membranes made of?

7. What is the chemical framework of life called?

8. Where is the Golgi apparatus found in the plant cell?

9. How do vesicles work?

10. Explain how grana work in chloroplast when the light intensity changes.

Suggested Readings

Ferguson, T. 2000. *Discovery windows on science*. New York: Van Nostrand Reinhold.

Turnbull, C. 2006. *Plant architecture and its manipulation*. Oxford: Blackwell Publishing.

Twenty First Century Science. 2006. Science Educators Group of York, Nuffield Curriculum Centre.

Internet

Internet sites represent a vast resource of information. The URLs for Web Sites can change. Using one of the search engines on the Internet, such as Google, Yahoo!, Ask.com, and MSN Live Search, find more information by searching for these words or phrases: **plant cell structure, prokaryotic cells, eukaryotic cells, plant primary cell wall, plant secondary cell wall, membranes, organelles, mitochondrion, plastids, chloroplasts, endoplasmic reticulum, ribosomes, Golgi apparatus, and microtubules.**

Growth: Cells to Tissue

CHAPTER 7

All of us, at some time during our lives, have wondered why there are so many different kinds of plants and animals. Why are they different? Do they ever change? These are commonly asked questions that have been answered only in recent years.

Understanding how plants grow, develop, and reproduce and do so with predictability is the key to understanding the science of plants. Why do oak trees always produce acorns that in turn always grow into oak trees? Why do all other organisms display similar growth and reproductive sequences? Since this predictable repetition always includes a single-celled stage at some point in the process, it has become obvious to scientists that the controlling mechanism is completely contained within a single cell and subsequently transmitted to resulting cells as growth proceeds. Therefore, this awesome "responsibility" must surely be housed in some molecule that is present in all cells.

In addition, scientists have long been aware of the visibility and activity of chromosomes during cell division. It was only logical to suspect the **chromosomes** of being responsible for containing and passing on the information that controls the growth and development and the inheritance of the organism's traits to the next generation. But of what are chromosomes made?

Objectives

After completing this chapter, you should be able to:

- ■ **Discuss what deoxyribonucleic acid (DNA) does**
- ■ **Describe DNA as a double helix that unwinds**
- ■ **Comprehend how DNA replicates itself**
- ■ **Describe mitosis (nuclear division) and cytokinesis**
- ■ **Recognize the differences between mitosis and meiosis**
- ■ **Recognize the different phases of mitosis and meiosis**
- ■ **Identify the types of simple tissue**
- ■ **Explain the difference in primary growth and secondary growth**
- ■ **Draw and label parts of the dicot stem and monocot stem**
- ■ **Describe the difference between a dicot root and monocot root**
- ■ **Discuss how the vascular tissue works in plants**
- ■ **Understand how a leaf is involved with the vascular system**

Key Terms

chromosomes	enzymes	somatic cells
hemoglobin	transcription	chromatin
nucleotides	translation	genome
double helix	RNA polymerase	developmental biology
X-ray crystallography	codon	gene regulation
gene	anticodon loop	lactose
triplets	interphase	glucose

operon
structural genes
promoter
histones
hormones
osmotic agents
temperature shock
    treatments
N-formylmethionine
leucine
tryptophan
point mutations
lethal
benign
silent mutations
mutagen
LSD (lysergic acid
    diethylamide)
X-rays
ultraviolet light
heavy metals
pesticides
mitosis
cytokinesis
replicated
interphase
metabolic activity
spindle fibers
chromatids
centromere
homologues
homologous pair
diploid
nucleoli

prophase
metaphase
metaphase plate
anaphase
identical genetic
    makeup
equatorial plate
meristematic regions
apical meristems
lateral meristem
secondary tissue
cell expansion
cytoplasm
plasma membrane
cellulose microfibrils
zone of elongation
zone of differentiation
parenchyma
collenchyma
sclerenchyma
sclereids
primary xylem
primary phloem
vessel
elements
tracheids
fibers
sieve tube members
companion cells
sink
epidermis
trichomes
bulliform
guard cells

stoma
gas exchange
primordia
ground tissue
internode
epidermis
cortex
pith
vascular bundles
vascular cambium
fascicular cambium
secondary xylem
secondary phloem
anchorage
absorption
synthesis
organic compounds
root hairs
pericycle
root meristem
stele
casparian strip
palisade cell
spongy parenchyma
veins
vascular bundles
bundle sheath cells
kranz leaf anatomy
secondary growth
wood
growth rings
cork cambium
periderm

DNA Replication

One early rational suggestion was to specify the complicated characteristics displayed by any single organism or species surely the molecule of inheritance would have to be protein. Only here would there be the possibility of thousands and thousands of different combinations of amino acids. With 20 amino acids, the number of possible sequences is enormous, and surely each sequence could specify the characteristics for a particular organism. The critical question was how the same molecules could be passed on from generation to generation. Carbohydrates or lipids, with their identical repeating units, could not specify complicated codes. The only possible solution seemed to lie in the proteins. But such was not the case. Beginning in 1951, James Watson and Francis Crick began piecing together information about the other molecule that was a constituent of the chromosome—DNA.

Monday-morning quarterbacking is easy when studying the history of the molecule basis of inheritance. The solution may seem obvious to a student of modern biology, but the following advances within the past century have been remarkable indeed, given the information at hand.

1. In 1869, Johann Friedrich Miescher identified a weakly acidic substance of unknown function in the nuclei of human white blood cells. This substance would later be called deoxyribonucleic acid, or DNA.

2. In 1912, Physicist Sir William Henry Bragg and his son, Sir William Lawrence Bragg, discovered that they could deduce the atomic structure of crystals from their X-ray diffraction patterns. This scientific toll would be key in helping Watson and Crick determine DNA's structure.

3. In 1928, a public health official studying bacterial pneumonia found that something called a transforming factor could be passed from dead to live bacteria and cause a change in their hereditary characteristics. In 1943, a scientist at Rockefeller University discovered that the transforming factor was DNA.

4. In 1941, George Beadle and Edward Tatum of the University of Chicago performed ingenious experiments with bread mold to show that a single gene is responsible for the synthesis of a single protein. Nobel Prize winner Linus Pauling used this information to show that sickle-cell anemia is caused by a mutation of the blood protein hemoglobin, and the only difference between normal- and sickle-celled anemia is caused by a mutation of the blood protein **hemoglobin**, and the only difference in normal and sickle-celled hemoglobin, is the substitution of two of the 600 amino acids.

5. In 1944, Oswald Avery, and his colleagues Maclyn McCarty and Colin MacLeod, identified Griffith's transforming agent as DNA. However, their discovery was greeted with skepticism, in part

because many scientists still believed that DNA was too simple a molecule to be the genetic material.

6. At the same time as the Watson-Crick study, other scientists were able to show that a bacteriophage, a virus that attacks bacterial cells, can change the genetics of the bacterial cell by injecting its own hereditary material. Using radioisotopes, they proved that the hereditary material was DNA and not protein.

7. Many different studies showed that the amount of DNA in body cells is two times the amount found in sex cells.

8. The convincing bit of evidence giving the basis for the Watson-Crick study was that DNA is composed of the 5-carbon sugar deoxyribose, some phosphate, two purines (adenine and guanine), and two pyrimidines (thymine and cytosine). The nitrogen bases (purines and pyrimidines) do not occur in equal proportions, but the proportions are fixed for a given species and are different between species. The amount of adenine always equals the amount of thymine, and the amount of guanine always equals the amount of cytosine.

9. When James Watson went to Cambridge to work at the Cavendish Laboratory with Frances Crick, they had access to a great deal of information concerning the basic chemistry of DNA. They knew not only that the molecule had to carry precise information that could be transferred from generation to generation, but also that the quantity of specific information needed would be enormous for complex eukaryotic organisms. Precision in information transfer was all-important, but if a mutation occurred, the new information transfer was all-important, and new information would have to be transferred with the same precision as the original information.

   When their work began, they already knew from the work of Rosalind Franklin that the molecule consisted of a multiple helix with bases inside. They also knew of Chargaff's experiments to show that the amount of guanine in an organism was always the same as the amount of cytosine, and the amount of adenine was equal to the amount of thymine. Considering all the chemical and X-ray crystallography data, they set about to construct a tin model of the molecule. They knew if DNA was indeed the correct molecule, it would have to have the correct configuration, size, and complexity to code for the vast amount of information that had to be processed. Working within the physical dimensions imposed by purines, the sugar deoxyribose, and phosphate, they deduced that the molecule had to be not a single helical structure as in proteins, but a double helix.

   In unraveling the complex structure, they discovered that the ends of the two strands were different. The phosphate group that joins two sugar molecules is attached to one sugar at the fifth carbon position (the 5' position) and to the other sugar at the third carbon position (the 3' position). This gave one strand with a 5' end and the other with a 3' end. In other words, the strands were opposite in direction, or antiparallel. It became apparent later on that the decoding information could be done only from one direction, leading to the concept of a sense strand and a nonsense strand.

10. In 1966, the genetic code is deciphered when biochemical analysis reveals which codons determine which amino acids. The four nucleotides of DNA are formed by a chemical bonding of the purine or pyrimidine to a molecule of deoxyribose and a phosphate group.

The four **nucleotides** of DNA are formed by a chemical bonding of the purine or pyrimidine to a molecule of deoxyribose and a phosphate group, as seen in Figure 7-1. These nucleotides are paired with the purines and pyrimidines facing each other as the sugar and phosphate protrude toward the outside. The purines adenine and guanine both have two rings and are thus the same size and shape (only a single ring substitution is different). The pyrimidine base thymine is a double ring with a single ring—all nitrogen base pairs that are the same dimension and fit into the double-stranded molecule.

Figure 7-1

A nucleotide consists of a nitrogen base—either adenine (A), guanine (G), cytosine (C), thymine (T), or uracil (U)—connected to a 5-carbon sugar and s phosphate group. In DNA, the sugar is deoxyribose and the bases are A, T, G, and C; in RNA the sugar and the bases are A, U, G, and C. Courtesy of Heidi Kamp.

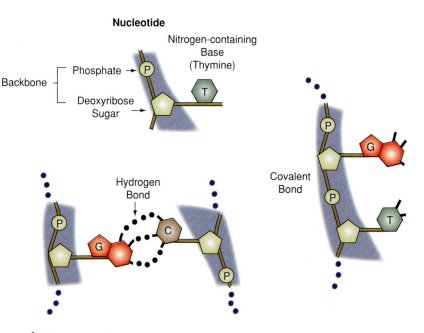

The complementary pairing of purines and pyrimidines is accomplished by hydrogen bonding between H-O and H-N. Only two such bonds occur between adenine and thymine, but three hydrogen bonds occur between guanine and cytosine. Therefore, it is not possible for adenine to pair with cytosine nor guanine with thymine.

The Watson-Crick model proposed two right-handed polynucleotide chains coiled in a helix around the same axis. This is often referred to as a **double helix** (see Figure 7-2). The molecule might be thought of as a ladder, twisted into a coil like a spring. The purine and pyrimidine bases of each strand are stacked on the inside of the double helix with their planes parallel to each other and perpendicular to the long axis.

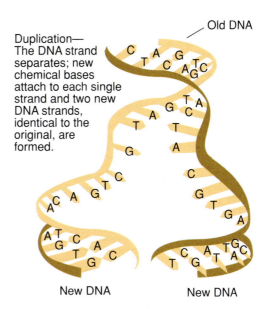

Duplication— The DNA strand separates; new chemical bases attach to each single strand and two new DNA strands, identical to the original, are formed.

Old DNA

New DNA   New DNA

Figure 7-2

The double-stranded DNA molecule separates so that each strand may replicate with its complementary strand. At the completion of this process, each chromosome consists of two identical chromatids with the same genetic information encoded.

Courtesy of Rick Parker.

The diameter of the entire helix is 2 nm. The space occupied by a purine is more than 1 nm, but the space occupied by a pyrimidine is less than 1 nm. Consequently, a purine-pyrimidine pair joined by hydrogen bonds occupies exactly 2 nm. The distance between nucleotide pairs in the coil is 0.34 nm, and there are 10 nucleotide pairs in one complete coil of the helix ($0.34 \times 10 = 3.4$ nm). All these dimensions were revealed by **X-ray crystallography**. Thus, constraints are imposed on the size and shape of the purines and the pyrimidines themselves and the way they fit into the molecule. In a single strand of the DNA molecule, one might encounter any possible sequence involving thousands of the four nitrogen bases, but in the matching strand complementary base pairing would always occur.

11. In 1970, Hamilton Smith, at Johns Hopkins Medical School, isolated the first restriction enzyme that cut DNA at a specific nucleotide sequence. Over the next few years, several more restriction enzymes would be isolated.

12. In 1972, Stanley Cohen and Herbert Boyer combined their efforts to create recombinant DNA. This technology was the beginning of the biotechnology industry.

All living organisms and therefore each of their cells must have one thing in common: the ability to replicate themselves exactly so that those characteristics remain those characteristics. For example, cow characteristics remain cow characteristics, and all other species are perpetuated with exceptional accuracy. Knowing that the molecule of inheritance is DNA, one needs to know exactly how it manages to replicate itself and pass on its exact molecular structures to succeeding generations.

The total number of nitrogen base pairing in a given chromosome varies, but for such a chromosome to replicate itself exactly and pass on this precise sequence to a new cell, it must "unzip" at the hydrogen bonds holding a new complementary strand. For example, the nucleotide containing guanine attracts another nucleotide containing cytosine in the nuclear sap to pair with it. The synthesis continues until each of the two original strands is bonded to it through the nitrogen bases, as shown in Figure 7-2.

Once the process is completed, the resulting chromosomes are no longer one double-stranded helical molecule, but two identical such molecules. A cell in which all its chromosomes have so replicated is ready to make two genetically identical cells.

Protein Synthesis

A typical chromosome has many thousands of subsections of its total nucleotide sequence, each controlling the production of a specific protein. Each of these subsections of the chromosome is called a **gene**. Since a given gene might include a few hundred to several thousand nucleotides in sequence, and since proteins are composed of many amino acids linked together, the message contained in each gene must somehow be put to work building a specific protein with a set sequence of amino acids.

Both strands of the DNA do not function in this process; only one does so. Suppose, for example, one strand of the DNA molecule read in part: adenine-cytosine-guanine (A-C-G). Then the complementary pairing would have to read: thymine-guanine-cytosine (T-G-C). Such three-part sequences are referred to as **triplets**, and they specify the location of a specific amino acid. Actually, the triplets of the DNA molecule do not translate directly into the code but serve as a template from which the code will be made. In the preceding example, A-C-G might be the template for specifying the correct message.

The process by which proteins are synthesized is rather complicated (see Figure 7-3). Since proteins are needed by the living cell for structure, storage, and most important for thousands of different kinds of **enzymes** with near-perfect specificity, it stands to reason that the process of forming proteins must be an exceptionally precise sequencing of amino acids. Enzymes, in particular, must have a primary structure that is accurate, predictable, and repeatable. The right kind and amount of enzymes must occur not only in the right place but also at the right time. Sometimes the linkage of a complicated series of biochemical events fails if even one enzyme is missing. A cell that fails to make a particular enzyme at a particular time might not survive.

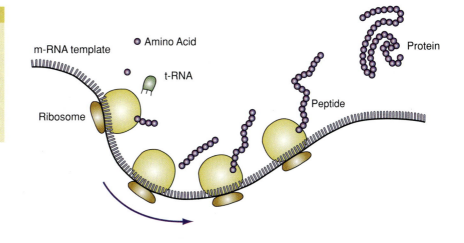

Protein synthesis is a sequential progression involving DNA, RNA, amino acids, and ATP. This "master plan" involves two basic processes: **transcription** and **translation**.

Transcription

In addition to the process of DNA replication, the DNA molecule within the nucleus provides the template for protein synthesis. Just as in the replication process, the double-strand DNA molecule "unzips" (the bonds are actually broken by an enzyme) for a gene portion of its length, and one strand becomes the template for making a single-stranded messenger RNA (mRNA). Since the strands of the DNA molecule are antiparallel, the enzyme responsible for forming a new mRNA strand can read the messages only from one direction, thus, eliminating any confusion about which strand should be read (see Figure 7-4). A particular triplet that becomes exposed by the unraveling and unzipping provides the genetic code for a new complementary nitrogen base pairing.

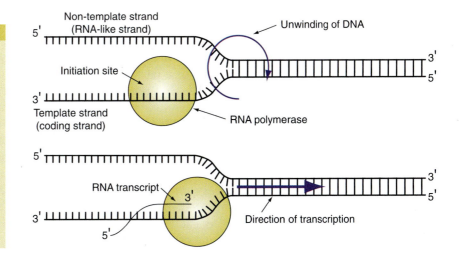

A special kind of enzyme called **RNA polymerase** causes this triplet to pair with nucleotides previously synthesized and present in the nucleus. Returning to our original DNA triplet sequence: adenine-cytosine-guanine, the complementary pairing within the new mRNA would be: uracil-guanine-cytosine. All forms of RNA uracil substitutes for the thymine. This new synthesized mRNA triplet is called a **codon** and is the code for a particular amino acid. Although triplets occur in both DNA and RNA, the term *codon* applies only to the triplets of mRNA. Subsequently, the DNA molecule unravels further, exposing more of the genetic code from the DNA. Each triplet accepts complementary pairing for making other codons, and the process proceeds until the entire DNA sequence (one gene) has been transcribed. Thus, the single-stranded DNA carrying the code for the protein to be made—a very large molecule— and the mRNA molecule anneals by allowing the hydrogen bonds to re-form. This process is a transcribing of the genetic code in the form of a series of many three-letter sequences (see Figure 7-5). The codons can be read only from one direction; to do otherwise would specify a different amino acid.

In eukaryotic cells, the newly formed mRNA molecule undergoes a modification or processing in which segments of the molecule are eliminated. Specific enzymes cut out or edit those portions of the message, which are garbled or fail to specify the correct protein sequence, as shown in Figure 7-5. These intervening sequences have a role in gene regulation or impart additional variability in genetic diversity.

Once the mRNA molecule has been formed and processed, it migrates out through a nuclear pore and becomes attached to a ribosome. Even though the molecule may be very large, perhaps an average of some 500 codons, the nuclear pores are sufficiently large to allow movement from the nucleus to the cytoplasm.

Translation

Once in the cytoplasm, the large mRNA molecule attaches itself to a ribosome; the same time, many amino acid molecules are being activated by an energy-dependent (requires ATP) attachment to a rather small transfer RNA (tRNA) molecule. There is a separate tRNA molecule for each of the 20 amino acids, and each one is specific because of a particular **anticodon loop**, a sequence of three nitrogen bases that read as a complement to the codon at the site of the ribosome-mRNA. Thus, a codon of mRNA aligns with the anticodon of tRNA and in the process positions the appropriate amino acid. Then, the next codon is pulled across the surface of the ribosomes, read, and matched with its appropriate anticodon loop, moving the second amino acid into the proper position. This process continues until the entire mRNA coding sequence has been translated (see Figure 7-5).

As each amino acid is being positioned in proper sequence, an enzyme causes a molecule of H_2O to be released between the –COOH group of one amino acid and the $-NH_2$ group of the adjacent amino

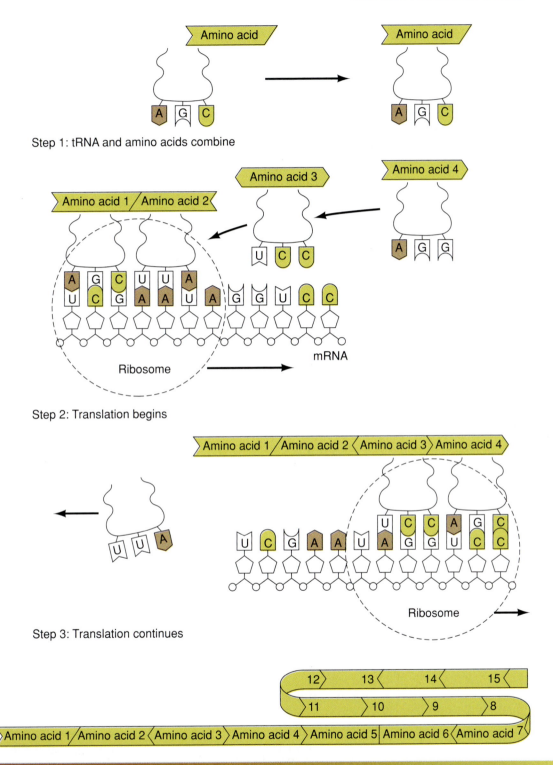

Step 1: tRNA and amino acids combine

Step 2: Translation begins

Step 3: Translation continues

Figure 7-5

Protein synthesis usually occurs by the pulling of the mRNA molecule through a series of ribosomes known as polysomes. This allows the same protein to be made at each ribosome. Translation occurs as the codon of the mRNA molecule is matched with the anticodon of the transfer RNA molecules. A previously activated tRNA-amino acid complex moves into position of the anticodon loop which codes for the three. Courtesy of Rick Parker.

acid, forming a peptide bond. As each new amino acid in the chain is added, it is detached from its tRNA, which is then free to move back into the cytoplasm to activate and deliver another amino acid when "summoned" by an appropriate codon. Thus, at the termination of the translation sequence, all amino acids are in the proper order and held together by peptide bonds. Proteins are also sometimes called polypeptides because of their structure. Technically a protein may consist of more than one polypeptide chain, but most protein is a single chain (see Figure 7-5). In most cases the mRNA is degraded in a short time and is not used as a template for the production of more than a few identical proteins, possibly only one. Thus mRNA is usually thought to be short-lived.

Multiple copies of the same protein can be made at approximately the same time by directing the same mRNA molecule along the surface of several ribosomes. Such groups of ribosomes are called polysomes and apparently allow the message to be used repeatedly before it is degraded.

The Gene

The gene represents a segment of DNA that codes for a particular protein. In prokaryotic organisms, the DNA is a single, circular molecule much like a very long rubber band. It is not associated with protein, it never condenses into a chromosome, and it is never visible under ordinary light microscopy. In eukaryotic organisms, however, each chromosome consists of a single link thread of DNA associated with both basic and acidic nuclear proteins. During **interphase**, the molecule is uncoiled and difficult to see, even with appropriate staining. During mitosis, however, the DNA condenses and coils so that the chromosomes do become visible under light microscopy.

In eukaryotic organisms, the cells of all tissues except sex cells are called **somatic cells**. They contain two sets of chromosomes that represent the inheritance from each of the parents. Since cells are generally small and since the nucleus is even smaller, chromosomes are visible only under rather high magnification. Even when viewed with a light microscope at 1000× magnification, some chromosomes are barely visible at their most highly condensed state. A study of their detail structure requires further magnification.

It is important to realize that a chromosome is made up of some nuclear proteins plus the DNA molecule. What one actually sees through the microscope is a stained preparation of the protein and DNA, the so-called **chromatin**. Only during mitosis or nuclear division does the chromatin become visible as distinct chromosomes. Organisms vary in their number of chromosomes, and the number has absolutely nothing to do with the complexity of the organism. Each chromosome represents a large amount of DNA, and the molecules are coiled and doubled back on them much like a tangled kite string. The "string" is nothing more than a very long DNA molecule consisting of thousands and thousands of nucleotide pairs. From an inheritance point of view, a gene represents a heritable characteristic, one that can be passed on from generation to generation. From a molecular point of view, a gene is a piece of a DNA molecule that carries the genes that are responsible for the entire code.

Not all genes are decoded or "turned-on" at the same time. In fact, geneticists say that not more than 5% to 10% of the **genome** (the entire genetic complement) is turned on at any one time. The kinds of proteins needed at the time of seed germination—those associated with breaking down storage macromolecules, beginning rapid metabolism and the synthesis of a great deal of new structural material during meristem development, and the ultimate production of new leaves and roots—may be entirely different from the proteins needed after the plant is fully grown and ready for sexual reproduction. There are many examples in molecular biology that show that transcription occurs in genes only when it is needed, but the regulatory system must be controlled ultimately by a complex interaction of the environment and internal factors such as hormones.

The overall problems of how an organism determines which genes to turn on and off is the basis of **developmental biology**, the combined synthesis of genetics, physiology, **gene regulation**, and morphology. With the completion of the mapping of human genome has come a much better understanding of gene regulation. This will lead to answers we all have had about the genome. In plants, many different ones have had their genome mapped and this is leading to improvement of them and will help to provide the world with better food.

Gene Regulation

The first information about genetic regulation (what turned them off and/or on) has come from studies with bacteria. Because the cell cycle is so short in such organisms, they lend themselves to elaborate experimentation in a short period of time.

In the organism *Escherichia coli*, a common bacterium in the digestive tract of healthy humans, the milk sugar **lactose** is split into **glucose** and galactose by the enzyme B-galactosidase. When lactose is present, the enzyme is present in large quantities; when lactose is absent, only traces of the enzyme can be found. Since lactose induces the enzyme to be formed, apparently turning on a gene, it is called the inducer and the enzyme is called an inducible enzyme. Some substances inhibit enzyme production and are called corepressors; such enzymes are called repressible enzymes.

The model developed for E. coli to explain gene regulation includes an **operon**, a group of **structural genes** that are functionally related and are aligned along a segment of DNA known as the **promoter**. The RNA polymerase attach to the promoter at the beginning of transcription. An additional DNA segment called the operator acts as the on-off switch for the promoter. In the "on" position, mRNA can be transcribed; in the "off" position, no mRNA can be made. Yet another gene, the regulator, determines the position of the on-off switch. The regulator is controlled by a repressor, which may bind to the regulator and keep the entire system in the "off" position. It would appear that inducers or corepressors control the repressor. In the lactose operon, lactose controls the repressor.

In eukaryotic organisms, chromosomal proteins play an important role in gene regulation. All genes are present in all chromosomes of all cells all the time. Complicated, multicellular organisms produce many different kinds of cells at different times during the life cycle and at different places in the organism. A majority of the genes are thus turned off or "silenced" most of the time.

Both acidic and nuclear proteins and basic nuclear proteins, the **histones**, are important in turning genes on and off in eukaryotic organisms. Various **hormones**, vitamins, **osmotic agents**, and **temperature shock treatments** have been used to induce transcription in eukaryotic organisms under laboratory conditions.

You may be curious by now about the size of a gene, and where, if the DNA molecule is a continuous thread, one gene stops and another begins. The secret lies in the genetic code itself. From the results of hundreds of experiments to determine which nucleotide triplet codes for which amino acid, we know that some triplets do not code for amino acids at all, but are stop signals to indicate the end of the code for a particular gene. The stop codons are U-A-A, U-A-G, and U-G-A. Since thousands of genes are all connected on the same chromosome, there must also be a start signal, and we know that the codon is always A-U-G. When this codon is at the beginning of the gene, it will code for **N-formylmethionine**, which we now know must be the first amino acid in the polypeptide chain forming any protein. Thus, the presence of N-formylmethionine as the first amino acid coded for by a gene allows for the start of the formation of the remainder of the protein. When A-U-G occurs anywhere else within the gene, it simply codes for methionine and does not function as a start codon.

If one considers that there are only four possible nucleotides put together in groups of three, then the possible numbers of codons is 4^3 ($4 \times 4 \times 4$, or 64). That may seem like a surprisingly small number, but since there are only 20 amino acids in protein, 64 are more than sufficient. As a matter of fact, all possible combinations are used. Some amino acids are coded by more than one codon. Six separate codons, for example, code the amino acid, **leucine**; one codon translates for **tryptophan**. By analyzing duplicate codes; it becomes apparent that the first two nucleotides in the code are rather "fixed" whereas the third nucleotide is variable (geneticists refer to the third nucleotide as being "wobbly" or imprecise).

Time Required for Protein Synthesis

No one knows just how long it takes for the translation process to occur at the surface of the ribosome, but it would be possible to calculate the time required to make a particular protein, for example,

1 of 2,000 amino acid units. By using precise radioisotope labeling experiments, it has been shown that only a few codons are translated per second, perhaps about 10. If that figure is accurate, then the calculation 2,000/10 = 200 shows that it takes 200 seconds to make the single protein molecule. That may seem like a long time. But remember that each cell may contain thousands of ribosomes, and they could all be coding for the same protein at the same time. In such experiments, protein synthesis usually occurs (that is, a polypeptide chain is completed) in about 5 to 10 minutes.

One must also bear in mind that hundreds of different proteins could be synthesized in the same time, provided that different genes were being transcribed from the chromosomes at the same time.

Accuracy of the Code of Mutations

The 64 codons are the only possible codes for incorporating amino acids into protein. The grand scheme for passing inherited traits from one generation to another seems to follow a precise plan. Occasionally, however, mistakes are made, and the overall consequences of those mistakes (some good, some bad) are discussed in the section on evolution. It is important to emphasize here, however, that **point mutations** occur because changes are somehow incorporated into the primary structure or amino acid sequence during the synthesis of a protein. Following are types of point mutations that can occur as a result of miscoding.

Transitional Mutants

In the DNA molecules, one purine-pyrimidine base pair may be replaced by another. For example, A-T may be replaced by G-C.

Transversional Mutants

A purine-pyrimidine pair may be replaced by a pyrimidine-purine pair. This type of point mutation is very common.

Insertion Mutants

This is sometimes called a frame shift, in which an extra base pair shifts the sequence out of phase by one pair. In reading all subsequent codons, the code is misread by one base in advance. This type of mutation may be caused by acridine dyes and of course has serious consequences.

Deletion Mutants

Here a base pair is deleted, which also causes a frame shift; the code is misread by one base delay. Deletion may also be quite serious and can be caused by high pH levels, high temperature, and by various chemical compounds, such as proflavin.

Not all mutations are **lethal** to the organism. Transitions and transerversions are relatively **benign**—that is, they have little effect on the quality of the protein. They are often referred to as **silent mutations**. On the other hand, insertions and deletions are usually lethal. Occasionally, deletion will occur in multiples of three, so there is no frame shift, but an entire amino acid is deleted.

Mutagen

Any substance that causes a mutation to occur is called a **mutagen**. The effects of industrialization have caused concern about the number of possible mutagens in the environment. Sometimes human society continues to use them, even though the possible consequences are obvious. High concentrations of caffeine, for example, may act as a mutagen, and the drug **LSD (lysergic acid diethylamide)** can have mutagenic effects. Likewise, gamma rays, **X-rays**, **ultraviolet light**, **heavy metals**, and **pesticides** are known to be highly mutagenic.

Cell Division: Mitosis and Cytokinesis

Scientists must be concerned not only with what goes on in a single living cell, but they must also consider how those single living cells grow to produce multicellular organisms. If all organisms maintained the cell uniformity found in most prokaryotes, there would be no opportunity for complicated, multicellular plants and animals with hundreds of different kinds of specialized cells in the same body. Sometimes a cell divides to produce two identical daughter cells that remain similar throughout their life span. Such similar cells make up a tissue. On other occasions, the cells specialize to form new tissues.

The process of a cell dividing to produce two new cells occurs repeatedly in single-celled organisms and in specific tissues of multicellular organisms. In single-celled organisms, the mother cell grows and then divides to produce two daughter cells that are essentially identical in size, structure, and genetic composition. They are, however, only about half the size of the original mother cell. Each of these two cells then begins a growth period until it reaches maximum size, then each repeats the division process.

The frequency of these cellular divisions varies from one species to the next; some divide every few minutes and others less often, up to once every day or so. *Escherichia coli* can replicate every 12.5 minutes. Given enough space and food, such organisms could produce incredible numbers in a short time. Available space and food do act as limiting factors to such growth, however, as they do for multicellular organisms. The "balance of nature" is kept intact by such controls.

All multicellular organisms contain eukaryotic cells with organized nuclei and chromosomes. The process of cell division involves two separate but interdependent steps. **Mitosis** is the process of nuclear division resulting in two nuclei; both genetically identical to each other and to the single nucleus that gave rise to them. **Cytokinesis** is the division of the parent cell into two new cells, each of which contains a full complement of all the different organelles and cytoplasmic constituents and one of the two newly formed nuclei.

Once initiated, the division of the nucleus is an essentially continuous process. Prior to the actual sequence of mitotic events is a period during which the cell makes extra cytoplasmic material, including all the organelles. During this period, the chromosomes also are **replicated** (see Figure 7-6).

Figure 7-6

DNA strands divide, and bases attach to the new strands to form identical genes for new cells. DNA structure coiled inside a replicating chromosome. Structures may be partially described by the position of the centromere. Courtesy of Rick Parker.

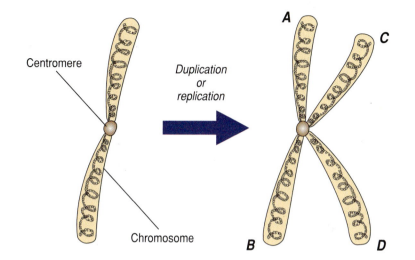

The time when cells are not dividing is called **interphase**, an in-between stage. It actually represents a time of intense **metabolic activity**; even though the chromosome cannot be seen through a microscope.

Interphase can be divided into three stages: two gaps (G1 and G2) separated by a synthesis (S) stage. The sequence is G1, S, and G2. The first (G1) is a time for growth of cytoplasmic contents, including numbers of organelles. At this time, substances are synthesized that either promote or inhibit the subsequent stages of divisions. In some plant cells, as in those of nonmeristematic regions, specific levels of inhibitor apparently act as a permanent block to mitosis, and these cells never divide again.

During the synthesis (S) stage, the DNA is replicated by the process described earlier. Purines, pyrimidines, sugar, and phosphate molecules that had been synthesized earlier are present in the nucleus and become the building blocks for the new DNA.

During the G1 stage, the mitotic apparatus responsible for the division process is synthesized and assembled, including the **spindle fibers**.

Mitosis

Although a continuous process, mitosis has been divided into four phases of activity for easier comprehension (see Figure 7-7). Since these phases are designated for convenience and do not reflect any true separation or pauses in the ongoing process, their application in the following description of mitosis is meant to be supplemental.

Figure 7-7

Even though mitosis is an ongoing integrated process. It is possible to recognize the developmental stages: (1) Interphase—period between one division and the next. Chromosomes cannot be seen but the nuclear membrane is visible. (2) Prophase—the nuclear membrane is disappearing and the chromosomes are tightly coiled. (3) Metaphase—spindle fibers appear and attach to the centromere. (4) Anaphase—chromosomes move into central position in the cell, the chromatids have separated and become new chromosomes. (5) Telophase—chromosomes arrive at the poles, and the nuclear envelope begins to reform. (6) Return to inter-phase. Courtesy of Rick Parker.

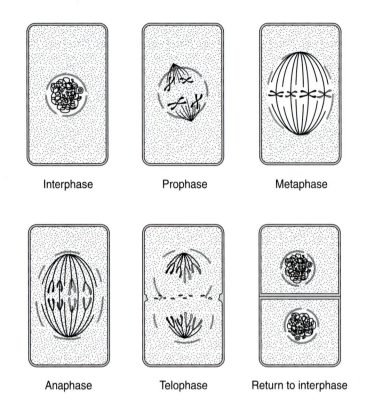

Interphase   Prophase   Metaphase

Anaphase   Telophase   Return to interphase

Mitosis begins with the chromosomes gradually changing from elongated, thin strands winding throughout the nucleus to a shorter, thicker structure that finally becomes visible with the use of a microscope and appropriate stains. Beginning with DNA replication, each chromosome has been composed of two genetically identical, parallel double strands. As the chromosomes reach this most condensed and shortest condition, these double strands become visible as two **chromatids** connected by a noncondensed area called the **centromere**. The position of the centromere varies from the center of the chromosome in some (metacentric position), to off-center locations (submetacentric position), to end positions of the chromosome (telocentric position).

The position of the centromere aids in specific chromosome identification because of the relative lengths of the two-chromosome arm on each side of it. Chromosome identification is important in many genetic studies and enables scientists to match up the two chromosomes in each nucleus that are identical in relative length, arm-length ratios, and most important in the traits genetically controlled by the genes of that chromosome pair. Even though the actual genetic message may be different on the two genes controlling each trait, these look-alike chromosomes are called **homologues**.

Essentially all plants have an even number of chromosomes in each cell because they exist in these **homologous pair**. With the exception of cells that are produced in reproductive organs, all eukaryotic cells normally contain a **diploid** number of chromosomes.

Diploid literally means "two sets" of chromosomes, each complete set coming from two parents that originally gave rise to the new organism through sexual reproduction. The diploid (2n) number varies in the plant kingdom from only 4 chromosomes (2n = 4) to over 1,200 chromosomes.

A given species, however, normally has the same diploid number of chromosomes in all its individuals. Sometimes multiples of these sets of chromosomes are produced, leading to polyploidy. Table 7-1 gives a few representative examples of chromosome numbers for different species.

Table 7-1

Chromosome Numbers of Different Organisms	
Organism	**Diploid chromosome numbers**
Haplopoappus gracilis (small desert plant)	4
Mosquito	6
Onion	16
Cabbage	18
Corn	20
Sunflower	34
Cat	38
Human	46
Plum	48
Dog	78
Goldfish	94
Ophioglossum reticulatum (fern)	1260

Courtesy of Rick Parker.

Prophase

The nuclear division begins with the condensation of the chromosomes to a shorter, thicker, and microscopically visible configuration. As the chromosomes are changing, the nucleolus or **nucleoli** gradually disappear from the nucleus, and the nuclear membrane (envelope) then begins to break up, finally disintegrating completely. Once the new visible chromosomes are fully condensed and free in the cytoplasm, long slender microtubules begin organizing into spindle fibers. The appearance of these structures signifies a translation from **prophase** to **metaphase**.

Metaphase

The spindle fibers continue to form opposite ends (poles) of the cell, ultimately meeting in the middle of the cell, three-dimensionally, and these fibers form a spindle- or football-shaped structure. Fully condensed chromosomes move to the center of the cell as the spindle is being formed and are attached at their centromeres to a pair of fibers, one from each pole. Not every fiber has a chromosome attached to it, but every chromosome, each appearing as a pair of chromatids, is attached to a spindle fiber. All the chromosomes are finally aligned in the center of the spindle and attached to fibers, forming what has been termed the **metaphase plate**. At this point in the process metaphase is complete, and the next sequence of events is grouped under **anaphase**.

Anaphase

Thus far, each chromosome has been a single unit because there has been only centromeres holding two genetically identical chromatids of each together. As soon as all chromosomes are attached to spindle fibers in the center of the cell, their centromeres divide, allowing the two chromatids of each chromosome to separate. Current theory suggests that the spindle fibers attached to these centromeres begin to shorten and pull the newly divided centromeres and their chromatids apart. The two chromatids of each chromosome are now called daughter chromosomes as they move apart through the cytoplasm to the opposite poles of the cell. Since the cytoplasm is an aqueous medium and the contracting spindle fibers pull on the centromeres, the arms of each daughter chromosome trail back from an equal number of V-shaped strands aiming toward each pole. Anaphase thus consists of the short time from separation of the chromatids at the metaphase plate until the two groups of identical daughter chromosomes reach the opposite poles of the cell.

Telophase

As the two identical sets of chromosomes reach the opposite poles of the cell, several events begin to occur almost simultaneously. These events are essentially the reverse of prophase. A nuclear membrane begins forming around each of the two groups of chromosomes, and the spindle fibers completely disassociate and disappear. The chromosomes also become less distinct as they change into long, thin strands again. Nucleoli re-form, attaching to special nucleoli-organizing areas of certain chromosomes. The end result of mitosis, then, is the formation of two complete nuclei—both genetically identical to each other and to the nucleus that gave rise to them. Since each nucleus is normally a cell (cytokinesis), the significance of their identical genetic makeup is clear. The growth, protein synthesis, and direction of all cellular functions in these two new cells must be the same as the parent cell, and it is the process of mitosis that ensures the **identical genetic makeup** of all new daughter cells.

Cytokinesis

Although discussed separately here, mitosis and cytokinesis usually occur simultaneously to produce two daughter cells, each having genetically identical nuclei. Cytokinesis is the division of the original cell and its cytoplasm into two cells, each containing an approximately equal amount of all cytoplasmic materials. Occasionally, division results in daughter cells of unequal size and content. Cytokinesis begins during the early telophase activities of mitosis. As the newly formed daughter nuclei pull apart and move away from the **equatorial plate** (metaphase plate) region of the cell, small vesicles migrate to that area and associate with microtubule fibrils found there. These fibrils may be derived from disassociated spindle fibers left in this region of the cell. The vesicles are produced by Golgi bodies and contain pectin, which forms the middle lamella between the cells. The vesicle-fusing process begins at the middle of the cell and continues toward the outer walls, building the cell plate, which completely separates the original cell into two halves when completed.

Once the cell plate is formed, both newly formed daughter cells form a cell membrane across the plate to completely enclose their respective cellular contents. Primary walls are then laid down outside the cell membrane, and two new daughter cells are complete although not yet full sized. The plasticity of the primary wall allows for cellular growth until each of the two daughter cells reaches maximum size; the secondary wall materials are added to provide rigidity and structural strength. It is during this period of cellular growth that the nuclei of each cell are said to be in mitotic interphase. During this period, the cell synthesizes extra cytoplasmic material and organelles, the chromosomes replicate, and finally specific material necessary for the initiation of mitosis, such as extra microtubules, are produced. The two daughter cells are now mother cells in their own right and are ready to undergo mitosis and cytokinesis.

Not all cells follow this sequence of interphase activity because not all cells divide more than once. In plants, new cells are produced only in apical meristem and lateral meristem. Those cells, which do not remain in these meristematic areas are thought to go through the G1 portion of interphase activities but are thought not to replicate their chromosomes or produce specific materials for mitosis.

Those cells, which do divide repeatedly, vary in the amount of time required for mitosis and cytokinesis. Some take a little more than an hour, and others take up to 3 or more hours, but they do not do so by the process of mitosis, since they do not have organized nuclei with true chromosomes.

Sometimes nuclei divide without cytokinesis taking place. The consequence is a cell with more than two sets of chromosomes, or a polyploidy. The genetic implications of polyploidy are discussed in the section on genetics and evolution.

Meristems

In multicellular organisms, actively dividing cells result in either new growth or replacement of old cells for overall tissue maintenance. Both growth and maintenance functions occur in animals, whereas plants only produce new growth. Growth regions in plants are called **meristematic regions**, and they occur in the tips of growing shoots and roots of all plants. Woody plants also have two kinds of lateral meristems: the vascular cambium and cork cambium.

Apical meristems

Whenever plants grow taller or whenever side branches grow longer, they do so as the result of the shoot meristem, a region of intense mitotic activity at the apex of the shoot (see Figure 7-8).

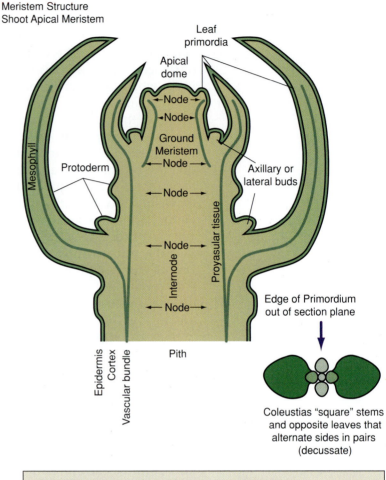

Meristem Structure
Shoot Apical Meristem

Coleustias "square" stems and opposite leaves that alternate sides in pairs (decussate)

No cap for protection-leaf primordia serve sufficiently

Apical dome, leaf primordia tips, lateral buds-interrupted zones of division

Internodes-interrupted zones of elongation (nodes quiescent)

Maturation also mosaic because of interspersed meristematic regions

Lateral buds

Lateral buds are also sites of meristematic potential and may or may not be activated at a particular time.

Roots

Roots also grow as the result of meristematic activity at their tips. Root and shoot meristems are essentially identical, except that shoot meristems are protected by young leaves being initial at the apex; root meristems are protected by a root cap, a group of cells that arises from the apex and provide a loose covering for it. These cells are constantly being sloughed off and replaced by new ones.

Shoot and root meristems

Shoot and root meristems (see Figure 7-8) that cause a plant to grow vertically (higher in the air and deeper in the ground) comprise the **apical meristems**, giving rise to the primary tissue of a plant.

Lateral meristems

Plants that grow for a relatively long time, and particularly those that survive the winter, do so as the result of **lateral meristem**, regions of meristematic activity that cause a plant stem or root to increase in diameter and give rise to **secondary tissue**. By far, the most important lateral meristem is the vascular cambium, a cylinder of cells in the roots and stems of woody plants that give rise to secondary xylem or phloem. In addition, some woody stems develop a phellogen or cork cambium that gives rise to bark. In general, monocots do not have lateral meristems, although some semiwoody monocots, such as palm trees, develop regions of cambial activity.

Accessory meristem

Although apical meristems and lateral meristems account for most growth, other pockets of meristematic activity sometimes persist for a long time. At the base of the blade and at the base of the sheath of grasses, for example, an intercalary meristem provides continuing mitotic activity long after the leaf is fully developed. This feature has adaptive significance because if a herbivore (or a lawn mower) cuts off the leaf, it continues to grow from the base and ensures survival of the plant. In such plants, the shoot meristem remains near the ground and thus prevents destruction of the plant. Certain meristem persist for some time in the margins of leaves, particularly in very elaborate or deeply lobed ones.

🌍 Tissues

Since cell division in plants occurs only in meristematic regions—specific, localized pockets of mitotic activity—cells adjacent to those regions begin to change and eventually become the specialized tissues that make up mature organs. Thus, one finds a lineage of cells trailing away from the meristem of the shoot or the root. The meristematic region is quite small, perhaps only 10 or 20 cells across and occurring in a cone of tissue. The meristem depth is also only 10 or 20 cells, and directly behind the apex is a group of cells that have recently divided and are in the process of becoming mature. They do so by going through a process of **cell expansion**, in which the vacuole expands greatly, pushing the **cytoplasm** close to the **plasma membrane**. The physical forces exerted on the cell wall dictate the direction of expansion. Even though these new cells have very thin cell walls and are quite extensible, the direction in which the **cellulose microfibrils** are laid down influences the relative strengths and weaknesses in the wall. Some directions "give" more than others, and instead of a perfect sphere, most cells begin to assume a shape characteristic of the tissues they will eventually comprise. The region immediately adjacent to the shoot or root apex is usually referred to as the **zone of elongation**, although in fact the cells may also be expanding laterally, and not all the growth is true elongation. Small meristematic cells, originally only 20 or 30 μm in diameter may suddenly elongate to perhaps several hundred micrometers.

During the elongation phase and as the cell begins to reach its ultimate shape, subtle internal chemical changes may begin to occur. This regional change of cell composition, thickness of the wall, and function is directly behind the zone of elongation and is called the **zone of differentiation**. The specialized nature of the cell, which contributes to the separation into tissue with specific functions, begins here.

Even though all cells start out the same as daughter cells produced in the mitotic process, they begin changing in the zone of elongation and continue to do so in the zone of differentiation. Here, they take on structural and functional characteristics that are to remain until they decompose as litter. Some plant cells die early in life, but they still function in tissues for a number of purposes. Vessels, for example, specialize and die early, but they continue to function in water and mineral transport.

Simple Tissues

Cells of the ground tissue are of three types: **parenchyma, collenchyma**, and **sclerenchyma** (See Figure 7-9). Each tissue type has specific functions, and parenchymal cells, for example, perform many functions. Parenchymal tissue in the stem or root may perform a task very different from that in a leaf. The ground meristem gives rise to the cortex, which is actually the foundation on which the stem and root are made. The vascular tissues are embedded within the ground tissue, whereas the epidermis surrounds it. The ground tissue of the leaf is called mesophyll, and all these systems are continuous. In cross section, the bulk of the developing stem is ground tissue. These primary functions of the ground tissue are synthesis support and storage.

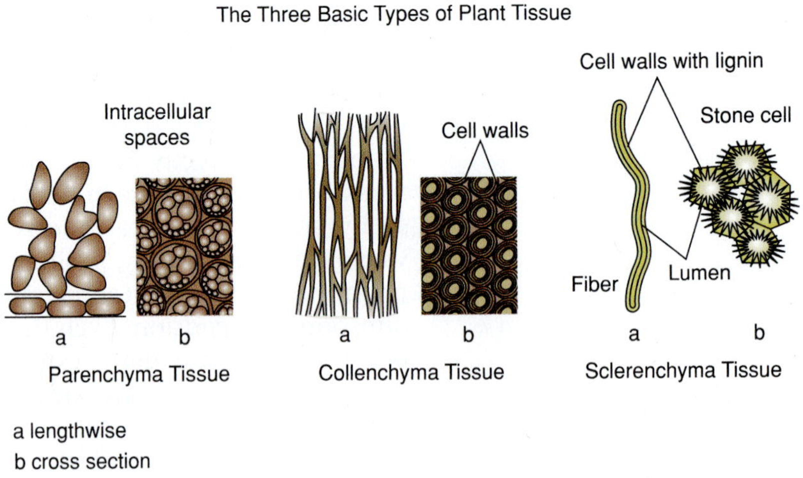

Figure 7-9

The basic tissue type is parenchyma. The living cells are thin walled and isodiametric, and there is considerable intercellular space. (B) Collenchyma so much less common but easily recognized by the thickening of the primary wall between the cells. (C) Sclerenchyma may be either fibers or sclereids. (D) These thick-walled support cells are almost always dead at maturity.
Courtesy of Carolina Biological Supply.

The Three Basic Types of Plant Tissue

Intracellular spaces

Cell walls

Cell walls with lignin

Stone cell

Fiber

Lumen

a   b
Parenchyma Tissue

a   b
Collenchyma Tissue

a   b
Sclerenchyma Tissue

a lengthwise
b cross section

Parenchyma

Parenchymal cells have relatively thin primary cell walls; they are often isodiametric; that is, they have essentially the same diameter in all planes. They generally retain all membranes and organelles and appear very much like what one might expect a typical plant cell to be.

Parenchymal tissues remain alive and functional at maturity, and they occur commonly in the cortex of stems and roots, in the pith of stems, in leaves, and in fruits. They also occur in certain regions of the primary and secondary vascular tissues and as ray cells in secondary vascular tissues. Parenchymal cells have large vacuoles and are often involved in storage. Many parenchymal cells are capable of cell division and are therefore important in wound healing and regeneration. They also initiate adventitious roots on stem cuttings and function in various tissues for photosynthesis, storage, respiration, protein synthesis, and secretion.

Collenchyma

Collenchyma cells usually occur as discrete groups of cells that form a tissue directly beneath the epidermis in stems and petioles, and they often border the veins in dicot leaves. The cells are generally elongated and unevenly thickened in the primary walls, which are capable of giving exceptional strength to young actively growing tissues and organs. The walls of these cells stretch readily and offer little resistance to elongation. Collenchyma cells, like parenchyma, are alive at maturity.

The uneven wall thickening characteristically occurs at the junction with two other cells, forming a "corner" with thickened deposits of cellulose, pectin, and water. This additional thickness provides mechanical support for young stems and leaves when subjected to a stress such as wind. One excellent place to observe collenchyma is in the petioles of celery or beets. In windy regions, celery may form a great deal of collenchyma, which tends to make the celery stalk (actually the petiole) tough.

Sclerenchyma

Sclerenchyma is characteristically by cells with very thick walls. Sclerenchyma generally falls into two categories of cells: fibers and **sclereids**, or stone cells. Fibers are usually elongated, have lignified secondary cell walls, and are important for strength and support in plant parts that have completed elongation. Fibers generally occur in strands or bundles, and they are very important in the consumer industry. Hemp and jute fibers are used to make twine, flax fibers are used in making linen, cotton fibers (technically epidermal hairs of the seed) are converted to thread and woven into a myriad of products. Other plants provide fibers that are processed into twine and rope. Originally most kinds of rope were made from hemp, but since this plant also produces marijuana, its cultivation is no longer legal in the United States.

Sclereids

Sclereids are variable in shape, some being highly branched like an octopus, others shaped like a bone, and still others relatively spherical. Compared with fibers, they are short cells, and they may occur singly or as groups of cells throughout the ground tissue. They are the primary component in seed coats, nutshells, and the hard pit of stone fruits, and they give a ripe pear its gritty texture. At maturity, sclerenchyma cells are dead. In dying they reach the ultimate specialization.

The strength of a fiber or sclereid depends on the thickness of the secondary cell wall. Prior to death, that cell must synthesize a large number of cellulose and lignin molecules, which impart strength to the cell. As indicated elsewhere, the secondary wall thickness is one of the important quality considerations in the cotton fiber and can be a major factor in the farmer's price. The thickness and number of wood fibers determine the strength and quality of the lumber.

Complex Tissue

A single type of cell characterizes the complex tissue (the so-called simple tissue just described). The ground tissue may be composed of parenchyma, collenchyma, or sclerenchyma. In contrast, complex tissues consist of more than one cell type, and these are involved in the vascular and dermal systems of the plant.

The provascular tissues give rise to the primary vascular system of the plant—the **primary xylem** and **primary phloem**. The cells of these tissues are highly specialized; xylem transports water and nutrients, whereas the phloem transports only organic solutes produced within the plant. These pipelines are the support system for higher plants; without them the stature of land plants would be much smaller than it is.

Xylem

Five basic cell types make up the primary xylem: **vessel, elements, tracheids, fibers**, and parenchyma. Although they all have specialized functions, vessel elements in angiosperms are the most important cells in water and nutrient transport. Vessel elements combine end to end, much like sections of a stovepipe; the vessel may run throughout the body of the plant. Vessels transverse the plant from small roots up through the center of the root system, through the stem, out through the branches, into petioles, and terminate near the margins of the leaves. Vessels are very large in diameter, and therefore the flux of water through them can be enormous. As the cells destined to become vessel members reach maturity, they take on a characteristic shape and size, and the protoplast, or living part of the cell, dies. At the time, the end walls are dissolved, and the vessel elements are connected end to end to form a continuous pipe, much like a drinking straw.

Tracheids

Tracheids also function in water transport but are less efficient in doing so because they are smaller in diameter, and the end walls are not totally dissolved as the cell reaches maturity. Therefore, resistance in water movement is greater, and not as much water can be moved along a given cross-sectional area of xylem. Obviously, those plants that have more vessels and few tracheids are more efficient in water transport. In surveying all the plants that have a vascular system (the tracheophytes), it can be seen that most of the angiosperms have vessels, most gymnosperms have none, and neither do most ferns. Angiosperms have both vessels and tracheids, and each contributes to the overall transport.

In addition to these water-conducting cells, xylem consists of fibers that are usually dead, thick-walled, and nonfunctional for transport. Their obvious function is in support, although occasionally fibers remain living and functioning for storage purposes. The parenchymal cells associated with primary xylem function for storage and synthesis of certain metabolites.

During the formation of the secondary cell wall in primary xylem, the wall is laid down in rings or in a spiral pattern, much like the reinforcing fibers in a plastic garden hose. This thickening gives strength to the vessel elements and tracheids while allowing for expansion of the cell.

Phloem

Primary phloem tissue is made up of **sieve tube members, companion cells**, fibers, and parenchyma. They transport organic solutes, primary sugars, over long distances within the plant. Students are often taught that xylem movement is upward from the soil, and phloem movement is downward from the leaves. Such comments are over simplifications since a great deal of transport is lateral or opposite of what we have been taught. Water moves to a point where the water potential is most negative, and sugars move to the point where the sugar deficit, or **sink**, is greatest. Transport of sugars and other organic solutes occurs through sieve tubes, formed by sieve tube members connecting end to end. The term *sieve tube* is appropriate because the end walls, unlike the perforated end walls of vessel elements, are left with portions of the wall remaining to form a sieve, much like a spaghetti strainer. The organic solutes dissolved in the protoplast of the sieve tube elements must be connected by holes in the sieve. Although the pores are large in terms of molecular size, they provide some resistance to movement of the viscous fluid, and transport through the phloem is considerably slower than water movement through the xylem. In gymnosperms, a more primitive type of sugar-conducting cell, a sieve cell, performs essentially the same function, although the efficiency of transport is reduced.

When wounding of the phloem occurs, the sieve tubes apparently synthesize a polysaccharide called callose, which plugs the sieve plate and reduces the flow. Such a mechanism stops the loss of organic solutes from a wound and thus enhances the plant's chance of survival.

In the process of differentiation of the sieve tube members, a unique development process occurs. Generally, whenever a cell loses its nucleus, the regulatory processes are uncoordinated, and the cell dies. In sieve tube members, however, maturity leads to a cell that is enucleate (without a nucleus), but otherwise the cell is alive and functions normally. The only other living cell known to be enucleate and functional at maturity is the red blood cell in mammals.

How a cell could function without a nucleus has long puzzled scientists, and we now believe that the close association of a sieve tube member with a companion cell allows for its regulation. The two cells have a great deal in common, having been derived from the same mother cell at mitosis. They are functionally connected, and the companion cell appears to have a great deal to do with the loading and unloading of solutes into and out of the sieve tube. Although much smaller in size than the sieve tube element, companion cells have all the normal cell components, including a nucleus. The function of fibers and parenchyma is the same in phloem and in xylem.

Dermal Tissues

The protoderm, the meristem tissue that gives rise to the epidermis, produces all the dermal, or outer, tissues of the plant. The **epidermis** produced by the shoot apex is made of a layer of cells that cover all parts of the plant, including the stems, both upper and lower surfaces of the leaves, all surface of flowers and fruits, and the root system. This "skin" functions like animal skin, acting as a protective barrier to reduce desiccation and prevents the entrance of harmful agents such as bacteria and fungi. The epidermal cells are living parenchymal cells that fit together tightly, sometimes looking like a jigsaw puzzle in face view. The outer surface secretes through the plasma membrane and into the cell wall a waxy layer called the cuticle, which reduces evaporation.

Sometimes the cuticle can become very thick on old leaves, even though it may develop cracks through which water can escape. There is no cuticle covering the epidermis of the root system.

The epidermal layer may also have various types of appendages, called **trichomes**, which may be specialized for protection, secretion, and other functions. Trichomes come in many shapes and sizes; they may be part of an epidermal cell, or they may be a series of cells attached to the epidermis. Generally, the epidermis is only a single layer thick, but in some plants such as *Ficus elastica* (rubber tree), the epidermis may be several layers thick and serve as a water storage tissue. Some epidermal cells on the upper surface of monocot leaves may enlarge greatly and function in water storage. These **bulliform** cells lose water rapidly during periods of stress and cause the leaf to curve upward, eventually rolling up as the wilting process progresses and thus helps the plant to conserve water.

Interspersed throughout the epidermal cells in dicots and indefinite rows in most monocots are **guard cells**—two elongated cells capable of flexing at the ends to form an opening between them. The two guard cells, together with the pore, which they form between them, are called a **stoma** (pl., **stomata**). Through regulation of the turgor pressure in these guard cells, the size of the pore may be controlled so that the pore is fully open, fully closed, or any stage in between. The stomata function is **gas exchange**, allowing water vapor to escape from the leaf surface and carbon dioxide and oxygen to enter and leave the leaf. Other gases also come and go, but CO_2 and O_2 are of primary importance.

Tissues and Development

Multicellular organisms assume a particular shape and function because the genetics of the organism dictate that a certain group of cells occur in a particular region, function as a tissue, and provide some service to the organism as a whole. Some plants and animals have many different kinds of tissue, and all of them are essential to the well-being of the organism. Three basic tissue types arise in the primary growth of vascular plants. Later, if the plant becomes woody, new secondary tissues arise from a different region of mitosis, the lateral meristems.

Primary Growth of the Stem

A stem arises from cells produced in the shoot meristem. A longitudinal section through a typical dicot stem reveals a dome of tissue at the apex, the apical meristem itself, flanked by lateral protrusions that eventually become leaves. The smaller **primordia** occur nearest the apex, and as one moves progressively away from the apex, the leaf primordia become larger. At the point of attachment of each of those leaves, a new zone of mitosis can be observed, which gives rise to a bud. Between the leaves are regions of **ground tissue** supporting the connecting vascular tissue, and this region becomes the **internode**. The internode makes up the bulk of the stem and elongation of internodes causes a plant to grow taller.

The Dicot Stem

If the young dicot stem is cut in cross section, stained properly, and viewed under a microscope, the tissues can be seen in proper orientation. The outside of the stem is covered by the **epidermis** with its waxy cuticle; directly inside is the **cortex**. Separating the cortex from the **pith** (both ground tissues) is a ring of **vascular bundles** (see Figure 7-10), the number of which is determined by the complexity of the tissue.

Some are square, for example, and may have only one vascular bundle in each corner. Other round stems might have as many as 15 or 20 bundles in a concentric ring. Each vascular bundle repeats the pattern of primary xylem toward the inside and

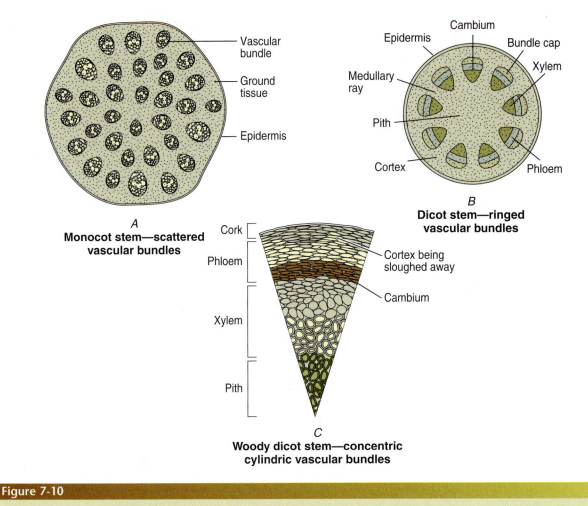

Vascular bundle

Ground tissue

Epidermis

A
Monocot stem—scattered vascular bundles

Cambium

Epidermis

Bundle cap

Xylem

Medullary ray

Pith

Cortex

Phloem

B
Dicot stem—ringed vascular bundles

Cork

Cortex being sloughed away

Phloem

Cambium

Xylem

Pith

C
Woody dicot stem—concentric cylindric vascular bundles

Figure 7-10

Cross section of a root tip showing apical meristem. Root and shoot meristems are essentially identical at their tips, root protected by the cap. Courtesy of Rick Parker.

---

primary phloem toward the outside of the stem. Both tissues may include large numbers of fibers that lend additional support to the stem.

As the stem becomes older, and if the dicot becomes woody, a **vascular cambium** begins to form between the xylem and phloem of each bundle. This cambium is called the **fascicular cambium**, and it joins with additional meristem, which begin to develop in the adjacent ground tissue and is called interfascicular cambium. When the lateral cambium joins completely, it forms a solid cylinder capable of laying down new **secondary xylem** toward the inside and **secondary phloem** toward the outside.

The Monocot Stem

Monocot stems are usually uniform in thickness, from the top to the bottom of the plant. In contrast, dicot stems are usually tapered from bottom to top. The meristem may extend from the apex a short distance, giving rise to upward divisions, providing an increase in elongation, and laterally allowing for an increase in stem diameter. In cross section, the vascular bundles are randomly scattered, as opposed to the ring of bundles in a dicot. Since they do not occur in a circle, it is impossible to distinguish pith from cortex in monocots (see Figure 7-10). It is simply called ground tissue, or ground parenchyma. Some monocots have hollow stems, however, and their vascular bundles may occur in a ring, as in the dicots.

The functions of roots in higher plants are **anchorage**, **absorption**, storage, and **synthesis** of new **organic compounds**. The patterns of cell and tissue development are highly predictable within closely related groups. In all plants that have **root hairs**, they consist of single-cell extensions of epidermal cells and are always derived from the epidermis.

At various points, lateral roots develop from the main root. This development begins at the **pericycle** as a pocket of meristem activity that quickly organizes into a new **root meristem**. As the lateral root grows in size, it breaks through the endodermis, cortex, and epidermis.

The Dicot Root

In most dicots the vascular cylinder, or **stele**, arises in the center of the root. The pattern is always with primary xylem in the center, radiating like spokes of a wheel. The number of spokes varies according to species. A common form has four spokes and is called a tetrarch pattern; those with six spokes are called hexarchy, and so on.

The primary phloem occurs in pockets between the spokes. Surrounding this region is a single layer of cells called the pericycle. Directly outside the pericycle is a single layer of cells called the endodermis. The endodermal layer is easily recognized by staining the **Casparian strip**, a suberized layer with walls so impermeable to water and solutes that substances must move instead through the plasma membrane.

Some plants do not develop primary vascular tissue in the center of the root, but rather in a well-defined cylinder of primary xylem and phloem. In that case, the ground tissue is separated into pith inside the vascular system with the cortex outside the vascular cylinder (see Figure 7-11).

The Monocot Root

Most monocot roots have pith in the central core surrounded by the protoxylem (first xylem) points, which establish the pattern for each species Most monocots have more than five protoxylem and phloem in the same pericycle, endodermis, cortex, and epidermis.

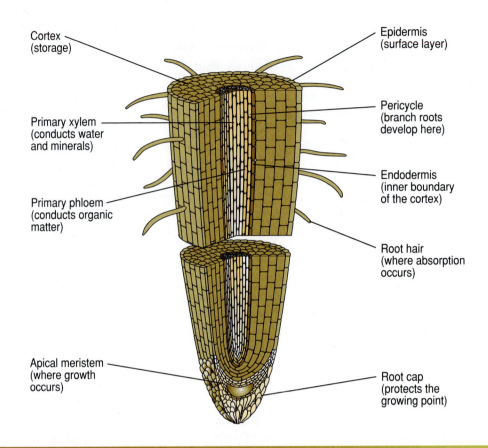

Cortex (storage)

Epidermis (surface layer)

Primary xylem (conducts water and minerals)

Pericycle (branch roots develop here)

Primary phloem (conducts organic matter)

Endodermis (inner boundary of the cortex)

Root hair (where absorption occurs)

Apical meristem (where growth occurs)

Root cap (protects the growing point)

Figure 7-11

Cross sections of a monocot stem. The vascular bundles are scattered throughout the ground tissue of plant stems. In the dicot stem, the vascular bundles form a ring of tissue in the ground parenchyma. Woody dicot stem has concentric cylindrical vascular bundles. Courtesy of Rick Parker.

Anatomy of the Leaf

Leaves are figuratively the heart of the photosynthetic green plant. One might think of roots and stems as nothing more than organs for support, absorption, and transport, but leaves contain practically all the chlorophyll and are responsible for essentially all photosynthesis, the exception being a few plants with green stems and few or no leaves.

The larger the leaf blade, the greater the photosynthetic surface for trapping sunlight. Size and numbers of leaves can be deceiving (see Figure 7-12). The total leaf surface on a large pine tree may be just as great as a similar size oak tree, but the number of pine leaves or needles is much greater.

The typical leaf blade is constructed of an upper and lower epidermis, either or both of which may possess guard cells. Generally, the lower epidermis contains a larger number of stomata than the upper epidermis, but sometimes the positions are reversed. Some plants, like begonia, have no stomata at all on the upper surface.

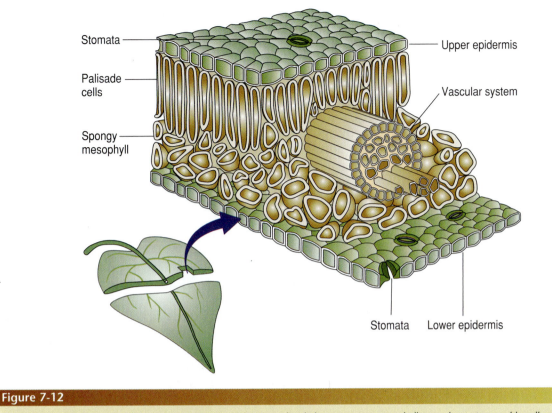

Stomata

Palisade
cells

Spongy
mesophyll

Upper epidermis

Vascular system

Stomata   Lower epidermis

Figure 7-12

Cross section of a leaf shows the upper and lower epidermis, palisade layer, spongy mesophyll, vascular system and bundle sheath cells, and the stomata. Courtesy of Rick Parker.

Below the upper epidermis, the typical leaf has a layer of vertical, parenchymatous cells called **palisade cells**. They have large numbers of chloroplast and are the site of photosynthesis. Epidermal layers, on the other hand, tend to be translucent and have few or no chloroplasts, except in the guard cells. Most leaves have a single layer of palisade, but some species have two or even three layers of palisade parenchyma; some environmental conditions may lead to this phenomenon.

Below the palisade layer, and making up the bulk of most leaves, is the **spongy parenchyma**, a layer of loosely packed cells containing chloroplasts and many intercellular spaces for water and gas exchange. The thickness of the spongy layer is also related to environmental conditions, and shade leaves tend to have less spongy parenchyma than do sun leaves on the same plant.

Embedded in the palisade or spongy layer are the **veins** or **vascular bundles**. Viewed in cross section, they appear just like the vascular bundle in a stem. The midrib may be a very large vascular bundle with much fiber reinforcement in both the xylem and phloem. In the smaller veins, the number of cells becomes smaller as one progresses from the midrib to the edge of the leaf in dicots. Net venation is typical in dicots.

Each lateral vein branches again and again, finally terminating with a single vessel or sieve tube in the parenchymal tissue. Water entering the tiniest root must make its way through a massive pipeline system to the main root, through a main stem, out through the branches, and finally from a stem into a petiole and its blade. All parts of the system must be highly integrated and connected; there can be no breaks, or the tissue will die. Likewise, the last sieve tube terminating in a leaf must be the site of "loading" of sugars produced in the process of photosynthesis. Once loaded into the phloem pipeline, those sugars may be transported to any region of the plant in which they are needed.

In some atypical leaves, a layer of cells called **bundle sheath cells** surround the vein. These cells, too, are densely packed with chloroplast, and such plants are said to have **Kranz leaf anatomy**, or the Kranz syndrome. This anatomical anomaly is associated with C4 dicarboxylic acid metabolism.

The structure of monocot and dicot leaves is not very different, except that monocots have parallel venation, and all veins terminate near the tip of the leaf. Since there is no lateral transport, damage to the middle of the leaf tends to lead to death of tissues beyond the transport system, whereas the net venation in dicots allows for diversion of water and nutrients, even if a portion of the leaf is damaged.

The blade of most leaves is connected to the stem by a petiole, a more or less round structure through which all the vascular connections to the blade must pass. The cross section reveals major or minor veins, and the number of cells in cross section is indicative of the leaf area that must be supported by the water, nutrients, and sugars flowing through those cells. The veins of the petiole are embedded in ground tissue and surrounded by the epidermis.

Secondary Growth

If one were to categorize the basic shape of plants as compared to the shapes of the animals, one would think of the similarities of fields of crops and trees of forests and orchards as being relatively tall and slender, althrough lateral growth. This thickening around the middle occurs only in woody stems and roots; the leaves do not have **secondary growth**.

Vascular Cambium

Plants that live for a long time, and especially those dicots and gymnosperms, which become woody and survive from year to year, do so as a result of lateral meristems. These cylinders of intensive mitotic activity give rise to secondary growth: tissues laid down toward the center of the plant become the secondary xylem, and those laid down toward the outside of the plant become the secondary phloem. Both tissues arise from the same single layer of cells, and no one is quite sure what causes one daughter cell to become xylem and another phloem. More cells end up as secondary xylem than as secondary phloem, and essentially all **wood** and secondary xylem.

Secondary Xylem

The basic cell types comprise the secondary xylem and another phloem. More cells end up as secondary xylem. Vessel elements, tracheids, fibers, and parenchyma are produced toward the inside of the stem by the vascular cambium. Each year's growth brings new xylem tissue at the cambial region, the older wood begins closer to the center. Most of the strength of the wood comes from thick-walled vessel elements, tracheids, and fibers. Some woody plants produce large vessel elements in the early spring growth than later in the season, and this pattern is termed ring-porous wood. Trees with ring-porous wood are easily recognized in cross section because the **growth rings** are quite distinct; each one representing a year under normal growing conditions. In some climates, however, more than one growth ring may be produced in any given year, leading to a misinterpretation of the age of the tree. Other trees produce the same size vessel elements throughout the growing season, and their wood is termed diffuse-porous. Interpretation of growth rings is more difficult in these trees.

The thickness of the secondary cell wall and degrees of lignification in fibers are important factors in the strength of wood and the quality of paper made from it. Some pulping woods have fibers of a length and thickness that are desirable for a particular purpose. Parenchyma in secondary tissue is primarily confined to rays, groups of cells that radiate from the center of the tree and function in lateral transport of water and nutrients. The patterns of the rays can be seen in the grain of polished woods used for furniture.

Secondary Phloem

The cells comprising secondary phloem are the same types as those of primary phloem. Again, the vascular cambium produces new sieve tube elements, companion cells, fibers, and parenchyma toward the outside of the stem.

Secondary phloem is important in long-distance transport of organic solutes from the "source" to the "sink," but the thickness of the tissue is never very great. Old primary phloem and secondary phloem are gradually sloughed off as part of the bark. Only the newly formed secondary phloem remains functional.

Secondary Growth in Roots

Regardless of the position of the vascular tissue, secondary growth comes about in some roots because cambium develops to separate the primary xylem from the primary phloem. Shortly after the cambium develops around each spoke of the patterned primary xylem, it soon rounds out to form a circle, laying down new secondary xylem to the inside and new secondary phloem to the outside. Some old roots may become very woody and achieve substantial diameters. As the size of the root becomes greater, the epidermis and cortex becomes stretched and may crack at irregular intervals. At approximately the same time that these outer tissues begin to slough off, new regions of cambial activity occur in the region of the epidermis. This new meristem, just as in stems, is called phellogen and leads to the production of cork, a dead tissue that protects the inner tissue from desiccation, mechanical injury, insects, and disease. The same basic cell types that occur in stem xylem and phloem also occur in the root.

Bark

The production of a rough, thick, mostly dead protective outer layer is common to most of the trees used for wood. Bark is the general term for all the tissue outside the vascular cambium. Bark is not wood but results from the activities of two lateral meristems: the vascular cambium, which produces phloem toward the outside, and the phellogen or cork cambium, which produces the cork and inner parenchymatous layer called phelloderm. Collectively the **cork cambium** and the cells it produces—the cork cells and phelloderm—make up the **periderm**, a tissue that replaces the epidermis as the protective outer covering. The dead cells outside the periderm are called outer bark, and the living cells to the inside are called inner bark. During the first year's growth, before any secondary phloem, has been produced by the vascular cambium, bark consists only of primary tissues. At the end of the first year's growth, the bark includes from (inside to outside) secondary phloem, primary phloem, cortex, periderm, and epidermal tissue. Each year of growth of the plant, more secondary phloem is produced. Since some phloem cells are soft walled, the old secondary phloem is sloughed off by the growing woody core pushing to the outside, and these cells cease to function in the conduction of organic materials. Only the most recently produced phloem cells are functional in capacity.

The first periderm formed grows during irregular periods of cork cambial activity, often keeping pace with the lateral growth of the stem for many years. Not all cork cambium activity is uniform, and the cambial layer usually is not a continuous cylinder around the stem. In many barks, new periderms form to the inside of the first formed periderm, entrapping phloem between the layers and resulting in an overlapping series of scales of plates. The rough barks (elm, maple, pine) are much more common than smooth barks (birch, sycamore), which are formed by periderms that occur as essentially continuous cylinders and form concentric rings of new periderms every few years. The combination of periderm formation patterns and continued increase in the girth of the trunk produces a variety of bark appearances, from an essentially smooth surface to deeply furrowed or ridged bark.

The pattern of secondary growth is essentially the same in roots and stems. Even the stretching of the epidermis and cortex is the same, being replaced by a phellogen that gives rise to the cork. Some kinds of woody plants have a great deal of phellogen activity, giving rise to very thick barks. Others are only a few cell layers thick, and therefore the bark is very thin. The most extreme example of cork thickness is *Quercus suber*, the cork oak. This tree did produce essentially all of the corks for wine bottles; some wineries are using plastic corks now. But more expensive wineries are still importing the cork from Portugal. These dead cells have special properties of gas permeability, which allow the wines to breathe during storage and aging.

Summary

1. All living organisms are able to reproduce and maintain their characteristic morphology because of a precise replication system inherent in deoxyribonucleic acid (DNA). This macromolecule occurs in nearly all organisms and consists of a double helix of nucleotides.

2. DNA is replicated in a precise fashion as the double coil unwinds and separates exposing nitrogen bases of the nucleotide that can attach to a complementary nucleotide.

3. In addition to replicating itself, DNA serves as the template for protein synthesis. This process occurs in two steps: transcription and translation. Transcription involves the uncoiling of the DNA molecule and exposing the template, allowing it to synthesize a new single-stranded messenger RNA. The complementary pairing ensures that each nitrogen base of the DNA pairs with its complementary base in the mRNA. After detaching from the DNA, the mRNA is edited and migrates out of the nucleus and into the cytoplasm, where it attaches to a ribosome, specific transfer mRNA is threaded through the ribosome, specific transfer RNAs for each amino acid match at the anticodon loop with the message and place the tRNA amino acid complex in the correct position. Finally, a peptide bond is formed between adjacent amino acids, ensuring the proper sequence for each specific protein.

4. A gene consists of a piece of DNA responsible for the coding of one particular characteristic. Genes are regulated by switches that turn them on and off.

5. Cell division in eukaryotes consists of mitosis, or nuclear division, and cytokinesis. The carefully controlled developmental sequence proceeds from prophase, to metaphase, then to anaphase, and finally telophase. After division, the nucleus goes

into interphase, which consists of two gaps separated by a synthesis phase in which the DNA is replicated. Interphase is a time of intense metabolic activity.

6. Cell division in plants occurs in meristems. Apical meristems of the shoot and root lead to primary growth, and lateral meristems of the cambium and phellogen lead to secondary growth.

7. The three simple tissue types are parenchyma, collenchyma, and sclerenchyma. Tissues are made up of various combinations of these cells. In primary growth, dermal, ground, and vascular tissues are produced by the meristem. The complex tissue types are xylem, phloem, and the epidermis.

8. In secondary growth, lateral meristems produce secondary xylem toward the center of the stem and secondary phloem toward the outside of the stem (or root). Leaves do not have secondary growth.

9. The herbaceous dicot stem has a vascular pattern of a single ring of vascular bundles. As the stem becomes older and if it becomes woody, the fascicular cambium between the xylem and phloem of the bundles link with a newly formed interfascicular cambium to produce a ring of dividing cells called the vascular cambium. All secondary growth arises from this region and the commercial products referred to it, as wood is technically secondary xylem. The monocot stem is characterized by vascular bundles scattered throughout the ground tissue, and consequently there is no distinction between cortex and pith. Except in rare cases, there is no secondary growth in monocots.

10. The dicot root begins with a cylinder of vascular tissue surrounded by the cortex and dermal tissues. Primary xylem occurs in characteristic patterns of radiating arms, and primary phloem occurs in pockets between the arms of xylem. As secondary growth begins, cambial activity between the xylem and phloem gradually links with interfascicular cambium outside the xylem to form the vascular cambium. Subsequent growth and development is similar to that in the stem; old roots and old stems have similar anatomy. The monocot root forms a vascular cylinder as a ring of xylem and phloem, with pith toward the inside and cortex toward the outside. The vascular cylinder of both monocot and dicot roots includes the pericycle, a single cylinder of cells capable of meristematic activity and the source of lateral roots. Surrounding other pericycle is another cylinder of cells called the endodermis, characterized by the Casparian strip, which forces water and solutes to cross living membranes to reach vascular tissue.

11. Leaves are organs modified for maximum area, usually consisting of a blade and petiole. The blade is composed of an upper and lower epidermis, enclosing a palisade parenchyma, spongy parenchyma, and veins of xylem and phloem. These veins

unload water through terminal vessels and load sugars through terminal sieve tubes for transport to other parts of the plant. Some leaves have Kranz anatomy, a specialized leaf structure in which the palisade layers are not well developed, and a bundle sheath of chlorophyllus cells surround the vein. This type of anatomy is associated with the C4 dicarboxylic acid form of photosynthesis.

Something to Think About

1. What is DNA?

2. Why is DNA so very important to all organisms?

3. A double helix means what?

4. How does mRNA work in the plant cell?

5. What part does DNA play in the genetic code of organisms?

6. Draw and name the phases of mitosis.

7. Name the three simple tissues.

8. Diagram a dicot stem.

9. Diagram a monocot stem.

10. Show a cross section of a leaf blade and label the parts.

Suggested Readings

Chawla, H. S. 2002. *Introduction to plant biotechnology*. Science Publishers, Mexico City.

Fleming, A. J. (Ed). 2005. *Intercellular communication in plants, annual plant review*, volume 16. Oxford/Boca Raton, FL: Blackwell Publishing/CRC Press.

Stille, D. 2006. *Plant cells: Building blocks of plants*. Mankato, MN: Compass Point Books.

Trigioavo, R. N. and D. J. Gray. 2004. *Plant development and biotechnology*. New York: CRC Press.

Internet

Internet sites represent a vast resource of information. The URLs for Web sites can change. Using one of the search engines on the Internet, such as Google, Yahoo!, Ask.com, or MSN Live Search, find more information by searching for these words or phrases: **cell structure, DNA, mRNA, plant tissues, cell division, mitosis, meiosis, dicot, monocot, primary tissue, and secondary tissue.**

CHAPTER 8

Wood

In overall botanical importance to human existence, only food plants rank above wood and wood products. In early human history, wood had even greater importance than the food plants, as a fuel and for weapons and tools. Today, over 4,500 products come from wood—the secondary xylem tissue (wood) of forest trees. Wood is used for housing, furniture, fuel, paper, charcoal, distillation by-products, and synthetic materials such as rayon, cellophane, and acetate plastics.

After completing this chapter, you should be able to:

- Know the chemical composition of wood
- Learn the difference in cellulose hemicellulose, and lignin
- Describe softwoods and what they are used for
- Describe hardwoods and what they are used for
- Count growth rings in wood
- Understand what reaction wood is
- Gain knowledge of the science of trees, dendrochronology
- Describe the many uses of wood
- Discuss the current and future productivity of the forest

Key Terms

tannin	sapwood	charcoal
heartwood	reaction wood	destructive distillation
lumens	tension wood	rayon
hardwoods	radiocarbon dating	cork
softwoods	board feet	veneer
balsa	transverse cut	plywood
relative density	radical cut	fiberboard
specific gravity	quarter-sawed	biological control
resin canals	parchment	controlled burning
early wood	sulfite process	monoculture
latewood	hemp	

🌐 Wood Structure and Secondary Growth

Wood is composed of the secondary xylem tissue laid down by the vascular cambium. Generally, the vessels and tracheids function in water and mineral transport, and fibers are support cells. Additionally, rays provide for lateral transport throughout wood tissue.

Chemical Composition and Properties of Wood

Many of the structural properties of wood depend on the arrangement of the component cells and on the chemical composition of the secondary cell wall. The three major cell wall constituents are cellulose and hemicellulose (polysaccharides and lignin, a chemically complex polymer of phenolic substances). Initially the secondary cell wall consists of cellulose embedded in a hemicellulose matrix. These polysaccharides are later cemented together by the deposition of lignin, which terminates any further cell growth by making the cell wall too rigid for additional expansion.

Cellulose

The most abundant naturally occurring organic compound is cellulose. It is a long molecule of several thousand glucose molecules linked end to end, which provides for the physical organization of the cell wall and approximately half the polysaccharides, found in cell walls. Like cellulose, they are linear molecules, but they may branch to provide a porous matrix around cellulose. Hemicellulose is usually found in greater amounts in woody angiosperms than is gymnosperms.

Lignin

The second most abundant component in the cell is lignin providing rigidity to the cell wall. Lignin is not found in all plant cell walls, but when present it is especially important and abundant in cells having a supporting function. The deposition of lignin normally occurs after the cell has reached its maximum size, beginning in the middle lamella, then in the primary cell wall, and finally in the secondary cell wall. Although lignin is often present in all three of these layers, lignification is most characteristic of the secondary cell wall.

The relative amounts of each of these materials control some of the physical properties of wood. Since there is more hemicellulose present in angiosperm wood, angiosperms generally contain more moisture. The greater lignin content in gymnosperm wood makes it more stable and less prone to warping.

The commercial value of wood depends on a combination of characteristics that make certain woods better suited to different uses. The specific gravity, figure, grain, cuts, and knots of wood are some of these properties. The following properties are also considerations for the best use of woods.

Durability

The resistance of wood to decay, wear, and insect damage is especially desirable in wood that is to be used structurally or that comes in contact with moisture. Since fungal decay is the most common form of wood destruction and since fungi thrive in warm, moist conditions, wood in contact with damp soil or subject to frequent rain and high humidity is more likely to decay. Fence posts, railroad, telephone poles, greenhouse table, coastal, or tropical structures and mine timbers require wood of exceptional durability. The natural preservation found in many trees is often toxic or unpalatable to decaying organisms and insects. This is especially true of **tannin**, which is found in amounts up to 30% in some woods. The most durable and resistant woods include redwood, cedars, black walnut, junipers, chestnut, bald cypress, black locust, and catalpa.

Color, Luster, and Polish

In wood used for furniture or cabinetry, the color of wood, especially the **heartwood**, is important. In addition, the natural luster, or ability to reflect light and the ability to take a polish, are functions of cell wall structure and types of cut. Some woods known to polish well are cedar, white pine, cherry, maple, walnut, holly, and some oaks.

Moisture and Shrinkage

The amount of moisture in the wood of a freshly cut tree varies from less than 10% of green weight in some species to over 75% in others. Some moisture is found in the **lumens** of the vessel elements and evaporates readily without causing any shrinkage in the wood. However, the cell walls of green wood comprise approximately 25% to 30% water, which is removed with more difficulty. As this water is removed from the cell walls, the wood shrinks. Most wood shrinks between 10% and 20% in volume if all the water is removed by oven drying. Nearly all this shrinkage is across the grain and greatest in the tangential direction across the width of flat-saw (tangentially cut) lumber. There is less than 1% shrinkage in the length of the boards.

To prevent uneven shrinkage and resulting warping, most commercial lumber is "seasoned" or "cured" by drying the wood under controlled conditions. This is especially important in hardwoods because of their use in furniture. In addition to prevent warping, proper drying reduces shipping weight and cost, increases strength, and improves the wood's ability to be glued, painted, stained, and polished.

Drying is done either in the open air, in drying sheds, or in kilns. The control of humidity and the application of heat are important variables in preventing uneven drying, which causes warping, cracking, twisting, and other distortions. Rapid air circulation helps regulate these factors, as does proper stacking with spacers to keep the boards straight and to allow for proper air circulation. Although it is economically important to dry wood as efficiently as possible, too rapid drying can result in stress, which may cause checks to form on the surface or allow distortions during manufacture or in use. Properly dried hardwoods have a moisture content of 6% to 8%, whereas most softwood for construction has from 15% to 19%.

Heat Conduction

As amazing as it might sound, fire protection often involves the use of wood. Solid wooden doors conduct heat slowly and help prevent the spread of a fire from one room to the next. Dry wood is a poor conductor of heat, and generally lighter woods conduct heat more slowly than heavy woods.

Acoustical Properties

For wood used in musical instruments, the resonance depends on a combination of elasticity, density, thickness, and cut. The soundboard is a piano, responsible for resonance and tonal quality, and is best made of spruce; laminated hard maple holds the metal tuning pegs tightly. The various woodwind instruments, such as clarinet, oboe, and bassoon, and the string instruments, such as violin, guitar, mandolin, and bass, all depend on the acoustical resonance properties of the woods used in their construction. A master instrument craftsman must know woods as well as music. In addition, the reeds of the clarinet, saxophone, and oboe are tapered strips of the woody cane from a large tropical grass, *Arundo donax*. Reeds from this plant have been used for woodwind instruments since at least 3000 BC.

Hardwoods and Softwoods

One of the most commonly used distinctions made when discussing wood is the categorization of all woody dicots as **hardwoods** and all coniferous gymnosperm as **softwoods**. Although there is some justification for such an oversimplification, there are a number of exceptions. One of the softest (lightest) of woods is **balsa** (*Ochroma lagopus*), a dicot; whereas slash pine (*Pinus elliottii*), a conifer, is harder than many hardwoods. What is actually being measured is **relative density**. Generally, the less dense or lighter a wood the softer and weaker it is.

Relative density is determined by measuring the **specific gravity** of the wood, which depends on cell size, cell wall thickness, and the number of different kinds of cells. For example, fibers can be thick walled and have small lumens and can be packed closely together, providing for a very dense (high specific gravity) wood with little air space. Conversely, fibers with thin walls and large lumens produce a wood with lower specific gravity. Vessel elements have relatively thin walls and large lumens, so a high vessel volume results in decreased specific gravity.

Specific gravity is determined by weighting a paraffin-coated block of wood (to prevent water absorption), immersing it in water,

FAMOUS TREES

Giant redwoods (*Sequoia sempervirens*) from Humboldt County, California, measured at 111.6 m. Although giant redwoods are generally considered to be the tallest tree species, unsubstantiated claims exist of a eucalyptus tree (*Eucalyptus reganus*) in Australia measuring over 140 m. Certainly there are a number of eucalyptus trees validated to be taller than 91 m, so they are easily the second tallest tree species in the world.

The tule tree (*Taxodium mucronatum*) found outside the city of Oaxaca, Mexico, and as shown in Figure 8-1, is believed to have the world's largest trunk circumference (42 m).

Only 40 m wide this tree is over 2,000 years old and was visited by Cortez, the Spanish Explorer that defeated and conquered the Aztec Empire, after he heard stories of its size. The tule tree is a youngster compared with the bristle cone pine (*Pinus longaeva*) found in the White Mountains of California, and in Nevada, Utah, and Colorado. Several of these remarkable trees are known to be over 3,500 years old; one of the oldest living ones, named Methuselah, is approximately 5,000 years old.

Figure 8-1

The tule tree (Taxodium mucronatum) is believed to have the worlds largest trunk circumference (42 m). Cortez is reported to have traveled to see this tree, located just outside of Oaxaca City, Mexico.

and weighing the displaced volume of water (1 cubic centimeter of water = 1gram).

$$\text{Specific gravity} = \frac{\textbf{Dry weight of wood}}{\textbf{Weight of displaced volume of water}}$$

Generally, a specific gravity of 0.41 or less is considered softwood, and above 0.41 hardwood. Most coniferous gymnosperms are softwoods and most dicots are hardwoods, but there is overlap. Several woods are above a specific gravity of 1.0, which makes them denser than water, so they sink.

Several structural differences exist between coniferous wood (softwood) and dicot wood (hardwood). Conifers are more homogenous and less complex than dicots. They contain no vessels or fibers, and approximately 90% of the woody tissue is composed of tracheids with little parenchyma. Some of the parenchyma present is associated with **resin canals**; long intercellular spaces present in the longitudinal system of cells and in some of the rays. Resin canals are found only in pine, spruce, larch, and Douglas fir. They are absent in the other conifers. Parenchymal cells surround the resin canal and produce the resin, which is secreted into the canal. Most parenchymal cells in coniferous woods are in the rays.

The rays, which provide the lateral or radial transportation capabilities of the wood, are uniformly smaller in conifer, being normally 1 cell wide and 1 to 20 cells high. Dicot rays are usually larger, up to 20 or more cells wide, and can be up to several hundred cells high. In addition, the ray volume in hardwoods ranges from 5% to 30% of the wood tissue; softwood ranges from 5% to 10%. The variability of the rays is also much greater in hardwoods, adding to their more heterogeneous structure.

Dicot wood has not only more and larger rays, but also a greater variety of cell types in the longitudinal system. In general, the presence of vessel elements, which are the primary water conducting cells, is the main difference, although certain dicots do contain some tracheids (oaks, for example). Dicot woods also contain fibers and parenchymal cells, and different species contain varying amounts and kinds of fibers. Thus, the kinds of cells, the relative amounts of each kind, and, most important, the presence of vessels distinguish the more complex structure of dicot wood from conifer wood.

Growth Rings

Although in a cross section of a log the phloem does not normally appear as discernible rings (because of the compacting of the soft cells), the hard-walled xylem cells form a new growth ring of wood each time the vascular cambium becomes active. In temperate regions, such activity usually occurs only once during the growing season each calendar year, and thus these are termed annual rings, as seen in Figure 8-2. The rings are a sectional view of the annual growth layer that form the full length of the tree.

Since water availability is one of the most important environmental factors controlling plant growth, a drought may cause early cessation of growth followed by a second burst of growth after subsequent rainfall. Infrequently, then, false rings may form, resulting in two or more apparent growth rings in one year. Such sporadic growth patterns are generally restricted to areas with unpredictable climatic patterns, such as arid and semiarid regions.

Woody growth appears as visible concentric rings because of the contrast in cell diameter and cell wall thickness from the early to the late part of the growing season. In typical years in temperate regions, growth begins in the early spring, while plenty of soil moisture is available from winter rains and snow. The cells produced are large and thin walled, making them less dense than the xylem produced in the summer. The less dense **early wood**, therefore, appears lighter than the smaller, thick-walled **latewood**. Whereas the morphology of cells formed from the vascular cambium change gradually with

Figure 8-2

This cross section of a tree trunk clearly shows the darker heartwood (inner rings) and lighter sapwood.

the growing season and does not present a sharp contrast from early wood to latewood, the interface between the latewood of one year's ring and the early wood of the next year's ring is clearly delineated. The annual rings are therefore distinctly visible.

A complex of environmental factors besides total water availability affects the width of the growth rings. Temperature, length of growing season, time of precipitation, disease, and soil fertility are among the variables that work together to control the patterns of lateral growth from year to year. A wide ring generally reflects a long, wet, and moderate growing season, whereas narrow rings usually reflect some kind of environmental stress. An extremely dry, cold winter followed by a hot dry summer could even prevent any lateral growth for that year.

Sapwood and Heartwood

With increased age, the center of the trunk ceases to function in the transport of water and minerals because it accumulates resins, tannins, oils, gums, and other metabolic by-products. Since trees do not have the ability to remove these compounds, they remain in the plant body. The centralization of all these materials allows the outer wood to remain unclogged and functional. These central wood rings, as shown in Figure 8-3, are often darker in color, more resistant to decay, and sometimes aromatic. This central heartwood is usually visibly distinguishable from the lighter, functional **sapwood** in fresh cut or cured and processed wood.

Figure 8-3

The growth rings are slightly irregular showing that the environment was unstable during its growth.

The natural preservation properties of some heartwoods, such as cedar, cypress, redwood, black walnut, and mahogany, makes them especially valuable for fine furniture. The ratio of heartwood to sapwood varies from one species to the next, as does the degree of visible differences between the two woods. Some characteristics of heartwood increase its commercial applications.

Reaction Wood

An interesting response of trees that have lost their vertical position is the production of **reaction wood**. For instance, if a tree has been bent by another tree falling against it or by a boulder rolling down on one side of it, reaction wood will form along one side of the trunk and bend the trunk back to a vertical position again. In young seedlings this can occur in a single growing season, whereas it takes many years in older trees. This phenomenon of plant movement is completely different from phototropic response because it involves the lateral meristems in the already elongated cells of the trunk, not the apical meristematic growth of the plant tip.

The reaction wood of dicotyledonous angiosperms is called **tension wood** and is produced along the upper sides of leaning stems and branches, causing reorientation by contraction on the upper side pulling the stem or branch back into a normal growth position. The fibers of tension wood resist cutting and project from the surface of boards sawed from hardwoods. This wooliness is an even greater problem in the planing of the boards during the finishing process. Some of these hardwood trees split at the cut end immediately after felling due to the release of the internal stresses caused by this tissue. Less of the particular log can be used in the production of board lumber than a log from trees without tension wood.

The reaction wood of gymnosperms is called compression wood and is produced along the lower side of the leaning stems or branches, causing straightening by expanding and pushing the trunk upright again. Compression wood is denser (harder) than normal wood. This wood is undesirable for commercial lumber because of its poor nail-holding characteristic.

Dendrochronology

Dendrochronology is the science of interpreting past conditions by studying the growth of wood. Since climatic conditions, especially water availability, influence the width of growth rings, and since there is normally only one ring produced each year, patterns of annual ring production directly reflect past climates. By counting the number of rings a given tree has, the age of that tree can be determined. The oldest trees are the bristlecone pine (*Pinus longaeva*) found in the White Mountains of California at above 3000 m elevation. Because of the short growing season and small amount of rainfall at elevation, bristlecone pines grow very slowly and have narrow rings, as many as 1,000 or more in less than 13 cm of lateral growth. Because of this slow growth, bristlecone pines are much smaller in both trunk diameter and height than the massive giant redwoods (*Sequoia sempervirens*), shown in Figure 8-4.

By matching ring patterns from the wood of a living tree with the wood of an older tree containing a partially overlapping growth ring sequence and continuing with the partially overlapping older woods from trees long dead, a continuous chronology of climatic patterns can be established for a particular region. Such a chronology has been established for the White Mountains, extending back over 8,000 years. Because bristlecone pine wood, dating back about 9,000 years, has been found, it is probable that this chronology can be extended even further back. Interestingly, by using radiocarbon

Figure 8-4

*The giant redwoods (*Sequoia sempervirens*) are considered to be the world's tallest trees, indicating consistently favorable growing conditions.* Courtesy of National Forest Service.

dating techniques on wood that has a known age through ring sequencing, it has been determined that **radiocarbon dating** is increasingly inaccurate for material over 1,000 years old. Carbon-14 dating of 1,600 and 3,300 years were found for bristlecone pine specimens having 2,000 and 4,000 annual rings, respectively. With continued cross dating, a more accurate age estimate for archaeological studies could be made, thus, improving our understanding of their cultural beginnings. Wooden utensils, tools, ornaments, and building timbers from archaeological digs can help provide information about the climate as well as the age in which a particular culture existed.

The study of tree rings also can tell a much more complete story about an individual tree's history than just its age and the general climatic conditions during its lifetime, especially when ring patterns from other trees in the region tell a slightly different story for the area in general. Damage due to fire, landslide, insects, and other natural occurrences can be seen in the patterns of annual ring formation.

Wood Uses

There are many uses of wood, some obvious and some less well known. The uses of wood and wood products are expanding because trees are a renewable resource. With proper management and use, the forests of the world will be able to provide this valuable raw material in more than sufficient quantities as long as it is needed.

Lumber

One of the most obvious uses for trees is the production of lumber for building and furniture. Many millions of **board feet** of certain softwoods are used each year for home construction because of their physical properties (a board foot is 1 foot long, 1 foot wide, and 1 foot thick). Trees such as white pine have a soft, uniform texture and an even grain that can be machined easily; does not shrink, swell, or warp significantly; is strong; and holds nails well. Not all gymnosperms have equally desirable qualities, although several other species are used for construction because of the ease with which nails can be driven into them. Framing a house out of oak, walnut, maple, or hickory, on the other hand, would be quite a task because their wood is so hard that boards would have to be drilled and screwed together instead of nailed.

Because of the grains, colors, and durability of hardwoods, they are most often used in furniture making. Some softwood, especially certain pines, is also used for furniture. The use of different woods depends on several properties, which are partially a function of the type of cut and the part of the tree used.

Cuts and Grains

In addition to the kinds of cells and the width of annual rings in a given wood, the appearance of a finished wood surface is a function of how the log has been cut. In a **transverse cut** or cross section, the annual rings appear as concentric circles. This cut is not commonly used to produce a commercial piece of lumber, although transverse slices that included the bark can make a unique and beautiful tabletop. A **radical cut** is made longitudinally through the center of the log. In a radial section of wood, the annual rings appear as parallel lines running the length of the flat board surface with the rays running at right angles across them. Since this cut is made through the center of the log, only a few boards can be cut from each log. Radial cuts are also said to be **quarter-sawed**.

The most common board surface is a log cut longitudinally but not through the center. This tangential cut results in the annual rings appearing as wavy bands with ends of the rays scattered out throughout. Tangential cut lumber is also said to be flat-cut or plain-sawed, and it is the most common cut because so many more boards can be sawed from a log than with the radial cut. The design resulting from these different cuts is the figure of the wood, not the grain. The grain of a wood technically refers to the direction of the fibers, tracheids, and vessel members. The density and size of the cells and the way they are grouped affect the texture of the wood surface more than does the surface pattern of the growth rings.

Knots

Another textural and structural consideration in woods is the presence of knots, which are the bases of branches that subsequent lateral growth has covered over. A higher proportion of knots are in the center of the trunk because the tree was younger and smaller with lower branches then. As the tree grows and ages, lower branches cease to be formed, and the old ones die and eventually break off, as shown in Figure 8-5. Just as the lateral growth will eventually cover the base of old branches, so will it cover scars from fire and other injury. You may have noticed that a nail or eye screw put in a tree trunk to hold a clothesline or bird feeder many years ago is not only partially, if at all, showing. It is the same lateral growth that ultimately turns two closely adjacent young trees into a "single" double-trunked tree. Occasionally, two different species growing in proximity results in the faster growing tree gradually enveloping the lower trunk of the other. Tree trunks have even been found with rocks in them. In many of the hardwood forests of France, no tree old enough to have been present during World War II is harvested because the metal bomb fragments and bullets that are often hidden well within the wood will quickly destroy a sawmill blade.

Fuel

More than one-third of the world's total population depends on wood for heating and cooking. The significance of wood as a fuel is greatest in developing countries, where more than 86% of all wood consumption is for fuel.

Approximately 1.5 billion people derive at least 90% of their energy requirements for cooking and heating from wood and charcoal, and another billion people depend on wood for at least 50% of their needs. These are impressive statistics, especially since these are also the countries with the fastest growing populations.

Because of the dependence on firewood in these countries, it is estimated that 50% of all wood used worldwide each year goes to fuel. In developed countries with stabilized population sizes and appropriate management and use practices, wood is a renewable resource that should never be in short supply. Theoretically, the developing countries, which depend much more heavily on wood, could produce enough through replanting and appropriate management to meet their needs. Practically, however, it may already be too late in some areas, since this resource has already been depleted to a critical level. The people are now depending on dried animal dung and crop residue for fuel. This removes much-needed nutrients from the soil, which will in turn be less productive. The vicious cycle continues as new land is cleared for more crop production to feed the growing population in these countries. Clearing without replanting removes even more of the wood that could have been used in the future for fuel.

In the United States, although only slightly over 10% of our total wood use is for fuel, over 1 million homes now use wood for their primary heat source. Since 1933, the wood fuel use growth rate has been about 15% annually. More recent trends are even greater because of the renewed interest in wood-burning stoves.

Even at this rate of increase, wood use for fuel is critical only in the developing countries, where the "energy crisis" is much more serious than ours, even with our petroleum shortages. If the developed countries fail to solve the energy crisis, however, we could well join the rest of the world in our dependence on wood as a primary fuel.

Paper

Paper production accounts for over half of the wood that is pulped through mechanical or chemical means. The balance of the wood pulp produced each year goes into cardboard and fiberwood. Paper is basically composed of separate plant fiber cells that have been slurred and then spread into thick sheets and dried. True paper was first produced around 105 AD in China. Before the Chinese developed a way of making paper, they wrote on strips of wood with a stylus and later on woven cloth, especially silk. Also in pre-Christian times, Egyptians made a type of paper by beating laminated stems of papyrus until they were very thin. The only other nonpaper material used prior to the second century was **parchment**, which is made from animal skins. The Chinese guarded the secret of paper making for over 500 years, establishing themselves as the only supplier of paper. Papermaking was a slow hand-labor process until the first machines were invented around 1800.

In 1865, the **sulfite process** for removing unwanted lignins from the pulp was developed. Lignin causes paper to turn yellow and become brittle with age. For economy, newsprint production does not include this lignin-removing process, and so newsprint eventually becomes yellow and brittle. To add weight and body to paper and to produce a smooth ink-impervious or "hard" surface, given fillers are mixed with the pulp while it is in the liquid slurry stage. Clay, alum, and talc are common fillers to add weight and stiffness to hard surfaces for writing purposes. To make colored paper, dyes are added at this pulp-slurry stage. Over 90% of all paper comes from wood; the rest is made from the fiber of other plants such as flax (paper money and cigarette paper), cotton, and **hemp**.

Charcoal

Charcoal is made by partially combusting hardwood blocks in the presence of very little air. Today charcoal is made by **destructive distillation** of hardwoods, and vapors are collected. From these distillation vapors wood alcohol (methanol), methane gas, wood tar, acetic acid, acetone, and hydrogen gas are separated. The distillation of softwoods, especially pines, yields turpentine and rosin, both valuable in the paint industry. Rosin is also used on musicians' bows and baseball pitchers handle a rosin bag to gain a better grip on the ball. Boxers and ballerinas shuffle their feet in a low-sided square rosin box to improve their footing.

Synthetics

Not all synthetics are made from petroleum by-products. **Rayon** was the first commercial synthetic fiber and is made from dissolved cellulose material of wood pulp. Other products include cellophane, acetate plastics, photographic film, and other molded plastics used as handles for tools. Although synthetics from petrochemicals are much more common and popular now, because wood is a renewable resource, rayon and other wood-pulp synthetics will be revitalized in the future. In fact, rayon blends are increasingly common in clothing.

Cork

Cork is a nonwood product of trees, specifically the "bark" of the cork oak (*Quercus suber*). Native to Europe, most cork is produced in Portugal because the climate conditions there are most conducive to rapid outer periderm growth. The cork cells are naturally impregnated during growth with a wax suberin, which makes them watertight and produces incredible buoyancy. A cork oak can be stripped of its outer bark after the first 20 years of growth and then every 10 years thereafter until the tree is approximately 150 years old. However, cork is being replaced by plastics for various items that at one time were unheard of (wine corks, fishing floats).

Other Wood Uses

The application of wood as lumber includes plywood, particleboard, veneers, and paneling as well as board lumbers for construction and furniture. In fact, the use of plywood for covering large surface areas is now preferred over board lumber in construction framing. Particleboard is especially popular in Europe, as is fiberboard, made in a similar manner. The manufacture of these materials was not economically feasible until the development of modern wood processing techniques and glues that would permanently bond wood and wood particles.

A **veneer** is a sheet of wood sliced from a log. Most commonly, the log is rotated against a stationary blade that moves inward, producing a continuous sheet of desired thickness. **Plywood** is made by cutting the continuous sheets to desired size and bonding them together under pressure. Alternate sheets are laid with their grains at right angles to each other to increase strength and reduce warping. When three or seven sheets are used, the center sheet is twice as thick as the outer ones so an equal amount of wood has the grain running in each direction. Different veneering techniques allow the thinner and smoother slices, which can be used as the "paneled" surface for a finished wall or for the facing on furniture.

Master craftsmen have produced veneered furniture for hundreds of years, and veneer has been known since 1500 BC: evidence of this has been found in the tombs of Egyptian pharaohs.

Undoubtedly the items painstakingly made by hand were highly prized because of the beautiful figures on the surface of the fine furniture. Although thickness of as little as 0.23 cm (1/110 inch) is possible, most veneer is about 0.084 cm (1/30 inch) thick.

Gluing wood chips together under pressure makes particleboard. Fiberboard is made by the same treatment, but the wood fiber is obtained by various chemical or mechanical pulping methods. Their use is primarily as insulation board or for building containers. However, when faced with a finished veneer or vinyl overlay, they are used for paneling and furniture.

Raw wood treated with preservatives is used extensively for fence post, telephone poles, mine support timbers, boat dock pilings, and timbers that are buried in unstable soils under building foundations. Most of the preservatives used are inorganic chemicals toxic to decaying organisms or toxic organic oils that prevent bacteria and fungi from decomposing (rotting) the wood tissues. If free from such decomposition, wooden post, poles, and timbers are incredibly strong and durable.

Forests and Forestry

The total U.S. land surface area of the 50 states is approximately 2.3 billion acres. About one-third of that amount, 760 million acres, is forested, and almost two-thirds of that forestland, 482 million acres, is classed as commercial timberland.

Management Policies

Proper management of existing forest includes thinning inferior trees to allow for more rapid growth of healthy ones, removing litter and diseased trees, which increase fire hazards, and employing the most efficient harvesting and restocking methods. In addition, studies show that some 168 million acres of commercial timberland would yield greater returns from the application of such intensive management techniques than other lands (see Figure 8-6). Economically, it would make sense to concentrate improvement efforts on the most productive lands.

Usage

With cost-effective technologies developed, those parts of harvested trees relegated in the past to waste could be used for pulp fuel biomass or synthetic extraction. Inferior, diseased, and damaged trees that are now culled from the harvestable strands could also become a resource instead of a liability.

Reducing Losses

An average of over 4 billion cubic feet of timber is lost each year to such destructive agents as insects, diseases, storms, and fires. Biological control of insect pests and fungi may effectively supplement existing efforts of controlling these destructive organisms. Controlled burning to remove ground fuel in the form of dead

Figure 8-6

With proper management, the commercial timberlands of the United States can provide for the wood and wood products needed for decades to come. Courtesy of National Forest Service.

trees, dry underbrush, and dead limbs has been used as a management tool in many of the federally controlled forest lands. However, after several summers of devastation from forest fires started by controlled burning, the U.S. government is rethinking this management practice.

Super Trees

New "super strains" of trees are being produced that will increase productivity in many future forests. By using cuttings, grafts, and tissue cultures to speed up the reproductive cycle, new strains are being developed that might be twice as productive as wild genetic strains. A concern with creating a **monoculture** that might be wiped out by a single catastrophic event is voiced by some researchers, but others are convinced that the development of super trees is the only way production will keep up with demands of the future. The United States is already the largest importer of timber in the world, and many economic experts warn against any increase in the dependence on imports.

Obviously, wood is incredibly important especially since it is renewable. With the current and future emphasis on wood and wood products, intelligent long-range planning is a must if we are to keep supply and demand in balance.

Summary

1. Over 4,500 different commercial products come from wood or secondary xylem. The giant redwood is the world's tallest tree, and the bristlecone pine is the oldest living organism. Wood is composed of cellulose, hemicellulose, and lignin. Properties of

wood that depend on the relative composition of these materials include durability, color, luster, polishability, shrinkage, heat conduction, and acoustics.

2. The terms *softwood* and *hardwood*, although somewhat misleading, refer to coniferous gymnosperms and woody dicots, respectively. Hardness actually refers to specific gravity or density.

3. In most trees, a single annual growing season results in the formation of a growth ring. These rings are xylem tissues, and as the tree ages, the inner (earliest) rings usually accumulate metabolic by-products and take on a different appearance. This heartwood no longer conducts water and minerals as the outer sapwood layers do.

4. Reaction wood allows trunks or branches to straighten up from a leaning position. Tension wood of dicots and compression wood of conifers differs in their mode of action, but they both produce the same results.

5. Although not wood, bark is an important part of most tree trunks. Cork is the bark of the cork oak (*Quercus suber*). The study of the past by using tree rings is called dendrochronology. The number of annual growth rings establishes the age, and the widths of the rings reveal the climatic conditions for that growing season. A continuous chronology for over 8,000 years has been established using dead tree trunks of bristlecone pines.

6. Building lumber can be categorized according to density, texture, and the part of the tree used. Transverse, radial, and tangential cuts result in different grains. The presence of knots is also a consideration when selecting building lumber. Knots are old branches covered over by the expanding trunk girth or diameter.

7. Wood used for fuel is an essential natural resource in most of the developing countries. The use of wood as fuel is increasing in the United States and in several other developing countries as well.

8. Paper production is a significant commercial use of wood, as is charcoal. Rayon is a synthetic fiber produced from wood, not petroleum by-products.

9. Other uses for wood include veneer making and plywood manufacture. In addition, some woods have natural preservatives that allows them to be used as timbers and pilings in wet environments.

10. Although the United States has adequate wood resources, some countries are not properly harvesting and caring for their forests. Proper management policies mean adequate timber production in the United States for many years to come.

Something to Think About

1. What is the world tallest tree?

2. List 5 softwood trees.

3. List 5 hardwood trees.

4. What are growth rings made up from?

5. Compare tension wood of dicots and compression wood of monocots.

6. Explain what dendrochronology is.

7. How are knots in wood made?

8. Rayon is made from what?

9. List some commercial uses of wood.

10. Proper management polices of the forest will mean what for the United States?

Suggested Readings

Bowyer, J. L., R. Shmulsky, and L. Haggreen. 2002. *Forest products and wood science*. Malden, MA: Blackwell Publishing.

Core, H. A., W. A. Côté, and A. C. Day. 1979. *Wood structure and identification*. Syracuse, NY: Syracuse University Press.

Wood and Fiber Science. 2005.

Internet

Internet sites represent a vast resource of information. The URLs for Web sites can change. Using one of the search engines on the Internet, such as Google, Yahoo!, Ask.com, and MSN Live Search, find more information by searching for these words or phrases: **chemical composition of wood, tule tree, cellulose, lignin, heartwood, sapwood, hardwood, softwoods, relative density, specific gravity of wood, growth rings, reaction wood, compression wood, dendrochronology, radial cut lumber, knots, renewable resource, paper, fiber board, plywood, and forest management.**

Function and Control

Plant-Soil-Water Relationships

CHAPTER 9

Of all the necessities for plant growth, water can be considered the most important. This may seem like an overly bold statement, since the absence of any one component essential for life will result in the organism's demise. In that sense, all essential components are equally important. Plants require water, carbon dioxide, oxygen, light, and certain minerals. Certainly a few environments exist in which light availability can be a problem; for example, a forest with closed canopies or houseplants in dark rooms. Additionally soils vary in the presence and amounts of certain minerals.

After completing this chapter, you should be able to:

- Discuss the importance of water to plant growth
- Describe how water moves through a plant
- Explain what turgor pressure does for the plant
- Describe the hydrogen binding of water molecules
- Discuss what soil is and where it comes from
- List the basic properties of soil
- Explain the physical process of diffusion and osmosis
- Discuss the chemical potential of water
- Outline the basic process of evapotranspiration in plant growth
- Describe how environmental factors affect water stress
- Explain how to make intelligent decisions concerning the growth of plants
- Define hydroponics

Key Terms

ionize	soil profile	deplasmolysis
cations	bedrock	water potential
anions	porosity	gradients
macroelements	macropores	soil-plant-air continuum (SPAC)
microelements	micropores	
co-factor	weathering	transpirational pull
osmoregulator	field capacity	endodermis
igneous	permanent wilting point	apoplastic movement
granite	kinetic energy	active transport
basalt	diffusion	symplastic movement
sedimentary rock	osmosis	pericycle
metamorphic rocks	plasmolysis	root pressure

guttation
cuticular transpiration
stomatal regulation
cohesive
adhesive
flaccid

radial micellation
crassulacean acid
    metabolism (CAM)
cuticle thickness
pubescent
xerophytes

compost
root bound
hydroponics

Water and Nutrition

The role of water plays a large role in the activities of minerals and soils. The minerals required by plants are taken in from the soil and dissolved in water, and the CO_2 in the atmosphere is dissolved in water at the cell surface. Even the stomata, through which CO_2 and H_2O enter and leave the plant, are controlled by water availability to the cells "guarding" these openings. Photosynthesis, too, begins when the water molecule is split by light energy.

Importance of Water in Metabolism

Plants must have access to water all the time. An aqueous environment is absolutely essential for all metabolic function; without it, life systems cease. Besides acting as a medium for dissolving solutes, water provides the turgor pressure for cells and participates directly in several biochemical reactions.

Water is therefore critical to every plant function. For approximately 95% of all plant species, this requirement is limited to fresh water. Marine plant species have an advantage, since about three-fourths of the earth is covered by oceans. Terrestrial plants, besides having to support their own weight, have another problem: Since they are not surrounded by water, as aquatics are, they need a more complex method of obtaining and distributing water (see Figure 9-1). Water must come to the plant; the plant cannot go to the water.

If life as we know it exists on any other planet, it also requires water. Unmanned space exploration always includes investigation of whether there is or ever was any evidence of surface water or atmospheric water vapor.

Figure 9-1

In some closed-canopy forests, water is not the limiting factor for growth, but low light intensity determines the kinds of understory that can grow there. Courtesy of National Forest Service.

Water is unique among the various liquids on earth. Because of hydrogen bonding patterns—the continual breaking up and re-forming—water is a liquid; but the cohesive strength of those bonds is so great that water strongly resists the separation of its molecules from each other, and they can withstand the equivalent of 100 atmospheres of tension before breaking. Thus, a capillary-sized column 100 m tall with a powerful suction applied at the top could easily pull a liquid column to the top. The actual height to which a column of water could be pulled is probably reduced because of impurities in most water that interfere with the perfect cohesive bonding of pure water. In addition, friction from the sides of the tube also affects the potential height reached. The final message, however, is the same: The great cohesive strength of water is an integral property in the movement of water through plants.

The adherence of water molecules to other surfaces is also a function of hydrogen bonding. This property of water allows capillarity, the process of climbing up walls of a narrow diameter tube. The resulting meniscus, or curved surface, of the water in such a tube is one of the first phenomena a student observes in a laboratory where accurate liquid measures are stressed. In plants, xylem vessels or tracheids exhibit capillary action.

Ionization, Dissociation, and pH

If minerals are to be taken up from the soil, they must first be dissolved in the relative acidity of the soil water. When salts are dissolved in water, they **ionize**, which simply means that they dissociate into electrically charged particles. For a given salt molecule such as table salt (NaCl, sodium chloride), the molecules are neutral with respect to changes when it is a dry salt. However, when the salt is dissolved in water, the molecule dissociates and become $Na+$ and $Cl-$. Ions with positive charges are called **cations**, and those with negative charges are called **anions**. Not all salt molecules dissociate in water; some do so almost completely, others hardly at all. The degree of dissociation is influenced by a number of factors, including temperature. The number of cations and anions present is important to the plant because the root system must take up the nutrients in the ionized form.

One of the most important contributing factors in dissociation is the relative acidity of the soil solution. The acidity of any solution is determined by a number of charged hydrogen atoms or protons (H^+) present. The term *pH* refers to the hydrogen potential and is defined as the negative logarithm of the hydrogen ion concentration (see Figure 9-2). This simply means that pH is a scale from 1 to 14 that describes the relative concentration of protons. The more protons present in solid solution, the greater the acidity. On the pH scale, a value of 7.0 is neutral; pure water has a pH of 7.0. Values less than 7.0 are considered acidic, and measures greater than 7.0 are basic or alkaline. Anything dissolved in water that decreases the number of protons increases the alkalinity because the scale is a negative logarithm, the smaller the number, the greater the H^+ ion concentration.

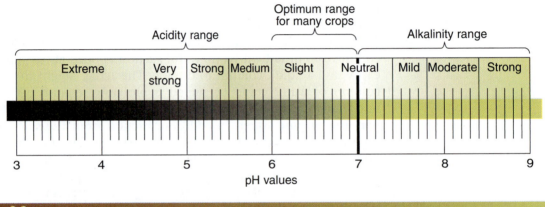

Figure 9-2

The pH scale goes from 1 to 14—this describes the relative concentration of hydrogen ions. The more protons present in the soil the greater the acidity. On a pH scale, value 7.0 is neutral. Values less than 7.0 are considered to be acidic, and if the values are greater than 7.0, the pH is basic. Courtesy of Rick Parker.

The pH of the soil solution determines the solubility of various soil salts. It is important to remember that all the essential soil nutrients do occur as salts; therefore, fertilizers are salts that can be good or bad, depending on their chemical makeup and concentration. Even potassium nitrate, a perfectly good fertilizer, can be bad salt if it is applied in too high a concentration.

Essential Plant Nutrients

Nutrition is just as important in plants as it is in animals. Cellular metabolism leading to the production of the organic molecules characteristic of all life requires only a few essential elements. In the surface layer of the earth, there are some 92 chemical elements. In addition, scientists have synthesized several artificial elements in recent years. In spite of this chemical and geological diversity, only 16 elements are considered essential for all plants. Just as certain molecules are needed in greater amounts than others, so too are elements needed in greater or lesser quantities. Strictly on an arbitrary and approximate scale, essential plant nutrients are divided into those required in greater concentration, the **macroelements**, and those required only in minute quantities, the **microelements**, some of which are shown in Table 9-1. The designation has nothing to do with the size of the element but only the quantity required for normal growth and reproduction.

Table 9-1

Sixteen Essential Plant Nutrients							
Nonmineral		Primary		Secondary mineral		Micronutrients	
Name	Symbol	Name	Symbol	Name	Symbol	Name	Symbol
Carbon	C	Nitrogen	N	Calcium	Ca	Boron	B
Hydrogen	H	Phosphorus	P	Magnesium	Mg	Chlorine	Cl
Oxygen	O	Potassium	K	Sulfur	S	Copper	Cu
						Iron	Fe
						Manganes	Mn
						Molybdenum	Mo
						Zinc	Zn

Courtesy of Rick Parker.

Each element, whether a macroelement or microelement, does have a specific function in plant metabolism; most are found as part of an organic molecule. Thus, if the element is missing, the molecule cannot be made in the cells. The only macroelement not found in organic molecules is potassium and its function is thought to be primarily as a **co-factor**, or chemical helper in enzymatic reactions, and as an **osmoregulator**. It is required in large quantities to act as a solute in stimulating the movement of water across a membrane. Although all carbon and some oxygen come from water, all the other elements must come via the same pathway. The "trick" in metabolism is to bring together all the essential nutrients in the right cell at the right time so that the myriad of enzymes packaged in specific organelles can join them together with various chemical bonds to form the molecules necessary for life.

A brief introduction to the specific functions of some of the elements is found in Table 9-2. Unquestionably, the single element most often deficient from soil over the face of the earth is nitrogen. That nitrogen exists in the atmosphere as the most abundant gas and in a very stable form is no assurance that nitrogen is plentiful for plants. Very few organisms have the ability to convert nitrogen gas to a form that can be taken up by plants. Certain microorganisms can do it, but higher plants have no ability by themselves to incorporate nitrogen. Even after nitrogen is in the soil in a usable form, it tends to be readily lost to volatilizing or leaching. One needs look only at a few different kinds of organic molecules to realize the importance of nitrogen. It is found in all amino acids and therefore all proteins, in chlorophyll, in nucleic acids, and in ATP, to name a few.

Table 9-2

	Functions of Nutrients in Plants and Their Deficiency Symptoms	
Nutrient	**Function**	**Deficiency symptoms**
Nitrogen	Promotes rapid growth; chlorophyll formation; synthesis of amino acids and proteins.	Stunted growth; yellow lower leaves; spindly stalks; pale green color.
Phosphorus	Stimulates root growth; aids seed formation; used in photosynthesis and respiration.	Purplish color in lower leaves and stems; dead spots on leaves and fruits.
Potassium	Increases vigor, disease resistance, stalk strength, and seed quality.	Scorching or browning of leaf margins on lower leaves; weak stalks.
Calcium	Constituent of cell walls; aids cell division.	Deformed or dead terminal leaves; pale green color.
Magnesium	Component of chlorophyll, enzymes, and vitamins; aids nutrient uptake.	Interveinal yellowing (chlorosis) of lower leaves.
Sulfur	Essential in amino acids, vitamins; gives green color.	Yellow upper leaves; stunted growth.
Boron	Important to flowering, fruiting, and cell division.	Terminal buds die; thick, brittle upper leaves with curling.
Copper	Component of enzymes; chlorophyll synthesis and respiration.	Terminal buds and leaves die; blue-green color.
Chlorine	Not well defined; aids in root and shoot growth.	Wilting; chlorotic leaves.
Iron	Catalyst in chlorophyll formation; component of enzymes.	Interveinal chlorosis of upper leaves.
Manganese	Chlorophyll synthesis.	Dark green leaf veins; interveinal chlorosis.
Molybdenum	Aids nitrogen fixation and protein synthesis.	Similar to nitrogen.
Zinc	Needed for auxin and starch formation.	Interveinal chlorosis of upper leaves.
Carbon	Component of most plant compounds.	
Hydrogen	Component of most plant compounds.	
Oxygen	Component of most plant compounds.	

Courtesy of Rick Parker.

Unlike animals, plants do not require any organic molecules for nutrition. They make their own, and many of these are required as essential molecules for animals. Take, for example, the requirement for vitamins. Most vitamins are fairly complex organic molecules made by plants. Since humans can synthesize only a few of the essential vitamins, we depend on plants to provide them for nutrition and good health. Green vegetables do provide certain nutritional components that you cannot get any other way except through dietary supplements.

The essential nutrients must be available in the soil solution in the right place and at the right time. Placement in the soil is critical, for if a root does not come within a few millimeters of an ion, no uptake can occur. There is very little movement in the soil itself, and the roots certainly do move downward in the soil, but not in response to any magical "pull" from plants. Most movement is strictly gravitational. This movement is influenced by many factors, but perhaps the most important is the amount of rainfall or irrigation. Excessive water in the soil will tend to leach its nutrients as it moves downward below the root zone. This is a problem in many parts of the world particularly in regions of high rainfall. Even improper irrigation management in regions of low rainfall can aggravate the situation. Loss of nutrients through leaching is a valid concern to developing agriculture in the moist tropics, and nitrogen availability is the biggest problem.

Some nutrients are far more soluble in water than others, and the pH factor is extremely important in determining solubility. Plant nutrients taken up as ions tend to be most soluble in a slightly acidic solution, at a pH of about 6.5. In regions of high rainfall, CO_2 in the atmosphere is dissolved readily in H_2O to form carbonic acid (H_2CO_2). Even though carbonic acid is relatively weak; it too ionizes to produce a proton (H^+) and a bicarbonate ion (HCO_3^-). The high concentration of H^+ reaches the soil as acidic rainfall and displaces many of the ions, but H^+ can do a good job of replacing almost everything. Such soils in high rainfall regions are referred to as acid soils, and it is almost impossible to add enough fertilizer to grow a crop; leaching removes the desirable nutrients almost as quickly as they are added. Nutrient leaching is a major problem in high rainfall, tropical regions.

The problem in arid regions is quite different. The rainfall there is so scanty that few salts are ever leached. Consequently, over time the problem of slat accumulation can become enormous unless there is sufficient rainfall to produce some leaching. Many western states in the United States suffer from this problem.

Diagnosing plant nutrient deficiency symptoms is not easy. One must learn a great deal about the plant in question before drawing too many conclusions about nutritional disorders. Sometimes nutrient deficiencies are mistaken for insect and disease problems, drought, heat stress, cold stress, and salt stress. After some experience, it becomes easier to spot specific deficiencies in some plants. In other plants, deficiencies are exceedingly difficult to diagnose, and both plant and soil analysis may be required before the diagnosis is certain.

Information concerning nutrient deficiencies in specific plants is obtained by growing the plants in a medium of known chemical composition, preferably water. But withholding one chemical element at a time, the nutritional disorder can be determined within a few weeks.

WATER: THE STUFF OF LIFE

Water covers three-fourths of the earth's surface and represents a major component of the bodies of plants and animals. For example, a one metric ton load of fresh cut alfalfa would weigh only about 236 kg with all the water removed.

This miracle liquid forms from two gases—hydrogen (H) and oxygen (O); two atoms of hydrogen and one atom of oxygen combine to form water,

H_2O. The water molecule is bipolar having charged poles like a magnet, giving water unique properties.

Depending on the temperature, water exists in three forms. It is a liquid between 0° and 100°C. It is a gas vapor at temperatures above 100°C, and water becomes a solid (ice) at temperatures below 0°C.

Of all the naturally occurring substances, water has the highest specific heat. This makes it a good coolant in biological systems and makes it resist rapid temperature changes. Specific heat is the amount of heat required to raise the temperature of a substance 1°C.

Water is the universal solvent, dissolving almost everything. It is powerful enough to dissolve rocks yet gentle enough to hold an enzyme in a fragile plant cell. As a solvent, it acts as a medium for biochemical reactions carrying products of metabolism and nutrients.

Within any temperature zone, the availability of water is the most important factor in determining which plants will grow and how productive they will be. Indeed, the future development of crop lands depends on the availability of water.

Soil

The earth is approximately 13,000 km in diameter, but only a very shallow outer layer of the land surface sustains life. At one point in the earth's history, that outer layer was essentially solid rock formed by gradual cooling of the **igneous** material that still comprises most of the mass of the earth. An approximately 80 km thick crust of cooled **granite** now overlays a less solid (semiviscous) layer of **basalt** that gradually becomes molten farther from the surface.

Granite occurs only where there is land; the ocean basins are composed of basalt covered by whatever sediments have slowly accumulated on the ocean floor. In addition to the igneous material, primarily granite formed by the cooling of the earth's crust, **sedimentary rock** is formed by cementation and solidification of sedimentary deposits weathered from granite. These deposits may include shale, sandstone, and limestone. Finally, a third type of rock can develop from either granite or sedimentary rocks subjected to extreme temperature and pressure inside the earth. These **metamorphic rocks** are different structurally, although made from the same basic components. Quartzite may form from sandstone, slate may form from shale, and marble may form from limestone, for example.

Rocks are composed of many different minerals, some existing as single elements, such as iron, phosphorus, potassium, copper, sulfur, and magnesium; other minerals, such as quartz (SiO_2), are combinations of elements (Si, silicon; O_2, oxygen). Minerals found in rock are referred to as inorganic (not from living organisms).

Soil Development

Weathering of solid rock occurs through both physical breakage and chemical breakdown. Both freezing-thawing and heating-cooling cycles produce the expression and contraction that crack and break solid rock into progressively smaller particles. Water runoff, root penetration, and wind also produce physical weathering of rock. Carbon dioxide and other gases combine with water to produce a weak acid that dissolves some minerals to chemically aid the weathering process. As these inorganic particles become smaller, more individual minerals become free and available.

The results of weathering are greatest near the surface, and soil formation is therefore limited to this relatively shallow zone. Soil is more than these weathered inorganic materials; it is a combination of inorganic particles, organic materials, and pore space occupied by water and air. In any case, these three basic components of soil play an important part in soil-plant-water relationships.

Soil scientists make a clear distinction between soil and dirt, defining the latter as "misplaced soil." In other words, someone sitting on bare soil at a picnic will end up with dirt on his shorts when he gets up.

Soil Profile

A vertical section of soil is called the **soil profile**, and it reveals more or less distinct horizontal layers—the soil horizons, as shown in Figure 9-3. The uppermost layer contains the majority of the organic matter, the result of biological decomposition. Climate, the activity of animals, and even cultivation has a great effect on the development of this layer. This is the A horizon and is usually noticeably darker because of the presence of organic compounds. The A horizon varies in depth from a few centimeters to as much as a meter. Because most of the biological activity occurs here, it is the richest horizon and varies in depth from a few centimeters to as much as a meter. Because most of the biological activity occurs here, it is the richest horizon in terms of total nutrients and is referred to as the topsoil. The maintenance of the A horizon is an important consideration in modern agriculture.

In addition to the organic materials present, the A horizon is composed of smaller inorganic particles. This is the result of the higher weathering activity near the surface. The abiotic (wind, water, temperature) and biotic (roots, decomposition, earthworm, and other organismal activity) components of weathering are more effective near the soil surface.

The B horizon is the next soil layer. It is characterized by having very little organic matter, being somewhat coarser in texture due to larger inorganic particles, and allowing deeply penetrating root

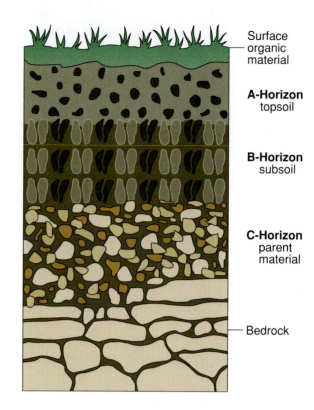

Surface organic material

A-Horizon topsoil

B-Horizon subsoil

C-Horizon parent material

Bedrock

systems. What organic matter is present is predominantly found near the top of the B horizon, close to the transition zone between A and B. The B horizon, therefore, is much lighter in color than the A horizon. Certain mineral components accumulate in greater amounts in the B horizon as well. Iron, calcium, carbonate, gypsum, clays, and aluminum oxides are commonly found in this layer.

The C horizon is typified by very large inorganic particles (rock sized), essentially no organic materials, and only occasional roots. This layer of the soil profile is actually a transition between the upper zones and the solid **bedrock** below. At this depth in the soil profile, the effects of weathering are minimal. The bedrock is sometimes called parent material because it is the source of the inorganic components of the soil. It is often impervious to root and water penetration.

Soil Texture

How course or fine the texture of the soil is depends on the size range of inorganic particles present. The particle sizes depend on the amount of weathering a given parent rock has undergone. The actual particle diameter determines whether it is generally classified as clay, silt, sand, or gravel.

The texture of a given soil is a direct measure of that soil's **porosity**, or the space between particles. That space contains either air or water, both essential to plant growth. Larger pores are termed **macropores**, and they drain water not held by capillarity. The smaller

micropores hold water against gravity, so most of the water available to root systems is found there. Texture, then, dictates the water-holding capabilities of the soil; water that percolates downward out of the root zone is replaced by air. Generally, the smaller the particle sizes, the closer they can pack together, thus lowering the porosity. It is important to remember that the adhesive property of water results in the surface of each soil particle being covered by water molecules. The greater the total surface area for a given group of particles, the greater the amount of water held in that type of soil. A 1 m cube of granite has 6 m^3 of surface area; breaking it in half results in the same volume of granite now having 8 m^2 of total surface area. If each of those halves is also broken in half, the total surface area exposed becomes 12 m^3. Each subsequent break results in smaller particles with increased total surface area for the same original 1 m^3 or rock; the process is called **weathering**. Weathering breaks down the mineral into smaller particles, increases the total surface area of a given volume of soil. The smaller the component particle sizes for a given volume of soil, then, the more water that can adhere to the surface area available to hold water ranges from less than 100 cm^2 per gram in course sand to over 1 million cm^3 per gram in clay.

The size of clay particles largely determines the physical and chemical properties of mineral soils. A pure clay soil, composed exclusively of clay-sized particles, therefore, can contain a large quantity of water. In addition, water molecules adhering to soil particles are tightly bound, which means they are not pulled away by gravity or plant root systems. Clays, then, can hold a lot of water, but it is not readily available to plants; only the water molecules held in the small capillary pore spaces are easily removed.

UNIQUENESS OF CLAY

Gravel, sand (very coarse to very fine), silt, and clay make up the soil. Clay is the smallest of these soil particles. An international system of soil classification based on diameter compares clay with the other soil particles:

Name of Particle	Diameter limits (mm)
Gravel	above 2.00
Coarse sand	1.00 to 0.50
Fine sand	0.25 to 0.10
Silt	0.05 to 0.002
Clay	below 0.002

Clay particles are plate-shaped and are composed of complex compounds such as kaolinite, illite, and montmorillonite. For example, the chemical formula for montmorillonite is:

$$(Na,Ca)_{0.33}(Al,Mg)_2Si_4O_{10}(OH)_2 \cdot nH_2O.$$

Clay is sticky and capable of being molded—highly plastic—when wet and very hard and cloddy when dry. Because of the small particle sizes of clay

and because all particles have the same negative charge, clay possesses some unique characteristics.

Clay in water forms a colloid—a mixture in which one substance is divided into minute particles and dispersed throughout a second substance. Since all clay particles have the same kind of electrical charge, they repel one another. Clay particles are closely packed and cannot move freely but maintain the same relative position. If a force is exerted upon clay, then the particles slip by each other. This permits clay to take on a shape as in pottery, brick making, or art objects, and even musical instruments such as the ocarina. When the molding force is removed, the particles retain their new positions because the same electrical forces as before act upon them. Clays become permanently hard when baked or fired.

Clay is also used in many industrial processes, such as paper making, cement production, and chemical filtering.

Pure sand, on the other hand, has less total surface area for the same total soil volume and, therefore, less moisture adhering to the particles. Because of the larger size and irregular shape of the particles, however, sand does not pack as tightly together as clay soils, resulting in greater porosity. When water is added to a sandy soil, a great deal of it is readily removed by root systems and gravity, and it dries out rapidly. It should be noted that, except for the surface particles, few soils, even desert sands, completely dry out; there is always a layer of water molecules held tightly to the particles, but it is not available to plants.

Silt-sized particles provide water-holding and porosity capabilities between those of clay and sand. The best soil for plant growth is a loam. Loam contains clay, silt, and sand-sized particles that provide a combination of water-holding capability (against gravity) and porosity (see Figure 9-4). Plant root systems must have available water and air as well as the main categories of soils as determined by texture.

The upper limit of water available to plants is a soil's **field capacity**. This is determined by first oven drying a given volume of soil to remove all moisture. Next, water is applied to the surface of this soil much as a heavy, steady rain or irrigation to the surface of this soil might do. As the water percolates downward, it replaces air in the soil pores from top to bottom. The container holding the soil must have holes in the bottom to allow excess water to drain off. After the soil is completely saturated with water, application of water at the surface is stopped. There is a period of continued downward water movement out of the soil in response to gravitational pull, this downward water movement out of the soil responds to gravitational pull. Once this

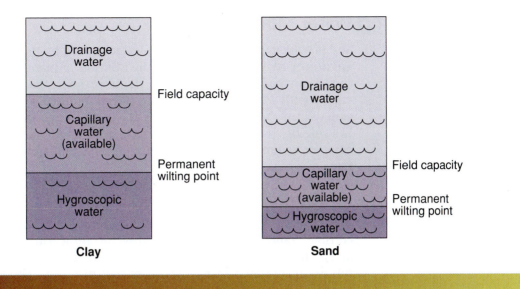

Clay   **Sand**

Figure 9-4

(A) Molecules are constantly in motion due to their kinetic energy. They move in straight lines until they collide with another mole-cule. (B) If large molecules are placed on one side of a container and small molecules on another, the colliding molecules at the interface gradually begin to intersperse. (C) And finally, they come to equilibrium as the large and small molecules become equal in concentration at all locations. Courtesy of Rick Parker.

downward flow has ceased, the soil is said to be at its field capacity. The soil is weighted at this point. The difference between dry weight and field capacity is there is some air in the larger soil pores, but other pores are filled with water that is available to plants.

In nature, a soil at field capacity loses its water through both surface evaporation and plant uptake. This occurs rapidly at first and then slows as the dry surface soil acts as a barrier to continued rapid surface evaporation. Eventually the plants lose water through daytime transpiration faster than they can take it up from the soil. For several days, plants wilt earlier in the day but become turgid again overnight as the root system continues to remove as much available water from the small soil pores as it can. Ultimately, the plants are unable to rehydrate overnight and become permanently wilted. Although the plant is not yet dead, if additional water is not added to the soil, death follows. At this point, the soil still contains a significant amount of water, but it is held in the smallest pores and on the particle surfaces so tightly that the plant's root system cannot remove it. The soil moisture content in such situations is termed the **permanent wilting point**.

Different soils have different field capacities and wilting points. A large field capacity is no guarantee of an abundant supply of water for plants, nor is a low field water holding capacity of soil necessarily too dry for plant growth. Different plant species are adapted to different soils and use the available water in the most efficient way possible. Still, a loamy soil that possesses the best combination of porosity and water retention is generally the most productive soil for plants.

Water Movement

Soil provides the reservoir for the water needed by plants. No single environmental factors affect the distribution of plants over the face of the earth as water. If you have the impression that we are preoccupied with water, your impression is well founded. We reemphasize water's importance in many places as a constant reminder of the critical rule of water in living organisms. Since all life takes place in an aqueous medium, it is important to understand how water moves from one area to another.

Kinetic Energy

Kinetic energy results from the motion of all molecules that are in constant random movement. Even if you fill a bathtub with water and let it become perfectly still, each individual water molecule in the tub is still in constant motion. If several drops of dye are carefully added to one end of the tub, the color slowly spreads throughout the water. Initially the water at the end where the dye is added becomes intensely colored, and the other end decreases as the color spreads toward the other end. Ultimately the water in the tub is equally colored throughout, and all this takes place without stirring the water at all. What has happened is an example of the process of diffusion.

Diffusion

Diffusion is defined as a random movement of particles from a region of high concentration to a region of lower concentration. Diffusion occurs because of the kinetic energy of matter, and it continues in the "downhill" direction until equilibrium is reached—that is particle concentration is equal in all regions.

Consider the example of the dye in the filled bathtub. If you examined any given individual dye molecule at the beginning of diffusion, you would not be able to detect that the net movement is from the application end toward the other end. Molecules move in all directions, and there is a net movement in one direction only because there are more dye molecules on one end. Thus, a greater percentage of them are able to move toward the other end. Once equilibrium is reached, the water and dye molecules are equally distributed throughout the tub. Although they are still in constant random motion, there is no net directional movement due to a concentration gradient.

Diffusion occurs in living organisms as well as in bathtubs and in the air around us. But since living organisms are cellular in construction and since water and other molecules need to move from cell to cell within these systems, the process of diffusion allows for movement within living organisms.

Osmosis

Osmosis is a special kind of diffusion in which water molecules move across a selectively permeable membrane in response to a concentration gradient. In other words, water moves across a membrane from a region in which its molecules are greatly concentrated to a region in which its molecules are less concentrated. Many students attempt to explain osmosis based on the concentration of salt dissolved in the water, but to do so would confuse the definition. All one needs to remember is that water molecules are most concentrated when there are no interfering or dissolved substances. As soon as some salt is put in pure water, the distance between the water molecules is increased and the water concentration is reduced.

If two adjacent cells have water concentrations of 90% and 92%, respectively, the 90% cell has 10% other molecules mixed in with the water, whereas the 92% cell has only 8% of other molecules to lower the water concentration (92%) to the lower concentration (90%). Equilibrium is reached when both cells have the same water concentration.

If the concentration of solutes outside the cell is greater than the concentration on the inside, water may be pulled from the cell out into the surrounding solution. As the turgor pressure falls below zero and becomes negative, the cell will begin to collapse. Even though the cell wall may impart rigidity, water will be lost from the vacuole and cytoplasm, causing the cell contents to collapse. This process is called **plasmolysis** and is, of course, disastrous for the cell. If water is quickly added to the external solution so that the solute concentration again becomes higher inside the cell, water will flow into the cell and bring about **deplasmolysis**. Plasmolysis occurs when salts in the soil solution, perhaps caused by overfertilization, become so concentrated that they create an inhibition of water absorption. Overfertilization in garden or houseplants will first be indicated by severe wilting, and if water can be applied quickly enough, the plants may revive.

In living systems, membranes are said to be selectively permeable because they allow some water-soluble materials to pass through while not allowing others. The natural selectivity enables cells to act as barriers to certain kinds of substances.

Water Potential

Water flows from one region to another because of a difference in potential energy. This potential energy is referred to as the **water potential**. Water melting from snow on a mountain, for example, has a great amount of potential energy, which it gives up in its travel down the mountainside. The potential energy might be used to turn a turbine to generate electricity as the water passes through, and the potential energy would be converted to a usable form of energy.

Cell water, too, moves in response to **gradients** in water potential, and in living systems the water potential consists of the following three components:

1. The solute potential comes from the activity of particles dissolved in water. Water molecules interact with each other through hydrogen bonding, and they perform their best work when interacting

with each other. If other kinds of particles are placed in the water, the distance becomes greater, and the water potential is reduced. The water potential of pure water is arbitrarily set at zero, and the addition of any solute to water (the solvent) reduces its water potential, and the value is negative.

2. The matrix potential is developed by the interaction of water molecules adhering to a wettable surface. The force required to remove water from that surface is called the matrix potential, and it, too, has a negative value.

3. The pressure potential is the force created by a real pressure against a membrane. The turgor pressure caused by water inside the vacuole and the cytoplasm pressing against the plasma membrane and cell wall develops a positive pressure potential.

The osmotic, matrix, and pressure potential combine to make up the water potential for a part of any system, and comparing the value with the components for another part of the system results in a gradient. The water always moves to the direction in which water potential is most negative. There are no exceptions, and the movement is strictly a physical process. Water movement *never requires metabolic energy*. Energy may be required to change the concentration of solutes in different parts of the system, but the water moves strictly by diffusion and osmosis, which are physical processes.

The Soil-Plant-Air Continuum

One can best conceptualize the movement of water in plants by considering the **soil-plant-air continuum (SPAC)**. This is simply a consideration of water movement as a series of linked pulling forces that extend all the way from the soil, into the roots, into the xylem, up through the stem, out through the branches and petioles, and finally into the blades of the leaf, as shown in the cross section of the leaf in Figure 9-5. The transport cells of the xylem can be thought of as part of a city water distribution system. As the main pipe moves water further toward the edge of the city, the size of the pipeline decreases in accordance with demand. In the leaf, vessel elements terminate in the mesophyll and supply water to a few living cells. The water movement continues out through the stomata and finally into the atmosphere. From root to leaf, water moves as a liquid column, and there is no gaseous phase. From leaf to atmosphere, water changes to vapor. Water moves along the entire distribution system from the reservoir (the soil) along gradients of water potential, always moving to the most negative point. In almost every case that means from soil, through the plant, and out into the atmosphere, providing the water needs of the plant along the way. The major driving force for this system is the **transpirational pull**, a tremendous wick-like force at the surface of the leaves that sucks the water into the atmosphere. It is no wonder that some 99% of all

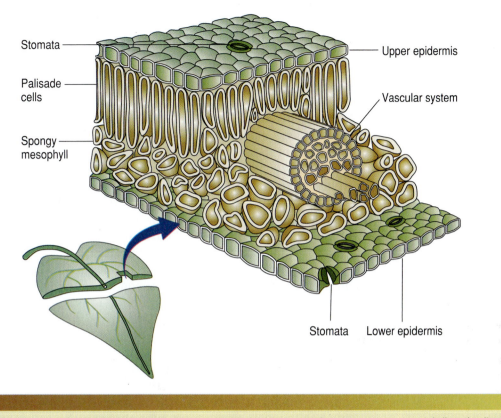

Stomata

Palisade cells

Spongy mesophyll

Upper epidermis

Vascular system

Stomata   Lower epidermis

Figure 9-5

Cross section of a typical leaf, showing the upper and lower epidermis stomata, vascular system, spongy mesophyll, and palisade cells. CO_2 enters the leaf through the stomata, and water exits the leaf at the same place. Courtesy of Rick Parker.

water taken up by plants ends up back in the atmosphere as a result of transpiration.

When there is precipitation and water percolates down into the soil pore spaces already occupied by the root system, water availability is no problem. The root zone must be constantly growing and branching, exploiting newly available soil water to guarantee an adequate supply between periods of precipitation. It is this new root growth, the youngest part of the root system, that is responsible for most water absorption. On this new growth, the epidermal layer greatly increases its total surface area in the form of root hairs, which begin developing a few millimeters behind the root tip. As these young roots push their way between soil particles and come into contact with water in soil pore spaces, the water passes through the cell membranes of the epidermal cells by osmosis. Water also has no problem moving through cell walls, either at the root or internally.

Normally, soil water has fewer ions in solution than does the cytoplasm of root epidermal cells; cells have a more negative solute potential than the matrix potential of the root external surface. Thus, water is taken from the soil into the roots as a result of the more negative water potential at the roots.

From the epidermal cells, water and solutes move inward toward the center of the root, where the vascular tissues are located. Water moves from epidermal cells through the cortex of the root to the **endodermis**. Water movement is primarily along cell walls without having to pass through cell membranes. Water movement along cell walls without passing through membranes is termed **apoplastic movement**. Once water reaches the endodermis layer, apoplastic water must pass through endodermal membranes because of a gasket-like band of suberin around each endodermal cell. Suberin is a waterproof waxy substance that seals off the route of water between cells. If you were to visualize the endodermis as a smokestack made of bricks, the mortar between each brick would be analogous to the Casparian strip around each cell (brick). This column completely surrounds the vascular tissues, and this is the first time that much of the water and its soluble ions are required to pass through a membrane since entering the epidermis. The endodermal cell membranes are selectively permeable, as are all cell membranes, and act to screen ions.

In addition to diffusion, mineral ions are brought into the root epidermal cells by the expenditure of metabolic energy. These ions are taken up by **active transport**. Once inside the epidermal cells, these ions can pass from cell to cell through the cortex to the endodermis through plasmodesmata. This pathway is called **symplastic movement**. Whether water and mineral ions move apoplastically (between cells) or symplastically, everything entering the vascular tissue must pass through the cell membranes of the endodermal cells.

Once past the endodermis, water moves through the **pericycle**, a layer of cells from which the new lateral roots originate. Inside the pericycle, water enters the xylem and becomes a part of a continued transportation network that terminates in the leaves. Through this network, water moves from the roots to incredible heights without the use of a pump. The movement of water is triggered by transpiration at the leaves.

Even when there is no transpirational water loss, the roots continue to absorb water and to accumulate ions in the xylem. If water is not moving up into the stems and leaves, the ions accumulate in gradually increasing concentrations in the xylem, causing the solute potential of the xylem to gradually become more negative. This more negative water potential in the xylem causes water to move by osmosis into the xylem from surrounding cells, ultimately forcing water and soluble ions upward. This force produces **root pressure**, which in turn forces water to be extruded at the leaf tips and margins through small pores of short plants.

The process of root pressure forcing water out of the leaves is termed **guttation** and only occurs when stomata are closed and when water is readily available to the root system, usually overnight Although capable of forcing water up the xylem when all the stomata are closed, when they are open, root pressure is a negligible component of water movement.

Water Movement throughout the Plant Body

Xylem is the tissue in which water moves in plants. It is made up of elongated cells having secondary cell walls with holes in them. Water moves through the xylem via vessel elements and tracheids in angiosperms and tracheids in the gymnosperms. Both tracheids and vessel elements are nonliving at maturity, and both may contain pits in their cell walls. Vessel elements also have larger holes in their secondary end walls, and occasionally in cell walls. These perforations are very important in allowing unimpeded water passage from one vessel element to the next.

Vessel elements are arranged end to end and form long, continuous strands throughout the plant. These strands are analogous to a long capillary-sized hollow tube because there are no cell membranes to retard water movement from cell to cell as there would be in living cells. Vessels move more water efficiently than do tracheids because the perforations in their end walls are larger than the tracheid's pits and because the cross-sectional area is much greater.

Transpiration

Unlike higher animals, which have circulatory systems that recycle fluids through the body, plants lose the vast majority of the water taken out of the soil. Transpiration is technically the evaporation of water from any portion of the plant body into the atmosphere. Actually, of the total water volume transpired by plants, so little of it is lost from plant surfaces other than the leaves that there is an understandable tendency to think of transpiration as exclusively a leaf-related phenomenon. In addition, the greatest bulk of water transpired by the leaves is lost through stomata. Although small, there are an amazing number of stomata on the leaf. For example, the lower surface of a typical cotton leaf averages 18,000 stomata per square centimeter.

Even though water does leave the plant and is lost in that sense, the term *lost* is misleading. In truth, the water that is transpired from stomata provides a critical service. The atmospheric CO_2 required for photosynthesis enters the leaf cells in an aqueous solution. It is the water at the surface of cells directly in contact with the atmosphere that facilitates CO_2 being taken in. It is also this water that evaporates into the atmosphere (transpires).

Transpiration occurs whenever the stomata are open, and the cell surfaces inside the leaf are exposed to the atmosphere A small amount of water is also lost directly through the cuticle and is termed cuticular transpiration. The reason water moves out of these cells is simple. The atmosphere normally has a lower water concentration than the cells that are in contact with it. Water passes through the cell membrane by osmosis and evaporation. These cells have a lower water concentration than adjacent interior cells not in contact with the atmosphere.

As water evaporates into the atmosphere, new water molecules are pulled up to take their place. Water moves along from cell to cell based on water potential gradients. In the xylem, the primary component of the water potential is the tension created by open stomata and transpiration pull. In the living mesophyll cells, the water potential is created primarily by the solutes and surfaces inside the living cells. Whether the water will remain in the leaf or dissipate through the stomata depends entirely on the overall water potential and the region where it is most negative. **Stomatal regulation** is a constant compromise for the plant. On the one hand, the stomata should remain open so the CO_2 can enter and participate in sugar production. On the other hand, closing stomata conserve water and may allow the plant to adjust to a lack of water in the soil.

Cohesion-Adhesion Transpiration Pull

Water moves from the roots to the leaves through a combination of the **cohesive** strength of water-to-water bonds, the **adhesive** attraction of water to vessel walls, and the transpiration at the leaves that pulls on the column of water below. It is the loss of water through transpiration that triggers the pulling tension on water in the xylem. As water evaporates through the open stomata, either from an open intercellular space in direct contact with the terminal xylem trace or from the outer surface of a cell, a negative water potential results, as compared to that of water in the xylem. Water in the xylem automatically moves out into the leaf in response to this negative water potential and pulls water molecules below up in the xylem tract to replace it. The negative water potential gradient produced in an actively transpiring leaf results from a combination of negative solute potential in the cells losing water, a negative matric potential (surface adhesion) on the cell wall and membrane surfaces, and especially a negative pressure potential caused by water being pulled out of the leaf into the atmosphere through evaporation.

Once the water starts moving from xylem cells toward open stomata because of this negative water potential, the phenomenal pull of the solid column of water in the xylem is transferred all the way down to the xylem originating in the roots. The great cohesive strength of water results in the column being pulled intact from bottom to top (root to leaf) without breaking.

There are few instances in which the water transport system is reversed and water moves into the soil. In fertilization burn, the concentration of solutes in soil is such a completing factor that the overall water potential is most negative in the soil. The atmosphere is pulling water one way, the soil is pulling water the other, and the direction with the most negative water potential wins.

The plant has the ability to modify water movement in various ways with a series of linked resistances; the most important of which is the guard cells on the leaf surface, which exerts stomatal regulation. The ability to close stomata when placed under stress is like turning off a faucet. If the fertilizer concentration is too severe, so much water will be pulled from the plant that wilting will occur, and unless the water flow is reversed, the plant dies.

One interesting case of water potential gradient occurs in the Atacama Desert of northern Chile. Rainfall in this desert is almost nonexistent, and no measurable precipitation may occur for a period of 10 years. As one might suspect, vegetation is sparse, but the dominant feature of the landscape is a mesquite tree (*Prosopis tamarugo*). This tree manages to survive because deep taproots mine water from underground aquifers. Few plants have such extreme adaptations to water stress.

Stomatal Regulation

The two guard cells bordering each stomatal opening determine whether the stoma is open or closed. The control is a physical response in the stomatal opening that is produced by the adjacent guard cell walls pulling apart when the cells are turgid and coming together when the guard cells are **flaccid**, as shown in Figure 9-6. Current theory suggests that the movement of water in and out of guard cells is an osmotic response governed by potassium levels in the guard cells. If potassium concentrations in guard cells increase, more negative water potential is created there and water moves osmotically into the guard cells, making them turgid. If the potassium levels drop, the water potential becomes more positive and water osmotically exits the guard cells, making them flaccid. A plant hormone, abscisic acid, is known to be somewhat involved in the changing of potassium levels.

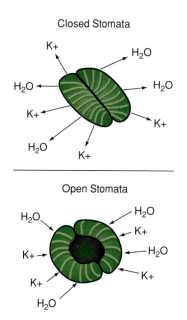

Figure 9-6

The stomatal apparatus consists of two identical guard cells, the inner walls of which are slightly thickened. Whenever turgor pressure increases inside these cells, they bend outward exposing the stomatal pore. The bending is also assisted by wall reinforcement known as radial mycellation. Courtesy of Carolina Biological Supply.

Closed Stomata

Open Stomata

The guard cells are structurally unique. They have slightly thicker adjacent cell walls than outer cell walls, and each has **radial micellation** of the cell wall with the bands farther apart in the outer cell wall and closer together in the inner or adjacent cell wall. When the guard cells become turgid, the outer cell walls expand more than the inner ones, pulling the inner walls apart. As the cells lose turgidity, the space between the adjacent inner walls come into contact again, preventing further transpiration and CO_2 uptake.

Because water loss and CO_2 uptake both occur when stomata are open, there is always a balancing act in progress. Plants must have CO_2 for photosynthesis, but on the other hand, only a given amount of water can be transpired and not replaced before the plant begins to wilt. The factors affecting this balance include both environmental and physiological controls.

Factors Influencing Transpiration

The single overriding control of stomatal activity is water availability. If there is not enough water available to the plant, all the guard cells become flaccid as the leaf wilts. Any time water availability fails to keep up with water loss and guard cells become flaccid, they close the stomata, preventing further transpiration until the roots can absorb additional water. Thus, any factor that increases transpiration rate relative to water uptake rate will result in stomatal closing.

Temperature

One of the most important regulatory factors is increased temperature. Above 30° to 35°C, stomata usually close automatically whether there is available water or not. At temperatures of (10° to 25°C) an increase of 10°C results in a doubling of the rate at which water evaporates (transpires) from the plant, thus, affecting stomatal regulation by causing more rapid water loss than uptake. Many plants, especially those in arid and semiarid regions of the world, regularly close their stomata during the middle of the day when the temperature is the highest. Such plants allow CO_2 uptake only in the morning and late afternoon, restricting their photosynthetic output.

Wind

By itself, but especially in combination with increased temperatures, wind movement across leaf surfaces results in increased water loss by physically pulling away water vapor molecules from the open stomata. Hot, windy days can impose a water stress on plants very quickly, resulting in early and sustained stomatal closure. On the other hand, wind may exert a cooling effect at the leaf surface and tend to offset the negative effects.

Humidity

A third environmental variable affecting transpiration rates is the amount of water vapor in the atmosphere—humidity. Since evaporation occurs by water molecules physically escaping from the water-atmosphere interface, the net number that escapes partially depends on how many water molecules are already in the atmosphere. Evaporation can be considered in terms of diffusion. When there are very few water molecules in the atmosphere (low humidity), the net flow of randomly moving water molecules is from leaf to atmosphere. Although the atmosphere will never have as many water molecules as a liquid surface, the closer the two densities become, the slower the net movement from water to air (evaporation). High humidity, then, slows the transpiration rate; low humidity increases it. A hot, windy, dry day, therefore, places a large stress on plants because of the combined effect on increasing the transpiration rate.

CO₂ Concentration

High internal concentrations of CO_2 cause stomata to close, and low CO_2 concentrations cause stomata to open. Although these factors may be relatively unimportant on a normal day, if stomata have closed prematurely for other reasons, fixation of carbon into sugars may stimulate the reopening of stomata.

Diurnal Stomatal Closing

The vast majority of plants close their stomata at night even though their transpiration rates decrease significantly in the absence of the daytime heat. Since there is reduced photosynthesis activity at night because of the absence of the light energy needed to drive critical reactions, the need for CO_2 is less, and continued stomatal activity is not required.

Conversely, there are plants that open their stomata at night and keep them closed all day. Cacti and other succulent plant groups use a different metabolic scheme to secure CO_2 for photosynthesis. The basis for this scheme is the conversion of CO_2 to certain organic acids at night, when water loss is slowest. During the day, these organic compounds release when sunlight energy becomes available to drive photosynthesis, the CO_2 required. Photosynthesis can occur then with minimal water loss. This system is termed **Crassulacean acid metabolism (CAM)** and will be discussed in more detail in Chapter 10. It is worth noting that CAM plants are essentially all desert-dwelling plants, and the development of the CAM pathway is a very successful water-conserving adaptation for their hot, windy environment.

Adaptations to Reduced Water Loss

The problem of balancing water loss via transpiration against the requirement for open stomata for CO_2 assimilation is most severe in hot, dry, and often windy environments. Plants that have successfully adapted to stress have done so by the evolution of several morphological and physiological features. These adaptations are basically of two strategies: (1) reduction of water loss through open stomata and (2) different mechanisms for CO_2 utilization that avoid having stomata open as much.

Morphological Adaptations

The first group of adaptations are structural and have evolved in direct response to the primary environmental features affecting rates, temperature, wind, and humidity. One modification actually does not involve the stoma at all, but rather affects the remainder of the leaf surface—**cuticle thickness**. The thickness of the waxy layer covering all the epidermis of leaves varies from so thin as to be almost nonexistent to so thick that you can scrape the leaf surface with your fingernail and accumulate wax (cutin) underneath. Generally, the more arid the environment, the thicker the cuticle, which minimizes as much water loss as possible. With the water loss essentially eliminated, the greatest amount of water lost from plants passes through open stomata. However, the evolution of a number of modifications has helped prevent excessive water loss in certain plants.

One of the most common modifications is the absence of stomata from the top surface of the leaf. The top surface is exposed directly to the sun, which increases the temperature significantly over that of the bottom surface. The higher the heat load, the faster transpiration occurs, so this modification lowers the total transpiration rate. The leaf surface is also very often highly **pubescent** in arid zone plants (**xerophytes**). The dense coverage of the hairs is thought to add reflectance, protect the stomata against wind, and produce increased humidity between the hairs and the leaf surface. The extra shading lowers the temperature while protecting against the wind, and the layers of increased humidity helps slow transpiration.

Another modification that increases humidity just outside the stomatal opening is the development of sunken stomata. Instead of being flush with the epidermal cells, the guard cells are recessed. The result of a sunken stoma is protection from the wind and the formation of small pockets of high humidity immediately outside the stoma, which slows transpiration rates.

To provide shade and possibly create a zone of higher humidity, some leaves roll downward at the margin. This creates essentially a recessed lower leaf surface that is partially protected from the wind and better shaded than flat lower leaf surfaces. The results are very similar to those in plants with sunken stomata, but for the entire lower leaf surface. Some plants use additional methods to avoid excessive transpiration. *Mimosa pudica* and other "sensitive" plants have leaflets that fold together in times of stress or disturbance, and other plants fold their large leaves up at night.

Grasses have modified upper epidermal cells, bulliform cells, which are enlarged for water storage. In periods of water stress, the bulliform cells lose their stored water, causing the grass leaf to roll up edge-to-edge, limiting further water loss from the upper epidermis.

Physiological Adaptations

Some plants have alternate pathways of carbon fixation and photosynthesis that provide for more efficient water use. Physiological changes have developed that allow such plants to survive in periods of low rainfall and in certain edaphic (soil) situations. The development of succulent leaves on plants that normally do not have them is yet another method of handling high water stress, this time by storing extra water.

🌀 Applications

All the aforementioned adaptations to survive water stress occur naturally and develop gradually over a long period. On a short-term basis, a plant not so modified cannot handle water stress well. Many houseplants have large leaves and are native to tropical regions, where abundant rainfall, high humidity, and moderate temperatures have allowed plants to thrive without the development of any morphological or physiological water-conserving modifications. When such plants are introduced into a lower humidity and/or hotter climate as a houseplant, their inability to conserve water often results in minimal or no growth, dehydration, wilting, and death in extreme situations. Watering heavily does not solve the problem; such plants can spend most of their time with closed stomata and very little new growth.

The care of houseplants is not so difficult. The key to successfully growing houseplants is to understand the specific water, light, and fertilizing needs of each and to try to imitate as closely as possible the environmental conditions of their native habitat.

Water

Houseplants vary in their water needs, but there are several general categories that apply to most common houseplants. Some must be kept well watered at all times; their soil must never be allowed to dry out. Very few plants thrive in a saturated soil, however. The key is to keep the soil moist but not flooded.

Other plants require an alternation of thorough watering followed by a soil-drying period. This group of plants is most susceptible to overwatering. More houseplants are killed by overwatering than by not watering enough (see Figures 9-7a and b). Overwatering can result in root rot, which is encouraged by insufficient root aeration. When a potted plant is watered, the entire root zone should be soaked. Shallow watering is the next most common mistake. When only enough water is applied to wet the top few centimeters of the potting soil, the plant looks well watered for a couple of days because the surface is damp. If the entire root zone does not receive water, however, the plant will ultimately die because the new growth areas of the roots will finally dehydrate. A good rule of thumb or "rule of finger" in this case, is to stick your index finger into the potting soil at least to the second knuckle to test for soil moisture. If the soil is dry to that depth, the plant probably needs watering.

Cacti and other succulents form a third watering category. They should be thoroughly watered, but much less frequently than other plants. In addition, they must have a well-drained soil because they are especially susceptible to root rotting. In prolonged periods of cool or cloudy weather or in areas of high humidity, their watering schedule should be even less frequent.

Fertilizer

Each species has specific nutritional needs and should be fertilized accordingly. Unless a plant has an unusual requirement for given specific minerals, however, a general-purpose houseplant fertilizer will contain appropriate levels of macroelements and microelements when mixed in the proper concentration. Do not fall into the trap of thinking, "if a little is good, a lot will be great" when it comes to fertilizer concentrations. As mentioned earlier, fertilizing with too high a concentration adds too many salts to the water in the soil, producing more negative water potential in the soil than in the roots. Water is pulled from the roots in such a situation, producing fertilizer burn.

For safe concentration and no loss of effectiveness, fertilize with half the recommended strength twice as often as is recommended. For many plants, in fact, such a fertilizing regimen actually produces better results.

Light

Different plants also have different light requirements. Some must have full sunlight for at least half the day; others need only indirect or reflected light, and some require very low light intensities. Leaves that receive too high a light intensity will literally sunburn. The resulting bleached areas followed by dry brown splotches on the leaf surface are different from the gradually spreading brown leaf margin or apex that results from fertilizer burn or insufficient water.

Soil

Potting soil for houseplants can be purchased already mixed and packaged or can be mixed at home with equal results. Recalling soil composition will clarify what one needs to use in mixing potting soil: organic material, inorganic minerals, and porosity for air and water. For the vast majority of common houseplant species, the following formula will produce a potting soil that provides these needs: equal parts of medium-coarse sand, peat moss, and local soil. The sand will provide porosity, the peat moss holds water well, and the native soil will add needed minerals. If local soils contain too much clay, are abnormally alkaline or acid, or are otherwise unsuitable, rich compost should be substituted (see the next section). Local soils should be sterilized by oven heating to 350°F for one hour or buy good quality potting soil. To ensure proper packing in the pot, dampen the mixture to preset the peat moss.

Because peat mass will ultimately compress and harden, houseplants need to be repotted every two to three years. A sure sign of peat moss packing is when water runs all the way through the pot very quickly after watering. The water will not absorb because the peat moss fibers are too tightly packed together. Instead, water will often run down between the wall of the pot and the compacted block of soil. The replacement of fresh potting soil improves water availability, aeration, and organic and inorganic nutrient availability.

Cacti and other plants native to dry regions should have an adjustment made in their potting soil formula. At least 50% medium-coarse sand should be mixed with local soil and peat moss. Cacti do best when they have a high-porosity soil that allows water percolation and proper aeration. Many other plant groups are also better adapted to sandy rather than loamy soil. Care should be taken not to overwater cacti even though they may be in an almost totally sandy soil. The top surface of sandy soil dries out rather dramatically in only one or two days after watering, even though the root zone has adequate moisture. Many people water based on the surface appearance. Cacti are very susceptible to root rot, even in a porous soil, if watered too often.

Compost

The addition of **compost** to a potting soil or garden enriches the soil with nutrients and texture. Compost, as shown in Figure 9-8, is made by placing organic materials in a pit or small mesh screened enclosure and providing an optimal environment for decomposition to occur.

Although animal tissues, woody plant parts greater in diameter than a pencil, and most Bermuda grass clippings are not to be used, essentially all other organic wastes can be composted. Leaves, garden pruning, kitchen vegetable waste, eggshells (after washing them inside), coffee grinds, orange rinds, and an endless list of other plant parts can be placed in compost. It is best to begin with an enclosure that provides a foot or two of depth. Adding a shovel full of soil to provide the decomposing organisms and even a little fertilizer mixed in with the composting material improves results. The top of the compost should be covered, and water should be added as necessary to keep the composting environment moist.

Figure 9-8

An ideal compost consists of a "mature" usable compost.

At the beginning of each year, the compost should be turned. Remove the bottom layer, which should be ready to mix with potting or gardening soil. Periodic turning or loosening to improve aeration during the warm part of the year prevents compaction and the resulting suffocation of the organisms carrying out the decomposition. Regular addition of new materials to the top grades the layers of activity, with the bottom the most completely converted. Although composts are easier to develop in warmer and wetter climates, which also usually have longer growing seasons, with proper planning and care a compost can be successful anywhere.

Potting

When potting or repotting a houseplant, you must keep in mind several rules of green thumb. First, do not use a pot with a top diameter more than 5 cm wider than the root zone of the plant. Second, use a pot with a drain hole in the bottom, preferably a clay pot for most plants because it allows better soil aeration. Third, when roots are forced to grow in the shape of the pot, often growing through the drain hole, it is time to repot; the plant is probably becoming **root bound**. Note that some plants need to be root bound to thrive. To repot those that do not, place one hand on top of the pot, turn the pot upside down, and strike the bottom of the pot firmly with the other hand until the root-soil mass slides out. Without unnecessarily damaging the root structures, select an appropriately larger container and repot.

The new pot must first have a drain hole covered. A piece of broken clay pot (shard) that is larger than the drain hole is most commonly used, although rocks and other materials are acceptable. Also a pot with a reservoir for water will work as well. Cover the hole, but do not seal it; this prevents the new soil from washing out but allows water to drain freely. To hold the shard in place as well as provide a new growth area for the plant roots, add several centimeters of freshly mixed (damp) potting soil to the bottom of the new pot before inserting the plant. Finally, suspend the plant in the pot with one hand such that the stem will clear the top of the soil with 2 to 4 cm of space left between the soil and the top rim of the pot. Add the soil and gently pack by tapping the pot on the tabletop; the soil should not be tightly handpacked. Thorough watering will help settle the new soil around the root system. An addition of soil to the surface may be necessary after several waterings have settled the soil and lowered the surface below the desired position.

Pests

The most dangerous houseplant pest is a plant owner who does not know the needs of their plants. Overwatering, overfertilization, underwatering, sunburning, and other problems of many houseplants are all a result of improper care. If these are eliminated, there are still several common insect pests that can cause problems. The following list describes the pest, its mode of damaging action, and recommended remedies that also work but are not always desirable in the enclosed atmosphere of your home. A registered systemic insecticide can be obtained from your local nursery, however, if you feel chemical control is warranted. There are more and more biological controls available. The mode of attack of each of these pests is the same: They pierce the plant tissue and suck out the plant juices.

Mealy Bugs

Mealy bugs are white or light colored and form clumps on all parts of the plant.

Appearance

Mealy bugs are white or light colored, often with pinkish hues, this oval insect grows to 3 or 4 cm in length and has a "scruffy" appearance; some look like small fury "pill bugs."

Location

Mealy bugs tend to clump or aggregate and are found on all parts of the plants, top to bottom (including roots), but more commonly on lower leaf surfaces and on stems. They are usually discovered in significant numbers before they noticeably damage the plant; however, if the bugs are not removed, the plant will look generally unhealthy at first. Then, localized discoloration and tissue damage follow.

Treatment

Mealy bugs can be removed mechanically by squashing them with your fingers or they can be sprayed with a 3:1 water-alcohol mixture. If a sprayer is unavailable, a cotton-tipped swab or cotton ball can be used for direct application.

Aphids

Aphids are soft-bodied insects that come in many colors and all shades. They form clumps or dense groups on most new growth of the plant. Courtesy of USDA Photo Gallery.

Appearance

Aphids are small (1 or 2 cm), winged or wingless, soft-bodied insects. Color can vary: They are green, gray, yellow, black, and all shades in between.

Location

Usually found in clumps or dense groups on the tenderest parts of the plant, such as the stem just below flowers or around new leaves. They deposit sticky "honeydew" on the plant surface (exudates of plant sap passed through their bodies), which is an excellent substrate for growth of sooty mold. They are normally noticed before significant plant damage is done; however, if unnoticed or not treated, they produce local discoloration followed by wilting of plant parts above the clumps of insects.

Treatment

Mechanically remove or squash with your fingers or wash with mild soapy water weekly. Regular washing is one of the best deterrents for all common houseplants.

Spider Mites

Spider mites are too small to see until their webs are produced on leaves and branches.

Appearance

They are too small to see readily without magnification by a land lens (less than 1 mm). They produce webbing at the leaf and branch axils, which then accumulate dust and can be seen. Spider mites are not an insect but a true mite that spins silk into a web, sometimes called red spiders.

Location

Found primarily on lower parts of plants at axillary branching points or lower leaf surfaces. They produce streaked discoloration of leaves, yellow to brown, followed by desiccation and an overall dried-out appearance.

Treatment

Since they thrive in dry environments, they can be controlled by misting the plant daily, increasing the humidity of the plant's immediate environment by keeping water in pot saucers or using room humidifiers. Use of soapy water once a week also works well. If webbing is detected, you will need something else to control them. There are parasite mites that will kill them, but you still have bugs around. They are not harmful to the plant only the spider mites.

White Flies

White flies can be detected when there seems to be a small cloud of smoke coming from the plant. They, like Aphids, secrete a honeydew as they feed and that is a perfect place for a secondary problem to arise, Black Sooty Mold. Control the White flies and control the Sooty Mold.

Appearance

White flies are tiny (1 or 2 mm), white flying insects that look like minute moths. They also have a quiescent period, a small (1 mm), oval, black scab-like feeding stage. This insect is the most serious greenhouse pest.

Location

White flies are found primarily on the underside of the leaves from the top to bottom. When the plant is disturbed, they swarm into the air and then return to the leaves, where they look something like dandruff on the lower surface. They are usually very noticeable before they damage the plant enough to cause symptoms.

Treatment

There is no satisfactory chemical treatment that will control them because they became resistant. Therefore, the life cycle must be broken. New biological controls are being used such as parasitic wasp and other types of parisites. Physical removal to prevent the spread to other plants by isolating infected ones is the best treatment. Buy healthy plants! Look for both the white flying stage and the quiescent stage before bringing a new plant into your home. These insects are very common in commercial greenhouses, and you should be careful to prevent their introduction into your house. Carefully check "vacationing" plants or plants held in greenhouses over the winter before bringing them back into your home.

Scales

Appearance

They look like nonliving scales, wounds, or markings and vary from white to brown. You will never see the stages with legs because the insect is too tiny.

The one you see is flat against the surface with its feeding parts embedded in the plant tissue. The insects secrete a material that camouflages its body to look like a nonliving scale or scab. Heavy

Scales look like nonliving scales, or markings, and vary from white to brown. They are found mostly on stems but can be on leaves.

infestations are fairly evident and are usually noticed prior to the plant suffering significant damage. Scales ultimately produce local tissue discoloration and finally wilting with heavy infestations. They are more common on yard ornamental plants.

Location

Scales are found mostly on stems but also on leaves especially along the midrib of the blade. They are fairly easy to see but are often overlooked because they do not appear to be alive.

Treatment

If the infestation is not too great, they may be scraped off with a fingernail or knife blade. Be careful not to damage the tissue by pressing too deep. For severe cases, dispose of the plants or isolate them from healthy plants.

Hydroponics

Plants do not need soil per se, but only what soil provides: anchorage, support, water, air, and nutrients. If all these needs could be met without the use of soil, the plants would thrive anyway. **Hydroponics** is the growing of plants in a liquid culture (see Figure 9-9). Water is obviously available, and depending on the specific needs of a given plant, air is available in moving water or aeration by bubblers, if necessary. A complete nutrients solution is added to the water that provides the corrected concentration of all the macronutrients and micronutrients. The Hoagland's solution can be used as a regular fertilizer solution for potted plants or for plants growing hydroponically.

Figure 9-9

Hydroponics is growing plants without soil. The plants are grown in a liquid culture.

A number of common houseplants can be rooted in a glass of water, and some, such as ivy, can be maintained hydroponically.

Commercial hydroponics greenhouse owners suspend many plants with their roots in a trough that contains gently flowing water and the nutrient solution. Because these plants are in the greenhouse, anchorage against wind is not a problem; support, water, air, and nutrients are all provided, and soil is not necessary. This system removes the time-consuming soil preparation, potting, and repotting, and the expense of the pots themselves. Hydroponically grown tomatoes are one of the most widely successful crops that can be grown out of season; hydroponic greenhouses are much more expensive to operate than the traditional method of growing plants.

Summary

1. Water is the most important factor in plant growth. It provides the solvent system for the distribution of nutrients and organic molecules, gives turgor to cells, and participates in a few biochemical reactions. The special properties of water are due primarily to the hydrogen bonding between water molecules and other hydrophilic substances.

2. Soil develops from igneous, sedimentary, and metamorphic rock material in the earth's crust. Chemical and physical weathering causes large mineral particles to be broken into smaller particles, giving rise to sand, silt, and clay. Combinations of these materials create loams, the soils in which plants grow. The highly weathered upper portion of the soil profile also contains organic matter and considerable air and water. The finer textured soils hold more water than do the coarse, sandy soils.

3. Diffusion is the physical process by which particles in random motion move from a region of higher concentration to a region of lower concentration. Osmosis is a special case of diffusion in which water molecules move through a differentially permeable membrane in response to a concentration gradient.

4. The chemical potential of water is called the water potential, and the concept is used to explain how water moves from one part

of a system to another. In the soil-plant-air continuum, gradients in water potential are developed by the components of solute, matrix, and pressure potentials. They combine to determine water movement to the most negative water potential.

5. Transpiration is the evaporation of water from a leaf surface. It accounts for approximately 99% of the water absorbed from the soil. The control of transpiration is exerted by the cuticle and by stomatal regulation, the opening and closing of the pores between two guard cells. Various environmental factors influence stomatal regulation, and plants cope with water stress by various morphological and physiological means.

6. An understanding of soil-plant-water relationships allows one to make intelligent decisions concerning the growth of plants—when to water, how to fertilize, how to mix soil properly, when to repot, and how to maintain healthy plants.

Something to Think About

1. What is the most important factor in plant growth?

2. What causes the special properties of water?

3. Where does soil develop from?

4. Explain chemical and physical weathering of the soil.

5. What do the combinations of these materials produce?

6. Where are the organic matter, air, and water located in the soil?

7. What is a special case of diffusion?

8. What is the name for the chemical potentials of water?

9. In the soil-plant-air continuum, gradients in water potential are developed by which components?

10. Describe hydroponics.

Suggested Readings

Benjamin, J. G., D. C. Nielsen, and M. F. Vigil. 2003. *Quantifying effects of soil condition on plant growth and crop production.* Centre National DeLa Recherche Scientifuqice, 116, 249

Feenstra, Gail, et al. 2002. *What is sustainable agriculture?* Davis, CA: UC Sustainable Agriculture Research and Education, University of California. Journal Geoderma, Elsevier, Amsterdam

Kraner, P. J. 1970. Plant and soil water relationship: A modern synthesis. *Quarterly Review of Biology, 45(2),* 218.

Hatfield, J., and K. Douglas. 1994. *Soil water interaction.* Boca Raton, FL: CRC Publishers.

Internet

Internet sites represent a vast resource of information. The URLs for Web sites can change. Using one of the search engines on the Internet, such as Google, Yahoo!, Ask.com, and MSN Live Search, find more information, by searching for these words or phrases: **water, soil, environment, sustainable, diffusion, osmosis, water potential, evapotranspiration, fertilizer, hydroponics, pressure potentials, and matric potential.**

CHAPTER 10

Energy Conversions

Essentially all biological energy emanates from the sun. The energy from the sun, if it is concentrated, may be used to an incomprehensible intensity. Even though only a fraction of that energy reaches the surface of the earth, it is equivalent to 1 million Hiroshima-sized atomic bombs per day. Some of the light directed at the earth is screened out or reradiated back into space before it reaches the ground.

The thermonuclear explosions of the sun represent one type of energy that is converted into various other types of energy, notably heat and light. In addition, energy may exist as **electrical energy, chemical energy,** motion, sound, and molecular forces that hold together the atoms within molecules. Energy is really the ability to perform work.

Scientific questions are constantly being put to the test. Does the experiment work every time? What conditions modify the results? Can many investigators in any part of the world repeat it? After such careful scrutiny, a number of natural laws have withstood the test of time and are now recognized as absolute. Two of these laws concern how energy flows through the solar system.

After completing this chapter, you should be able to:

- Relate the laws of thermodynamics to plant science
- Describe how energy throughout the world performs work of various kinds
- Discuss how energy is stored in chemical form
- Relate the oxidation-reduction reactions in plants to that in animals
- Describe how photosynthesis is based on two reactions
- Outline how respiration is the reversal of the chemical reaction for photosynthesis
- Recognize that aerobic respiration is glycolysis
- Discuss how energy is stored as ATP
- Describe the Krebs cycle

Key Terms

electrical energy
chemical energy
exergonic
endergonic
entropy
anaerobic organisms
aerobic organisms
oxidation reduction
oxidation
NAD
electron transfer
cytochromes
light and dark reactions
photons
chlorophyll

photosynthetic bacteria
fluorescence
phycobilins
anthocyanins
B-carotene
chloroplasts
cytoplasmic streaming
white light
electromagnetic
   spectrum
hill reaction
photophosphorylation
photosynthetic unit
photosystem I (PSI)
photosystem II (PSII)

Z-scheme
cyclic
noncyclic
   photophosphorylation
calvin cycle
paper chromatography
aspartate
C4 pathway
photorespiration
calorie
compensation point
monoculture
krebs cycle
flavin
facultative anaerobes

🔅 Laws of Thermodynamics

The first law of thermodynamics states: *Energy can neither be created nor destroyed, but it can be converted from one form to another.* It is possible; for example, to convert the energy locked in a hydrogen bomb to heat, light, short wave destruction radiation, sound, and other forms of energy. Within the atoms, a potential energy is held, capable of causing an incredible amount of "work" (and here the term *work* is used loosely because many of us prefer not to equate widespread destruction with work). Just the same, the potential energy of the atoms and molecules is put to work. The energy itself is not lost but is converted from one form into several.

Some forms of energy are used more efficiently in machines, but efficiency is nothing more than the percentage of energy that is converted into the human concept of work. This remainder is left in the original form, or converted into some type of energy that is not as desirable. Unlike water and nutrients, which cycle through ecosystems and are used over and over again, energy does not cycle; it changes, and eventually it ends up in a form that is of no tangible benefit to the system.

The second law of thermodynamics states: *When energy is converted, the potential energy of the final state will always be less than the potential energy of the initial state provided that no energy enters or leaves the system.* The law helps us to understand that water at the top of a waterfall has potentially more energy than water at the bottom of the waterfall, as shown in Figure 10-1.

Some of the energy is given up as the water spills over the face of the rocks. If a turbine happens to intercept the falling water, part of the potential energy may be turned into useful work. The system is never 100% efficient, and some of that energy is lost from the system as heat. Heat can be an acceptable form of energy; if it is concentrated, it too may be used to generate steam and turn the same turbine. In many cases, however, the heat is so diffuse that it may not be practical to turn it into work.

Consider all the electrical machinery used in factories. Engineers measure efficiency in terms of work produced per unit of electricity consumed. Electricity not converted to work eventually becomes heat, which is of no benefit to the factory. In fact, it means that the ambient or air temperature will increase, and the factories will have to be cooled to a greater extent than would be required if the machines were more efficient. The analogy holds in biological systems, although in warm-blooded animals and in plants the heat is dissipated rapidly and totally lost from the system. The energy was not destroyed but merely changed to a form that was of no use to the system. What eventually happens to the heat? It warms the earth and is eventually lost into space. You have probably never thought of a tree or rose bushes as warm, but in fact these plants produce body heat just as humans do, although at a much-reduced rate. Lack of insulation in plants allows the heat to escape. It is possible to measure the interior temperature of

large trees; it is several degrees higher than outside temperatures. But small plants seldom show a differential. In addition, some plant parts may produce a greater amount of heat during reproduction, but the heat is quickly lost to the atmosphere.

Chemical reactions that occur spontaneously are called **exergonic** reactions (energy yielding), and the potential energy at the end of the reaction is less than the potential energy at the beginning of the reaction. An **endergonic** reaction is one in which energy must be added to the system for the reaction to occur. Endergonic reactions can never occur spontaneously.

One concept of energy occurs in its organization of orderliness. Although difficult to define, **entropy** describes the degree of orderliness. As the energy is converted and work proceeds, the matter becomes less structured and less orderly, and the randomness becomes greater.

The heat energy of the sun remains as heat, warming the earth and driving the hydrological cycle and thus allowing life to exist.

light energy provides the potential energy for conversion in biological systems. Green plants begin the conversion process.

In biochemical reactions, as in factories and power plants, no energy transfer is 100% efficient. Energy transfer in organisms that cannot use atmospheric oxygen, the **anaerobic organisms**, is relatively inefficient; organisms that do use atmospheric oxygen, **aerobic organisms**, transfer that energy with far greater efficiency.

Oxidation Reduction

Energy changes occur in chemical systems because of **oxidation reduction**. Oxidation may be defined as a *loss of electrons*, and reduction may be defined as a gain of electrons. In most biological systems, the loss of electrons and reduction may be defined as *a gain of electrons*, in most biological systems the transfer of electrons is also accompanied by the transfer of protons. Thus, when energy transfer is described for biochemical systems, the changes in structure are usually written for the loss of gain of protons, rather than the electrons, although the electron transfer is implied. In this text and in others you will often see an electron-transfer molecule indicated as being oxidized (no special designation) or reduced (+2 H). For example, the primary electron acceptor molecule for photosynthesis is nicotinamide adenine dinucleotide phosphate (NADP). Whenever you see NADPH2, the designation is for the reduced form. The NADPH has gained electrons to become reduced; however, this gain is indicated by the addition of hydrogen, or protons. In other words, the addition of H^+, a proton, automatically means a reduction. Any oxidation of one molecule must be accompanied by a reduction. One cannot occur without the other. Energy is transferred from one molecule to another, demonstrating the law of conservation of energy. Each step of photosynthesis that involves electron transfer must have a reduction accompanied by an oxidation, yet the net result produces an overall reduction of CO_2 to produce a sugar. By the same token, respiration involves oxidation reduction that also occurs in distinct steps, but the overall result is to produce a net **oxidation** of sugar by losing electrons and protons. The primary electron acceptor in respiration is not NADP but **NAD**; the terminal phosphate is missing.

In both photosynthesis and respiration, **electron transfer** occurs in small steps with only slight changes in energy levels (see Figure 10-2). This is made possible by an integrated chain of electron acceptors; the most important of which are a group of proteins called **cytochromes**. The cytochromes all have a similar structure, with an atom of iron at the center of the molecule. Since iron (Fe) can exist as F^{++} or Fe^{+++}, it is possible to change from one cytochrome to another simply by transferring in electrons on the iron atom. Cytochrome structure is very similar to the central proton of the hemoglobin molecule, which transports oxygen in red blood cells. The electrons are "carried" by these protein transfer agents in much the same way that the oxygen is carried by hemoglobin.

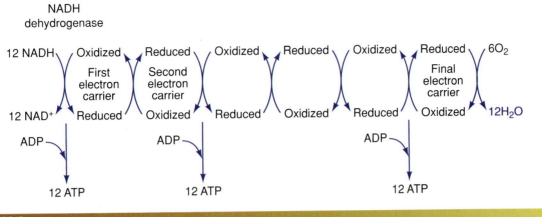

Photosynthesis

Little attention seems to have been given to photosynthesis and gases in the atmosphere prior to 1774, when the English chemist Joseph Priestly discovered oxygen and conducted experiments to show that animals cannot live without it. Later, he was able to show that if one places a green plant in a closed container with an animal, the animal will live. He concluded that green plants growing in light were able to replenish the "poison air" (CO_2) by producing a substance called oxygen. A few years later, a Dutch physician discovered that only green tissue produced oxygen, and then only in the light. When placed in the dark, plants consume oxygen just as animals do. The pieces were all finally put together in 1840 when Nicolas-Théodore de Saussure presented the formula for photosynthesis:

$$CO_2 + H_2O \xrightarrow{\text{Light}} CH_2 + O_2$$

This equation states that light energy is used to enable carbon dioxide and water to combine to form a sugar, giving off oxygen as a byproduct. The light energy is thus stored as chemical energy in the sugar. Although the concept is simple, the exact mechanism by which photosynthesis occurs is complicated and the process involves many steps.

For convenience and clarification, the process is divided into **light and dark reactions**. The light reactions must have light energy to initiate them, but the dark reactions will go to completion without light. The dark reactions can take place in light or dark conditions but do not require darkness. Possibly they should be named the "light independent reactions."

The light energy arrives at the earth's surface having traveled the entire 150 million km in a matter of 8 minutes. The light energy is contained in discrete bundles called **photons**. Some photons contain more energy than others, and the shorter the wavelength, the greater the energy per photon. Thus, a photon of blue light (about 450 nm) has a great deal more energy than a photon of red light (about 650 nm). Wavelengths shorter than the visible spectrum, including ultraviolet light and X-rays, contain considerably more energy than do the photons of the visible spectrum.

Pigments

Energy-capturing molecules that are specific for certain wavelengths are called pigments (see Figure 10-3), although biological systems have only a few of these, plants do contain some significant radiation-capturing pigments.

Figure 10-3

The absorption spectrum for chlorophyll a, chlorophyll b, and carotenoids. Note that the chlorophyll absorbs in both the red and blue portions of the spectrum, whereas the carotenoids absorb only in the blue and blue-green portions.

Courtesy of National Weather Service.

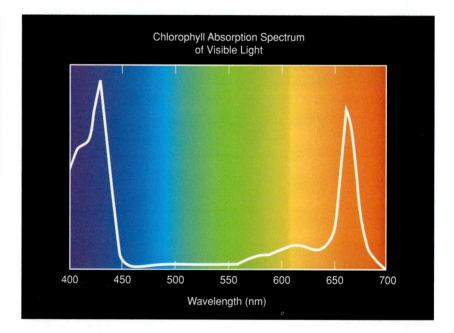

Chlorophyll Absorption Spectrum of Visible Light

400   450   500   550   600   650   700

Wavelength (nm)

Chlorophyll

The basic green color of plants comes from the pigment **chlorophyll**. Chlorophyll's *a*, *b*, and *c* are slight chemical variations of a basic structure (according to species), but they perform essentially the same function. Higher plants have varying ratios of chlorophyll a and chlorophyll b, with chlorophyll a always present in the higher concentration. Chlorophyll a is the primary pigment, and it occurs in all photosynthetic organisms except the **photosynthetic bacteria**. These "primitive" prokaryotes have a special kind of chlorophyll called bacteria chlorophyll.

Both chlorophyll a and chlorophyll b absorb in the red and blue portions of the visible spectrum. Yellow-green is not absorbed but is reflected, and the human eye perceives it as green. Not all animals see the same colors. Dogs are relatively color-blind, honeybees see only in the short wavelength of the visible spectrum and into the ultraviolet, and butterflies see only in the red or far-red portion. Such specificity has allowed coevolution of certain plants and animals.

The structure of the chlorophyll molecule allows for a resonance, or vibration of electrons, from one atom to another within the molecules. Electron movement, or excitation, is brought about by a photon of the correct wavelength (red or blue) striking the molecule. When the electron does become excited, the light energy becomes available to be (1) trapped and converted to chemical energy in the process of photosynthesis, (2) emitted at a longer wavelength with loss of energy as **fluorescence**, or (3) lost as heat. The first law of thermodynamics applies; light energy is converted into chemical energy in a complicated series of reactions called photosynthesis. In the intact leaf, fluorescence and heat loss are relatively unimportant. However, if chlorophyll is extracted by grinding leaves in an electric blender with acetone, and the extract is poured into a test tube and a bright incandescent light directed at it, the extract will appear dark red rather than green. This demonstration of fluorescence happens because the photosynthetic apparatus has been disrupted. The incandescent light continues to excite the chlorophyll molecule, but there is no electron acceptor to trap the energy. Therefore, the light energy is emitted as light, but always at a longer and less energetic wavelength than originally absorbed. In the case of chlorophyll, the wavelength emission is in the red portion of the spectrum. For other pigment molecules, fluorescence may occur as green, yellow, blue, purple, pink, or other colors. Fluorescent paints that glow under black light or ultraviolet light function in exactly the same manner. Anytime fluorescence occurs, the additional energy originally held by the photon at that time it is stuck is lost as heat. In other words, part of the energy is retained in a photon with a lower energy level, and part of it is converted to heat. Again, note that energy has not been created or destroyed, but merely converted to a different form.

Other Photosynthetic Pigments

In addition to chlorophyll, two accessory pigment systems perform in photosynthesis. These groups of molecules are the carotenoids and **phycobilins**. Carotenoids tend to be red, orange, and yellow, whereas phycobilins are red and blue. Both pigment systems function in electron captures and transfer. The phycobilins are restricted to cyanobacteria and the red algae. All these pigments can be excited by a photon of light of appropriate wavelength and then transfer that energy to a chlorophyll a molecule capable of continuing with the photosynthesis process. In other words, carotenoids and phycobilins act as antenna pigments, whereby they gather in the energy over a relatively large area and deliver the energy to a central molecule of chlorophyll a. Carotenoids also help protect the chlorophyll molecule from photo oxidation or bleaching. When shade-adapted leaves are placed in bright sunlight, they often bleach because the chlorophyll loses structure and there are few accessory pigments. Accessory pigment, then serves two important functions: (1) a transfer of light energy to a chlorophyll a molecule and (2) protection of chlorophyll from photo oxidation.

The accessory pigments are less subject to photo oxidation than chlorophyll and tend to last longer in the fall. When days become shorter and nights become cooler, orange and yellow pigments (which had been masked by the massive numbers of chlorophyll molecules) are displayed in the traditional fall colors when chlorophyll pigments eventually bleach. Certain other pigments, including **anthocyanins** of red and purple, add the spectrum of color that floods the countryside during autumn in most temperate parts of the world. Carotenes serve several other important functions in nature unrelated to photosynthesis. **B-carotene**, for example, is the precursor for vitamin A and is exceedingly important in animal diets.

One of the standard measures of feed quality for livestock and wildlife is to measure the concentration of B-carotene. Carotene also gives egg yolks their characteristic yellow and affects the pink pigmentation of flamingo feathers. The standard diet of flamingos includes algae-rich carotene. When dry food is used to feed flamingos in captivity, and when the carotene content is low, the pink pigmentation fails to develop, and feathers are almost white.

The Chloroplast

In prokaryotic cells that carry on photosynthesis (photosynthetic bacteria and cyanobacteria) there are no **chloroplasts**. The chlorophyll molecules are attached to membranes in the cytoplasm, and the entire process takes place there. In all eukaryotic photosynthesis cells, however, the processes are neatly packaged into organelles, and the chloroplast functions in the conversion of light energy into chemical energy. The chlorophyll molecules are attached to membranes of the chloroplast. The living chloroplast is readily visible through a light microscope because of the green chlorophyll. The nucleus, which is larger, is rather transparent and visible only with special lighting and optical techniques. Chloroplasts are saucer shaped, or flattened and elliptical, and bound by a double membrane. They are approximately 5μm in diameter, and they move freely in the cytoplasm. **Cytoplasmic streaming**, indicated by the movement of chloroplast, tends to speed up considerably as light intensity increases and slow as light intensity decreases. The number of chloroplasts in a cell varies from about 20 to more than 100.

A thin section of a chloroplast can be viewed through the electron microscope to reveal an outer membrane and an inner membrane. A three-dimensional network of stacked thylakoids makes up the grana. Strands that hold the entire system in place connect adjacent stacks of membranes. The thylakoids support the enzymes associated with the light reactions. The matrix between the grana is called the stroma, and the enzymes involved in the dark reactions are located there.

The Light Reactions

Light appears to saturate an area simply because the photons are so close together in time and space. In natural sunlight, photons of all energy levels (that is, all colors of the rainbow) strike the earth's surface at the same time and give the illusion of blended or relatively **white light**. Remember that the photosynthesis spectrum is only a tiny portion of the entire **electromagnetic spectrum**, and regardless of the energy level of each photon, only those in the red or blue wavelength are capable of exciting a chlorophyll molecule with energy levels correct for the photosynthetic process.

Whenever a red or blue photon strikes the chlorophyll molecule, it sets off a chain of reactions. Excitation causes an electron in the molecule to be raised to a higher energy level. The excitation is short lived, and the electron tends to give up the energy in a matter of milliseconds. It is important, then, for the chlorophyll molecule to be located near the proteins that will allow the energy to be captured and converted to chemical energy. This is accomplished by an electron transport system that passes the electron downhill as the energy is released. The electrons released by the chlorophyll molecule are replaced by those from a molecule of water. These are two basic reactions in photosynthesis: photolysis, or the so-called **Hill reaction**, and **photophosphorylation**.

Photolysis

Photolysis is an overall reaction to the energy gained from the excitation of the chlorophyll molecule that is used to split a molecule of water (*photo*, "light"; *lysis*, "to split"). This is one of the few instances in which water actually enters into a biochemical reaction:

$$H_2O \;\overset{\text{Light}}{\longrightarrow}\; 2H + \tfrac{1}{2}O_2$$

In this reaction, as in all chemical reactions, the beginning components on the left are called the substrates and the resulting components on the right are called products.

Oxygen is shown as half a molecule here because atmospheric oxygen is always O_2. In other words, this reaction must occur two times before a molecule of oxygen can be produced:

$$2H_2O \;\overset{\text{Light}}{\longrightarrow}\; 4H + O_2$$

Both products are exceedingly important in this case. The oxygen goes into the atmosphere as a gas which becomes the source of oxygen for all aerobic organisms. If one extrapolates backwards, oxygen can exist only if photosynthesis occurs. The ultimate conclusion is that aerobic organisms could not exist until photosynthesis occurred. Atmospheric oxygen has increased through geological history from none approximately 3.4 billion years ago to 21% of the air around us at the present time. This level of oxygen ensures that aerobic respiration can occur at a rate to produce sufficient energy compounds (primarily ATP) necessary for life.

The $2H^+$ produced in the splitting of water become the reducing source for driving the process of photosynthesis. The positively charged hydrogen atoms reduce NADP to $NADPH_2$. The electrons are used to replace those given up by the chlorophyll molecule. Thus, electrons "flow" from one acceptor to another, and the overall process is called electron transport.

Photophosphorylation

The substances that participates in biochemical reactions are called metabolites (a molecule involved in metabolism), and for many molecules to participate they must be phosphorylated. By examining a biochemical pathway, you can see that essentially all the intermediate metabolites are phosphorylated. It seems there is a great deal of phosphorus moving around in living cells. Where does it all come from, and how is the transfer made?

Recall that the "energy currency" for all living cells is ATP. These particular molecules function so effectively in their roles because the bonding of their (third) phosphate group creates a tremendous amount of stored energy—approximately 700 calories/mole. (A mole is the molecular weight, in grams, of any substance.) Few other biological molecules can store that much energy. ATP is like a super battery with exceptional storage capacity. In accordance with the law governing the conservation of energy, all of that stored in ATP, and essentially all usable energy, comes from the sun. It is stored directly in the ATP molecule by photophosphorylation—ATP formation using light energy mediated by the chlorophyll molecule. This is the second general light reaction of photosynthesis. The reaction proceeds as follows (Pi is an unattached, inorganic phosphate):

$$\text{ADP} + \text{Pi} \xrightarrow{\text{Light}} \text{ATP}$$

Since 7,000 calories/mole of energy are stored in the terminal phosphate group of this molecule, that much energy (actually more because of the inefficieny of their system) had to be transferred from light energy coming from the sun.

$$\text{Photolysis } H_2O \xrightarrow{\text{Light}} 2H + \tfrac{1}{2}O_2$$

$$\text{Photophosphorylation ADP} + \text{Pi} \xrightarrow{\text{Light}} \text{ADP}$$

The thylakoids of chlorophyll contain layers of proteins and various pigments, each **photosynthetic unit** being made up of approximately 300 to 400 pigment molecules (chlorophyll and carotenoids in higher plants). Photons of light may strike any of these accessory pigments to begin the chain reaction that eventually culminates in sugar formation. Only a single molecule of chlorophyll a is necessary to complete the photosynthetic process. Thus, each photosynthetic system acts like a giant solar collector, and efficiency of the system is greatly enhanced.

Evidence indicates that there are two different kinds of photosystems: **photosystem I (PSI)** and **photosystem II (PSII)**. In PSI, the central chlorophyll a molecule absorbs most efficiently at red wavelength of 700 nm and is usually called P700 molecule. In PSII, the central chlorophyll a molecule absorbs most efficiently at a red wavelength of 680 nm; hence, the designation P680 nm. Even though both are chlorophyll a, the central molecules or reaction centers absorb at slightly different wavelengths. The diagrammatic representation of electron flow in these light reactions is often referred to as the **Z-scheme**, in which the events of PSII occur before those of PSI.

Light entering PSII is absorbed directly (by the reactive center) or indirectly (by the antenna molecules), and the excited electron is transferred to an unidentified electron acceptor molecule. At the same time, photolysis of water occurs in which H_2O is split, giving its electrons to the electron-deficit P680 molecule. In the process, oxygen is given off into the atmosphere, and the photons become the reducing source to be used later in the series of reactions.

The excited electrons are accepted by an unidentified oxidization acceptor molecule, which becomes reduced. They are passed downhill, releasing energy through a series of protein molecules that comprise an electron transport chain. Cytochromes make up a proton of this chain. The energy captured by ADP and Pi to form ATP through photophosphorylation. At that point, another photon causes a similar excitation of the P700 reaction center as PSI, and again the excitation energy is transferred to electrons and transports them through a similar electron transport scheme. This time the terminal acceptor is NADP and the electrons are delivered to that molecule. The protons to complete the reduction come from the molecule of H_2O that was split at the time of the initial PSII activation.

It is possible for PSI to work independently of PSII, causing electrons to be raised to a high energy state, received by the other acceptor, and then passed downhill by the electron transport scheme, forming ATP along the way through **cyclic** photophosphorylation, as shown in Figure 10-4. Some prokaryotes and presumable all primitive photosynthetic organisms formed ATP this way, although no H_2O is split, no O_2 evolved, and no NADP reduced. Eukaryotic organisms can perform both cyclic and noncyclic photophosphorylation. Only in **noncyclic photophosphorylation** does the production of $NADPH_2$ occur, allowing CO_2 to be reduced to a carbohydrate.

Dark Reactions

Even though photosynthesis is often considered a light-requiring process, the two processes just described are actually the only ones that require light. If, for example, one had the products of the light

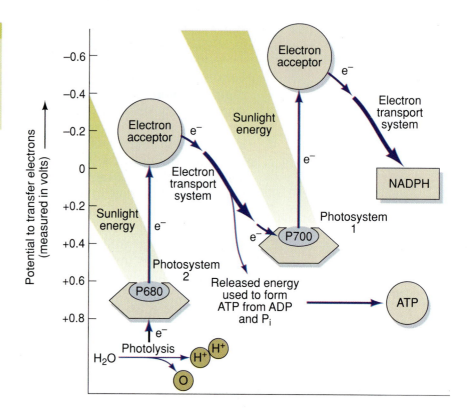

reaction, some chloroplast with enzymes, and some carbon dioxide in the correct reaction mixture, the remaining aspects of photosynthesis could go on in a test tube in a dark room. In truth, the so-called dark reaction usually takes place in the presence of light, but light is not required.

C2 Metabolism: The Calvin Cycle

Melvin Calvin and his coworkers at the University of California, after the discovery of radioisotopes, grew the green algae *Chlorella* in a complete nutrient medium in a large round flask and provided light from all directions. The efficiency of the sugar production was quite good. The flask was equipped with a port near the top where radioactive carbon dioxide ($14C_2$) could be injected with a hypodermic needle. At the bottom, a stopcock allowed the contents of the flask to be withdrawn rapidly and dropped into boiling methyl alcohol, where the cells would be killed instantly. The actual metabolic procedure, as shown in Figure 10-5, has been named the **Calvin cycle**. Then sugars were extracted and spotted onto large sheets of filter paper for a testing process called **paper chromatography**. Certain solvent systems caused the paper to be used as a wick; the chemical spot moved along with the solvent at a rate dependent on the physical and chemical properties of the substance in the spot. Some moved rapidly, others moved slowly, and by the time the solvent front reached the top of the paper, the chemicals had moved to various positions. The paper was dried, turned 90 degrees, and subjected to another solvent system. Further separation occurred, and many spots appeared when the paper was sprayed with a suitable stain. "These separate spots were then compared with known standards for identification." Melvin Calvin, 1950. The same procedure forms a basis for many

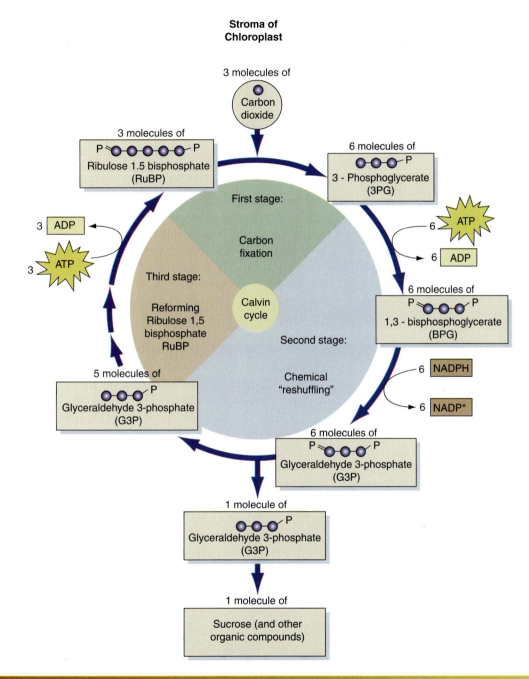

Stroma of Chloroplast

3 molecules of

Carbon dioxide

3 molecules of
P ⌕O-O-O-O-O P
Ribulose 1.5 bisphosphate
(RuBP)

6 molecules of
O-O-O P
3 - Phosphoglycerate
(3PG)

First stage:

Carbon fixation

6 ← ATP

6 → ADP

3 ← ADP

3 ← ATP

Third stage:

Reforming Ribulose 1,5 bisphosphate RuBP

Calvin cycle

6 molecules of
P O-O-O P
1,3 - bisphosphoglycerate
(BPG)

6 NADPH

Second stage:

Chemical "reshuffling"

6 → NADP⁺

5 molecules of
O-O-O P
Glyceraldehyde 3-phosphate
(G3P)

6 molecules of
P ⌕O-O-O P
Glyceraldehyde 3-phosphate
(G3P)

1 molecule of
O-O-O P
Glyceraldehyde 3-phosphate
(G3P)

1 molecule of

Sucrose (and other organic compounds)

Figure 10-5

The Calvin cycle explains how CO_2 is incorporated into ribulose 1,5-bisphosphate to form a 6-carbon unstable intermediate. The first stable product is 3-phosphoglyceric acid (PGA). By incorporating six molecules of CO_2, the cycle generates a 6-carbon sugar while regenerating the RuBP.

sophisticated biochemical techniques used extensively in laboratory research. However, with the advent of the latest biotechnological procedure, other methods have also been developed.

When Calvin injected $14CO_2$ as a source of carbon and allowed the cells to carry on photosynthesis for a few minutes, he found that many spots had been labeled (becomes radioactive) with 14C. This detection of radioactivity areas was made possible by placing an X-ray film on the areas where the radioactivity appeared as black spots on the negative. These spots were gradually reducing the time of exposure to radioactivity, Calvin was able to work backward and gradually develop a pattern of labeling based on the rate of incorporation of various compounds.

This was not an easy task, and it finally became clear that the first labeled compound was the 3-carbon compound 3-phosphoglyceric acid (PGA). Since CO_2 has only one carbon atom, and three carbons appear in the product; the compound of incorporation would appear to be a carbon compound. However, after many months of searching no 2-carbon compound was found, and the molecule eventually turned out to be a 5-carbon compound ribulose 1,5- bisphosphate (RuBP) When CO_2 is combined with the 5-carbon molecule, it produces a very unstable 6-carbon intermediate, which immediately splits into two 3-carbon molecules, detected as PGA. Calvin went on to elucidate the entire pathway, a series of steps that involved a rearrangement of 3-, 4-, 5-, 6-, and 7-carbon sugars to produce a final product hexose (6 carbon sugar) while regenerating the RuBP.

As CO_2 enters the leaf, it is absorbed on the wet surfaces of the mesophyll cells. The CO_2 crosses the plasma membrane, enters the cytoplasm, and crosses the chloroplast membrane, where it combines with the RuBP. The remainder of the cycle is merely a breaking down and restructuring of sugars so that hexose sugar is produced as a by-product of the cycle, and RUBP is regenerated. The $NADDPH_2$ produced in the light reaction provides the reducing source (H_2) needed for this process.

The enzyme responsible for causing CO_2 to condense with RuBP is RuBP carboxylase. Not all enzymes are equally efficient in catalyzing their specific reactions, and this one is notoriously inefficient. RuBP carboxylase does not attach to CO_2 very well and the only reason that carbon metabolism works as well in nature as it does is because large quantities of enzymes are synthesized by the plant. In some species, this so-called fraction 1 protein makes up more than 50% of the entire leaf protein. As world food shortages become greater, more attention is being devoted to the use of leaf proteins. In most cases the bulk of the leaf protein will be the enzyme RuBP carboxylase.

The C4 Photosynthetic Pathway

The Calvin cycle, which has also come to be known as the C3 pathway because the first detectable product is a 3-carbon compound,

NOBEL PRIZE FOR PHOTOSYNTHESIS

Melvin Calvin (April 8, 1911–January 8, 1997) was born in Saint Paul, Minnesota. He was the son of Jewish immigrants from Russia. His father was Lithuanian and his mother Georgian. Calvin earned his bachelor of science from the Michigan College of Mining and Technology (now known as Michigan Tech University) in 1931 and his PhD in chemistry from the University of Minnesota in 1935.

Calvin joined the faculty at the University of California, Berkeley, in 1937 and was promoted to professor of chemistry in 1947. Little was known about photosynthesis until about 1950. An understanding of the way in which carbon dioxide is fixed into sugars came about primarily through the efforts of Melvin Calvin and his coworkers.

After the discovery of radioisotopes in conjunction with World War II defense efforts in the United States and elsewhere, certain isotopes became available for scientific purposes. The single isotope, which has proven to be most effective and useful in biological research, is carbon 14, a radioactive form of the carbon atom. The stable isotope and the one found in greatest abundance in nature is 12C, but 14C is most useful in tracing the pathways of the carbon atom in metabolism. This carbon isotope emits beta rays and is relatively safe in experimentation.

Using the carbon-14 isotope as a tracer, Calvin and his team mapped the complete route that carbon travels through a plant during photosynthesis, starting from its absorption as atmospheric carbon dioxide to its conversion into carbohydrates and other organic compounds. In doing so, the Calvin group showed that sunlight acts on the chlorophyll in a plant to fuel the manufacturing of organic compounds. He and Andrew Benson received the 1961 Nobel Prize in Chemistry. The process has come to be known as the C3 pathway or the Calvin cycle.

In his final years of active research, he studied the use of oil-producing plants as renewable sources of energy. He also spent many years testing the chemical evolution of life and wrote a book on the subject that was published in 1969. Calvin also researched organic geochemistry, chemical carcinogenesis, and analysis of moon rocks.

is not the only means by which green plants fix carbon or incorporate it into sugars. In the early 1960s, workers at the Hawaiian Sugar Planters Association found that if $14CO_2$ was taken up by sugarcane leaves, the first detectable products were not 3-carbon molecules but the 4-carbon organic acid malate and the 4-carbon amino acid **aspartate**. This system of carbon fixation is now known as the **C4 pathway** of metabolism. It occurs in certain groups of plants and represents quite a departure from the C3 pathway (see Table 10-1). In fact, C4 metabolism still uses the Calvin cycle, but only in the later stages of fixation.

Table 10-1

Some Photosynthetic Characteristics of Carbon Fixation			
Characteristic	**C3**	**C4**	**CAM**
Leaf anatomy	No distinct bundle sheath of photosynthetic cells	Well-organized bundle rich in organelles	Usually no palisade cells, large vacuoles in mesophyll cells
Carboxylating enzyme	Ribulose bisphosphate carboxylase	PEP carboxylase, then ribulose bisphosphate carboxylase	Dark: PEP carboxylase Light: mainly ribulose bisphosphate carboxylase
Transpiration ratio (gmH$_2$O/gm dry weight increase)	450 to 950	250 to 350	50 to 55
Requirement for Na$^+$ as micronutrient	No	Yes	Unknown
CO$_2$ compensation point (ppm CO$_2$)	30 to 70	0 to 10	0 to 5 in dark
Optimum temperature for photosynthesis	15° to 25°C	30° to 40°C	35°C
Dry matter production (tons/hectare/year)	22 + 0.3	39 + 17*	Low; high

*Some C4 plants are less efficient than C3 plants, particularly at lower temperatures and lower light intensities.
Black 1973, Sailsbury and Ross 1978.

Anatomy of C4 Plants

The leaf anatomy of C4 plants is quite different from that of typical C3 plants. The vascular bundles of C4 plants are surrounded by a group of cells called bundle sheath cells. The density of chloroplast is much greater there than in the other spongy mesophyll cells. This is termed Kranz anatomy, and it occurs in almost all C4 species.

Fixation of C4 Plants

In C4 metabolism CO$_2$ enters through the stomata and is absorbed on the surface of the wet mesophyll cells. Since CO$_2$ is readily soluble in water, it forms carbonic acid and immediately dissociates into H$^+$ (which makes the cell sap more acidic) and HCO-3, the bicarbonate ion. The bicarbonate ion actually enters into carbon fixation. Oxygen is released and may exist in the stomata. The key to efficiency in C4 metabolism is the initial enzyme that captures the bicarbonate, PEP carboxylase. The enzyme connects carbon to PEP (phosphoenolpyruvate), an important intermediate in respiration. PEP carboxylase is found almost exclusively in the spongy mesophyll cells, so carbon is fixed there as malate and aspartate. These two 4-carbon compounds are then transferred in to the bundle sheath cells, where they are decarboxylated (carbon is dropped off). Then this carbon is fixed just as it is in the normal Calvin cycle metabolism.

The advantage here is that a very efficient carbon-capturing enzyme, PEP carboxylase, is present in mesophyll cells to collect great quantities of CO_2 and feed it into bundle sheath cells surrounding the vascular system. Once the sugars are finally made in the bundle sheath cells, they are readily loaded into the phloem, which is directly connected to the bundle sheath cells. The sugars can then be transported through the sieve tubes to parts of the plant where they are needed.

C4 Efficiency

Seemingly, C4 plants have evolved a cumbersome mechanism for absorbing CO_2 from the atmosphere. Why shouldn't they use the C3 method, which is shorter and simpler? Many experiments and measurements in nature, however, have shown that on the whole C4 plants perform better under conditions of high light intensity, high temperature, and low soil moisture than do C3 plants. Many weed species are C4 plants; they are more competitive than many crop plants for the reasons just mentioned. On the other hand, one must consider the environmental conditions for each particular situation before making broad statements about competitive ability. In the dark forest with adequate soil moisture, C4 plants are not competitive at all. Under these conditions, a typical C3 weedy plant species would have every advantage. Evidence suggests that C4 metabolism evolved in the dry topics, where plants are often subjected to high light intensity, high temperature, and drought.

The efficiency of C4 plants arises from their ability to concentrate carbon even under conditions of low CO_2 in the atmosphere. They also have the ability to maintain respiration rates at a uniformly low rate, in both light and darkness. In C3 plants, respiration rates increase in the light, a process termed **photorespiration**, and this increased use of photosynthesis decreases the efficiency of carbon fixation. The first contact of CO_2 with enzymes after entering the leaves is at the spongy parenchyma, where PEP carboxylase predominates. Once the initial fixation has occurred, it is possible to channel high concentrations of CO_2 into the bundle sheath adjacent to the phloem pipelines, which can readily load and transport the sugars. C4 metabolism is like a special kind of partitioning in space: The cells with the correct enzyme are in exactly the right place in the leaf.

The CAM Photosynthesis Pathway

A third type of carbon fixation has evolved for some plants, and the mechanism has many of the biochemical aspects of C4 metabolism. Instead of partitioning metabolites in space (between mesophyll and bundle cells), CAM plants use PEP carboxylase to fix the carbon into malate, which is then stored until the next day, when the carbon is released and refixed in normal Calvin cycle metabolism. This kind of metabolism is often thought of as being partitioned in time rather than in space. Location of the cells is not so important, but the day/night time interval is very important.

Originally discovered in the *Crassulaceae* and subsequently named Crassulaceae acid metabolism (CAM), this system is known to occur in many succulents and plants that grow in xeric (desert) environments. You might wonder why plants would absorb CO_2 at night, fix it into malate, the then refix it into Calvin cycle sugars the next day. The answer lies in the stomatal control of these plants. Unlike normal C3 and C4 plants, CAM plants open their stomata at night, when the relative humidity is higher and transpiration losses are lower. When sunlight reaches the plant, the stomata close, and new CO_2 cannot enter the plant. Although this slows the overall photosynthesis efficiency, it allows plants to live in more arid regions than would otherwise be possible. CAM plants manage to survive in the desert and in semiarid conditions, but the total biomass production is not as great as that of other plants. The best CAM plant with commercial importance is the pineapple.

Ecological Aspects of Carbon Fixation

The unit of energy used to quantify metabolic demands and consumption is the calorie. A **calorie** is the amount of heat (or heat equivalent) necessary to raise the temperature of 1 gram of water to 1°C. The same unit of energy is used to measure the energy value of foods and feed stocks for combustion.

The energy in any organic matter, such as a slice of bread, is determined by totally combusting the material in a bomb calorimeter (calorie meter) and measuring the calories of heat given off. The calorie used by nutritionists is actually 1 Kcal or 1,000 calories.

Each year the sun delivers 13×10^{23} calories to the surface of the earth; knowing how long the sun shines at a given location for all seasons of the year, it is possible to calculate the number of calories per square meter per unit time. That value is the upper limit of productivity, since it is not possible to get more energy out of that square meter of earth's surface than the sun put into it. Therefore, we often state that the upper limit on crop yield is limited by the amount of light striking that surface. In fact, the actual upper limits of productivity are far less than the theoretical limits because of the inefficiency of the system. About one-third of the sunlight is reflected back into space, and much is absorbed by the earth and converted to heat, which becomes the driving force for operating the water cycle in nature. Green plants are able to convert only a small percentage of the total light available. Recall that many of the photons are not the correct wavelength to be absorbed by the chlorophylls or the accessory pigments. On an annual basis, perhaps only 0.1% of the light energy that is capable of being converted to chemical energy is considered smaller. The opportunities are greatest where (1) the leaves are close together but not overlapping and (2) the leaves are exposed to light for the greatest part of the year. Such conditions exist more often in the tropics, and the theoretical productivity is greatest in those regions.

The Leaf Area Index

The leaf area index (LAI) is defined as the ratio of total leaf area to a unit of ground area. It might seem that if leaf area were exactly equal to ground area ($1m^2$ leaf area/$1m^2$ ground area = LAI of 1.0), maximum efficiency could be achieved. Actually, fully grown plants often have LAIs that exceed 1.0 and in fact can reach values of 10 to 20. These plants grow where the leaf canopy has many layers, and the bottom layer is very shady. The shade leaves have little opportunity for photosynthesis.

Do shaded leaves benefit the plant? Part of the answer depends on the type of metabolism for that particular species. If the plant carries on C4 metabolism, in which high light intensity is a major factor in photosynthesis efficiency, then a lower, shady leaf might be a liability or use more energy than it makes. On the other hand, the lower leaf of a C3 species in a dense forest with little light might be perfectly adapted to low light intensity and carry on photosynthesis at a lower but acceptable rate.

Respiration is constantly being balanced against photosynthesis in both nature and agriculture. The light intensity at which CO_2 being fixed via photosynthesis exactly equals the CO_2 being released by respiration is called the compensation point, and it has important ecological implications. A leaf just at the **compensation point** is contributing nothing to the buildup of metabolites or storage of material in a plane. It follows that photosynthesis must exceed respiration by a considerable amount if plants are to accumulate biomass. In some cases, senescent or shaded leaves carry on so little photosynthesis that the leaf falls below the compensation point and becomes a liability. When that happens, hormonal changes usually stimulate abscission, which helps the overall productivity budget for the plant.

The objective in agriculture is to provide as much productivity as possible. Farmers attempt to achieve a photosynthesis/respiration ratio of about 40:1. Productivity is not always measured in total biomass, since only the selected biomass is important to the farmer. With wheat, for example, selected biomass is the only reproductive structures; the roots, stem, leaves, and portions of the fruit are often discarded. With current emphasis on more comprehensive use of biomass, other parts of the plant have significance for an overall increase in productivity.

Farmers are always concerned about how patterns and density of plants maximize productivity. In temperate climates, much of the caloric value from the sun is lost because during certain seasons no leaves exist to carry on photosynthesis. In winter months and even in early spring, when the seeds first germinate or when the buds on the trees are just breaking, fields may be bare. The LAI is so low that efficiency is poor. Later, leaves begin to expand, and new leaves are produced on many new shoots. The LAI may reach a value of 1.0 fairly early in the season. Then rapidly reach a value of 6 or 8 by the time leaves have been produced. Such values are considered acceptable for most crops and would imply that the spacing of the plant is correct. If the plants are placed too close together, they compete for light, water and nutrients. If they are planted too far apart, much of the water and nutrient reservoir in the soil may not be used. Understanding crop ecology is important to the entire concept of productivity, and much is to be gained by knowing the nutritional and water requirements, pattern of rooting, type of carbon fixation and other factors that contribute to yield.

Net Primary Productivity

Just how much biomass of different types does vegetation produce? This is an important question both ecologically and agriculturally (see Table 10-2). Measurements have been made for various crop plants (**monoculture**) in various types of natural ecosystems, including aquatic ones. At one time it was thought that the marine biome could be producing as much as 80% of the world's oxygen and therefore carrying on 80% of the world's photosynthesis. Those estimates have been revised downward, and ecologists currently believe that the oceans are responsible for about 35% of the world's total photosynthesis. Most of that is produced by the single-cell phytoplankton that lives primarily in the surface waters. Depths below that level are too dark for effective photosynthesis by most organisms. The giant kelp varies on photosynthesis at a rapid rate, but their distribution is limited almost exclusively to coastlines. Algal beds and reefs are exceptionally productive, as are tropical rain forests, swamps, and marshes. Notice that monoculture of agriculture production is not particularly productive, and because water is limited, desert biomes are notoriously poor in biomass production. In many cases, including a number of catastrophic factors and human factors, ecosystems do not reach their potential production. There can be no question the human activity,

Table 10-2		
Maximum Photosynthesis Rates of Major Plant Types under Natural Conditions		
Type of plants	Example	Maximum photosynthesis (mg CO_2/dm^2/hr)
CAM Plants grow in arid regions and have stomata open at night and closed during the day to conserve water.	*Agave Americana* (century plant)	1 to 4
Tropical, subtropical, and Mediterranean evergreen trees and shrubs; temperate zone evergreen conifers	*Pinus sylvestris* (Scotch pine)	5 to 15
Temperate zone deciduous trees and shrubs	*Fagus syvatica* (European beech)	5 to 20
Temperate zone herbs and C3 pathway crop plants	*Glycine max* (soybean)	15 to 30
Tropical grasses, dicots, and sedges with C4 pathway	*Zea mays* (corn or maize)	35 to 70

primarily in the past 170 years, has accelerated the loss of vegetation and greatly reduced the productivity of both the terrestrial and aquatic ecosystems. In some grasslands of the central United States, overgrazing has led to the degradation of the landscape, elimination of grass species, and the invasion of weeds such as prickly pears, cactus, and broomweed. The "usable" productivity is almost devoid of grass species, and the substituted species grow at a reduced rate or replacement vegetation fails to establish, leaving a bare soil.

Productivity and how improper human decisions, pollution, changes in climate, and other factors affect it, is poorly understood. Many of the world's ecosystems are being and will continue to be upset or destroyed. Estimates place the loss of tropical rain forest at more than 20.00 km^3 per week, or more than 500,000 km^3 per year. Tropical rain forest currently produces about 22% of the world's net primary production (and oxygen). At the current rate of depletion, productivity loss on the world basis is about 5% every 3 years, and all tropical rain forests will be gone in approximately 35 years. Not only will the loss negatively affect oxygen and net primary productivity, but the tropical rain forest might also be a source for new crops and medicinal plants. Thus, future gains for human welfare may never be realized unless their destruction is halted.

Respiration

Respiration is the overall process by which the energy stored in carbohydrates is gradually released then transferred to ATP, the energy currency molecule. Unlike the burning of a bonfire, which results in combustion of the wood and the release of heat, the fuel of carbohydrates is "burned" at a slow and controlled rate so that the energy released can be partially captured and stored for later use. You will recall that glucose is the building block for starch and cellulose. Since cellulose is hard to digest, it makes an excellent structural material. However, digestion of starch will yield glucose. Glucose is also the chief transport carbohydrate in animals, but instead of storing it as starch, it is stored as a very similar macromolecule, glycogen. From this word is derived the term for the first phase of respiration—glycolysis. The process of glycolysis and aerobic respiration is identical in the plant and animal system.

Aerobic respiration proceeds in three major phases: (1) glycolysis, (2) the Krebs cycle, and (3) electron transport. Any aerobic organism utilizes sugar in this manner, usually combusting the sugar molecule completely to form CO_2 + H_2O. Note that these products of respiration are the substrates of photosynthesis. In other words, respiration is an approximate reversal of photosynthesis.

$$CH_2O + 2 \longrightarrow CO_2 + H_2O + \text{energy}$$

This time, instead of storing light energy as chemical energy, the energy level already stored in the sugar molecule is released to form ATP. Also, remember that all reductions must be accompanied by an oxidation; when as photosynthesis was a reduction process in which CO_2 was reduced by hydrogen produced in the photolysis of water, the burning of sugar while utilization oxygen is an overall oxidation process. All burning or combustion processes are oxidation. In the bonfire analogy, the energy released by the wood is slowly released and partially recaptured by ATP, although some is lost as heat. Again, the process is far from 100% efficient. Actually, the degree of efficiency varies according to a number of factors, which will be discussed later.

The Mitochondrion

The organelle containing the package of enzymes that function in respiration is the mitochondrion. Barely visible through the light microscope, mitochondria are rounded to rod shaped, 1 or 2 μm in diameter, and approximately the same size as a typical bacterial cell. The number of mitochondria varies in different tissues, depending on rates of respiration and need for ATP. Far more numerous than chloroplast, they may number a thousand or more per cell. Although the origins of both the chloroplast and mitochondrion are subjects of debate, they both have double membranes and both contain DNA, which is responsible for part of their replication.

The internal membrane of the mitochondrion is involved to provide a large surface area, much like a stack of shelves. These folds are called the cristae, and the total extent of the cristae is an indication of respiratory activity, although the number of mitochondria in a cell is probably a better indication. The matrix between the cristae holds the enzymes responsible for the Krebs cycle, and the surface of the cristae holds the enzymes responsible for electron transport. Thus, even within a single organelle, the positioning of enzymes allows for a partitioning of functions. Mitochondria occur in all eukaryotic cells, both plant and animal. All living cells respire constantly, but the rate of respiration depends on many genetic and environmental factors.

Glycolysis

The first distinct series of reactions related to respiration—Glycolysis—does not take place in the mitochondrion at all, but rather in the cytoplasm. The enzyme and reaction function equally well under aerobic and anaerobic conditions. Glycolysis, as shown in Figure 10-6, begins with glucose, which must first be phosphorylated by the transfer of ATP energy.

Glucose + ATP ⟶ Glucose phosphate + ATP

Once "energized" by this phosphorylation process, glucose 6-phosphate undergoes conversion by splitting into 3-carbon sugar and finally, after partial oxidation, ending up as pyruvic acid. Pyruvic acid (also called pyruvate) becomes a key intermediate, which can function as a substrate under both aerobic and anaerobic conditions.

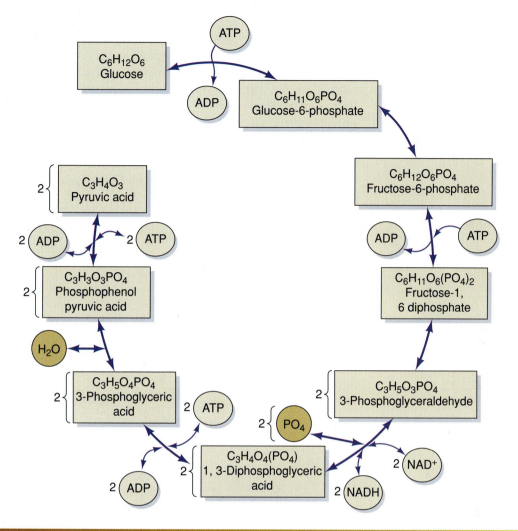

Figure 10-6

Glycolysis always begins with a molecule of phosphorylated glucose. In a series of reactions, this 6-carbon sugar is broken down into two molecules of 3-carbon pyruvic acid (pyruvate). One carbon is lost from each molecule of pyruvate, and the remaining 2-carbon fragment enters the Krebs cycle.

The Krebs Cycle

Pyruvate, which can readily penetrate the mitochondria membrane, enters into the **Krebs cycle**, as shown in Figure 10-7, the second phase of aerobic respiration named after the Noble Prize winner Sir Hans Krebs. This regenerating cycle is composed of a series of 4-, 5-, and 6- carbon organic acids. Prior to entering the cycle, the 3 carbon pyruvate is decarboxylated to give a 2-carbon acetate fragment. The fragment is activated by continuing with coenzyme A to become acetyl-coenzyme A. The carbon is lost as CO_2 and can be used as one measure of the rate of respiration. The acetate (2-carbon) fragment condenses with the 4-carbon oxaloacetic acid to form the

6-carbon citric acid, which subsequently loses another CO_2 to produce a 5-carbon organic acid, which subsequently loses another CO_2 to produce 4-carbon organic acid. After a series of conversion and oxidations (loss of electrons and protons) the 4-carbon oxaloacetic acid is regenerated to combine with another acetate fragment, and the cycle begins anew. Since one carbon is lost from each pyruvate just before entering the cycle, and since two carbons are lost in one turn of the cycle, then two complete turns of the cycle are necessary to completely lose all carbons from a single molecule of glucose (6-carbons):

1 glucose ⟶ 2 pyruvate ⟶ Krebs cycle ⟶ 4CO₂

Figure 10-7

In the Krebs cycle, acetyl CoA condenses with oxaloacetic acid to form citric acid. This 6-carbon organic acid loses one CO_2, and later a second CO_2 is lost. Electron transport occurs at several locations during the oxidation process. Substrate level phosphorylation produces only one ATP, the FAD-electron transport generates two ATP, and the NAD-electron transport generates three ATP. Courtesy of Rick Parker.

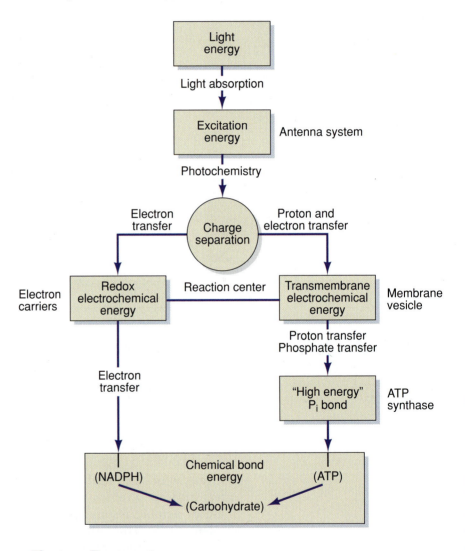

Electron Transport

Thus far, there has been no discussion about how oxidation is coupled to electron acceptors so that the energy is captured in an orderly manner, or about H_2O as a product of the respiratory process. The third and final phase of respiration is electron transport.

Electron transport, also sometimes called oxidative phosphorylation (as distinguished from photophosphorylation), is the process by which electrons are passed from the oxidation of Krebs cycle organic acid to the electron acceptor NAD and subsequently to **flavin** mononucleotide (FMN), coenzyme Q, cytochrome b, cytochrome c, cytochrome a, and cytochrome a_3 In this chain reaction, energy is transferred from the electron transport chain during the coupling of P_i, ADP to form ATP.

$$ADP + Pi \longrightarrow energy \longrightarrow ATP$$

This is precisely the same reaction described as one of the light reactions of photosynthesis. In this case the energy is provided by the oxidation of sugar rather than by light energy.

If a pair of electrons is passed along the entire length of the electron transport chain, ATP is made in three separate places. Finally after the electrons have lost most of their energy, they are transferred to molecular oxygen to produce H_2O.

$$2\bar{e} + 2H^+ + \tfrac{1}{2}O_2 \longrightarrow H_2O$$

Electron transport chains contain a group of cytochromes, the same iron-containing enzyme active in photosynthesis. These sites of electron transport exist anywhere electrons and protons become available during the oxidation process. Such opportunities occur at three sites of the Krebs cycle, enter into the electron transport chain at a later position, and make only two molecules of ATP. Finally, one molecule of ATP is made within the cycle itself, and this method of making ATP is called substrate level phosphorylation. The consequence of electron transport then are to synthesis ATP and deliver electrons and protons to oxygen required by aerobic organisms. If it were not for the need of oxygen to combine with the terminal protons and electrons in the electron transport system, there would be no reason for us to breathe and distribute oxygen to all cells.

Fermentation

Many microorganisms are able to carry on respiration without oxygen for their entire life. Such anaerobic organisms contain a special group of enzymes that permit a partial oxidation of sugars to produce ethanol. This process that occurs in animals is called anaerobic respiration. In anaerobic respiration, the end product is lactic acid.

Even in aerobic systems such as our own, the lack of oxygenation in a specific tissue can lead to the production of lactic acid. When humans exercise excessively without having gradually built up muscle tone, this results in muscle soreness from lactic acid accumulation in those tissues. The red blood cells cannot transport oxygen needed in electron transport.

In fermentation, the natural process used in making beer and wine, alcohol is produced from many carbohydrate sources. In any case, complex carbohydrates must be converted to glucose as the beginning substrate, just as in aerobic respiration. The process of glycolysis is exactly the same (recall that enzymes of glycolysis perform equally well with or without oxygen). Once pyruvate has been produced, however, the subsequent steps are unique, depending on whether conditions are aerobic or anaerobic. If no oxygen is present, pyruvate is decarboxylated, and the acetate fragment is oxidized directly to ethanol. This entire process goes on in the cytoplasm and the mitochondria are not involved. Some organisms called **facultative anaerobes** obviously have a great adaptive advantage. Some of these produce toxins and are exceedingly difficult to control.

Efficiency of Respiration

The total energy contained in a mole (1 gram molecular weigh) of glucose is 686,000 calories, 189 gm. When burned completely in a bomb calorimeter, all this energy would be released and the end products, as incomplete aerobic respiration, would be CO_2 and H_2O. In anaerobic respiration, a fairly large portion of that energy is captured and "repackaged" as ATP. It is thus possible to calculate the efficiency of aerobic and anaerobic respiration by knowing the number of moles of ATP that are in the process.

During glycolysis, two molecules of ATP are consumed in the initial stages of oxidation prior to the formation of two 3-carbon molecules, which undergo oxidation giving up electrons to NAD. This particular electron transport system is not complete, and each molecule of NAD releases only enough energy to make 2 ATP (a total of 4). Subsequently, each of the two 3-carbon molecules undergoes substrate-level phosphorylation on two different occasions. This gives an additional 4 ATP. At the end of glycolysis, pyruvate is formed, and the total ATP yield is –2 and +8 for a total of 6 ATP.

In the oxidation and decarboxylation of pyruvate, an additional pair of electrons is released to NAD, providing 3 ATP for each of the molecules, a total of 6 ATP. Subsequently, the Krebs cycle releases electrons and protons to NAD at three locations in the cycle, each forming 3 ATP, for a total of 9 ATP. In addition, an oxidation delivers the substrate level and 2 ATP as formed by an oxidation that delivers electrons to FAD (FAD is another electron acceptor that allows for only 2 ATPs to be made). Since the Krebs cycle must go around twice for complete oxidation of the glucose, all these values must be doubled for a total of 24 ATPs. The balance sheet looks like the one in the above diagram.

If each high-energy bond in ATP stores 700 calories/mole, then $36 \times 7,000 = 252,000$ calories; the remainder of the energy is converted to heat. The efficiency is thus 252,000/686,000, or almost 37%. Although seeming low, this is actually a very good efficiency of energy conversion. In an anaerobic organism, the efficiency of conversion is much poorer. In typical lactic fermentation, only 2 ATP molecules are produced, and the efficiency is 2%. In alcoholic fermentation, the energy yield is only slightly higher. It thus appears that primitive anaerobic organisms were terribly inefficient in preserving the energy trapped by organic molecules. Only after the advent of aerobic respiration did the process reach a respectable level of efficiency.

Substrate for Respiration

Once organic molecules have been formed, they function in biochemical reactions, as precursors for other metabolites, and in storage. Although one biochemical pathway may be involved in the synthesis of a compound, another may be involved in its breakdown. Large storage molecules are usually insoluble and become parts of the permanent plant structure. Others, such as starch, are stored as insoluble macromolecules into their smaller component. Each group of molecules has its own set of digestive enzymes.

starch —— amylases ——▶ glucose

protein —— proteases ——▶ amino acids

lipids —— lipases ——▶ fatty acid

Under most conditions, a plant or animal uses a reserve organic molecule for respiratory energy only after it is first converted to glucose. In the starch this is an easy one-stop conversion. But if the product is a lipid, lipases must break the lipid into fatty acid, then oxidation of the fatty acid, and then oxidation of the fatty acid produces 2-carbon acetate fragments that become available to the Krebs cycle.

Implications of Metabolism

Organisms that carry on photosynthesis produce for themselves and for the heterotrophs all the beginning organic molecules of life. Starting with simple sugars, photosynthesis provides a "carbon skeleton" from which all organic molecules are synthesized. It is easy to become so caught up in the details of photosynthesis, and respiration that one forgets how they interact to perform all the feats of metabolism. Macromolecules are contently being made from smaller molecules, and in a different part of the cell the macromolecules may be broken down into smaller molecules so that the carbon, hydrogen, and oxygen atoms may be rearranged to form an entirely different organic molecule. Some metabolic intermediates play key roles in providing large carbon pools from which other molecules are made. The Krebs cycle, for example, provides a 4-, 5-, and 6-carbon skeleton for the synthesis of many different organic molecules. One need only perform an animation reaction to add NH_2 (obtained from nitrogen in the soil) to one of these organic acids to form an amino acid. Purines and pyrimidines, lipids, hormones, sterols, pigments, alkaloids, starch, cellulose, proteins, nucleic acids, and many other compounds come directly from this backbone of photosynthesis and respiration. A more detailed discussion of secondary metabolism was presented in Chapter 6.

Catalyzed by enzymes, these biochemical reactions are carried out with speed and perfection. Compartmentalization is the key to packaging enzymes in specific organelles so that one metabolic process is not hindered or cancelled by another. You should begin to realize by now that the miracle of life must begin with the gene, for without the specificity of enzymes, biochemical reactions would not occur at a rate sufficient to sustain life. The living cell must surely be the most complicated system of operation that one could ever comprehend. The most sophisticated computer program should do well.

🌀 Summary

1. The laws of thermodynamics give important insight into the behavior of matter and its conversion into energy. The first law concerns the conservation of energy, and the second law allows one to make predictions about the probability that a reaction will occur. Energy is used throughout the living to perform work of various kinds.

2. Oxidation-reduction reactions in the biological system allow electrons and their accompanying protons to move from one molecule to another; every reduction must be accompanied by an oxidation. These reactions are all important in the energy conversion process of photosynthesis, in which CO_2 from the atmosphere is reduced to a carbohydrate through light energy.

3. A portion of that energy becomes stored in chemical compounds important to the cell and organism. Some of these stored sugars have been made; the atoms may be rearranged, substituted, deleted, or supplemented to form new chemical compounds important to the cell and organism. Some of these stored sugars are converted to glucose and pass through aerobic respiration to provide the energy for the cell and organism. Some of these stored sugars are converted to glucose and pass through aerobic respiration to provide the energy for the cell.

4. Photosynthesis consists of two basic light reactions—photolysis and photophosphorylation—followed by the dark reactions of the Calvin cycle. Some plants are especially adapted to high temperature, high light intensity, and water stress because the enzyme system for capturing CO_2 is different, and the processes are separated in time or space. Although such plants do still use the Calvin cycle of normal C3 plants, their additional enzymes cause them to be classified as C4 or CAM plants.

5. In its simplest form, respiration is the reversal of the chemical reactions for photosynthesis. Photosynthesis is an overall reduction process; respiration is an overall oxidation process.

6. The basic processes involved in aerobic respiration are glycolysis, the Krebs cycle, and electron transport. Glycolysis occurs in the cytoplasm under either aerobic or anaerobic conditions, must exist for further processing of the pyruvate in the mitochondrion. If oxygen is present, pyruvate is shuttled through the Krebs cycle, and the end products are CO_2 and H_2O. Much of the energy is stored as ATP. If oxygen is not present, anaerobic reactions lead to the formation of lactic acid in animals and ethyl alcohol in plants and microorganisms.

Something to Think About

1. What laws give an insight into the behavior of matter and its conversion into energy?

2. Which law concerns the conversion of energy? What does the second law allow?

3. Explain oxidation reduction in biological systems.

4. How does photosynthesis figure into this equation?

5. The energy is stored as what?

6. Once the sugars are made, what happens to the atoms?

7. What are the two basic light reactions involved with photosynthesis called?

8. Explain the Calvin cycle.

9. Discuss the difference in C4 or CAM plants.

10. Describe photosynthesis and respiration.

11. List the basic processes involved in aerobic respiration.

12. Where does glycolysis occur?

13. What happens when oxygen is present during glycolysis?

14. If oxygen is not present, what happens?

Suggested Readings

Baker, N, Ed. 1996. *Photosynthesis and the environment*. Netherlands: Kluwer Academic.

Black, C. C. 1973. Some Photosynthetic Characteristics of Carbon Fixation, Annual Review in Plant Physiology 24, 253–86.

Borror, D. J. 1960. *Dictionary of root words and combining forms*. Mountain View, CA: Mayfield.

Campbell, N. A., L. G. Mitchell, and J. B. Reece. 1999. *Biology: Concepts and connections* (3rd ed. and earlier editions). Menlo Park, CA: Benjamin/Cummings.

Lamber, H., and M. Ribas-Coubo. 2000. Plant Respiration Vol 18 Springer, Dordrecht, The Netherlands.

Lawlor, D. 1993. *Photosynthesis molecular, physiological and environmental processes*. Essex, UK: Longman Scientific, Technical.

Marchuk, W. N. 1992. *A life science lexicon*. Dubuque, IA: Wm. C. Brown.

Salisbury and Ross. 1978. Influence of light temperature, and carbon on the growth of plants, The Journal of Ecology 85, 3 359–372.

Speer, B. R. 1997. *Photosynthesis pigments*. Berkeley, CA: University of California, Berkeley Museum of Paleontology.

Internet

Internet sites represent a vast resource of information. The URLs for Web sites can change. Using one of the search engines on the Internet, such as Google, Yahoo!, Ask.com, or MSN Live Search, find more information by searching for these words or phrases: **law of thermodynamics, energy conversion, anaerobic respiration, aerobic respiration, oxidation reduction, photosynthesis, pigments, light reactions, ATP, ADP, Calvin cycle, Krebs cycle, stomata, glycolysis, fermentation, glucose, protein, lipids, and amino acids.**

CHAPTER 11

The Control of Growth and Development

Plant growth and development, and reproduction depend on the interaction of environmental factors with genetic makeup, or genotype. These two factors determine what an organism looks like, which is called the phenotype.

After completing this chapter, you should be able to:

- Explain the difference between phenotype and genotype
- Define growth and development as applied to plants
- Explain developmental biology
- Outline how growth and development are often limited to the environment
- Discuss how a plant's biological clock works
- Describe how reproductive cycles are attuned to biological clocks
- Recognize the primary receptor phytochrome (a protein-pigment)
- List and explain the mode of action of each of the hormones
- Describe plant response to different stimuli (tropisms), such as gravity, light, etc.
- Discuss why some plant movements may be temporary

Key Terms

phenotype	Pfr	indoleacetic acid (IAA)
genotype	red light	tryptophan
differentiation	far-red light	auxin threshold
diurnally	hormones	abscission layer
cold damage	gene activators	parthenocarpic fruits
water stress	stimulates	apical dominance
water quality	inhibitors	alpha-amylase
evapotranspiration	endogenous	aleurone layer
duration	exogenous	adenine
intensity	auxins	kinetin
phytochrome	abscisic acid	postharvest physiology
diurnal	ethylene	controlled atmosphere
seasonal cycles	plant growth regulators	epinasty
biological clocks	coleoptiles	growth retardants
photoperiodism	agar	callus

Key Terms, *continued*

florigen   thigmotropism   nyctinasty
geotropism   nastic movements   seismonasty
polarity   thermonasty

Principles of Growth and Development

Even if a farmer plants the best quality seed available, the crop will be a failure if sunshine, water, and nutrients are not available in the right amounts at the right time. A light frost or a heat wave could spell disaster. In nature, the **phenotype** responds to the biotic and abiotic environments. Soil factors, precipitation, and sunlight must be adequate for the plant to complete its growth cycle during the right season. Consider the consequences for a seed that happens to germinate at the beginning of winter. If that particular species were a warm-season annual, the seed would be killed by the frost. Another plant might survive the frost, lie under snow during winter, and complete its life cycle the following spring and summer. Why does one survive and the other does not?

The answer lies in the **genotype**. The winter annual must have genes that code for protection from the cold. The warm-season annual does not have such genes. Scientists have learned a great deal about the expression of specific genes that are modified by the environment. Even though experiments can be designed to look at specific gene-environment interactions, the phenotype represents the totality of hundreds of thousands of genes, which have been influenced by the environment. Phenotype must be viewed as a summation of gene-gene interaction and gene-environment interaction.

You may have wondered why two seedlings derived from the same parent plant could grow at different rates, flower at different times, and assume shapes that are quite different. There are two basic reasons: variability and environment. Sexual reproduction leads to variability; for example, you do not look exactly like your brother or sister. Even identical twins possess slight differences. The environment also affects genetic makeup. It is certainly possible to identify a single gene as one that gives rise to the synthesis of a particular protein, which may represent an enzyme responsible for catalyzing a particular biochemical reaction. However, the timing of that gene expression, the amount of enzyme produced, and many other factors depend on the environment. Some genes are expressed during periods of light and others in the dark; some genes are turned only when it rains; others are activated with the onset of autumn. All these environmental factors, acting in concert, bring about the gene expression, which is manifested as the phenotype.

The term's growth and development are often used imprecisely. Sometimes the two are even used interchangeably, which is certainly incorrect. Growth refers to an increase in size of volume of a cell, tissue, or organism. It happens because cell division is accompanied by an increase in cell size. One can even refer to growth in populations, that is, an increase in numbers. There are some problems with the definition. A germinating seed imbibes water and therefore increases in volume: Should this uptake of water alone be considered growth of the seed? Probably not, although the uptake of water is also accompanied by many metabolic changes but is not considered growth. Some involve digestion of storage macromolecules; others involve the synthesis of totally new molecules. The clarification, the definition for growth, is usually modified to state *an irreversible increase in size or volume*. Thus, permanent increases in dry matter are interpreted as growth.

Development is a summation of all the activities leading to change in a cell, tissue, and organism. The life cycle of an organism, for example, is a progressive change from a single fertilized egg, the zygote, through embryonic changes, and finally the changes that accompany full maturity of the organism. Growth, then, is a part of development. Development is an orderly sequence of events dictated by a precise set of genes turning on and off at every stage. Compare the life cycle of an organism with a piano keyboard (the chromosomes). Music is produced only when the correct keys (the genes) are struck in unison and the harmonic effect is pleasing to the ear. The correct sequence of genes has been turned on at the right time. Even a "good" chord must occur at the right time or it fails to integrate the musical composition.

Another important concept is **differentiation**—the chemical and physical changes associated with the development process. A meristematic cell can become a vessel element. In a short time, that cell changes its metabolism, shape, and wall thickness, and finally the protoplast dies. Thus, while developing, the cell certainly underwent growth and differentiation. The two parts of the system constitute the developmental process.

Developmental biology is the study of how organisms, their cells, and their tissues achieve a final predictable form and function. It embodies genetics physiology, biochemistry, biophysics, and many other disciplines, but gene regulation is the key in development.

Limitations of Growth and Development (Stresses)

Cells, tissues, organs, and organisms do achieve their predictable form and function as the result of the environment exerting its influence on the genome, the entire complement of genes for the species. Many factors combine to give the overall environmental influence, such as temperature, light, and moisture. Even if a particular gene has been turned on, it may not be expressed, or the expression may be highly modified if the environment fails to meet certain limits of tolerance. Those environmental factors outside the normal range of tolerance for a given environmental insult cause stresses from many sources that tend to limit growth and reproduction capacity. With all the new technology, many plant genomes have been mapped. Not only does this tell us about the environmental effects on a particular organism but helps to explain many of the unknowns associated with a genome.

The dry dormant seed lying in the soil has little resemblance to the organism that will spring from the embryo if the proper conditions are met. Seeds carry with them a "backpack" of stored molecules to nurture them in the early stages of germination. The early hours are precarious ones, and even the slightest desiccation at the wrong time may cause death.

Although a few kinds of seed will germinate on the soil surface (some actually require light and must lie on the surface), most seeds require the insulating and protective qualities of the warm, moist soil for germination. If stresses are not imposed and if the seed has imbibed water and oxygen, the machinery of germination is set in motion. Macromolecules of starch, proteins, and oils are broken down, enzymes already present begin to function and new ones are rapidly synthesized, and metabolism gears up quickly. Respiration rates, for example, increase dramatically within a matter of hours, and the embryo begins to undergo mitosis and cell enlargement rapidly. Some cells of the embryo grow faster than others; the radicale is almost always the first organ to emerge. Why the radical grows faster than the plumules is unclear. It is part of the entire development sequence that causes an organism to turn genes on and off at different times during the life cycle. Wheather the genes actually get turned on and off is, again, determined by stress. The mystery is being unraveled but it will still take time because the interactions are complex and the answers elusive. We know that growth and development occur only within certain physiological limits imposed by the environment.

Temperature

Fluctuations in temperature occur **diurnally** and seasonably, and in some parts of the world the fluctuations are more pronounced than in others. As a general rule, greater seasonal variations occur as one moves toward the poles. Temperature extremes may be modified by factors such as altitude, clouds, relative humidity, ocean currents, and high and low pressure areas. Temperature affects each biochemical reaction in a particular way.

Generally, as the temperature increases within physiological limits—the tolerance range for that species—the rate of the reaction doubles for each 10°C rise in temperature. These are obvious limits;

most biochemical reactions cease at approximately 43°C because the proteins are denatured and lose function at that point. This means that although the primary structure (amino acid sequences) may remain intact, the secondary, tertiary, and quaternary structures may collapse under several heat conditions. The enzyme then ceases all activity. Some enzymes can resist changes under intense heat. Consider for example the algae that manage to grow and reproduce in hot springs at or near the boiling point of water. Although there are exceptions, lower temperatures do not usually denature enzymes. **Cold damage** to plants may occur because the rate of the biochemical reactions is so slow that metabolic function is impossible. When temperatures fall below freezing, ice crystals may form and actually pierce membrane and cell walls. Although scientists do not fully understand how perennial plants are able to tolerate extremely low temperatures and survive, some believe the increasing viscosity of the cytoplasm of the cells to be a major factor in preventing ice crystal formation. Plant tissues also tend to dehydrate during the winter, and they rehydrate in the spring just as growth begins. Great gains in physiology and the biochemical makeup of the plant cells are presenting understanding of genes of various organisms and how they allow survival under extreme heat and cold while the genes of another organism fail to be regulated and the organism dies.

Water

Although it certainly is true that many modern cities import water through pipelines or canals for many miles, human migration and settlement has been historically tied to locals where fresh water is found. Although water has been discussed in various contexts in this book, it still deserves special attention as a factor in controlling plant growth (see Figure 11-1). As a medium for the support of all chemical reactions in living tissues, water dominates and controls the rates and combinations of reactions within living cells. **Water stress** in plant growth presents a major challenge for modern agriculture as growing human populations demand greater productivity. The stress problem is one of degree; each additional increment of water may increase productivity of plants a few percentage points, but the energy cost of water supply determines the amount of supplemental irrigation. Does one get enough in return to justify the additional cost?

Figure 11-1

Water stress in plant growth presents a major challenge for modern agriculture as growing human populations demand greater productivity. With the use of food crops for making fuel, water is even more important.
Courtesy of USDA Photo Gallery.

In natural ecosystems, the water stress problem is quite different. Here, survival is the key. Productivity may be exceedingly low in some desert ecosystems, but if the species can survive and manage to reproduce, even occasionally, then there is a good chance for relative success under those environmental conditions, provided that some outside force does not intervene. Plant geneticists, physiologists, and biotechnologists are particularly interested in understanding the mechanisms underlying drought tolerance. Drought tolerant plants are now being developed through genetic engineering by using genes from drought tolerant plants and putting them into less tolerant plants to create a drought tolerant plant.

Soils

One of the most serious stresses is imposed by salts, both naturally occurring and artificially applied; plant owners sometimes fertilize too heavily. Even though a little bit is good, a lot is not necessarily better. Fertilizer burn actually comes about because water is pulled out of the plant and back into the soil: The plant cannot take up water fast enough to compensate for losses from the leaf surface.

Salt damage is far more subtle than are the effects of overfertilizing. Damaging salts, often not a part of the essential nutrients, are found in most irrigation water. Such salts contribute to the overall problem of **water quality**. So-called fresh water used for irrigation contains some salts; even bottled drinking water contains some salts to keep the water from tasting "flat." Deionized and distilled water, on the other hand, have most of the salts removed. In nature, both underground water and surface water acquire dissolved salts, and these are transported along with the irrigation water. When water is applied to plants, it leaves the soil and leaf surface through the process of **evapotranspiration**, which takes up much of it. As the water molecules escape into the atmosphere, they leave the salts in the soil or in the leaves, where they accumulate unless removed by runoff or leaching below the root zone. Salt accumulation is generally not a problem in zones where adequate rainfall occurs; but in arid and semiarid regions, comprising approximately one-third of the entire land surface of the earth; salt accumulation is a major problem. Rainfall is not great enough to leach salts out of the root zone. These are the same regions to which water is transported for irrigation. The Imperial Alley of California is a good example of prime farmland plagued by salts. The ecological problem is enormous and worsening progressively.

Light

For any novice plant grower, the importance of light becomes apparent very quickly. One of the first questions likely to be asked of a nursery sales person is whether a new houseplant needs a sunny window or shade. Too many new "botanists" decide that the plant is not getting enough light and immediately transfer it from the dark corner to the hottest south window. The change in light might be too drastic for the plant, and bleached or dead leaves might be the consequence.

The process of photosynthesis involves the conversion of light energy into a stored chemical energy as sugars or other organic molecules. So light is absolutely essential for sustained plant growth, but the amount and kind can vary greatly. Plant physiologists are concerned with three factors of light **duration**, **intensity**, and quality.

The duration is important in photosynthesis because the conversion process is an accumulative one. For all practical purposes, a plant can accumulate twice as much sugar in 10 hours of sunlight as it can in 5 hours. But there are limits, and the accumulation process may slow during very long periods. The season and distance from the equator dictate the amount of natural sunlight to fall on any given spot, and it is quite predictable for any given date.

In a desirable climate at the equator, every plant has 12 hours of sunlight each day in which to accumulate photosynthesis products; season has little if any effect on the amount of light received. But as one proceeds toward the north and south poles, the effect of this photoperiod becomes greater and greater. At the North Pole during summer, daylight is almost constant, but during the winter there the days are quite short. Few plants are photosynthesizing at that time of the year. The seasons are reversed at the South Pole, and days are longest there when they are shortest at the North Pole.

As expected, this finely synchronized system has important implications for reproduction. Like animals, plants reproduce during advantageous seasons. To be in the process of flowering at the time of the first blizzard or onset of winter could result in death of a particular plant and even extinction for the species. Through the process of natural selection, species have ensured that the onset of reproduction will be triggered at an appropriate time of the year. The photoperiod system is a foolproof way to accomplish this. Its predictability is absolute. If the plant species continues to grow in the same place, the length of day is a certain, and thus ideal. Most plants have developed a mechanism to signal and change the developmental process from strictly vegetative growth to reproductive growth, and still allow time for the reproduction processes to be complete before the environmental conditions change. This process does not depend on the accumulation of a given amount of photosynthesis, storage capacity is certainly important in providing energy for the reproductive process. Instead, a protein pigment called **phytochrome** represents the trigger itself. More will be said about this remarkable chemical in the latter part of this chapter.

Intensity

Light intensity is important in the growth and development of all plants. Some plants are adapted to a shady, woodland environment, whereas others survive only under sunlight. Light intensity is mediated not only by the shading of other leaves, but also by local weather conditions, including clouds. Sugarcane in Hawaii, for example, is not as productive as agriculturists would like because too many clouds surround the island. As a C4 plant (see Chapter 10), sugarcane is most productive under conditions of full sunlight. Many houseplants fail to survive because of low light intensity, whereas others die from excessively bright locations within the room.

Quality

Light quality is important in plant growth and development primarily because of the absorption spectrum of the chlorophyll and carotenoid molecules. Visible light without red or blue components will undermine the plant growth and possibly inhibit flowering. Most sunshine contains a complete wavelength distribution, but artificial lights tend to be deficient in certain proteins of the visible light. Incandescent lamps produce light primarily in the red and far-red portions of the spectrum, whereas fluorescent lamps produce light primarily in the blue region. High-quality environment-control chambers where plant experiments are conducted are equipped with a mixture of both incandescent and fluorescent lamps to stimulate natural sunlight. With the correct mix of lamps, the quality can be duplicated rather well, but it is exceedingly difficult to get the intensity of natural sunlight. A special type of fluorescent lamp supplies more red light and therefore a more balanced spectrum than other fluorescent lamps.

Human Intervention

Each species has its own set of optimum conditions for growth and reproduction. By domesticating certain plants and animals, humans have learned about the environment best suited for those species. On the other hand, we really do not know the subtle requirements for optimum growth and reproduction for many species, even the domesticated ones. Each genome responds to the environment in a specific way. Change one or a few genes and the tolerance will change. One plant may be able to tolerate conditions near the south window very well (many of the succulent desert plants), but others do not have the right combination of genes to tolerate that level of heat and light intensity. Put the cactus in a dark corner and it will not thrive at all, yet a philodendron may grow very well there.

For some plants, one of the great stress factors is the human one. We may starve, overwater, underwater, overfertilize, overheat, freeze, desiccate, pollute, or crowd the plant to death. It certainly is not necessary for every person who ever looks at a plant to have a degree in agronomy, horticulture, or botany, but common sense concerning some of the principles of growth and development of plants can go a long way in making life greener, more pleasing, and rewarding.

ANIMAL HORMONES FROM PLANTS

Phytoestrogens are chemicals produced by plants that act like estrogens in animal cells and bodies. They are often trace substances in food. These chemicals mimic and supplement the action of the hormones, estrogen, from the animal or human body. Phytoestrogens are a comparatively recent discovery, and researchers are still exploring the nutritional role of these substances' metabolic functions.

Phytoestrogens were first discovered as causing "red clover" disease. This condition causes infertility in sheep that graze on red clover, a plant high in phytoestrogens. Why do plants produce phytoestrogens? They might be a form of plant defense, limiting predation of plant species by causing long-term infertility, increased risk of postnatal mortality, and other adverse reproductive effects on grazing herbivores. In the 1970s, phytoestrogens were first investigated as a treatment for postmenopausal symptoms.

Phytoestrogens mainly fall into the class of flavonoids: the coumestans, prenylated flavonoids, and isoflavones are three of the most potent in this class. The best-researched are isoflavones, which are commonly found in soybeans and red clover. Lignan has also been identified as a phytoestrogen, although it is not a flavonoid.

The estrogenic properties of these biochemicals have been shown to be due to their structural similarities to the hormone estradiol. Mycoestrogens produced by fungi also have similar structures and effects.

According to research, flaxseed contains the highest total phytoestrogen (lignan) content. Isoflavones are found in high concentration in soybeans and soybean products like tofu, whereas lignans are mainly found in flaxseed.

Today many over-the-counter products containing phytoestrogens are available to relieve menopausal symptoms.

Cycles

It is certainly not difficult to identify **diurnal** and **seasonal cycles** in our own lives. Work schedules, eating habits, body functions, and even psyches (many of us are cantankerous early in the morning!) are affected. A modern ailment is jet lag; rather uncomfortable physical emotional state brought about by passing through several time zones in a short period of time. Seasonal cycles, too, cause us to have changes in moods, and we eat and dress differently.

Plants too are carefully attuned to these **biological clocks**; the most important of which is the 24-hour clock. Perhaps you have never really thought about why our days are not 16 hours long, or maybe 32, or even 50. Would it make any difference to you if we had 8 hours of darkness year-round? Because of the physical design of our solar system, the earth rotates on its axis once every 24 hours to give us alternating exposure to the sun; at the same time each year, our planet follows an elliptical orbit around the sun. Because of the shape of the orbit, together with the tilt of the earth on its own axis, seasons are dictated for each precise location on the earth. Some other planet in some other solar system might have an entirely different biological clock. Ours is unique to the planet, and living organisms have systems for responding to the days and seasons.

The system certainly is not perfectly predicable. We cannot foretell what the exact temperature and relatively humidity will be on any particular day of the year from now, but we have learned to expect some limits. The plants and animals that thrive at any particular locality do so because their ancestors have survived a rigorous screening over many generations. Those that do not have the correct genes to allow for survival under those conditions are eliminated.

The day/night cycle allows a period of energy accumulation during the daylight hours; at onset, photosynthesis ceases. Daylight allows a plant to "recharge its batteries" and perhaps produce a little extra for storage. That's what agriculture is all about. Too much night, or even too much cloudy weather, could greatly reduce productivity. But is night actually necessary? Could plants photosynthesize in constant light? Experiments in controlled environmental chambers with constant light have shown that some species seem to perform perfectly in constant light, but others do not.

So how do plants use their biological clocks to maximum advantage? The seasonal cue is rather predictable in some localities, but not so predictable in others. Frost kill date and long-term meteorological averages can vary over several weeks. Thus, seasonal cues such as changes in precipitation and temperature are imprecise at best, but the photoperiod cue is absolute. Not all external control mechanisms are cued by day length increases. Nights become warmer, and the cue to the DNA calls for transcribing genes and synthesizing proteins responsible for renewed growth. Soon the buds break, cell division begins in the shoot apex, new leaves are produced, and a stem internode elongation occurs.

How a plant perceives changes in day length apparently has nothing to do with photosynthesis, and therefore chlorophyll is not involved. On the other hand, it is a light cue, so pigment—a molecule that becomes activated or excited—must be responsible.

The first clue about its identity came from two U.S. Department of Agriculture scientists in Beltsville, Maryland, in the 1920s. W. W. Garner and H. A. Allard were studying tobacco, which in Maryland normally flowers in the late summer. One plant in their tobacco field failed to flower and continued growing until frost time. Intrigued by the plant, the scientist took cuttings of it, which they grew in a greenhouse. Various experiments with fertilizers, irrigation, temperature, and other factors failed to induce flowering. Finally, these plants had somehow delayed flowering until winter. Seeds taken from the greenhouse were planted the following season and again failed to flower until winter. It was clear that the factor responsible for the initiation of flowering was the length of the day, all other factors being equal. The plants would simply not flower unless the length of day was shorter than a critical number of hours. Garner and Allard called this phenomenon **photoperiodism** and went on to work with other plants. Some species would flower only when the length of day was longer than some critical value and response to a change in the proportion of light and dark in a 24-hour day.

These scientists were able to categorize plants as long-day, short-day, and day-neutral type. Long-day plants flower in the summer, short-day plants flower in early spring or fall, and day-neutral plants will flower under a variety of light conditions. The absolute length of the light period is not the most important factor, but whether it is a longer or shorter period than some particular interval. Consider the common cocklebur (*Xanthium strumarium*), which will flower when the length of the light period is less than 16 hours. Many varieties of poinsettias, as shown in Figure 11-2a, as well as, chrysanthemums, as shown in Figure 11-2b, are also induced to flower when the day length is about 14 hours, or longer. Thus, wheat, as shown in Figure 11-3, is by definition a long day plant, as are many cereal crops.

Figure 11-2a

Some plants, such as poinsettia, require short days and long nights for flowering. The colors of poinsettia come from bracts (leaves) that color, not petals.

The photoperiodism is at least partially controlled by a protein pigment called phytochrome. Unlike the other plant pigments, phytochrome is a very large protein molecule capable of existing in two different forms, referred to as Pr (phytochrome red) and **Pfr** (phytochrome far-red.) The Pr form can be made to change into the Pfr form if a **red light** of 660 nm wavelength is shown on the pigment in solution. This conversion occurs rapidly, in a matter of seconds or at most minutes. Once converted to the Pfr form, it can be reversed to the Pf form of **far-red light** if a 730 nm wavelength is shone on

Figure 11-2b

Chrysanthemums also require short days and long nights to produce flowers. This was discovered by Yoder Brothers Greenhouses in the late 1800s, and they have been producing chrysanthemums year-round ever since.

Figure 11-3

Others, such as wheat, require long days and short nights to flower.

the solution. This reversal can go on indefinitely, and the pigment form at the end depends on the last light used to excite the molecule. The system can be dependent graphically.

Pr ⟶ red light ⟶ Pfr ⟶ Far Dark-Red

Note that once the pigment is in the Pfr form, it can gradually revert to the Pr form in darkness, even if no far-red light is present. This change is very slow, and the length of the night is a factor in determining the ratio of Pr to Pfr. In nature, the pigment is primarily in the Pfr form at the end of the day because there is more red light in sunshine than far-red light.

This knowledge of photoperiodism has allowed plant scientists to take advantage of seasonal flowering and modify the flowering process to suit the needs of the grower. For example, when it is known how long it takes to bring chrysanthemum flowers into production after initiation, the length of the day and the date for peak flowering can be precisely determined. This timing is particularly critical for specialized sale, not only in chrysanthemums, but also Easter lilies and Christmas poinsettias.

Further experimentation has shown that the time-measuring drive is not controlled only by the phytochrome. It has not been possible to draw a conclusion about the ratios of Pr to Pfr and explain the phenomenon of long and short day plants.

The latest research on phytochrome is that phytochrome is a 3-D structure of the light detecting protein. Phytochrome is twisted into a molecular knot, an uncommon shape for any protein. Scientists theorize that the knot helps give phytochrome an overall stability as it snaps back and forth between two different forms in response to changes in light color.

Knowing the 3-D structure of phytochromes will allow researchers to determine the specific switching mechanism plants use to respond to light and how nanotechnology may also find a light-activated switch useful as they develop novel microscope devices.

Plant Hormones

Scientist are learning more about chemical messengers called **hormones**, which are indirectly responsible for much of the control of growth and development. Although it is difficult to demonstrate that a specific hormone can turn a particular gene on or off, the evidence shows that hormones are intimately involved in the regulation process. In animal systems, some hormones are in the **gene activators**; others clearly are not. It is temping to suggest that some plant hormones are also gene activators, but there is no direct experimental evidence to prove the point. Hormones may be both **stimulates** and **inhibitors**, accelerating growth in one tissue and inhibiting it in another. Sometimes the same hormone can perform both functions depending on its concentration.

The regulation of growth and development by all the internal (**endogenous**) and the external (**exogenous**) factors are partially due to these chemical messengers (see Figure 11-4). By definition, a hormone is an organic molecule synthesized in very small quantities in one part of an organism, when the molecule exerts some profound physiological effect. You might think the definition sounds suspiciously like that for an enzyme, but enzymes are synthesized within the same cell in which they operate. Hormones may be transported over great distances before they reach a target cell or cells. The definition for plant and animal hormones is the same. Animal hormones tend to be very specific: In the human body, for example, literally hundreds of them have been identified. Plant hormones, on the other hand, tend to be very general, and each kind of hormone may perform many different functions.

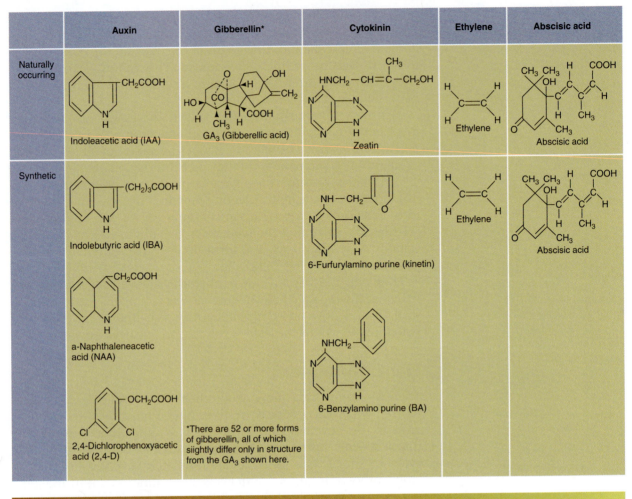

	Auxin	Gibberellin*	Cytokinin	Ethylene	Abscisic acid
Naturally occurring	Indoleacetic acid (IAA)	GA₃ (Gibberellic acid)	Zeatin	Ethylene	Abscisic acid
Synthetic	Indolebutyric acid (IBA) / a-Naphthaleneacetic acid (NAA) / 2,4-Dichlorophenoxyacetic acid (2,4-D)	*There are 52 or more forms of gibberellin, all of which siightly differ only in structure from the GA₃ shown here.	6-Furfurylamino purine (kinetin) / 6-Benzylamino purine (BA)	Ethylene	Abscisic acid

Figure 11-4

Chemical structures of various growth regulators. These naturally occurring growth hormones have been developed synthetically. Courtesy *of Rick Parker.*

Currently, there are five general classes of known plant hormones **auxins**, gibberellins, cytokinins, **abscisic acid**, and **ethylene**. These molecules tend to be rather small, much smaller than the giant proteins that act as enzymes. Most of the plant hormones have molecular weights in the range of 200 to 300, which allows them to move through tissues with relative ease. Some animal proteins do act as hormones, but none are known in plants. The simplest kind of hormone is ethylene: $CH_2=CH_2$. Even though it is a very simple gas, it exerts a tremendous physiological influence on plants.

Many chemicals not found naturally in plants exhibit growth-regulating properties similar to those of the hormones. Since they are not synthesized in the plant, they do not meet the true criteria for a plant hormone and therefore are called plant growth regulators. By definition all hormones are **plant growth regulators**, but not all plant growth regulators are hormones.

Auxin

Although most hormones research has been carried out since 1940, the history of interest in hormones is much older. The earliest recorded observation leading to the discovery of plant hormones is that of Charles Darwin. In 1881, he and his son, Francis, reported on experiments performed with grass and oat **coleoptiles** in *The Power of Movement in Plants*. They first described the phenomenon of photoperiodism, the bending of plants toward a unidirectional light source. The Darwins placed molded lead caps on the tiny shoot apex and noted that the shoot tip did not bend; when the lead cap was removed, the plant responded to the light. Their notes recorded that "when seedlings are freely exposed to the lateral light some influence is transmitted from the upper to the lower part, causing the latter to bend" (see Figure 11-5).

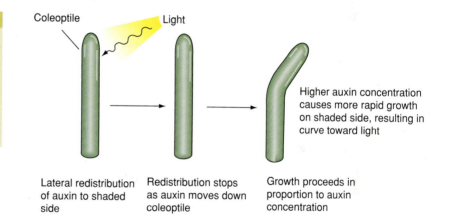

Figure 11-5

Darwin and son Francis's experiment with oat coleoptiles described the phenomenon of phototropism—the bending of plants to a light source. Courtesy of *Biological Sciences Department, NCA & T State University.*

Coleoptile Light

Lateral redistribution of auxin to shaded side

Redistribution stops as auxin moves down coleoptile

Growth proceeds in proportion to auxin concentration

Higher auxin concentration causes more rapid growth on shaded side, resulting in curve toward light

Essentially nothing was done about the Darwins' observations during the latter part of the nineteenth century and early part of the twentieth century. Then in 1926, the Dutch plant physiologist Frits W. Went, working in his father's oratory at the University of Utrecht, performed a remarkable set of experiments. Went found that the tip of the coleoptiles could be cut off and placed on a tiny block of **agar** (the gelatin-like material made from certain marine algae that maintains liquid in a semisolid state at room temperature). After about an hour, the coleoptile tip was removed and the agar block placed on one side of another decapitated oat coleoptile. In a short time, the upper part of the coleoptiles began bending away from the side of the agar block. These substances later diffused out of the block and into the second decapitated oat coleoptile and moved only down the side of the tissue cylinder directly below it, causing the cells to elongate more rapidly than those on the opposite sides of the cylinder.

This differential in the rate of growth on opposite sides of the cylinder of tissue caused the entire cylinder to bend. This experiment was particularly significant because it proved for the first time that the stimulus described by Darwin was a chemical one, rather than physical or electrical. The chemical substance was named auxin, after the Greek *auxin*, meaning "to increase." Although we now refer to any substance that will cause a similar bending response an auxin, the only naturally occurring one is **indoleacetic acid (IAA)**.

Many chemicals have auxin-like properties, and some, such as 2,4-dichlorophenoxyacetic acid. (2,4-D), are used as herbicides or weed killers.

A low concentration of 2,4-D is far more active than the same concentration of IAA in many plants, and small amounts can cause excessive respiration, excessive cell expansion, and finally tissue death; however 2,4-D is used extensively in plant tissue culture as a growth regulator. A mixture of 2,4-D is closely related to 2,4,5-trichlorophenoxacetic acid (2,4,5-T), which is Agent Orange—used as a defoliant in the Vietnam War.

IAA is synthesized from the common amino acid **tryptophan**; it is found as a by-product of both fungal and animal metabolism, and a particularly rich source is pregnant horse urine. However, these organisms neither use nor respond to it. Although no hormones are found in truly large concentrations, auxin seems to be concentrated at the site of synthesis in embryos, apical meristems, and in young leaves and fruits—all actively growing tissues. Auxin has directional or polar movement, from the growing tips of the plant toward the base of the plant: Hardly any moves in the other direction. Movement generally occurs through parenchymal cells rather than through vascular tissue.

As with other hormones that exert tremendous influence at low concentrations, it is difficult to predict a response for auxin. It behaves differently as the condition changes; many dicots are more sensitive than monocots, and the root is more sensitive than the shoot. From time to time and place to place, tissue sensitivity changes, apparent that very high concentration of auxin is toxic; hence the use of 2,4-D and 2,4,5-T as weed killers. If extremely low concentrations of 2,4-D are used, it can be a very good synthetic auxin source for stimulation of growth.

Auxin Effects

Unquestionably one of the most important actions of auxin is Darwin's observed effects of cell elongation. Occurring as it does in meristematic tissues, auxin is apparently responsible for the rapid growth and elongation of tissues, and in the shoot apex it diffuses downward and causes the stem to elongate. Young, tender stems are most readily affected; and their cells elongate rapidly. If the shoot apex is removed from the plant, apical auxin supply is eliminated and the shoot stops elongating. If auxin is added to the cut surface, growth resumes. It appears the intact plants already have an optimum amount of auxin being produced in the shoot tips. And extra application generally fails to produce a response. The **auxin threshold** is apparently saturated under most growth conditions; the same concentration of auxin that causes shoot elongation will cause an inhibition of root growth. This differential is probably due to a difference in tissue sensitivity.

Although leaf growth does not appear to be directly controlled by auxin production, leaves do contain auxin. If for some reason, including seasonal triggers, the level of auxin in the leaf falls, there is a good chance that an **abscission layer** will form at the base of the petiole. Enzymes form that break down cell walls and cause the petiole to become very weak. The leaf eventually falls from the plant. Of course, this is normal for deciduous trees in the fall, but severe stress during the growing season can also cause premature leaf fall. Such stresses may be induced by drought, flooding, high and low temperatures, light stress, and a number of other factors. Abscission zone also forms at the base of the peduncle and may lead to premature fruit shedding.

One of the important commercial uses of auxin is to stimulate the production of adventitious roots on stem or leaf cuttings. Again, depending on the type of tissue, age, and a number of other factors, various concentrations of auxin may be needed to stimulate lateral roots. In commercial horticulture practices, various synthetic auxins are used not only to speed the rate of production, but also to increase the total number of roots produced.

Auxin is also important in the growth of fruit. Developing seeds are a rich source of auxin, and that source is responsible for the growth of the fruit surrounding the seed. If ovules fail to be fertilized or if for some reason the embryo aborts, the fruit will normally fail to develop. One of the best examples is the strawberry. The achenes are produced on the outer surface to the receptacle, and each one is responsible for a region of growth necessary to produce a normal accessory fruit. If the achenes on one side of the receptacle are removed when the fruit is small, that side of the fruit fails to develop, and an abnormal strawberry is produced. This is apparently why strawberry shape is so varied. It is possible to substitute auxins for achenes and cause a normal fruit to develop. Such fruits that develop without fertilization are called **parthenocarpic fruits**. Greenhouse tomatoes and cucumbers are often produced in this manner.

Auxin is also responsible for the phenomenon called **apical dominance**. If the shoot apex is removed, the level of auxin drops in the main stem, and the adjacent lateral buds (at the next lower node) are released from inhibition and begin to grow (see Figure 11-6). Buds at successively lower nodes may also begin to grow, but the influence is greatest at the bud nearest the shoot apex. If auxin is applied to these buds, they fail to grow, proving that they are inhibited by auxin moving basipetally (downward) from the shoot apex. Plants in nature that have one central leader, such as trees with a typical conical shape, are said to have very strong apical dominances. An oak tree, on the other hand, with much branching and no central leader, is said to have weak apical dominance. The ultimate in apical dominance is exhibited by columnar palm trees, as shown in Figure 11-7, which have a single meristem located at the shoot apex. Such trees die if the shoot apex is ever removed.

Figure 11-6

Ponderosa pines exhibit strong apical dominance with a central leader.

Figure 11-7

Palms along a beach. Columnar palm trees have no lateral buds; therefore, the apical bud controls the growth of the entire tree.

Mode of Action

Obviously a plant hormone with so many different functions (only a few have been described here) must influence regulation of several different systems. Certainly one of the most important is cell elongation. Detailed studies by many physiologists and biochemists have shown that auxin increases the plasticity or "stretch ability" of the cell wall. Since living cells normally have turgor pressure that allows the plasma membrane to push against the cell wall, if the wall becomes loosened, then the cell can stretch until the wall again becomes a significant barrier. As cells become older, they fail to respond to auxin as do young cells, and the elongation finally becomes impossible. Some rigidity is characteristic of older, mature tissues and is the foundation upon which structural wood is built.

Gibberellins

At approximately the same time that Went was performing his classic experiments with auxin, Japanese scientists were investigating a fungal disease that caused excessive stem elongation, stem weakening, and finally stem collapse and death of the rice plants. The disease was known locally as the *bakanae,* or crazy-seedling, disease of rice. The organism was finally identified as *Gibberella fujikuroi,* and the band of active growth compound causing *bakanae* was named gibberellic acid. Because scientific communication was so poor between the Orient and the Western world at the time, the information was not "discovered" by Westerners until British workers reported in 1955 that minute quantities of gibberellin acid would cause genetically dwarfed pea plants to grow to a normal size. Within a few years, dozens of laboratories were experimenting with gibberellins. In both dwarf pea and dwarf corn, the amount of gibberellin applied was directly proportional to the amount of increased growth up to that for a normal plant. The obvious but incorrect conclusion was that the normal pea and corn plants contain plenty of endogenous (synthesized within the plant body) gibberellin whereas dwarf mutants somehow failed to synthesize the hormone. But dwarf plants have even more gibberellin than do normal plants. Experiments suggest that light represses growth by causing a sensitivity of the tissue to endogenous gibberellin, perhaps through a gibberellin inhibitor in the dwarf plants. Dwarf pea and corn are the standard plants for gibberellin; bioassay is a method of quantitatively determining activity of a substance using a living organism. (Both straight growth and curvature of oat coleoptiles are used as bioassays for auxin).

As one might expect from the original rice disease, gibberellic acid proved to be a growth promoter. There are now more than 50 slightly different compounds, all closely related chemically, classified as gibberellins. Less than half occur naturally, and most have no known biological activity. Gibberellin has been extracted from seeds, young shoot and root tips, and young leaves. Its concentration diminishes greater in older tissues. In the 1950s, botanists found that some plants that flowered only under long days could be induced to flower under short days if given gibberellic acid. This exciting news led to the incorrect conclusion that gibberellin was unequivocally identified as the flowering hormone. It does cause flowering in some species, but not in others.

A good example of how a hormone acts as a chemical messenger is the effect of gibberellin on the germination of barley seed. In the normal germination process, following imbibition of water, various enzymes begin to break down macromolecules. Barley seeds have a large starchy endosperm that must be broken down into sugars to provide energy for the germinating seed. This process is triggered by the synthesis of the enzyme **alpha-amylase**, which causes starch to be converted in a process that was speeded up dramatically with the addition of gibberellic acid, and later studies showed that the gibberellin caused the production of the enzyme in the **aleurone layer**, a group of cells surrounding the endosperm. Even if exogenous gibberellin is not applied, the process eventually occurs: The embryo itself produces the gibberellin, which causes the synthesis of the enzyme, eventually breaking down the starch. If the seed is split in half so that the embryo is removed, no alpha-amylase is produced. It is still difficult to prove that the gibberellin is the direct trigger that turns the gene on and off, but the circumstantial evidence is most convincing.

The brewing industry uses gibberellin to speed up the conversion of starch to sugar in barley seeds. In beer brewing, malting barley is used as the sugar source for making alcohol. The malting process is rather slow if seeds are allowed to germinate normally, but if gibberellin is added to the germinating seeds, the conversion to sugar is greatly expedited and the entire process becomes far more efficient. Increased efficiency translates into income for the brewer.

Gibberellin has also been used commercially to increase the fruit size and cluster uniformity in Thompson seedless grapes. The embryo in each of these grapes fails to provide the optimum amount of gibberellin, and if the clusters are sprayed with a dilute concentration of gibberellin very early in fruit development, the ultimate size may be doubled or tripled, and the quality of the product is not usually changed.

Cytokinins

In the 1940s, workers at various laboratories began studying plant tissue culture as a means of isolating specific tissues or organs, removing them from external influence, and then studying nutritional and growth factors one at a time. Although tissues taken from various plants, including tobacco and carrots, grew for a while in culture, the cells eventually stopped dividing. Many culture media were tried, but liquid coconut milk proved to have some factor that immediately caused the cells to divide. After years of attempts at isolation, the factor was finally identified as a derivative of the purine **adenine**, the same substance found in nucleic acids. The compound was named **kinetin**, and now the entire class of compound isolated from plants that causes cell division is called cytokinins. Since cytokinins are so effective in promoting cell division, it was thought this compound or a similar one might be responsible for the uncontrolled growth of cancer cells. During the 1960s, a great deal of research effort was devoted to trying to associate cytokinins with tumor development in both plant and animal cells, but no direct connection was ever made. Instead, cytokinins appear to be naturally occurring compounds responsible for the control of the cell division process in meristematic regions. Cytokinins also seem to slow the turnover of proteins in plant tissues, which retard the aging, or senescence, process. For example, if cytokinins are sprayed on lettuce leaves, the storage life is greatly enhanced, and the product can be shelved for long periods. This allows supermarkets to distribute fresh vegetables more efficiently and ultimately buffers price fluctuations in a higher volatile market.

Abscisic Acid

Just as auxins, gibberellins, and cytokinins are generally considered to be promoters of growth, others are generally considered to be inhibitors of growth. They reduce the rate of growth or induce dormancy at critical times, such as the onset of fall and winter. The primary molecule that controls these inhibitory processes is abscisic acid. Originally identified in dormant buds of ash trees and potatoes, it was later found to be the same molecule that caused the abscission of leaves, flowers, and fruits. When days become shorter in the fall and temperatures begin to decrease, levels of abscisic acid gradually rise in the abscission zones of petioles. Leaves fall, and the level of abscisic acid in the dormant buds remains very high throughout the winter. With the onset of spring, levels of abscisic acid decrease at the same time that levels of auxin, gibberellin, and cytokinin begin to rise. Buds are activated and stems begin to grow and develop for the cycle to repeat itself.

Whenever a water stress occurs in leaves, abscisic acid levels rise in the guard cells, which triggers the movement of K^+ out into adjacent cells, making the solute less concentrated and causing water to move out of the cells. When the pressure decreases in the guard cells, the stomatas close, and transpiration is greatly reduced. Thus, it appears that abscisic acid is intricately involved in plant water stress.

Ethylene

For many years greenhouses were heated with open space heaters. Occasionally an attendant would arrive at work after a cold night to find that all the leaves had fallen off the plants. Various toxic gases were suspected, but many years went by before ethylene was recognized as a major component of exhaust gases and the association made between leaf abscission and ethylene. Even more powerful than abscisic acid, this gas can trigger unusual growth and development responses even in trace amounts. The ripening of fruit involves a number of chemical and anatomical changes, and in the process of ripening storage molecules change to soluble sugars, as in bananas. This is part of the respiratory process and is accompanied by the release of large quantities of ethylene. As bananas ripen naturally, the dark flecks that appear on the peel are pockets of concentrated ethylene. One can increase that concentration and speed the ripening process by enclosing the fruit in a plastic bag so that the ethylene given off by the ripening fruit is concentrated and not allowed to diffuse. Green stalks of bananas are ripened in this manner at the supermarket, or they are placed in a small gassing room and treated with bottled ethylene gas.

The study of **postharvest physiology** is a major segment of the field of plant science and horticulture. Scientists study the environmental and biochemical factors that influence the storage and ripening process of fruit and vegetables. Some fruits, such as banana and avocado, produce large amounts of ethylene and ripen very suddenly, and this ripening is accompanied by a rapid rise in the rate of respiration of that fruit tissue. This sudden increase in respiration is called the climacteric rise and at one time was the subject of many laboratory studies.

The storage of apples has received particular attention because apples are a major crop and subject to rapid deterioration under improper storage conditions. Postharvest physiologists have learned that if apples are stored in a **controlled atmosphere** of temperatures just above freezing, high relative humidity, low oxygen, and high carbon dioxide levels, storage life can be increased dramatically. It is now possible to store apples year-round with essentially no reduction in quality. This finding has obvious implications for availability of fresh fruits for the consumer and the stabilization of prices.

Although the effects of ethylene are well documented, some of the effects are so interrelated with auxin that it is difficult to determine which is causing the effects. In addition to its effects on fruit ripening, ethylene causes leaves to abscise, chlorophyll to leach, flower pigments to fade prematurely, and leaf petioles to grow more rapidly on the upper side and therefore curve down. The petioles response to ethylene is called **epinasty** and should not be confused with the phototropism attributed to auxin.

Growth Retardants

Many commercial growth regulators are advertised as **growth retardants** and are sold for the purpose of reducing pruning or mowing costs, decreasing stem length in flowering plants such as azaleas and poinsettias, and decreasing biomass production to reduce fire hazards. It appears that several of these compounds act by blocking the synthesis of gibberellin. Others may act as antiauxins, and in fact several possibilities exist for counteracting the effects of hormones.

Hormone Interactions

It is probably apparent by now that hormone action is rather complicated. Change the concentration and the promoter may become an inhibitor. Both auxin and gibberellin act on cell elongation; both ethylene and abscisic acid cause abscission. It is important to remember that hormones never act alone in any tissue, and the bottom line of growth and development depends on the relative concentrations of all the hormones acting in concert to produce the final product. Tissue culture experiments have revealed a great deal about the interaction of auxin, gibberellin, and cytokinin (see Figure 11-8). If the ratio is changed slightly, one may get **callus** (undifferentiated tissue); change the ratio again, and only roots may be produced; and if the ratio is just right, both roots and shoots may be produced, eventually leading to a complete new plant from clonal tissue. Many exciting results have been obtained in understanding how the chemical messengers do their work.

The Flowering Hormone

Thousands of hours of research have gone into the search for the flowering hormone. Various laboratories have isolated an extract that will cause a vegetative shoot apex to become reproductive and produce flowers, but so far no one has ever chemically identified the compound. The floral stimulus is obviously transported from leaves to the shoot apex. For example, it is possible to take a short-day plant such as cocklebur (*Xanthium*) and place it under long-day conditions so that the plant remains entirely vegetative. If a single leaf is enclosed by a dark box so that it receives only short-day light, the plant will flower. In addition, if a cocklebur plant is growing under long-day conditions and if a leaf from a flowering short-day plant is grafted onto the vegetative plant, the chemical stimulus is transferred to the apex and the noninduced plant will flower. This elusive chemical stimulus has been given the name of florigen, although nothing is known about its true chemical nature or structure. It seems particularly ironic that, in the age of modern miracles and molecular biology, such a compound could not be isolated, but scientists are coming closer to the conclusion that **florigen** may be a combination of substances perhaps including gibberellin and other hormones.

Plant Movements

You have already learned that some plant tissues have organs that are capable of modifying their rate of growth so that there is a change in direction. Darwin's experiments with the oat coleoptile showed that cells on the dark side of the stem elongate more rapidly than those on the light side. Went later demonstrated that the auxin levels are higher in the cells that elongate more rapidly, and even if the rate of cell division is unchanged, the cylinder of tissue will curve toward the light. An explanation of this phenomenon is that light causes a migration of auxin molecules from the light side to the dark side of the stem. It has been shown that seedlings kept in constant darkness elongate much more rapidly than seedlings grown in light (from the time of emergence from the soil). Such plants are said to be etiolated, and they have tiny underdeveloped leaves and weak, pale internodes. If stem tissues are analyzed, the etiolated (dark-grown and spindly) seedlings contain much more auxin. Etiolation in the dark helps the seedling to force its way through the soil before expending all its energy. Plants in shaded conditions tend to elongate faster than bright light-grown plants, which allows them to compete more effectively for the light.

Plant leaves and flowers also move a great deal, often exactly tracking the sun. For instance, sunflowers, as shown in Figure 11-9, track the sun from early morning to sunset, and leaves of many other plants do the same. This allows the plant to readjust constantly so that only perpendicular rays from the sun hit the leaf surface, and therefore the photosynthetic efficiency is increased.

Figure 11-9

In this field of sunflowers, all the flower heads point in the same direction. These flowers track the sun from sunrise to sunset.

Geotropism

The earth's gravitational pull may cause growth responses. Roots grow down into the soil (positive **geotropism**), and shoots grow upward from the ground (negative geotropism). All these tissues start out from the same embryo, originally consisting of only one cell. As soon as the first division of the fertilized egg occurs, a **polarity** is established, and it is possible to determine which direction is heads and which is tails, that is, shoot and root. Tissues in the same organism react in its tails, this is shoot and root. Tissues in the same organism react positively at one end and negatively at the other and one theory suggests differential tissue sensitivity to certain hormones, powerhouse auxin. Gravity may cause a higher concentration of auxin on the lower side of the tissue than on the upper side. Thus, it is possible for a given auxin concentration to accumulate on the lower side of the plant embryo and stimulate those cells to elongate faster. That tissue would curve upward and become the shoot. At the other end of the organism, the same concentration of auxin might be inhibitory to the cells and slow the rate of elongation. Thus, cells on the upper surface would grow more rapidly, and the tissue would curve downward and become the root.

Thigmotropism

Thigmotropism is a word derived from the Greek *thigma* ("touch") and refers to directional growth caused by plants touching a solid object. This phenomenon is often exhibited by climbing plants such as English ivy, which send out aerial roots when a shoot comes in contact with a wall, as shown in Figure 11-10. The roots attach the plant to the surface so that it may continue to climb upward. Hormones are undoubtedly involved in the contact response and subsequent differentiation of tissues.

Figure 11-10

Ficus benghalensis *(Banyan tree) sends out aerial roots when it touches a solid object. This is called thigmotropism.*

Nastic Movements

Nastic movements are those made in response to some stimulus but are not oriented relative to the direction of the stimulus. Some of these movements are reversible and do not involve true growth. Others are true growth responses, as in the epinastic responses described for ethylene, and the changes are relatively permanent. Other nastic responses include **thermonasty**, as in the opening and closing response of tulips due to fluctuations in temperature; **nyctinasty**, the so-called sleep movements of leaves due to changes of turgor of certain cells located at the base of the petiole; and **seismonasty**, a response to shaking or some other mechanical disturbance.

The sensitive plant (*Mimosa pudica*) and the Venus flytrap (*Dionaea muscipula*), shown in Figure 11-11, both display leaf movements not related to differential growth (the response is much too rapid). They are due instead to turgor movements. Large, turgid cells act as a hinge for these organs. When the trigger is released, an electrical impulse is transmitted instantaneously to these storage cells, membrane properties are changed, and the water pressure is released. After a period of time, the water pressure is regained and the trap is reset.

Figure 11-11

Nastic movements are made in response to some stimulus. The Venus flytrap (Dionaea muscipula) responds in two ways: nyctinasty and seismonasty. Courtesy of Julia Blizen, Cove Creek Garden.

Summary

1. The genetic composition of a plant—the genotype—is modified by the environment to produce a phenotype, the physical and chemical features that characterize the plant. The genotype specifies the potential characteristics of the organism, but whether they are in fact expressed depends on the type of environment.

2. Growth is denied as an irreversible increase in size or volume. Development is a summation of all the activities leading to changes in a cell, tissue, organ, or organism. Development is an orderly sequence of events dictated by a precise set of genes turning on and off at every stage. Differentiation comprises the chemical and physical changes associated with the development process.

3. Developmental biology is the study of how organisms, their cells, and their tissues achieve a final predictable form and function.

4. Growth and development are often limited by environmental extremes that suppress genetic function. Temperature, water, salt, light, and other factors combine to limit gene expression by influencing the turning on and turning off of specific genes.

5. Plants are carefully tuned to biological clocks, the most important being the 24-hour and the day/night cycles. Seasonal changes provide environmental signals to make certain changes in growth and development. Reproductive cycles are usually attuned to these environmental signals to make certain changes in growth and development. Reproductive cycles are usually attuned to these environmental cues and the flowering process in many plants due to photoperiodism. The primary receptor in this system is a protein-pigment that senses a change in the day length.

6. The primary plant hormones are auxin, gibberllic acid, cytokinin, abscisic acid, and ethylene. They act as stimulators and inhibitors of growth by directly or indirectly influencing gene regulation.

7. Plants respond to a unidirectional light source by compensatory growth, causing the cells on the dark side to grow more rapidly than those on the light side. The curving response occurs because of change of growth rate of localized cells, but the change in growth rate is brought about by a migration of auxin from the light side to the dark side. This bending response to light is called phototropism, and other tropisms include a response to gravity (geotropism) and a response to touch (thigmotropism).

8. Other plant movements may be temporary and occur in response to a change in membrane properties so that the turgor in certain target cells changes rapidly.

Something to Think About

1. Compare phenotype and genotype.
2. Define growth.
3. What is the difference in growth and development?
4. What is development biology the study of?
5. List the limiting factors in growth and development.
6. Explain plant biological clocks and what they control.
7. Name the primary plant hormones and tell which are promoters and inhibitors.
8. Discuss tropisms.
9. List the nastic movements of plants.
10. How do these movements work?

Suggested Readings

Fosket, D. E. 1994. *Plant growth and development*. San Diego: Academic Press.
Larcher, W., E. Huber-Sannwald, and J. Wiese. 1997. *Physiological plant ecology* (3rd ed.). New York: Springer.
Raghavan, V. 2007. *Development biology of flowering plants*. New York: Springer.
Taiz, L., and E. Zeiger. 2006. *Plant physiology* (4th ed.). Sunderland, MA: Sinauer Associates, Inc.

Internet

Internet sites represent a vast resource of information. The URLs for Web sites can change. Using one of the search engines on the Internet, such as Google, Yahoo!, Ask.com, and MSN Live Search, find more information by searching for these words or phrases: **plant growth and development, photoperiod, phytochrome, plant biological clocks, short-day plants, long-day plants, day neutral plants, plant hormones, auxin, IAA, parthenocarpic fruit, apical, gibberellins, bioassay, cytokinins, abscisic acid, ethylene, postharvest physiology, flowering hormone, tropisms, and nastic movements.**

PART 4

Evolution
and Diversity

CHAPTER 12

Sexual Reproduction and Inheritance

Prokaryotic organisms, and some eukaryotic ones, reproduce asexually. Usually a single-celled individual splits into two genetically identical individuals. Even in this case, there can be slight genetic differences between the two offspring as a result of mutations, but there is no guarantee. Most eukaryotic organisms, on the other hand, produce offspring by sexual reproduction. Sexual reproduction has ensured that all individuals are at least slightly different. Identical twins may be an exception, but even here, single gene coding mistakes or mutations can result in minor differences.

Without the resulting variability among individuals that sexual reproduction produces, you and your friends would look alike, and you would not be able to distinguish your parents from your neighbors or your dog from any other. All other individuals of each kind of organism would look and function the same. There would be no cultivated food crops because there never would have been a larger tomato, a juicier grapefruit, or higher yielding wheat inflorescence to select as the start of a better crop plant.

Objectives

After completing this chapter, you should be able to:

- Outline the process of meiosis
- Identify where the meiosis process takes place in angiosperms
- Explain how a single-celled zygote is produced
- Describe the differences in the processes of meiosis and mitosis
- Discuss how mutations form
- Recognize the alternation of generations
- Explain double fertilization
- Diagram the inheritance patterns Mendel expressed
- Describe complete dominance and recessive genes
- Demonstrate an understanding of backcrosses
- Draw a Punnett square showing a dihybrid cross, involving two traits
- Discuss the implications of genetics engineering
- Outline how actual genetic code works
- Describe molecular evolution

Key Terms

meiosis
chromosomes
allele pair
bivalent
cytokinesis
interkinesis
anaphase II
telophase II
chiasma
sporophyte
gametophyte

alternation of
    generations
haploid
microsporogenesis
microgametogensis
megagamete
sporogenous
tapetum
exine
intine
integuments

microphyle
embryo sac
egg cell
synergids
antipoidal cells
endosperm
Gregor Mendel
pure parents
F1
dominant
recessive

313

hereditary factors
F2
principle of segregation
punnett square
heterozygous
homozygous
backcross
incomplete dominance
monohybrid crosses

dihybrid cross
trihybrid cross
linked
principle of
  independent
  assortment
epistasis
pleiotropic
multigenic

genetic engineering
recombinant DNA
protein sequencer
polymerase chain
  reaction (PCR)
taq polymerase
exons

Meiosis

The key to sexual reproduction is a process of nuclear and cellular division different from that of mitosis (Chapter 7). The critical features of **meiosis** are the production of daughter cells having only one set of **chromosomes**, half the number of the parent cells that began the process. Diploid cells have two sets ($2n$) of chromosomes, half that number, one set (n), is termed the haploid number. The fusion of two haploid cells during fertilization completes the sequence of events in the reproduction cycle, producing a genetically unique organism.

Most sexually reproducing organisms are diploid, which means that all the cells making up the body of the organism, that is, somatic cells, contain a nucleus with two sets of homologous, or similar, chromosomes. These homologues are similar in size, position of their centromere, and most important, their genetic composition. Each of the two chromosomes of a homologous pair contains genes for the same group of traits, and these genes occur in the same sequence from one end of the homologue to the other. It is important to remember that homologues contain genetic instructions for the same traits but not *necessarily the same instructions*. Thus, the expression of each trait is controlled by two genetic messages; two homologous chromosomes control one gene from each trait. These two genes are called an **allele pair** or alleles. Alleles are always in the same position (locus) on the two homologous chromosomes. All the functions of each diploid somatic cell, therefore, are genetically encoded twice, once per haploid set of chromosomes. A set of chromosomes is composed of one chromosome from each different homologous pair, as shown in Figure 12-1, and it does not matter which of the two homologues is present.

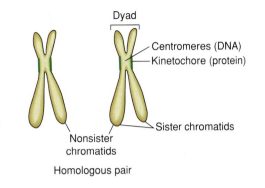

Figure 12-1

Two homologous chromosome pairs. Homologues are the same length, have their centromeres in the same position, and contain genetic messages for the same set of traits. Each of these chromosomes is composed of two genetically identical chromatids. Genes for the same trait are called alleles. Courtesy of Carolina Biological Supply.

The following analogy may help you visualize homologous chromosomes. Hold your hands together so that your palms are facing each other and your fingers are touching and pointing in the same direction. Your two thumbs and each of your four fingers are paired, with similar digits across from each other. This is analogous to five homologous chromosome pairs in that each member of each pair is the same size and length, and they look similar. The matching joint creases in your fingers are analogous to the loci for the two genes from the same trait—alleles. Meiosis is a process that occurs only in specific tissues of sexually reproducing organisms. In angiosperms, the total flower is the reproductive structure, but meiosis occurs in the anthers and ovules. In each of these tissues a different sequence of events takes place.

Like mitosis, the process of meiosis is a continuous series of activities, but it has been subdivided into phases for convenience of discussion (see Figure 12-2). Unlike mitosis, meiosis I is the reduction division because it is during this sequence of events that the haploid conditions are produced from a diploid parent cell. The second meiosis division, meiosis II, is similar to mitosis in its activities. For convenience, meiosis is discussed using the subdivision of meiosis I and II with each further subdivision into the phases of nuclear divisions presented for mitosis.

The importance of a reduction division preceding the fusion of two gametes is apparent if you consider what would happen without it. If a diploid organism produced gametes through mitosis, each would be diploid, just like all the somatic cells. Human gametes (egg and sperm) each would have 46 chromosomes, the diploid number for all somatic cells. Fusion of the two gametes would result in a single-celled zygote having 92 chromosomes. Two such individuals, each producing gametes without a reduction division, would have an offspring with 184 chromosomes in each cell. By the fifth generation, the offspring

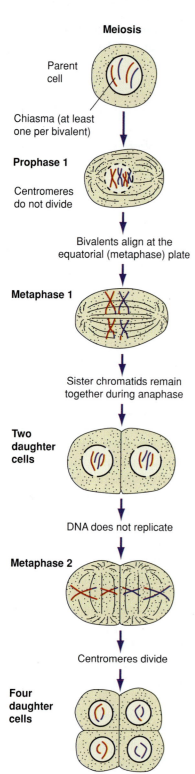

Meiosis

Parent cell

Chiasma (at least one per bivalent)

Prophase 1

Centromeres do not divide

Bivalents align at the equatorial (metaphase) plate

Metaphase 1

Sister chromatids remain together during anaphase

Two daughter cells

DNA does not replicate

Metaphase 2

Centromeres divide

Four daughter cells

Figure 12-2

Stages of meiosis, from the single diploid cell having six chromosomes through representative stages of the two meiotic divisions.
Courtesy of Carolina Biological Supply.

would have 1,472 chromosomes in every cell of every individual. Obviously, this doubling of chromosome number for each generation could not continue for long before there would be an unsolvable space problem. In addition, problems with multiple genetic messages would occur, especially in animal systems.

Most organisms within a given species, therefore, are constant in their chromosome number. As already stated, all humans have a diploid number of 46 chromosomes (23 pairs). By comparison, dogs have 78, horses 74, elephants 56, and chimpanzees 48 chromosomes. The diploid number of plants ranges from a low of 4 in a couple of species to a high of 1,290.

Meiosis I

As in mitosis, the cellular activities of interphase, including G1, S, and G2 phases, prepare the cell for division; these activities include DNA replication to produce chromosome to compose of two genetically identical chromatids united at the centromere. Prophase I is underway when the threadlike chromosomes first become visible. Next, the nuclear membrane starts to break down, and the spindle fibers begin forming. Later, the nucleolus becomes gradually less distinct.

As the chromosome coil becomes more condense and visible, the homologous pairs physically come together. This process of synapsis does not occur in mitosis. Synapsis produces an exact pairing of the two homologues with the same genetic regions side by side for the full length of the two chromosomes. The pairing is evident as soon as the chromosomes become visible. The pairing homologues, each composed of two chromatids, are called a **bivalent** (*bi*, "two"; *valent*, "combined, associated"). Thus, each bivalent contains four chromatids.

During bivalent formation, the nuclear membrane and nucleolus continues to disappear, the spindle forms, and the paired chromosomes become fully condensed and begin moving toward the center of the cell. Prophase I ends at this point.

Metaphase I

Metaphase I begins as the chromosomes arrive at the equatorial plane of the cell and their centromeres attach to spindle fiber. Typically, the bivalents are doughnut shaped because their centromere regions push away from one another during full chromosome condensation. As each bivalent lines up on the spindle, the two centromeres attach to fibers on opposite sides of the equatorial plane. For example, in a plant having a diploid chromosome number of 6, there are three bivalents attached to the spindle at this point in meiosis. Each of these three bivalents contains a chromosome from each of the two parents that formed the reproducing plant. Which member of the homologous pair lines up on which side of the equatorial plane is a random process.

The next event during meiosis results in the actual reduction in the number of chromosomes each daughter nucleus receives. The centromeres of each of the two homologous chromosomes paired during the first part of meiosis I are pulled apart and move toward opposite poles of the cell—this occurs in anaphase I. Each chromosome is still composed of two chromatids. Half the total number of chromosomes goes one direction and half in the other. It is important to remember that in each resulting group, there is one chromosome from each bivalent. Thus, each pole has a complete set of chromosomes with one genetic message for each trait. In a plant where $2n = 6$ represents the diploid condition, $n = 3$ represents the haploid condition that results from meiosis I.

As the separated homologues reach opposite ends of the parent cell, several events occur at once—what is considered telophase I. During telophase I the uncoiling of the chromosomes, the reformations of the nuclear membrane, the reappearance of nucleoli, and the complete disappearance of spindle fibers occur. Here, it is important to point out that in different organisms these events vary in their level of completion. In some organisms, telophase I events never really occur before the chromosomes begin the sequence of activities leading to the second meiotic division. In others, telophase I proceeds almost to completion prior to the start of meiosis II division. **Cytokinesis** can also occur completely, partially, or not at all, depending on the organism. In instances when telophase I events do take place to completion, there can be a period termed **interkinesis**, but it differs from interphase in that there is no DNA replication.

Meiosis II

Each of the two daughter nuclei resulting from meiosis I is haploid, having one set of chromosomes composed of two chromatids each. The events that these two haploid daughter nuclei go through are essentially the same as those in mitosis. In prophase II, the chromosomes become tightly recoiled and are visible as having two chromatids each. The nuclear membrane and the nucleolus gradually disappear, and the spindle fibers begin forming. Once the spindle is formed and the chromosomes are fully condensed, the chromosomes move the equatorial plane between the ends of the spindle. This is metaphase II. Each chromosome acts independently, with centromeres attaching to spindle fibers at the equatorial plane.

The centromeres then divide at the beginning of **anaphase II**, and the two chromatids separate and move toward opposite poles. As these newly formed chromosomes reach the opposite poles, **telophase II** events begin. At the end of telophase II, there are four resulting daughter nuclei, each one containing a single complete set of chromosomes and each one being genetically different from the others.

Crossing Over

Seemingly, the four haploid daughter nuclei produced by meiosis should be two pairs that contain genetically identical members. In other words, each of the two daughter nuclei resulting from meiosis I will be genetically different because of the random assortment of homologues, but each chromosome in those two daughter cells should have identical chromatids. These chromatids then separate in meiosis II. But the chromatids then separate are not identical, especially because of the process of crossing over, an exchange of genetic material during the early part of meiosis

This exchange is between adjacent chromatids on the two homologous chromosomes of a bivalent. As homologous chromosomes undergo synapsis during early prophase I, each chromosomes is composed of two genetically identical, or sister, chromatid arms of the bivalent condition adjacent to nonsister chromatid arms, which often cross over each other and associate with the outside nonsister chromatid arms of the bivalent for a portion of their length. As the chromosomes become more condensed and visible, this configuration can be seen as a crossover of chromatid segments and is termed a **chiasma**. Chromosomes average one chiasma per bivalent, which means there are, on average, as many chiasmata in every meiotic division as the haploid chromosome number.

As bivalents reach their most condensed state, the homologues begin to repel one another and physically separate, except at the crossovers. The crossed-over chromatid segments break from their original strand and fuse with the adjacent chromatid of the homologue. This results in an exchange of genetic material between two of the four chromatids in a bivalent. The exchange produces two chromosomes, each composed of two genetically different chromatids.

When the chromatids separate during meiosis II, therefore, the four resulting nuclei all are genetically different from each other. This genetic variability, combined with the mixing of genetic information that occurs as a result of random assortment and separation of homologues during meiosis I, provides the potential for a phenomenal number of original genetic combinations in gametes. To this genetic diversity, add the random possibilities of which two genetically different gametes will fuse during fertilization, and it becomes clear why no two sexually reproducing individuals are ever alike.

Mutations

Although the replication of genetic information normally occurs flawlessly, coding mistakes, or mutations, do occasionally occur. Most mutations are not lethal, and many are thought to have absolutely no effect at all; however, others produce changes in the appearance or function of an organism. Mutations are another source of variability in sexually reproducing organisms. Scientific dogma argued against such "transposable elements." Dr. Barbara McLintock suggested otherwise and won the 1983 Nobel Prize in Medicine.

Between the random assortment of homologues, chiasmata, mutations, and chance combinations of gametes during the process of sexual reproduction, variability is ensured. Variability is essential for the success of organisms as their environments gradually change over many generations.

The Angiosperm Life Cycle

The actual process of meiosis is essentially the same in all sexually reproducing organisms. What the resulting haploid cells (nuclei) are called and what they do next differ. In animals, the haploid cells usually differentiate to become gametes, which fuse during fertilization into a single-celled diploid zygote. That zygote can then begin dividing mitotically to form a multicellular organism that grows, matures, and reinitiates the entire cycle (see Table 12-1).

Alternation of Generations

In plants, however, different sequences of events usually occur. Plants have two generations: **sporophyte** and **gametophyte**.

The sporophyte generation produces spores, each of which develops into a gametophyte, which produces gametes. The gametes fuse to form a sporophyte, and the alteration between these phases continues in successive generation. This is called the **alternation of generations** in plants, as shown in Figure 12-3. The sporophytes are diploid, producing haploid spores by meiosis. These haploid spores divide mitotically to produce the multicellular, haploid gametophytes. Specific cells produced by the haploid gametophytes are gametes, which may fuse with other gametes to form a diploid zygote. The zygote divides mitotically to form the diploid sporophyte stage of the life cycle.

The gametophyte (**haploid**) stage or generation is the dominant form for lower (nonvascular) plants—the sporophyte begins less evident and smaller. With increasing complexity from the lower to the higher (vascular) plant forms, this relationship gradually changes in the angiosperms (the highest plants), where the sporophyte is the dominant, long-lived, independent group and the gametophytes are small short lived dependent groups of specialized cells that differentiate within the flowers of the sporophyte tissues of the flower, one in the anthers and the other in the ovules. In the anthers, **microsporogenesis** (the making of small spores) is followed by **microgametogensis** (the making of small gametes). In the ovules, megasporogenesis is followed by megagametogenesis. The spores produced in the ovules are larger than the spores produced in the anthers and are therefore termed megaspores. Likewise the egg (in the ovule) is larger, a **megagamete**.

Microsporogenesis and Microgametogensis

In the developing anther of a flower, there are four areas of **sporogenous** (fertile) cells, one in each of the four sacs comprising the anther. These cells initiate the development sequence to produce

Table 12-1

Mendel's Dihybrid Cross F2 Results

Traits	Number of plants*
Yellow-round	315
Yellow-wrinkled	110
Green-round	108
Green wrinkled	32

315.110.108.32 is approximately a (9:3:3:1 ratio).

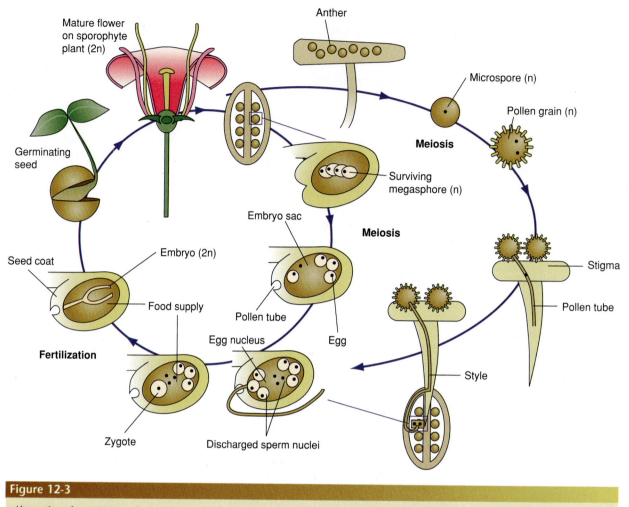

Figure 12-3

Alternation of generations; the alternation between the sporophyte and gametophyte stages in the life cycle of all plants. Courtesy of Carolina Biological Supply.

pollen grains. The **tapetum** is a layer of nutritive tissue that provides nourishment for the sporogenous cells as they divide. The tapetum, as well as the sporogenous cells, are diploid. The cells in the sporogenous areas that undergo meiosis are called microspore mother cells, or microsporocytes. Each diploid microspore mother cell produces four haploid microspores. In monocots, cytokinesis normally occurs as each of the two meiotic divisions occurs, whereas in dicots the four haploid nuclei are produced first, then cytokinesis forms all four microspores simultaneously at the end of meiosis.

Each of the four haploid microspores enlarges and develops into a pollen grain, as shown in Figure 12-4. The outer coat of pollen grain, the **exine**, is a tough layer that makes pollen one of the most persistent plant parts over the entire geologic fossil record of higher plants. The inside layer of the pollen grain, the **intine**, is not as resistant and is composed of cellulose and pectin produced by the microspore cell. Pollen grains vary in size from less than 20 μm to over 200 μm in diameter and in external morphology from smooth to very ornamented, with ridges, spines, and surface undulation. There are even pollen grains with winglike projections of the exine that aid in wind dispersal.

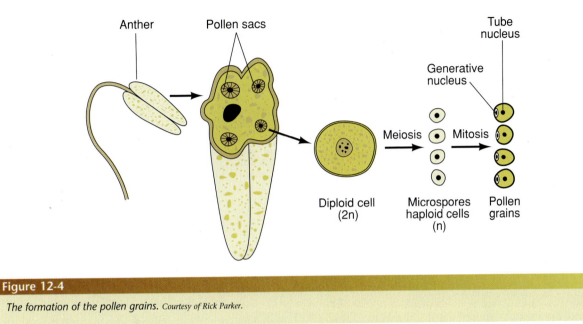

Figure 12-4

The formation of the pollen grains. *Courtesy of Rick Parker.*

Microgametogenesis, which occurs inside the developed pollen grain, comprises the gametophyte stage, the gamete-producing organism in the alternation of generations. The haploid microspore nucleus divides mitotically to produce two haploid nuclei: the tube nucleus and the generative nucleus. In some species, this binucleate microgametophyte is the mature pollen. In most species, however, the generative nucleus divides mitotically to produce two haploid gametes, or sperm nuclei, prior to pollen release.

If the pollen is released in the binucleate condition, the generative nucleus divides mitotically sometime during or soon after pollination to produce the two nonmotile sperm nuclei, since both must be present for double fertilization to occur. The tube nucleus is so called because after the pollen grain lands on a receptive stigma, the tube nucleus directs pollen germination through one of the pores in the exine and forms a pollen tube down through the tissue of the style. The tube nucleus produces an enzyme that dissolves intercellular tissue of the style and forms a small tube through which the tube nucleus and two sperm nuclei travel to the ovary.

Megasporogenesis and Megagametogenesis

The ovary of a flower may produce from one to hundreds of ovules, depending on the species. Each ovule is connected to the placental tissue of the ovary by stalklike funiculus. The ovule develops two outer layers, the **integuments**, which enclose the nucellus tissue within. There is a small opening, **microphyle**, at the end where the integuments come together. Through this opening, the pollen tube containing the sperm nuclei enters the ovule prior to fertilization.

In the nucleus tissue, there is a single diploid megaspore mother cell that undergoes meiosis and cytokinesis to produce four haploid megaspores (see Figure 12-5). At the conclusion of meiosis, these four megaspores are usually aligned in a chain. Normally the three cells nearest the micropyle degenerate, leaving only one functional megaspore to develop into the megagametophyte.

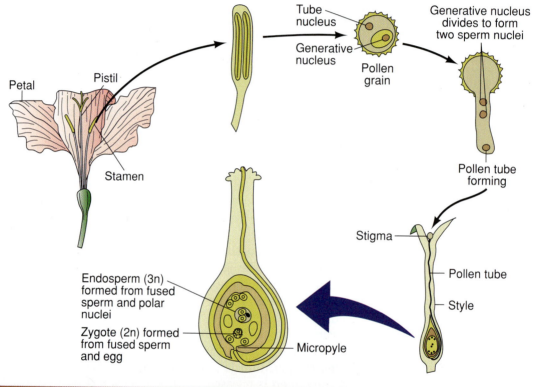

Figure 12-5

After pollination, the pollen grain germinates and a pollen tube forms intercellular down the style. The tube contains the tube nucleus and two sperm cells. Courtesy of Rick Parker.

This single haploid megaspore undergoes three consecutive mitotic divisions to produce two, then four, and finally eight haploid nuclei within a single cell in the ovule. This cell, the **embryo sac**, is the mature megagametophyte. The eight nuclei are normally arranged with four at the micropyle end and four at the opposite end of the embryo sac. Subsequently one nucleus from each group of four migrates to the middle of the embryo sac. These are the two polar nuclei. Cell walls are synthesized around the remaining three cells at the micropylar end, resulting in a middle **egg cell** flanked by two

synergids. The three nuclei at the other end of the embryo sac also form cell walls and are called the **antipoidal cells**. Both polar nuclei remain in the center of the original cell, becoming a binucleated central cell. This eight-nucleate, seven-cell condition is the megagametophyte, and single-egg cell is the gamete. Each ovule in a mature ovary contains a megagametophyte or gamete. Each ovule in a mature ovary contains a megagametophyte or gamete "plant" just as each pollen grain contains a two- (or three-) nucleate microgametophyte.

Fertilization

The culmination of sexual reproduction in angiosperms is a unique form of fertilization occurring in the ovule. The pollen tube normally enters the ovule through the micropyle opening. Once adjacent to the egg, it forms a pore in its wall to allow the two sperm cells to exit. Fusion between haploid egg and sperm nuclei occurs, and a diploid zygote cell is formed. This is true sexual fertilization. Angiosperms are said to have double fertilization because the other sperm nucleus fuses with the two polar to form a triploid cell (three sets of chromosomes). This second fertilization is one feature that makes angiosperms unique among plant groups (see Figure 12-6).

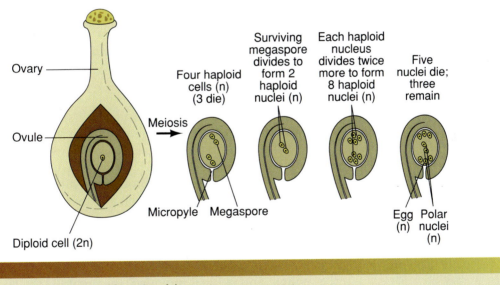

Figure 12-6

After pollination, next is fertilization and formation of the zygote. Courtesy of Rick Parker.

The zygote forms the embryo of the seed through repeated mitotic divisions, whereas the triploid **endosperm** tissue of the seed forms mitotically from the triploid nucleus. The seed coat develops from the ovule's integuments. The new individual becomes relatively dormant in its seed, ready to be dispersed, germinate, and grow to maturity to initiate the reproductive cycle again. It is worth repeating that the most important consequence of sexual reproduction is variability.

🌀 Inheritance

Sexually reproducing diploid organisms have two genetic expressions for each trait, one provided by each parent. These two alleles are located at the same position, or locus, on homologous chromosomes. The study of genetics entails an understanding of how these alleles are passed from one generation to the next and how their messages control the appearance of an organism. The appearance of an organism is termed its phenotype; the genetic information controlling it is the genotype. Only the phenotype can be seen; thus, the study of inheritance stems largely from studying the phenotypes of individual traits making up the organisms.

The first such extensive study in which the results were carefully recorded and published was in 1866 by an Austrian monk named **Gregor Mendel**. Mendel studied a number of traits of the common pea plant (*Pisum sativum*), each of which had two different expressions. Between 1856 and 1868, he made hundreds of carefully recorded crosses and observations about the inheritance of these traits through several generations. Mendel made these observations and drew sound conclusions about the control of inheritance without knowing about chromosomes and meiosis; these were first described later in the century.

In spite of the stability of gene position as described by classic Mendelian inheritance, it is possible for a gene or groups of genes to be transferred from one chromosome to another.

Mendel's Experiments

"**Pure Parents**" are those that produce offspring bearing only one expression for a given trait. When Mendel crossed plants that were pure for one phenotype of a given trait (for example, yellow seeds) with plants pure for the other expression of trait (for example, green seeds), the results were probably a bit unexpected. It seems logical that the offspring from such a cross would have some plants with green and some with yellow seeds. But, as Table 12-2 shows, the **F1**, or first filial (*filial*, "offspring"), generations for every such cross had only one of the two possible expressions appear. These results were consistent regardless of which parent variety provided the pollen and which provided the ovule. Mendel termed the expression of the trait that appeared in the F1 plants the **dominant** character, the other he called **recessive** character. Thus, yellow is dominant and green is recessive for pea color. This means that if the **hereditary factors** for both yellow and green are present, the expression of the trait will be yellow. The term *gene* was not proposed until much later, but Mendel's factors are equivalent to genes. Note that the seed characters are expressions of the genotype of the parent plants not that of the embryo (next generation plant) contained therein.

Table 12-2

Mendel's Experimental Crosses of Traits and Results			
Parental traits	F1 Result	F2 Result	F2 Ratios
Yellow × green seeds	100% yellow	6022 yellow; 2001 green	3.0:1
Round and wrinkled seed	100% round	5474 round; 1850 wrinkled	2.96:1
Red × white flowers	100% red	705 red; 224 white	3.15:1
Inflated × constricted pods	100% inflated	882 inflated; 299 constricted	2.95:1
Long × short stems	100% long	787 long; 277 short	2.84:1
Axial × terminal flowers	100% axial	651 axial; 207 terminal	3.14:1
Green × yellow pods	100% green	428 green; 152 yellow	2.82:1

When Mendel let the F1 plants self-pollinate, he noticed that in the next generation approximately one-fourth of the plants had green seeds. As can be seen from these results, the hereditary factor for the recessive expression of the traits reappeared in the **F2** generation after being absent in the F1 generation. So the recessive factor can be carried through a generation unchanged even when it is not expressed. This phenomenon held true for each of the other six character combinations studied; the recessive expression was totally absent from all F1 plants that were allowed to self-pollinate. Most important, these recessive characters appeared in the F2 generation in approximately a 25% frequency in every case.

Mendel concluded from these results that each plant has two hereditary factors for each trait studied. When gametes are formed, these two factors segregate and only one factor of each pair is included in each gamete. The formation of a new individual following fertilization has two of these factors coming together, one from each parent, to direct the expression on the trait in question. Mendel's **principle of segregation** explains how a recessive character can be hidden, but not lost, from one generation to the next.

By combining what is now known about genes, meiosis, and the separation of homologous chromosomes with Mendel's experimental crossing results, an explanation is possible as to how these traits are inherited and controlled. Here, letters represent the two possible alleles of a given trait, the capital letter for the dominant allele and the lowercase of the same letter for the recessive. The mathematical possibilities can be traced for the experimental crosses to show what happens to these genes. *Y* represents the allele (hereditary factor of Mendel) for yellow, *y* the allele for green. A pure parent producing yellow seeds would have two alleles for yellow, one on each of a homologous pair of chromosomes and are the same loci on those chromosomes. The genotype for this yellow-seed parent would have a *YY*. Similarly, a pure parent for green seed would have a *yy* genotype. During meiosis in each of these two parents, the two alleles for seed color segregate into different gametes produced containing one of the *y* alleles, 50% the other. In a yellow (*yy*) × green (*yy*) cross, therefore, all offspring produced during fertilization will have a *Y* allele from one parent and a *y* allele from the other. All the F1 plants thus produced would have a *Yy* genotype and a yellow phenotype. Although only the phenotype can be seen, the genotype of the F1 is known because the genotypes of the pure parents were known.

This cross also shows how the yellow allele, *Y,* has complete dominance over the allele for green seed color, *y.* When there is one of each of these alleles, the phenotype is always yellow. A **Punnett square** visually demonstrates how alleles for a given trait from two parents segregate. Fifty percent of the gametes receive one of them, and 50% receive the other. Joining each possible gamete type of one parent with each possible gamete type of the other parent is done by connecting each of the two letters across the top of the square with each of the letters down the side of the square. It can be seen that each union of gametes produces a *Yy* genotype. Four out of four genotype are *Yy*. And this is termed a **heterozygous** genotype (*hetero*, "different"). Both *YY* and *yy* genotypes are called **homozygous** (*homo*, "same"); *YY* is homozygous dominant and *yy* homozygous recessive. Only the homozygous recessive genotype, *yy*, will produce genotypes that do not include a dominant allele for yellow. (see Figure 12-7).

Figure 12-7

Punnett square: AA × aa monohybrid cross. The offspring are three with the dominant phenotype and one recessive giving the 3:1 ratio.
Courtesy of Carolina Biological Supply.

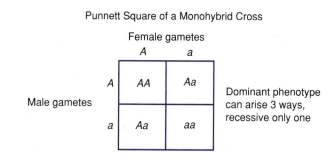

Punnett Square of a Monohybrid Cross

Female gametes

Dominant phenotype can arise 3 ways, recessive only one

When Mendel allowed the F1 plants to self-pollinate, he allowed plants all having the heterozygous genotype *Yy* to interbreed. Each F2 plant produces 50% *Y* and 50% *y* gametes because of the gene segregation during meiosis. The resulting F2 genotype frequencies are 1*YY*:2*Yy*:1*yy*. These genotypes produce a 3 yellow:1 green phenotypic ratio because *YY* and *Yy* will yield the same phenotype—yellow seed. This 3:1 ratio was approximated by all of Mendel's crosses. Of course, a Punnett square established the *theoretical* mathematical frequencies for each possibility, and the larger the number of crosses, the closer the observed ratio should approximate the expected ratio. Mendel's results are considered to be amazingly close to the expected, possible frequencies. Remember that Punnett squares reflect ratios, not the actual number of offspring.

Backcrosses

When an offspring is crossed with one of the two parental types that gave rise to it, backcross has occurred. For instance, if one of the

F1 plants has a *Yy* genotype and was backcrossed to the homozygous recessive parent with green seeds (*yy*), what would be the genotype and phenotypic ratios produced? A Punnett square setting up the **backcross** would indicate that the resulting progeny should have a 50:50 ratio of *Yy* and *yy* genotypes. Since all *Yy*'s will be yellow and all *yy*'s will be green, this will result in a 50:50 yellow/green phenotype *yy* ratio. The backcross between an F2 (*Yy*) with the yellow seeds of the F2 offspring and one of its parents (all *Yy* because they are F1 plants) is also possible; however, there is a problem with visually interpreting the results because there are two possible genotypes, *YY* and *Yy*, for yellow seed. All that can be seen is yellow seeds: The genotype cannot be seen directly. In this situation, there is a way to determine the genotype for a plant with a dominant phenotype (in this case yellow seed)—a testcross is performed.

Testcrosses

When the phenotype for a given trait is the dominant expression of that trait, the genotype could be either homozygous dominant or heterozygous. For the yellow-versus-green seeded pea plants just discussed, yellow could be either *YY* or *Yy*. To determine which genotype a plant possesses, a testcross is performed. Since the yellow seeded plant has the "unknown" genotype, a cross must be made with a plant with a known genotype. This makes possible offspring with different phenotype, depending on which of the two dominant genotypes the unknown plant possesses. A plant with the homozygous recessive genotype, a green-seeded plant, is used for the known genotype.

Thus, a testcross would be a green-seeded plant (*yy*) × a yellow-seeded plant (either *YY* or *Yy*). Establish two Punnett squares and work through the two possible crosses: *yyxYY* and *yyXYy*. You should find that the first cross of homozygous recessive × homozygous dominant would result in 100% yellow-seeded offspring. The second cross of homozygous recessive × heterozygous would result in a phenotypic ratio 50:50, yellow- and green-seeded offspring. Depending on the results of the testcross, therefore, the genotype of the yellow-seeded plant can be determined. Such a cross between plants with dominant and recessive phenotypes will work for any trait that is controlled by complete dominance, as were the seven pea plant traits studied by Mendel.

Incomplete Dominance

In all seven of the pea plant traits studied by Mendel, one allele always showed complete dominance over the other allele for that trait. For other traits, however, this is not always the case: The expression of many traits is controlled by the alleles always blending their contributions. Such genetic controls are said to be examples of incomplete dominance. In Mendel's pea plants, red flowers were completely dominant over white. A heterozygous genotype, therefore, would have a red flower. The genotype frequencies for incomplete dominance crosses are based on the same principle of segregation; only the phenotype expression of heterozygous individuals will differ. In a plant with **incomplete dominance** for flower color, such as snapdragon, heterozygous plants would be pink. Two pink-flowered plants cross-pollinating would produce 25% red-, 50% pink-, and 25% white-flowered offspring. The 1:2:1 genotypic ratio of homozygous dominant/heterozygous/homozygous recessive is directly reflected by the phenotype in plants with incomplete dominance, as shown in Figure 12-8.

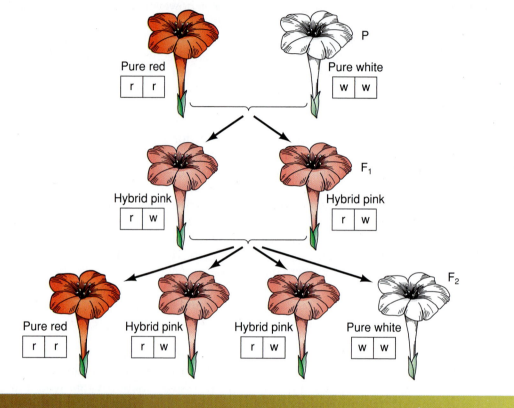

Pure red

| r | r |

Pure white

| w | w |

P

Hybrid pink

| r | w |

Hybrid pink

| r | w |

F₁

Pure red

| r | r |

Hybrid pink

| r | w |

Hybrid pink

| r | w |

Pure white

| w | w |

F₂

Figure 12-8

Sometimes the genes do not express dominant nor recessive traits. In this case, the gene effects are used together and come up with a different color. For example, red-flowered plants crossed with white-flowered plants. The offspring of a red gene and a white gene would give the flower a pink color. Courtesy of Rick Parker.

Dihybrid Inheritance

To this point, we have discussed the inheritance of only one trait at a time, or **monohybrid crosses**. To consider only a single trait is convenient and serves to demonstrate the basic mechanisms of Mendelian genetics, but it is not a realistic approach. A **dihybrid cross** involves inheritance patterns for two traits considered simultaneously, a **trihybrid cross** involves three simultaneous traits, and so on. There are two different situations possible when more than one trait is being studied. The alleles controlling one trait can be on a different chromosome pair from the alleles controlling the other trait, or both allele pairs can be on the same chromosome pair. In the latter case, the allele pairs can be on the same chromosome pair. In the latter case, the two traits are said to be **linked**. We discuss linkage later in this chapter, but first let us consider dihybrid inheritance of traits located on different chromosome pairs (unlinked traits).

If we were to select two of the morphological traits studied by Mendel, we would study their inheritance patterns together much as we did with a single trait. If a plant that is pure, or homozygous, for round yellow seed is crossed with a plant that has wrinkled green seed, we would have a cross similar to the single trait of seed color examined earlier in the chapter. Each parent plant would be homozygous—one homozygous dominant for both traits, for seed color, the letters *R* for round *r* for wrinkled will also be used. The homozygous dominate parent, then, would have a *YY/RR* genotype with the two *Y's* representing two alleles for yellow and the two *R's* representing two alleles for round. The other parent will be homozygous recessive for each trait and would have a genotype of *yy/rr*. Homologous chromosomes line up and separate during meiosis I independently of other pairs (see Figure 12-9). This is called the **principle of independent assortment** and is very important in the production of genetic variability.

Figure 12-9

Dihybrid cross of two pure parents, one homozygous dominate for two traits (AABB), the other homozygous recessive for those two traits (aabb). All F2s are heterozygous for both traits (AaBb) and all have the dominant phenotype of purple flowers and red leaves. Courtesy of Carolina Biological Supply.

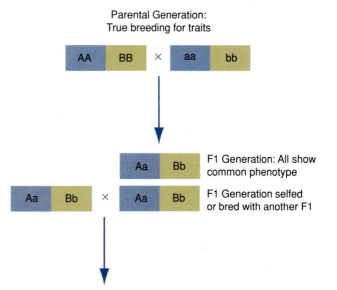

Parental Generation:
True breeding for traits

AA BB × aa bb

Aa Bb — F1 Generation: All show common phenotype

Aa Bb × Aa Bb — F1 Generation selfed or bred with another F1

F2 Generation with various phenotypes

The more homozygous pairs there are the greatest number of possible alignments. The formula for determining the possible independent alignments is *2n*, where *n* = the number of bivalents involved. Using the formula for figuring the possible combinations in a dihybrid cross, we can see that there are two or four possible combinations of alleles going to different gametes. Three bivalent (*2n = 6*) gives 23 or 8 possible alignments. There would be eight possible combinations when using a second color for each one of the homologues. Even though each gamete has the same genetic composition for this cross, it is important that the Punnett square be set up using all possible combinations. Of course, only one of the total number of alignments that could occur actually happens in each parent cell undergoing meiosis.

All the F1 plants resulting from this cross are genotypically heterozygous for both traits and all have yellow and round seeds, since both yellow and round expressions are completely dominant over their recessive counterparts. At this point, the dihybrid cross is essentially identical to a sample monohybrid cross, except with two traits. The next step, however—crossing F1 among them—is a different situation. The principle of independent assortment applies in determination of the possible genetic combinations of this cross.

The resulting phenotype is in a 9:3:3:1 ratio of yellow round/ yellow-wrinkled/green-round/green wrinkled. Again, these are ratios of expected phenotypes: The total of 16 does not represent the number of F2 offspring produced.

As complicated as this dihybrid cross was, imagine such a cross with characters displaying incomplete dominance or a trihybrid cross (three nonlinked traits). This use of the Punnett square becomes a bit unwieldy beyond dihybrid crosses, and even trying to represent visually the possible phenotypes in an incomplete dominance dihybrid cross is more confusing than clarifying. The point is that when the study of inheritance includes the more realistic situations involving multiple traits or expressions other than complete dominance, the complexity of such a study is much greater.

Linkage

It should be obvious that many genes occur on a single chromosome, and therefore at telophase I they must segregate as a group. Such genes or alleles are said to be linked. Linkage serves to stabilize the genome, allowing coordinated genes to continue acting together. If all genes sorted independently, there would be a significantly greater chance for unsuccessful recombinations to occur, resulting in poorly adapted individuals. Thus, linkage is important in the balance between genetic stability within a generation and the production of new recombinations and variability. Two linked traits will not produce the typical 9:3:3:1 ratio in the F2 generation of a dihybrid cross (as previously discussed) because the genes are not subject to independent assortment during meiosis. Until the mechanism of crossing over was elucidated by a geneticist working on fruit flies (*Drosophila*), such varied crossing results were difficult to understand. Now it is easy to see how, the linked traits can become "unlinked" by a crossover. The farther apart physically such genes are, the greater the likelihood that they will cross over; the closer together, the stronger the linkage. However, genes that are very far apart on a long chromosome arm may actually have a higher linkage frequency than others closer together because two crossovers on the same arm can relink them.

Chromosome Mapping

A genetic linkage map shows the relative locations of specific DNA markers along the chromosome. Each marker is like a mile marker along a highway. Any inherited physical or molecular characteristic that differs among individuals and is easily detectable is a potential genetic marker. Markers can be expressed DNA regions (genes) or DNA segments that have no known coding function but whose inheritance pattern can be followed. DNA sequence differences are especially useful markers because they are plentiful and easy to characterize precisely. Markers must be polymorphic to be useful in mapping; that is, alternative forms (alleles) must exist among individuals so that they are detectable among different members in the mapping population.

A mapping population is the group of individuals that will be evaluated for their "score" at a set of markers. This raw mapping data is analyzed by software that constructs the map by observing how frequently the alleles at any two markers are inherited together. The closer the markers are, the less likely it is that a recombination event (a crossover during meiosis) will separate the alleles, and the more likely it is that they will be inherited together. Thus, unlike other types of maps, the distance between points on a genetic map is not measured in any kind of physical unit; it is a reflection of the recombination frequency between those two points. This genetic map unit is measured in terms of centimorgans (cM, named after the geneticist Thomas Hunt Morgan). Two markers are said to be 1 cM apart if they are separated by recombination 1% of the time. The genetic distance tells you little about the physical distance—the actual amount of DNA separating the markers. This genetic to physical distance relationship varies between species, and varies between different spots within the genome of a single species.

A genetic map helps us understand the structure, function, and evolution of the genome. It can be an important tool for agricultural crop improvement. Recent work has shown that the genetic maps of many closely related species (for example, the grains) are quite similar with respect to the content and location of genes, and scientists are trying to determine how the genetic map of one species may be applied to others.

Gene Interaction

Up to this point, inheritance has been discussed in terms of a single pair of alleles controlling one trait. For purposes of demonstrating the mechanism of genetic control and inheritance traits, a one-gene/one-trait model is easy to understand. However, probably very few traits are exclusively controlled by a single allele pair. It is very likely that most genes do not act alone; rather, interactions of nonallelic pairs of genes control the expression of the phenotype. It is truly remarkable that Mendel was lucky enough to have chosen those seven traits in the garden pea that exhibit complete dominance, no linkage problems, and apparently single allelic control. Apparently they are single-gene traits because it is very possible that other gene interactions also are involved in the control or expression of these traits, even though there is no overt incidence for such action.

Epistasis

One of the most common examples of gene interactions is one gene having a masking effect on the expression of another, nonallelic gene. It is worth noting that this is not a case of dominance because **epistasis** genes influence nonallelic genes. Dominance is the effect of one allele over the other in an allele pair.

An interesting example of epistasis has been discovered in white clover (*Trifolium repens*). Hydrocyanic acid (HCN) is present in high concentrations in some strains of white clover, whereas other strains contain none. Some experimental crosses between positive (with HCN) and negative strains have resulted in an unusual F2 ratio for a species previously thought to have a single gene pair control of HCN presence with a positive dominance. For such a plant, the expected ratio of positive to negative in the F2 would be 3:1. When the following results were obtained, it was difficult for the researchers to accept a single-gene pair control:

P	POSITIVE × NEGATIVE
F1	100% positive
F2	351 positive:256 negative

The 351:256 results more closely resemble a 9:7 ratio, which would indicate a modified 9:3:3:1 ratio of a dihybrid cross, indicating two pairs of genes. Epistasis can operate in any cross involving two or more pairs of genes. The results of epistatic action would be a reduction in the number of expected phenotypic expressions in which two or more of the classes become indistinguishable from each other.

It was discovered that white clover has two different enzymes that affect the production of HCN in the following way:

GENE A	GENE B
Enzyme a	Enzyme b

Precursor compound ⟶ Cyanogenic glucoside ⟶ HCN

An individual, therefore, must have at least one dominant of each of two pairs of genes, A and B, to produce HCN from the chemical precursors available in all the plants. Without epistasis, a 9:3:3:1 ratio of enzymes a and b/enzyme a only/enzyme b only/neither enzyme would be normal for a dihybrid cross in the F2 generation. The 9:7 ratio would be enzymes a and b (HCN production)/no HCN because one or the other enzyme is absent, and biochemically all the other results are the same due to epistatic influence.

There are other examples of epistasis in corn, mice, chicken, and other animals. The actual effects of the epistatic genes often vary from the example just given, but the general control of such genes is predictable within any genetic system studied.

Pleiotropy

Not only do genes interact to control the expression of a single trait, but many genes affect more than one trait. Single genes that can affect the expression of several traits are said to be **pleiotropic**, because of the nature of all gene action—the production of proteins and resulting enzymes—it is probable that most genes have some influence on other traits during the development of the organism. Thus, not only does more than one pair often influence the phenotypic expression of a single trait, but also single genes can be involved in the control of many different characteristics.

Multigenic Inheritance

Most of the genetic controls of phenotypic characteristics discussed thus far have been discrete or discontinuous. There is no blending; rather, the phenotype represents one expression or the other. Flowers are red or white, seeds are yellow or green, and seed coats are round or wrinkled. Even considering the effects of incomplete dominance, epistasis, and pleiotropy as modifiers of genetic control, the possible phenotypes change only in degree or number of discrete possibilities. For a trait to vary continuously there must be **multigenic** or polygenic control.

Multigenic control is one of the combined additive effects of two or more allele pairs of genes. Height, color, and shape of plants and plant parts are typical examples of traits with continuous variations of expression. Fruit or inflorescence size in many plants, and human skin, hair, and eye color also vary in a continuous fashion. The more allele pairs involved in the multiple gene control of a trait, the more continuous is the blending of possible phenotypes. Of course, it is unrealistic to assume that all allele pairs affecting the same trait additively does so with equal influence. Environmental changes and modifier genes also can come to bear on final phenotypic expression. In spite of how little is really known about multigenic inheritance, we can assume that many traits are controlled in this manner, and considerable research effort has been directed at better understanding the phenomenon of multiple gene control of heredity.

Molecular Genetics

Since Mendel's pioneering work, the revelation of DNA structure must be ranked as the most significant discovery in the area of genetics. Since the mid-1970s, the area of molecular genetics has seen incredible breakthroughs in understanding gene action. Researchers study inheritance at the nucleotide level. They have learned to cut, splice, and insert genes into organisms and study the effects. Most of this **genetic engineering** started with bacteria and viruses but has progressed to the point that many plants have been transformed. For example, corn is now transformed with *Bacillus thunguriensis* to keep the corn ear-worms at bay. Roundup soybeans, are resistant to Roundup so the crop can be sprayed with the herbicide. This is only a few of the examples of the use of genetic engineering.

The engineered genes have been used to produce insulin, interferon, human growth hormones, various vaccines, a hormone that makes dairy cows produce more milk, and cloned sheep, calves, and cats. The genes that have been painstakingly produced for these and other experimental uses are often termed **recombinant DNA**. Basically, any genetic material that is produced synthetically or in a different organism and then introduced into the test organism is called recombinant DNA. The two areas given greatest attention by molecular geneticists are the production of new medicines and the improvement of domesticated plants and animals. At present, the world market for genetic engineering in agriculture alone could easily be $400 billion. Although the potential for introduction of new genes into food crops and livestock to enhance their productivity is great, these applications will be somewhat slower to develop than in areas of medicines. The safety of these products for human consumption has to be tested. Acceptance by the public will be the cause of the slowdown.

PUBLISHING GENOME SEQUENCES FOR PLANTS	DNA was first sequenced in 1977. The first free-living organism to have its genome completely sequenced was the bacterium *Haemophilus influenzae*, in 1995. Then in 1996, baker's yeast (*Saccharomyces cerevisiae*) was the first eukaryote genome sequence to be released, and in 1998, the first genome sequence for a multicellular eukaryote, a free-living nematode about 1 mm long (*Caenorhabditis elegans*), was released.

The "Gene Machine"

The production of a synthesized gene once took four to eight months of tedious biochemical manipulations. Since the advent of gene synthesizers in 1982, functional gene segments of one's choosing can be automatically reproduced in a single day; by simple typing in the desired genetic sequence on the gene machine's keyboard, geneticists can splice the appropriate nucleotides automatically, and a gene fragment is ready to be introduced into the DNA of an experimental organism. Existing sequences can be duplicated or modified, or researchers can even design totally new sequences.

By using this jointly with a **protein sequencer**, a machine that can read out the exact sequence of amino acids in a protein, researchers can now carry out sophisticated experiments rapidly.

The protein sequencer can analyze even tiny amounts of a useful enzyme (protein) in only a few hours. Which amino acids are in the protein, and in what order, can next be programmed into the gene synthesizer. This machine will produce a genetic fragment to be inserted into bacteria to direct the production of large quantities of the desired protein. There are many new instruments used in genetic engineering to simplify each step in the use of DNA production, and the **polymerase chain reaction (PCR)** is one of those.

Who would have thought a bacterium hanging out in a hot spring in Yellowstone National Park would spark a revolutionary new laboratory technique? The PCR, now widely used in research laboratories and doctor's offices, relies on the ability of DNA-copying enzymes to remain stable at high temperatures. No problem for *Thermus aquaticus*, the sultry bacterium from Yellowstone that now helps scientists produce millions of copies of a single DNA segment in a matter of hours.

In nature, most organisms copy their DNA in the same way. The PCR mimics this process, only it does it in a test tube. When any cell divides, enzymes called polymerases make a copy of the entire DNA in each chromosome. The first step in this process is to "unzip" the two DNA chains of the double helix. As the two strands separate, DNA polymerase makes a copy using each strand as a template.

The four nucleotide bases, the building blocks of every piece of DNA, are represented by the letters A, C, G, and T, which stand for their chemical names: adenine, cytosine, guanine, and thymine. The A on one strand always pairs with the T on the other, whereas C always pairs with G. The two strands are said to be complementary to each other.

To copy DNA, polymerase requires two other components: a supply of the four nucleotide bases and something called a primer. DNA polymerases, whether from humans, bacteria, or viruses, cannot copy a chain of DNA without a short sequence of nucleotides to "prime" the process, or get it started. So the cell has another enzyme called a primase that actually makes the first few nucleotides of the copy. This stretch of DNA is called a primer. Once the primer is made, the polymerase can take over making the rest of the new chain.

A PCR vial contains all the necessary components for DNA duplication: a piece of DNA, large quantities of the four nucleotides, large quantities of the primer sequence, and DNA polymerase. The polymerase is the **Taq polymerase**, named for *Thermus aquaticus*, from which it was isolated.

The three parts of the polymerase chain reaction are carried out in the same vial, but at different temperatures. The first part of the process separates the two DNA chains in the double helix. This is done simply by heating the vial to 90–75°C (about 165°F) for 30 seconds.

But the primers cannot bind to the DNA strands at such a high temperature, so the vial is cooled to 55°C. At this temperature, the primers bind or "anneal" to the ends of the DNA strands. This takes about 20 seconds.

The final step of the reaction is to make a complete copy of the templates. Since the Taq polymerase works best at around 75°C (the temperature of the hot springs where the bacterium was discovered), the temperature of the vial is raised.

The Taq polymerase begins adding nucleotides to the primer and eventually makes a complementary copy of the template. If the template contains an A nucleotide, the enzyme adds on a T nucleotide to the primer. If the template contains a G, it adds a C to the new chain, and so on to the end of the DNA strand. This completes one PCR cycle.

The three steps in the polymerase chain reaction—the separation of the strands, annealing the primer to the template, and the synthesis of new strands—take less than two minutes. Each is carried out in the same vial. At the end of a cycle, each piece of DNA in the vial has been duplicated.

But the cycle can be repeated 30 or more times. Each newly synthesized DNA piece can act as a new template, so after 30 cycles, 1 billion copies of a single piece of DNA can be produced! Taking into account the time it takes to change the temperature of the reaction vial, 1 million copies can be ready in about three hours.

PCR is valuable to researchers because it allows them to multiply unique regions of DNA so they can be detected in large genomes. Researchers in the Human Genome Project are using PCR to look for markers in cloned DNA. Many medically and industrially important substances are proteins. The availability of this technology will allow researchers to work much more efficiently and carefully toward ethically acceptable uses for recombinant DNA in genetic engineering applications. One of the major applications is in the criminal justice system. Many people serving time in jail have been proven not guilty using the DNA matches from evidence. Now it is possible to use modifications of existing genes, or copies of them, to be put into organisms in which such traits would be desirable. Existing genes and even completely new genes are used in these efforts. A less expensive method for inserting the gene construct into the plant's DNA can be done using *Agrobacterium tumefaciens* (a bacterium) to insert the gene constructs into the DNA of the desired plant.

Genetic Gibberish

The ability to biochemically disassociate the individual nucleotides of a gene and establish their sequence led to one of the most astonishing discoveries in genetics since the discovery of DNA. Molecular geneticists have found that the complete sequence of nucleotides in the DNA is not translated into the messenger RNA. Mixed in with the actual genetic code for a given protein are segments of genetic gibberish or strings of nucleotides within the nucleus—a long strand of heterogeneous nuclear RNA. This long RNA becomes edited to delete the gibberish segments, leaving only the sensible segments that are spliced together to form mRNA in the existing nucleus. The long RNA is a direct, one-to-one copy of the DNA message, including the gibberish material.

It is provable that the sensible portions are actually much shorter than the gibberish interruptions. The sensible segments have been called **exons** because they are expressed, whereas the gibberish parts are called introns because they are intragene segments. Data indicate that exons average about 100 to 300 genetic letters, whereas introns average 1,000, with some having up to 10,000 letters within a single gene.

A number of interesting hypotheses suggest why there are any intron segments at all, let alone so many more than exon segments. These long spaces improve the odds that, when errors occur or when fragments or genes are exchanged or lost during recombination, the breaks will occur in intron segments, leaving the exon intact. Thus, all the necessary genetic code for the production of a specific protein will be available in the form of mRNA once all the introns are cut out and the previously interrupted exons are spliced together.

Other molecular geneticists theorize that these intron segments are not just spacers but are actually involved in regulation, possibly instructing the DNA as to which protein should next be coded into the RNA strand. Still others would condone that exons and introns provide the genetic flexibility necessary to allow for the occasional production of novel new genes. In this way, point mutations and genetic recombination would be supplemented significantly as sources for new genetic variability. Since genetic variability is the raw material for evolutionary change, this area of molecular biology has been termed molecular evolution.

Undoubtedly, few fields of biology have ever moved forward as quickly or generated as much interest as the area of molecular biology. Especially intriguing is genetic control and the potential for manipulating that control.

Summary

1. Sexual reproduction ensures variability among offspring. The process of meiosis is the key to producing that variability. Meiosis produces haploid cells and forms a diploid cell. The fusion (fertilization) of two haploid cells produces a new, single-celled zygote with a unique combination of genes. In angiosperms, meiosis occurs in the anthers and ovules of the flower.

2. Meiosis is two sequential nuclear divisions: The first is a reduction division producing nuclei with only a single set of chromosomes (haploid); second is like mitosis, further producing genetically different cells. The sequence of nuclear and cellular events is different cells

or similar to that of mitosis in both meiosis I and II, except that in the former, homologous chromosomes pair up and then separate.

3. Crossing over, the formation of chiasmata, results in additional genetic recombination, thus, increasing variability. Mutations may also result in genetic variability.

4. All plants have alternation of generations between a haploid spore-producing (sporophyte) stage and a gamete-producing (gametophyte) stage. In angiosperms, both stages occur within the tissue of the flower. In the anthers, microsporogenesis is followed by microgametogenesis. In the ovules, megasporogenesis and then megagametogenesis occur. The resulting gametes fuse in a unique double fertilization to produce a new sporophyte stage in the form of a diploid zygote that will develop into the embryo of the seed, complete with a triploid nutritive tissue.

5. All sexually reproducing, diploid organisms have two genetic expressions (alleles) for each trait. The study of inheritance patterns in peas by Gregor Mendel helps elucidate the ideas of gene control of phenotypic expression in offspring. Traits with complete control of dominance can be homozygous dominant, homozygous recessive, or heterozygous dominant genotype. Experimental crosses between pure patterns producing F1 and then between F1 plants producing F2 offspring, can demonstrate the patterns of gene control.

6. Backcrosses between offspring and parent plants, especially a testcross between a dominant plant and a homozygous recessive plant, enables research to study gene control. Some traits have incomplete dominance, allowing for a blending of genetic control and intermediate phenotypic traits.

7. Dihybrid crosses involve two traits. When the alleles controlling those traits are on the same chromosome, they are said to be linked. Other gene interactions include the modifying effects of epistasis and pleiotropic and the complex control of multigene influence.

8. Molecular genetics is an area of science that studies the modification of genes. Genetic engineering has allowed the artificial production of recombinant DNA. Technology advancement in the form of gene synthesizers and protein sequences are enabling scientists to study gene function much more rapidly than ever before.

9. The actual genetic code found in messengers, RNA is now known to have genetic gibberish or intron segments interspersed between the exon segments that actually carry the protein-producing genetic sequence to the ribosomes. The study of molecular evolution is a promising new approach to the understanding of gene function and control.

Something to Think About

1. Define sexual propagation.

2. Diagram a complete flower and show how pollination and fertilization takes place; label all parts.

3. Compare and contrast meiosis and mitosis; determine each phase.

4. Describe what takes place in each phase.

5. What is crossing over?

6. Diagram and label the sexual life cycle of angiosperms; indicate where mitosis and meiosis occur.

7. Diagram and label the sexual life cycle of gymnosperms; indicate where meiosis and mitosis occur.

8. What is a Punnett square?

9. Make a Punnett square with pure parents producing F1s and then between F1 plants producing F2 offspring.

10. Who was the father of genetics?

11. What is dominance in genes, recessive genes, and intermediate genes?

12. Write the abbreviations for homozygous dominance, homozygous recessive, and incomplete dominance.

13. Define molecular genetics and explain how it is being used in your life.

14. What is a genome and why is it important not only in plants but also in animals and humans?

15. Explain how you feel about genetic engineering.

Suggested Readings

Charles, D. 2002. *Lords of the harvest: Biotech, big money and the future of food.* New York: Perseus Book Group.

Howell, S. 1998. *Molecular genetics of plant development.* Cambridge, MA: Cambridge University Press.

Leyser, O., and S. Day. 2002. *Mechanisms in plant development.* Malden, MA: Blackwell Publishing Professional.

Slater, A., N. W. Scott, and M. R. Fowler. 2003. *Plant biotechnology: The genetic manipulation of plants.* New York: Oxford University Press.

Steeves, T. A., and J. M. Sussex. 2004. Patterns in Plant Development (2nd ed.). Cambridge, MA: Cambridge University Press.

Internet

Internet sites represent a vast resource of information. The URLs for Web sites can change. Using one of the search engines on the Internet such as Google, Yahoo!, Ask.com, or MSN Live Search, find more information by searching for these words or phrases: **plant sexual life cycle, plant genetics, meiosis, mitosis, Gregor Mendel, F1 generation, F2 generation, phenotype, genotype, microsporogenesis, megasporogenesis, dihybrid crosses, molecular genetics, PCR, and recombinant DNA.**

Genetic Engineering and Biotechnology

CHAPTER **13**

Today, agriculture needs a new infusion of science and technology and new capabilities that will restore and enhance the competitiveness of United States' agriculture in the world marketplace. The products of biotechnology offer one of the most exciting opportunities to meet these urgent needs.

After completing this chapter, you should be able to:

- Define biotechnology, genetic engineering, and related terms
- Understand the basic processes of biotechnology research
- Recognize the degree of progress made in biotechnology research up to this point
- Identify the latest developments or applications resulting from biotechnology research in plant science
- Describe future impacts of biotechnology research and genetic engineering
- Discuss environmental, ethical, control, and conflict of interest concerns brought about by biotechnology research
- Name five plants altered by genetic engineering
- List four goals of genetic engineering in plants
- Describe the process of genetic engineering
- Explain how DNA controls formation of proteins
- Describe a transgenetic plant
- Diagram how the genetic code is translated
- Give three advantages of genetic engineering over traditional selective breeding

Key Terms

biotechnology
genetic engineering
genetic code
helix
messenger RNA (mRNA)
nucleotides

recombinant DNA (rDNA)
plasmids
cross-protection
biocontrol
transgenic

biopharmaceuticals
genomics
bioinformatics
proteomics

Biotechnology Defined

The term **biotechnology** refers to an array of related basic sciences that have as their centerpiece the use of new methods for the manipulation of the fundamental building blocks of genetic information to create life forms that might not ever emerge in nature—life forms that can expand and enhance the well-being of humans.

The American Association for the Advancement of Science called **genetic engineering** one of the four major scientific revolutions of this century, on a level with unlocking the atom, escaping the earth's gravity, and the computer revolution. The newfound ability to manipulate cellular machinery has been termed a biotechnology revolution. It could have as profound effect on our society as has the information revolution occurring alongside it. Many believe that the impact of biotechnology will be as great or greater in agriculture as in medicine.

The biotechnology revolution in agriculture is a part of an overall increasing sophistication of biological techniques for improving the production, processing, and marketing of food and fiber. Biotechnology, for instance, allows an acceleration of the process of selection and breeding that has been under way for over 100 years.

In the case of agriculture, as shown in Figure 13-1, biotechnology is built on a broad base of existing and ongoing scientific research that supports and enhances the use of the new methods in genetic engineering and related techniques. This science base helps define what should be genetically engineered and enables the products of fundamental research in the laboratory to be practically applied in the field.

Figure 13-1

Genetic engineering (the process of manipulating or inserting genes in an organism) is being used on more and more crops. The crops or products represented here have either already been genetically engineered or are involved in ongoing or planned transgenic studies. Courtesy of USDA/ARS.

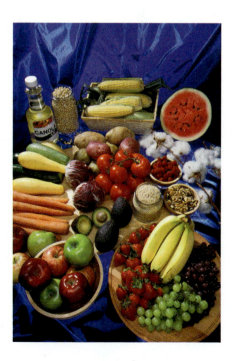

Genetic Engineering

Every cell in plants, animals, and microbes contains the genetic information to allow perpetuation of that cell or organism. The study of the structure, chemistry, and function of this genetic material has been the basis of understanding that has enabled the biotechnology revolution to come of age.

Genetic engineering, or manipulation, involves taking genes from their normal location in one organism and either transferring them elsewhere or putting them back into the original organism in different combinations. Its value to biotechnology is twofold.

First, scientists can take useful genes from plant and animal cells and transfer them to microorganisms such as yeasts and bacteria that are easy to grow in large quantities. Products that once were available only in small amounts from an animal or plant are then available in large quantities from rapidly growing microbes. One example is the use of genetically engineered bacteria to produce human insulin for treating diabetes.

The second benefit holds particular promise for plant and animal breeders. Genetic engineering allows desirable genes from one plant, animal, or microorganism to be incorporated into an unrelated species, thus avoiding the constraints of normal crossbreeding. A wider range of traits is available to the breeder, and these traits can be incorporated more quickly and more reliably into target species than possible with conventional methods.

Needed Knowledge

Genetic material is arranged in helical strands that contain the code for triggering the characteristic functions of that organism in succeeding generations. The discovery of the structure of the DNA helix in the 1950s (see Table 13-1), the unraveling of genetic code in the 1960s, and the development and refinement of the tools of genetic engineering in the 1970s has caused something fundamentally different to happen in biotechnology science. Using enzymes as "genetic scissors,"

Table 13-1	
Brief History of Genetic Engineering	
Year	**Event**
1944	DNA identified as genetic material
1953	Double strand DNA structure identified
1973	First transgenic bacteria prepared
1976	First genetic engineering company (Genentech) established
1980	First patent for genetically engineered microbe
1982	Approval of first genetically engineered drug
1986	First field test of genetically engineered plant
1987	Genetic engineering patent extended to higher life forms

the genetic structure of cells can be snipped apart and reconstructed in combinations impossible to achieve by natural reproduction (see Figure 13-2). Scientists can not only alter existing genes, but construct totally synthetic genes to cause the organism to perform a desired function.

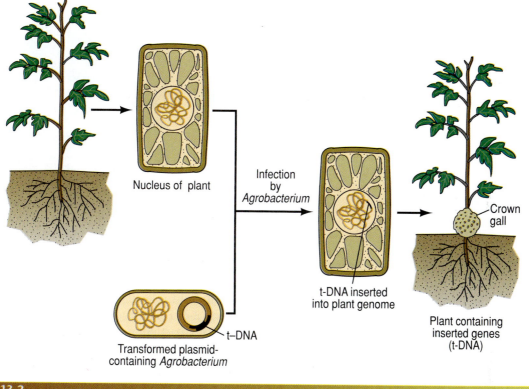

Nucleus of plant

Infection
by
Agrobacterium

t-DNA inserted
into plant genome

Crown
gall

t–DNA

Transformed plasmid-
containing *Agrobacterium*

Plant containing
inserted genes
(t-DNA)

Figure 13-2

Gene splicing transfers beneficial genes in one species and transfers them to another species.

To make genetic engineering work, methods are being developed to transfer genetic material into plants and animals and to make sure that the function that has been engineered is expressed at the right place and at the right time. But genetic engineering is only a part of biotechnology. The total picture includes understanding the physiology and biochemistry of the function of interest and knowledge of the existing genetic codes that regulate the process. This allows scientists to understand what to genetically engineer to produce a more desirable organism. Once such product has been created in the laboratory, a variety of techniques such as tissue culture (see Figure 13-3) are often needed to recreate an organism that can compete in a practical ecosystem. The techniques of plant breeding and development are used to take the final product back to the field.

Figure 13-3

A rose plant that began as cells grown in a tissue culture. Courtesy *of USDA/ARS.*

The Genetic Code

Before scientists could undertake such genetic manipulation, they had to unravel the secrets of the **genetic code**. They discovered that DNA is a long double-stranded molecule wound in a spiral called a **helix**, as shown in Figure 13-4. Each gene is a segment of the DNA strand that usually codes for a particular protein. Proteins, like DNA, are also long, chainlike molecules. They are constructed from 20 different amino acid building blocks. They are extremely versatile molecules, anywhere from a few dozen to several hundred amino acids long. Unlike the regular spiral formed by DNA strands, proteins fold and twist into an enormous variety of three-dimensional shapes. The plant and animal bodies possess thousands of different kinds of protein and each plays a specific role in life. That role can be structural or physiological—for example, the proteins involved in the photosynthetic processes.

The DNA code is translated into amino acid sequences in proteins, through an intermediary called **messenger RNA (mRNA)**—a single-stranded molecule similar to one side of the double-stranded DNA. To be able to control the protein manufacturing process, scientists needed to understand the detail of the DNA coding system.

The DNA Jigsaw

The DNA molecule contains subunits called **nucleotides**. Each nucleotide comprises a sugar component (deoxyribose), a phosphate component, and one of four different bases—adenine (A), guanine (G), thymine (T), and cytosine (C). Scientists discovered

Figure 13-4

DNA strands divide and bases attach themselves to the new strands to form identical genes for new cells.

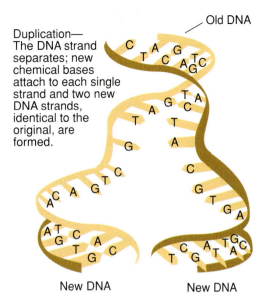

Duplication—
The DNA strand separates; new chemical bases attach to each single strand and two new DNA strands, identical to the original, are formed.

Old DNA

New DNA   New DNA

that DNA was formed from two strands of nucleotides, held together by the bonds between the bases on opposite strands. The entire structure is like a ladder. The sides are formed by the sugar and phosphate groups, and the rungs are the bases. The bases in the "rungs" are matched in pairs like pieces in a jigsaw. The two strands forming the ladder are then wound around one another to form the helix (see Figure 13-4).

These DNA molecules contain the blueprint for all proteins made in a cell. Each sequence of three bases along the DNA strand is a chemical code for 1 of the 20 amino acids—the building blocks of proteins.

Translating the Code

To make the proteins, the DNA molecule is unraveled, the strands separate, and the cell makes a copy of the relevant part, in the form of single-stranded messenger RNA. The mRNA then moves to the cell's "factories" called ribosomes, where it acts as a template for protein manufacture. The code for the protein is read off the base sequence on the mRNA, and the appropriate amino acids are added to the protein one by one, aligned against the mRNA code by small segments of RNA called transfer RNA (tRNA) (refer to Figure 13-2).

The coding system is universal. It is basically the same in all animals, plants, and microorganisms. A piece of DNA from a plant inserted into the chromosome of a bacterium makes perfect sense to the bacterial cell.

Scientists have known the detailed amino acid sequence of many key proteins for a long time. Once they also understood which base sequences in DNA were represented by which amino acids, they could identify the genes in the chromosome that coded for particular proteins.

Recombinant DNA Technology

Identifying the genes is not enough. The next step is to be able to copy the gene and insert it into other cells—cells that can be grown easily using existing microbiological techniques, or cells of other plants or animals where the protein is required (see Figure 13-5). To do this, scientists used new biochemical techniques, involving special enzymes, to break the DNA strand at chosen points, insert new segments, and "stitch" the strand back together again. The result, known as **recombinant DNA (rDNA)**, is DNA that incorporates extra segments bearing genes it had not previously contained.

Figure 13-5

Research Geneticist Ann Blechl and colleagues are the first to insert modified Fusarium chitinase and glucanase genes into wheat plants, which may lead to wheat plants that are more resistant to Fusarium head blight.
Courtesy of USDA/ARS.

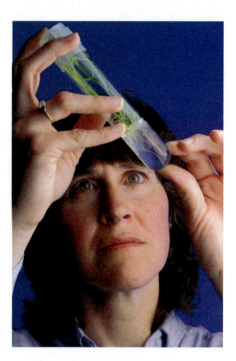

Insertion of genes into different organisms is made much easier by the existence of bacterial **plasmids**—small circles of DNA that are much smaller than the bacterial chromosome. Some of these plasmids can pass readily from one cell to another, even when the cells are far apart on the evolutionary scale. Using the special "cut and paste" enzymes mentioned earlier, scientists can insert genes from one organism into a bacterial plasmid, then insert the recombinant plasmid into a living microorganism, where it will direct the synthesis of the desired proteins. Human insulin for treating diabetes can now be produced in this way.

So far, scientists have used genetic engineering to produce the following:

- Improved vaccines against animal diseases such as foot rot and pig scours
- Pure human products such as insulin and human growth hormone in commercial quantities

- Existing antibiotics by more economical methods
- New kinds of antibiotics not otherwise available
- Plants with resistance to some pesticides, insects, and diseases
- Plants with improved nutritional qualities to enhance livestock productivity

Targeting Agriculture

In the past, agriculture has been an energy and labor intensive industry. Biotechnology offers the opportunity to reduce both these costs in future operations. Inherent resistance to pests and disease can reduce the use of chemical pesticides, reducing the cost of production and the potentially harmful environmental effects of such practices. The possible uses of biotechnology for agriculture are limited only by imagination and initiative, such as the peach and apple orchards in Figure 13-6.

Figure 13-6

Since World War II, American agriscience has progressed at a breathtaking pace. Here, a technician checks on futuristic peach and apple "orchards." Each dish holds tiny experimental trees grown from lab-cultured cells to which researchers have given new genes. Courtesy of USDA/ARS.

The total system for food and fiber production is extremely diverse and multifaceted, providing a broad range of potential applications of biotechnology. Biotechnology is not only enhancing the traditional enterprises in food and fiber production, it also is producing new high technology industries that are providing new jobs and producing new goods and services. Sometimes thinking of biotechnological applications is limited to production agriculture, where an exciting new array of scientific breakthroughs is being developed. Other exciting areas related to production agriculture include: the new application of biotechnology to food processing and manufacturing, to new methods for ecologically sound disposal of wastes, and to biochemical engineering where totally new products are being produced from agricultural residues using biotechnological tools.

Biological Factories

One of the early uses of biotechnology has been to use simple organisms such as bacteria and yeast as so-called biological factories to produce biologically active compounds. Genetic codes for these compounds are inserted into the genetic makeup of these simple organisms along with genetic instructions that cause the production of the desired material.

Through these techniques, for instance, human insulin is now produced and is replacing insulin from animal sources for treatment of diabetes in humans, as shown in Figure 13-7.

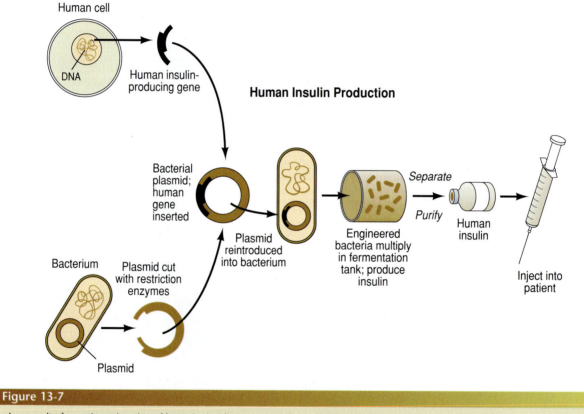

Human Insulin Production

Figure 13-7

As a result of genetic engineering of bacteria, insulin is now readily available and relatively inexpensive.

Interferon, a biological anticancer treatment and antiviral material previously available in only minute quantities, can now be produced inexpensively and in large quantities using biological factories. Hormones, such as bovine growth hormone, have been manufactured using this method. The product is being injected in dairy cattle to enhance milk production by some 20% to 30%. Diagnostic reagents and improved vaccines for animal and human disease are being produced using these techniques. Early progress has been rapid in this area because the genetics of these organisms is relatively simple.

Plant-Water Relationships

In many parts of the world, water is the limiting factor in food production. Biotechnology is being used to greatly enhance the production of plants with high drought-stress tolerance. These plants will maintain yields in environments with much less water, and there is the promise of developing plants that can use brackish water. These developments could have a profound effect on stretching water resources and will be crucial as water for irrigation becomes less available and more expensive.

Plant Productivity

Plant growth and development has been the subject of investigation for decades, but until recently remained poorly understood. Biotechnology makes it possible to isolate, characterize, and manipulate specifc genes. This new technology provides a powerful tool to understand plant growth and development and a way to directly manipulate the process. Opportunities in this field include altering chemical composition of the plant product, improving processing quality, producing plants resistant to stress or herbicides, altering plant size, changing the ratio of grain to stalk, and making an annual plant a perennial (see Figure 13-8).

Figure 13-8

Perhaps genes could be introduced to make wheat a perennial crop. Courtesy of USDA/ARS.

Nitrogen Fixation

The transfer of nitrogen-fixing genes to plants is possible with the techniques of biotechnology. Nitrogen-fixing genes might be transferred to plants that do not now have this nitrogen-fixing capacity. These genes would have to be in a form that could be incorporated into the plant genome, replicated, and expressed. The genes would have to be expressed in an environment amenable to nitrogen fixation where the enzyme, nitrogenase, could be protected from oxygen and where the enzyme system could tap into the sources of reductant and energy from the plant.

The full complement of nitrogen-fixing genes has been cloned from *Klebsiella pneumoniae* and transferred to and expressed in another bacterium, *Escherichia coli*.

Plant Disease Resistance

Genetic engineering offers an exciting and environmentally sound way of reducing the cost and increasing the effectiveness of plant pest control, through the development of genetic resistance to disease. Because disease resistance is controlled by relatively few genes, this area is among the most favorable candidates for early application of biotechnology to plants.

The search for resistant genes involves screening cultivars or species of plants to identify individuals that exhibit resistance to infection, replication, or spread of a pathogen. If the resistance trait is the result of expression of a single gene or genetic locus, plant breeders begin the task of introducing the resistance trait into a cultivar having desirable agroeconomic traits that will ultimately be released to the farmer. The plant-breeding process (see Figure 13-9)

Figure 13-9

Plant geneticist evaluates different genetic sources of alfalfa to identify plant traits that would increase growth and enhance the conversion of plant tissues into biofuel. Courtesy of USDA/ARS.

usually requires more than five years and considerable evaluation of progeny to eliminate plants that contain undesirable traits in addition to the disease-resistance trait.

Tobacco Mosaic Virus

Tobacco mosaic virus (TMV) causes the leaves of some important crop plants, including the tomato, to wither and die. Scientists have incorporated into the tomato plant a gene that protects it from infection. It has the same effect as a vaccine for humans. This approach is now being applied to other viral diseases in crops.

Manipulating Microbes

Plant microbes affect agriculture both detrimentally and beneficially. Economic losses and human suffering result from plant diseases caused by microbes, so much ecological research deals with plant pathogens (disease-producing agents). Unfortunately, with so many important crops, diseases, and agroecosystems, only a small proportion can be investigated intensively.

Scientists focus on microbial influences that maintain plant health. One of the successful applications involves cross-protection, in which the infection of plant tissues by one virus suppresses the disease caused by another closely related strain of the virus. The protecting strain must have negligible impact on the host. Such strains have been found naturally, but also can be created in the laboratory via biotechnology. **Cross-protection** has been used successfully in protecting citrus trees from severe strains of Citrus tristeza virus.

Another successful example involves the bacterium that causes crown gall of stone fruits and other plants: A nonpathogenic (or "friendly") strain produces an antibiotic that inhibits the pathogenic strain. Because the two strains are closely related, the nonpathogen survives in the same niches as the pathogen, and responds similarly to environmental fluctuations. Consequently, upon deliberate release of the biocontrol agent, close association of the two bacteria is assured. For crown gall, the biocontrol agent was naturally occurring, but with biotechnology it will be possible to engineer normal resident nonpathogenic microbes into biocontrol agents.

Some microbes can be used as biocontrol agents for weeds. For example, a fungus is used to control northern jointvetch in rice and soybean fields. Knowledge of the ecology and epidemiology of this fungus contributed to the development of a rational, effective biological control (**biocontrol**) approach. Even though the genetic and biochemical bases of pathogenicity are unknown for this pathogen, it is so specific in its actions and is relatively unable to be dispersed widely that it makes a desirable biocontrol agent. With biotechnology, other pathogens of weeds can be altered for use in biocontrol. The ecology of specific candidates will have to be well described to assure selection of those with the greatest potential for safe, effective use.

DARWIN DAY

Darwin Day is the anniversary of the birthday of Charles Darwin on February 12, 1809. Science and people in general have been deeply affected by Darwin's work. Now with over 200 years of evidence supporting his initial findings, modification and refinement continue to this day. For his contributions to the body of scientific knowledge and his commitment to scientific methodology, his birthday is celebrated globally. These celebrations of Darwin's work, and tributes to his life, have been organized since his death in 1882.

Darwin provided the first coherent theory of evolution by means of natural selection. His theory has had far-reaching implications in almost all disciplines—rocking the foundation of our knowledge base. Charles Darwin's *Origin of Species* (published in 1859) is a key work in scientific literature and in evolutionary biology. The book's full title is *On the Origin of Species by Means of Natural Selection, or the Preservation of Favoured Races in the Struggle for Life*. It introduced the theory that populations evolve over the course of generations through a process of natural selection. It was controversial because it contradicted religious beliefs of biology. Darwin's book was the result of evidence he had accumulated on the voyage of the *Beagle* in the 1830s and added to through continuing investigations and experiments after returning.

The book is readable and attracted widespread interest on publication. It also was controversial, generating much discussion on scientific, philosophical, and religious grounds. Here is how Darwin described natural selection in his book:

> *As many more individuals of each species are born than can possibly survive; and as, consequently, there is a frequently recurring struggle for existence, it follows that any being, if it vary however slightly in any manner profitable to itself, under the complex and sometimes varying conditions of life, will have a better chance of surviving, and thus be naturally selected. From the strong principle of inheritance, any selected variety will tend to propagate its new and modified form.*

> *Owing to this struggle for life, any variation, however slight and from whatever cause proceeding, if it be in any degree profitable to an individual of any species, in its infinitely complex relations to other organic beings and to external nature, will tend to the preservation of that individual, and will generally be inherited by its offspring. . . . I have called this principle, by which each slight variation, if useful, is preserved, by the term of Natural Selection, in order to mark its relation to man's power of selection.*

The scientific theory of evolution has evolved since Darwin first presented it, but natural selection remains the most widely accepted scientific model of how species evolve.

Some advocates would like to have a public holiday declared for February 12, 2009. The year 2009 will mark an important year for Darwin Day celebrations. It will be the 200th anniversary of Darwin's birth, and will also mark the 150th anniversary of the publication of Darwin's *On the Origin of Species*.

Controlling Plants

Mobile or transposable elements provide an opportunity to isolate and identify genes that would enhance crop quality and productivity. These mobile elements provide a direct link between plant characteristics, for example, disease resistance, plant height, organ shapes, and the DNA molecules that control the particular traits.

Transposable elements provide one way of physically isolating genes that control complex plant traits because they provide a direct connection between the observable trait and the DNA molecule that controls it. Even though all the intermediate steps in the process (the messenger RNA, the proteins produced by the genes, and the pathways) may be unknown, the gene responsible for the trait can be identified and physically isolated using a tagging procedure.

Nutritional Quality of Plants

Many plant foods are deficient in nutrients or lose nutritional value during storage. Some plants have other features that are not optimum for human or animal health. Genetic engineering can be used to both improve and retain nutritive value as well as to modify undesirable properties of plant products. For instance, through genetic engineering, the composition of dietary fats can be modified to reduce their possible contribution to cardiovascular disease.

Scientists are working to develop a sulfur-rich feed plant for sheep. Research has shown that sulfur supplements in the diet help sheep produce better quality wool fiber. Scientists believe it would be more cost effective to feed the sheep on pasture that was naturally sulfur-rich. Using biotechnology, scientists have developed alfalfa strains that produce a sulfur-rich protein in their leaves. They now plan to develop pasture grasses with the same characteristics.

Biological Control of Pests

Biological control exploits natural factors in the life cycle of harmful insects as a means of control. Some possibilities include use of highly specific insect pathogens (bacteria, viruses, fungi) to produce insect disease or death or to use unique viruses that interfere with the insect immune system, making it more vulnerable to disease. Also, insect chromosomes or genes with some ability to control population growth are under study as well as methods that interfere with normal growth and maturation. All these processes of biological control are potentially capable of being enhanced through the use of genetic engineering to improve the effectiveness of the crop insect's natural enemy. These processes are highly specific to single insect species and so are highly desirable environmentally as alternatives to chemical pesticides (see Figure 13-10).

For example, an insecticidal protein has been successfully incorporated into tomato plants to provide protection from some leaf-eating insects. The protein comes originally from *Bacillus thuringiensis,*

a naturally occurring bacterium that lives in the ground. Using genetic engineering techniques, scientists have inserted the gene for this protein into the plant's genetic material. When an insect eats the modified plant, the protein is released and the insect dies.

Another example is glyphosate, which is an environmentally friendly, widely used broad spectrum herbicide. It is easily degraded in the agricultural environment and works by interfering with an enzyme system that is present only in plants. Unfortunately, the herbicide kills crop plants as well as weeds, but scientists have now used genetic engineering methods to breed crop plants that are glyphosate resistant. By planting these modified crops, farmers can control weeds by spraying with glyphosate alone.

Transgenetic Plants

Transgenic plants and animals result from genetic engineering experiments in which genetic material is moved from one organism to another, so that the latter will exhibit a desired characteristic. Scientists, farmers, and business corporations hope that transgenic techniques will allow more precise and cost-effective animal and plant breeding programs. They also hope to use these new methods to produce animals and plants with desirable characteristics that are not available using current breeding technology. Table 13-2 lists genetically modified plants.

In traditional breeding programs, only closely related species can be crossbred, but transgenic techniques allow genetic material to be transferred between completely unrelated organisms, so that breeders can incorporate characteristics that are not normally available to them. The modified organisms exhibit properties that would be impossible to obtain by conventional breeding techniques.

Table 13-2			
Genetically Modified Plants			
Crops	**Vegetables**	**Flowers**	**Trees**
Alfalfa	Asparagus	Arabidopsis	Apple
Canola	Cabbage	Petunia	Pear
Corn	Carrot		Poplar
Cotton	Cauliflower		Walnut
Flax	Celery		
Potato	Horseradish		
Rice	Lettuce		
Rye	Peas		
Soybean	Tomato		
Sugar beet			
Sunflower			
Tobacco			

Although the basic coding system is the same in all organisms, the fine details of gene control often differ. A gene from a bacterium will often not work correctly if it is introduced unmodified into a plant or animal cell. The genetic engineer must first construct a transgene—the gene to be introduced. This is a segment of DNA containing the gene of interest and some extra material that correctly controls the gene's function in its new organism. The transgene must then be inserted into the second organism.

Making a Transgene

All genes are controlled by a special segment of DNA found on the chromosome next to the gene and called a promoter sequence. When making a transgene, scientists generally substitute the organism's own promoter sequence with a specially designed one that ensures that the gene will function in the correct tissues of the animal or plant and also allows them to turn the gene on or off as needed. Figure 13-11 shows a common transgenetic plant. For example, a promoter sequence that requires a light "trigger" can be used to turn on genes for important growth regulators (hormones) in plants. The plant would not produce the new hormone unless provided the appropriate trigger.

Inserting the Transgene

Unlike animals, plants do not have a separate germ line (eggs and sperm), and all cells of a plant retain the capacity to develop into a whole plant. This makes inserting the transgene much simpler. The transgene can be introduced into a single cell by a variety of physical or biological techniques, including using viruses or derivatives to carry the new gene into the plant cells.

Figure 13-11

Tomatoes—a transgenetic plant.
Courtesy of USDA/ARS.

Tissue culture techniques can then be used to propagate that cell and encourage its development into a transgenic plant, all of whose cells contain the transgene. Once the plant is produced, nature takes over and increases plant numbers by normal seed production.

Uses of Transgenic Techniques

Transgenic methods have now been developed for a number of important crop plants such as rice, cotton, soybean, and oilseed rape and a variety of vegetable crops like tomato, potato, cabbage, and lettuce. New plant varieties have been produced using bacterial or viral genes that confer tolerance to insect or disease pests and allow plants to tolerate herbicides, making the herbicide more selective in its action against weeds and allowing farmers to use less herbicide.

For example, a new variety of cotton has been developed that uses a gene from the bacterium *Bacillus thuringiensis* to produce a protein that is specifically toxic to certain insect pests including boll-worm, but not to animals or humans. (This protein has been used as a pesticide spray for many years.) These transgenic plants should help reduce the use of chemical pesticides in cotton production, as well as in the production of many other crops that could be engineered to contain the *Bacillus thuringiensis* gene. In another case, a gene from the potato leaf roll virus has been introduced into a potato plant, giving the plant resistance to this serious potato disease.

Transgenic technologies are now being used to modify other important characteristics of plants such as the nutritional value of pasture crops or the oil quality of oilseed plants like linseed or sunflower.

Novel Uses

With techniques similar to those used to make insulin-producing bacteria, it may be possible to develop animals that produce other useful **biopharmaceuticals**—drugs produced by living tissues. For example, researchers have developed transgenic animals such as cows and sheep that secrete economic quantities of medically important chemicals in their milk. The cost of these drugs may be much less than for those produced using conventional techniques.

Eventually it may also be possible to develop crops for non-food uses by modifying the starches and oils they produce to make them more suitable for industrial purposes, or to use plants rather than animals to make antibodies for medical and agricultural diagnostic purposes. In the cut flower industry, transgenic research may yield products such as blue carnations as novelty items.

Advantages over Selective Breeding

Transgenic technology is an extension of agricultural practices that have been used for centuries—selective breeding and special feeding or fertilizing programs. It may reduce or even replace the large-scale use of pesticides and long-lasting herbicides. Transgenic technology is still experimental and is still very expensive. If it could be made commercially viable, it would offer a number of advantages over traditional methods.

Compared with traditional methods, transgenic breeding is:

- More specific—scientists can choose with greater accuracy the trait they want to establish. The number of additional unwanted traits can be kept to a minimum.
- Faster—establishing the trait takes only one generation compared with the many generations often needed for traditional selective breeding, where much is left to chance.
- More flexible—traits that would otherwise be unavailable in some animals or plants may be achievable using transgenic methods.
- Less costly—much of the cost and labor involved in administering feed supplements and chemical treatments to animals and crops could be avoided.

Micropropagation (Tissue Culture)

Plant breeders already use micropropagation techniques in which whole plants are grown from single cells or from small plant parts for rapid multiplication of identical, disease-free plants, such as the papaya plantlets in Figure 13-12. If necessary, genetic engineering can be used to incorporate desired characteristics from other species into the cell prior to propagation.

Figure 13-12

Papaya plantlets raised in a petri dish by a process called micro-propagation. The plantlets will undergo genetic testing to see whether they will yield so-called perfect flowers, which will develop into fruit. Courtesy of USDA/ARS.

The Future

Current research will see the improvement and development of crops for specific purposes. Plants that require less water could be developed for countries with arid climates. Crop plants engineered to be tolerant to salt could be farmed in salt-damaged farmland or could be irrigated with salty water. Crops with higher yields and higher protein values are also possible. Biotechnology will help to do the following:

- Improve farming productivity
- Protect our environment by allowing reduced and more effective use of chemical pesticides and herbicides
- Reduce food costs

Table 13-3 summarizes the long-term goals of plant biotechnology.

Table 13-3

Long-Term Goals of Plant Biotechnology		
Agronomic traits	**Quality traits**	**Specialty chemicals**
Herbicide tolerance	Oil composition	Plastics
Insect control	Solids content	Detergents
Virus resistance	Nutritional value	Pharmaceuticals
Fungal resistance	Consumer appeal	Food additives

"Omics" Revolution

Following biotechnology and genetic engineering has been the development of three new fields of study: genomics, bioinformatics, and proteomics. This "omics" revolution in science has been very fast and, in many cases, borders on overwhelming.

Genomics

The term *genome* originated in 1930 and was used to denote the totality of genes on all chromosomes in the nucleus of a cell. Incredibly, DNA was not identified as the genetic material of all living organisms until 1944. The genetic code was elucidated in 1961; with these fundamental insights in hand, it was possible to contemplate the concept that biological organisms had a blueprint consisting of finite numbers of genes. The sequence of these genes encoded all the information required to specify the reproduction, development, and adult function of an individual organism.

Genomics is operationally defined as investigations into the structure and function of very large numbers of genes undertaken in a simultaneous fashion. Because all modern genomics have arisen from common ancestral genomes, the relationships between genomes can be studied with this fact in mind. The study of all the nucleotide sequences, including structural genes, regulatory sequences, and noncoding DNA segments, in the chromosomes of an organism plant or animal is called genomics. The massive interest and commitment to genomics flows from the generally held perception that genomics will be the single most fruitful approach to the acquisition of new information in basic and applied biology in the next several decades.

Despite the importance of plants, much basic research on plant functions still needs to be done. For example, scientists need to learn more about how plants grow, how they protect themselves from disease or insects, and how they respond to their environment. One of the best ways to do this is to study the information encoded in a plant's DNA, the material which makes up its genes.

An understanding of genomics means that information gained in one organism can have application in other even distantly related organisms. Comparative genomics enables the application of information gained from facile model systems to agricultural or nonmodel taxa. The nature and significance of differences between genomes also provides a powerful tool for determining the relationship between genotype and phenotype. Plant genomics hold the promise of describing the entire genetic repertoire of plants. Ultimately, plant genomics may lead to the genetic modification of plants for optimal performance in different biological, ecological, and cultural environments for the benefit of humans and the environment.

Recent scientific advances through private and public investments in studying DNA structure and function not only in humans but in other organisms such as plants can extend a new biological paradigm to improving the useful properties of plants that are important to humanity. Solutions to many of the world's greatest challenges can be met through the application of plant-based technologies. For example, the revitalization of rural America will come from a more robust agriculture sector; reductions in greenhouses gasses can be achieved from the production of plant biofuels for energy; chemically contaminated sites can be rehabilitated economically using selected plants; and worldwide malnutrition can be greatly reduced through the development of higher yielding and more nutritious crops that can be grown on marginal soil.

Molecular plant breeding efforts have received a boost in the past decade from massive amounts of gene and genome sequence information (genomics), which have been used in parts to generate molecular markers for marker-assisted breeding in species such as alfalfa—the number one forage crop—white clover, and tall fescue.

Functional genomics can produce massive amounts of data on the majority of genes of a sequenced organism. Functional genomics includes transcriptomics, proteomics, and metablolmics, which identify and quantify thousands to tens of thousands of gene transcripts (RNA), proteins, and metabolites, respectively, in cells, tissues, or organs. By revealing concerted changes in RNA, proteins, and metabolites during plant development, under optimal or stressed conditions genes determine important plant characteristics or traits.

Other disciplines involved in plant genomics are genetics, biochemistry, biophysics, molecular and cell biology, and physiology. They are also used to elucidate the precise biological function(s) of specific gene products. Ultimately by combining genomics, functional genomics, and other approaches, the aim is to identify genes with key roles in plant metabolism, growth, development, and response to the environment, which will be of value to plant breeding efforts to improve plant performance, yield, and quality.

Genomics and functional genomics approaches are used to identify genes and process that enable plants to respond and adapt to environmental challenges, such as pathogen attack (bacterial, fungal, or viruses), intensive grazing, or abiotic stress such as drought and soil acidity/aluminum toxicity. Another target of this type of research is plant secondary metabolism, which yields an amazing array of compounds of industrial and medicinal value. Secondary metabolism is crucial for the production of structural compounds in plants such as lignin, which affects digestibility of plant material.

Bioinformatics

The part of the "omics revolution" that has made it possible to analyze and interpret all the genomics data is bioinformatics; for example, genomics generates data, bioinformatics provides the analytical tools enabling those data to be interpreted.

Bioinformatics is the study of the inherent structure of biological information and biological systems. It puts together the ever increasing avalanche of systematic biological data with the analytic theory and practical tools of computer science and mathematics.

Genomics data can be viewed as the greatest encoding problem of all time. These bioinformatics problems open the way for biologists to collaborate with mathematicians and computer scientists, because it aims to translate "biology problems" into new challenges that are interesting to theoreticians: problems of information content, structure, and encoding, which inherently interest theorists. By contrast, the common view and practice of bioinformatics, as simply an application of existing mathematical and computer science to biological problems, translates biological questions into the language of information content, structure, and encoding, so the mathematicians and computer scientists are needed to help solve these problems.

Bioinformatics is a rapidly developing branch of biology and is highly interdisciplinary using techniques and concepts from informatics, statistics, mathematics, chemistry, biochemistry, physics, and linguistics. It has many practical applications in different areas of biology and medicine. Research in bioinformatics includes method development for storage, retrieval, and analysis of the data. Bioinformatics is a rapidly developing branch of biology.

Proteomics

Included in the "omics" revolution are a number of powerful tools including variations on the theme of proteomics. **Proteomics** aims to identify and characterize the expression pattern, cellular location, activity, regulation, posttranslational modifications, molecular interactions, three-dimensional structures, and functions of each protein in a biological system.

The potential of proteomics for plant improvements is outstanding considering that the developments in robotics and nanotechnology achieved in the last 10 years allow for the genome sequencing of many organisms. The new genomics science highlights our ignorance of the newly discovered sequenced genes.

In plant science, the number of proteome studies is rapidly expanding after the completion of the *Arabidopsis thaliana* genome sequence and proteome analysis of other important or emerging model systems. This analysis of plants is subject to many obstacles and limitations as in other organisms, but the nature of plant tissue, with their ridged cell walls and complex variety of secondary metabolites, means that extra challenges are involved that may not be faced when analyzing other organisms.

Proteomics is a complementary approach to solve many problems; from mutant characterization to genetic variability estimates, from the establishment of genetic relationship to the identification of the genes involved in the response to biotic or abiotic stresses, from the definition of genes coregulated or coaffected by a given molecule to the quantification of their amounts, proteomics, with its recent development in mass spectrometry and database management is today a very powerful new tool to deal with the global approaches of modern biology.

Proteomics is a promise of a great revolution. If genomes and microarrays gave us a glimpse into the blueprint of life, proteomics was going to unravel the working end of the cell: the protein machinery.

But, it has rapidly become apparent that proteins were not going to give up their secrets without a fight. Proteins are much more diverse in their properties than nucleic acids, a single protocol for example preparation or analysis is unlikely. There is no PCR for proteins, so the amount of starting tissue and detection sensitivity are critical limitations. Protein concentrations extend over a far greater dynamic range than nucleic acids. Proteomics must deal with differences in abundance of six to eight orders of magnitude, meaning that the few most abundant proteins often interfere with detection of low-level proteins. None of these problems are insurmountable, but they have slowed the appearance of the expected biological results as each problem needed to be solved individually.

With the sequencing of the Arabidopsis genome, there is a mature proteomics technology platform. Now is the time to bring the resources and tools together. Proteomics gives us the molecular mechanisms that will fill in the gaps that have existed in the defined genetic pathways. Proteomics will blend perfectly and powerfully with genetics. The revolution has begun and it continues at a rapid pace.

Biotechnology Policy, Public Perception, and the Law

While the opportunities for using biotechnology in agriculture are truly fantastic, they have triggered public concern and a corresponding need to develop methods to assure that biotechnology is used in an environmentally sound manner. Some people fear that widespread use of plants and animals with altered genetic characteristics may threaten the environment by disturbing the existing balance between organisms. This balance is a dynamic one. Since gene mutation and changes in gene position within chromosomes are normal events in all living organisms, organisms with new properties are constantly emerging. However, transgenic technology does expand the scope of these events. Careful examination of the properties of the transgenic organism is essential before it is studied outside the closed environment of the laboratory.

Agricultural researchers are enhancing traditional methods of manipulating plant and animal germplasm used for over 100 years as well as using the guidelines in recombinant DNA studies directed by human applications. In both cases, there is a sound track record of safety associated with research and its products. Recombinant DNA techniques have been safely employed since the mid-1970s through an essentially self-imposed series of safety guidelines and reviews within the scientific community. Formerly under the control of the National Institutes of Health, these safety procedures are now the responsibility of the U.S. Department of Agriculture. Methods and procedures are being completed to assure continued safety in applying recombinant DNA techniques for agricultural research and in producing biotechnology products.

🌐 Summary

1. Success in biotechnology and genetic engineering is based on our increasing understanding of the genetic code. Biotechnology research benefits agriculture, forestry, and industrial processing by diversifying crops and crop products, with increasing concern and care for the environment. Some plants have already benefited from genetic engineering.

2. Much current research focuses on understanding and developing useful promoter sequences to control transgenes and establishing more precise ways to insert and place the transgene in the recipient. Much still needs to be done to improve our knowledge of specific genes and their actions and of the potential side effects of adding foreign DNA and of manipulating genes within an organism.

3. Following biotechnology and genetic engineering has been the development of three new fields of study: genomics, bioinformatics, and proteomics. The "omics" revolution in science continues at a rapid pace.

4. A formal regulatory system exists to examine areas of risk or uncertainty before field testing is approved. The government, industry, and other interested groups are also considering more general questions on the uses of this new technology, including ethical questions and sociological consequences.

Something to Think About

1. Explain why to a cell, DNA is DNA whether it came from a plant or animal, or from a different species.

2. Name the four bases found in DNA.

3. List five crops that have been genetically engineered.

4. Name four of the goals of biotechnology.

5. Identify the role of the following: DNA, rDNA, tRNA, and mRNA.

6. Explain how DNA controls the formation of proteins.

7. Why are people concerned about biotechnology and genetic engineering?

8. What future impacts could genetic engineering have on agriculture?

9. Define genetic engineering, biotechnology, and transgenetic.

10. Describe the advantages that genetic engineering has over traditional selective breeding.

11. Do the benefits of genetic engineering and biotechnology outweigh the risks?

Suggested Readings

Halford, N., Ed. 2006. *Plant biotechnology: Current and future applications of genetically modified crops.* New York: Wiley.

United States Department of Agriculture. 1986. *Research for tomorrow. The yearbook of agriculture.* Washington, DC: USDA.

United States Department of Agriculture. 1992. *New crops, new uses, new markets. 1992 yearbook of agriculture.* Washington, DC: USDA.

United States Department of Agriculture. 1994. *Biotechnology and sustainable agriculture: A bibliography.* Washington, DC: USDA.

Vasil, I. K., Ed. 1990. *Biotechnology: Science, education and commercialization. An international symposium.* New York: Elsevier Publishing Company.

Internet

Internet sites represent a vast resource of information. The URLs for Web sites can change. Using one of the search engines on the internet, such as Google, Yahoo!, Ask.com, or MSN Live Search, find more information by searching for these words or phrases: **biotechnology, agroeconomic, agroecosystems, biocontrol, bioinformatics, biopharmaceuticals, genetic code, genetic engineering, genomics, DNA, RNA, micropropagation, proteomics, and transgenetic.**

CHAPTER 14

Diversity: Vascular Plants

No fossil evidence suggests that life on land existed prior to 450 million years ago. Various marine organisms must have repeatedly washed up on shore, but the rocky, barren environment was inhospitable to them. This period also coincided with the receding of the oceans and the exposure of algae to the land. The seawater offered protection from desiccation, and the sun's destructive ultraviolet radiation acted as a buffer so that temperature fluctuations were moderate and allowed organisms to absorb nutrients directly from the minerals dissolved in it. Marine organisms did and still do cope with the osmotic problems of a saline environment, but otherwise growth and development in the ocean was without many of the problems that were to be encountered on land.

After completing this chapter, you should be able to:

- Discuss the cycle that plants have gone through in the past 500 million years
- Name the primitive vascular plants
- Describe how these plants developed their vascular system
- Name which one of the four primitive vascular plant families actually gave rise to the living seedless vascular plants
- Discuss the progression of the Pterophyta, ferns that contain almost 12,000 living species
- Draw and label the fern life cycle
- Identify the differences between a non–seed bearing vascular plant and seed bearing vascular plant
- Compare and contrast the gymnosperms and angiosperms
- Discuss the co-involvement of flowering plants and insects
- List and tell how different insects and animals are involved in the co-involvement
- Explain secondary plant compounds that have strong physiological effects on humans

Key Terms

ozone	carboniferous period	seed coat
multicellular green alga	seeds	after-ripening dormancy
bryophytes	seed ferns	cycads
vascular plants	gymnosperms	ginkgos
antheridium	carboniferous period	conifers
archegonium	fronds	gentophytes
conduction systems	fiddlehead	angiosperms
sporophyte	ovule	trimerophytes
gametophyte	nucellus	progymnosperms
paleozoic era	integuments	cone (stobilus)

microsporangia
microspore mother cells
microspores
prothallium cells
tube cell
seed-scales complexes

polyembryony
maidenhair tree
mormon's tea
ephedrine
bronchodilator
paleobotanists

coevolution
protected ovules
inferior ovary
pseudocopulation
alkaloids
isozymes

Movement to Land

Although some oxygen was apparently produced by electrical discharges that split water molecules, it was not until cyanobacteria began photosynthesizing, and thus obtained their reducing source (H^+) from a photolysis of water that oxygen was put into the atmosphere in appreciable quantities. The oxygen molecule, O_2, automatically forms a small amount of **ozone**, O_3, which has the ability to absorb ultraviolet radiation. Although the proportion of ozone is relatively small compared with the concentration of oxygen, it is a significant factor in the Earth's atmosphere. The fossil record indicates but evidence suggests the significant factor in the Earth's atmosphere. The fossil record indicates that cyanobacteria began photosynthesis about 3 billion years ago, but evidence suggests that significant oxygen accumulation did not occur until about 1 billion years ago. It was not until approximately 450 million years ago that a sufficient concentration of ozone finally began to absorb enough harmful radiation that organisms could survive when exposed directly to the sun's rays.

Single-celled marine organisms existed as the sole life forms for a long time. Then true multicellularity evolved, first as a single-dimensional filament, then as a two-dimensional sheet of cells, and finally as three-dimensional organisms with considerable differences in the external and internal environment of the organism as a whole. Colonial organisms such as the green alga *Volvox* developed modified internal concentrations of nutrients and gases. As multicellular marine plants became larger and more complex, specialized cells formed to accommodate and alleviate problems of "communication" between cell tissues. Survival is more likely for organisms that evolve a system of coping with environmental changes.

No evidence exists that any of these prokaryotes or eukaryotes survived on land until a **multicellular green alga** established itself on land in the Silurian period, about 400 million years ago. Evidence suggests that it may have evolved symbiotically with a fungus, perhaps like the modern lichens. Probably thousands of algae were washed onto land before successful establishment was achieved. At best, the terrestrial habitat was precarious, but surviving offspring had improved structure/function relationships, and the landscape finally had its first occupants.

Adaptations of Land Plants

The fossil record suggests that early land plants were not significantly different from marine progenitors. Some of the earliest plants evolved as **bryophytes**, a group that could be considered an evolutionary dead end. Bryophytes failed to develop specialized transport systems for water and nutrients, which in turn limited their stature. They also failed to develop an effective cuticle, the root system did not become extensive enough to carry the plant through periods of stress, and gas exchange mechanism (stomata) failed to develop or regulate efficiently.

While the bryophytes followed one line of evolution, another group developed a level of specialization not before achieved on land. Elongated, thin-walled cells had already developed in some of the multicellular marine algae, and mutation of those cells gradually led to conducting cells that eventually evolved into xylem and phloem. At first the cells had obstructions in the end walls, and long-distance transport was rather slow. With a relatively inefficient cuticle and system of gas exchange, plants could not grow very tall, and the surface area of aboveground parts had to remain small. Such plants were branched and had almost no leaves. As xylem and phloem became more efficient, water and nutrients moved faster, leaves expanded, and the area available for capturing light energy increased; thus, more photosynthate could accumulate, creating greater biomass. Those plants that developed one efficient transport system for water and another for the transport of organic solutes are called **vascular plants**.

The vascular plants and bryophytes diverged long ago, shortly after the progenitor alga made the transition from ocean to land. Some bryophytes are fossilized and calculated to be more than 350 million years old; the ancient specimens are similar to the alternation of generations, it is believed that the green algal ancestor must also have had the characteristic.

The rate of evolution was rapid after the movement to land. Extreme environments lead to a more expeditious selection of variants than does a stable environment; therefore, fluctuations in temperature, changes in light conditions, nutritional scarcity, changing relative humidity, and unpredictable water availability promoted swift plant evolution on land. One of the critical features of survival on land is the protection of sex cells; organisms successful in making the transition must have gametes with an outer layer of protective cells. The development of the **antheridium** (male multicellular sex organ) and the **archegonium** (female multicellular sex organ) were important advancements in the success of land plants.

Early in the movement to land, organisms developed multicellular sporangia well protected by walls that prevented desiccation. Such structures could be dispersed really by wind. At the same time, plants were beginning to develop a cuticle to keep the plant itself from drying. Efficient **conduction systems** evolved from primitive transport cells hampered by friction to more streamlined cells with little or no frictional loss. Gradually the alternation of generations evolved an increasingly important role for the **sporophyte**; the **gametophyte** was much reduced in size and became more dependent on the sporophyte. In the flowering plants, the gametophyte represents only a tiny fraction of the life cycle.

During the latter stages of the **Paleozoic era**, the climate was stable and mild, similar to the climate of modern subtropical and tropical regions. Although mountains were forming in what is now the eastern United States, Texas, and Colorado, larger areas of the United States were flat, and the far western portion of the continent was still covered by the sea. In most places, the saline waters were not deep, some being vast coal swamps. Very stable, warm, humid regions gave rise to lush plants with a great deal of biomass. This era, known as the **Carboniferous period**, produced most of the vegetation that decomposed and formed vast deposits of coal, oil, and natural gas.

Evolution and Distribution of Vascular Plants

The vascular plants include those with and without **seeds**. A great deal has been said in older botanical literature about the **seed ferns**, a group of plants that is now extinct. The fossil record clearly shows that they were not ferns at all, but primitive **gymnosperms** and therefore seed bearing. Vascular plants do not produce seeds but reproduce by spores. There are three extinct (Rhyniophyta, the oldest; Timerophyta, fossil found 350 million years ago, and Zosterophyllophyta, early Devonian 375 millions years ago) and four living divisions (Lycophyta, a group of herbaceous line; Sphenophyta, horsetail's jointed stems; Psilophyta, epiphytes; and Pterophyta, the ferns). Pterophyta is currently the largest and the most evident group. There are 12,000 living species, and they are cosmopolitan (throughout the world). The age of the ferns in the fossil record dates from the **Carboniferous period**, a time in which they were the dominant vegetation. Water was plentiful, and the climate was warm, humid, and unchanging. In many respects, the world was like a giant greenhouse with ideal growing conditions for many plants. Some ferns grew to 8 m tall and had broad trunk bases with many aerial roots as props.

Most extant ferns are tropical, but some are found in temperate regions, including the mountainous regions of the United States; a few are adapted to aridity and inhabit deserts. Living species vary in size from the tiny aquatic *Azolla microphylla* with fronds or fern leaves less than 2 cm long, to the giant tree ferns, some of which may reach almost 25 m in height and 30 cm or more in diameter. All this stem tissue is primary in origin; and only one species of fern has a vascular cambium. Fern **fronds** come in myriad shapes and sizes, and many are highly dissected. Unlike the seed plants, ferns produce leaves from a coiled position. The frond expands by unrolling from base to tip with new growth produced at the apex; such new leaves are often covered with surface hairs. The coiled frond is called a **fiddlehead** because of its resemblance to the neck and head of a violin, as shown in Figure 14-1.

Figure 14-1

Fern leaves or fronds unroll from base to tip. In the coiled condition, they are referred to as a crosier or fiddlehead. They are called this because they resemble the neck and head of a violin.
Courtesy of National Arboretum.

Reproduction of Ferns (Nonseed Plants)

In almost every case, ferns are homosporous, that is, produce only a single kind of spore. These spores are born on the underside of fertile fronds, as opposed to the non-spore-bearing sterile fronds. Spore mother cells in the fronds undergo meiosis. The spores are produced in sporangia, and many sporangia are grouped together in a region to make up a sorus. The sori are arranged on the underside of the leaf in varying patterns, as shown in Figure 14-2, depending on the species, and are often mistaken for insect eggs.

Figure 14-2

Fertile fronds of ferns produce spores in sporangia on the underside of the frond.

Each sorus may be covered with a membrane-like protective layer called the indusium. When the spores mature, the indusium shrivels or breaks, and the spores are released. They are catapulted through the air for maximum dispersal.

If the spore lands on a suitable substrate, the spore starts to germinate; the haploid spore will give rise to a chain of cells called a protonema. Mitosis takes place, the result is a free-living bisexual gametophyte. The absorption of water and nutrients is achieved by rhizoids, which grow like root hairs on the underside of the haploid plant body.

The aquatic-dwelling ancestry of the ferns is revealed by a reproductive system in which motile sperm are produced within the antheridia, swim to the neck of the archegonium, and fertilize the egg to produce a new diploid zygote (see Figure 14-3).

MAINTAINING DIVERSITY

In 1990, the U.S. Congress authorized establishment of a National Genetic Resources Program (NGRP). It is the NGRP's responsibility to acquire, characterize, preserve, document, and distribute to scientists germplasm of all life forms important for food and agricultural production.

The Germplasm Resources Information Network (GRIN) Web server provides germplasm information about plants, animals, microbes, and invertebrates. This program is within the U.S. Department of Agriculture's Agricultural Research Service.

In terms of plants, the National Plant Germplasm System (NPGS) is a cooperative effort by public (state and federal) and private organizations to preserve the genetic diversity of plants. The world's food supply is based on intensive agriculture, which relies on genetic uniformity. But this uniformity increases crop vulnerability to pests and stresses. Scientists must have access to genetic diversity to help develop new varieties that can resist pests, diseases, and environmental stresses. The NPGS aids the scientists and the need for genetic diversity by acquiring, preserving, evaluating, documenting, and distributing crop germplasm important for food and agricultural production.

Since many important crop species originate outside the United States, the first steps toward diversity are acquisition and introduction. New germplasm (accessions) enter NPGS through collection, donation by foreign cooperators, or international germplasm collections.

GRIN provides NGRP personnel and germplasm users with continuous access to databases for the maintenance of passport, characterization, evaluation, inventory, and distribution data important for the effective management and utilization of national germplasm collections.

For more information, start at the Web site for Germplasm Resource Information System: www.ars-grin.gov.

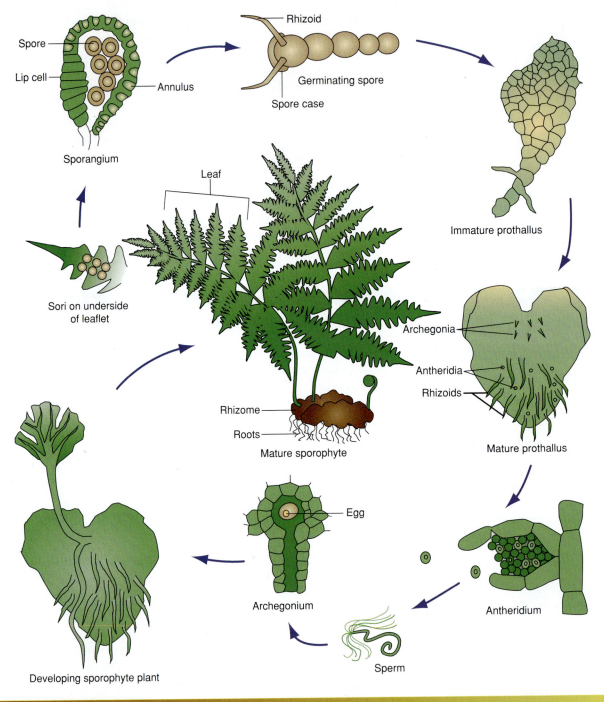

Spore

Lip cell

Annulus

Sporangium

Rhizoid

Germinating spore

Spore case

Immature prothallus

Leaf

Sori on underside
of leaflet

Rhizome

Roots

Mature sporophyte

Archegonia

Antheridia

Rhizoids

Mature prothallus

Egg

Archegonium

Antheridium

Developing sporophyte plant

Sperm

Figure 14-3

The fern life cycle resembles the life cycle of the angiosperm.

The Seed Plants

Spores produced by ferns and lower plants are adapted to many environmental stresses and therefore are able to survive under extreme conditions; however, such spores fail to exhibit the survivability of seeds. Seed plants represent a significant jump in adaptation to more extreme climates. The oldest seeds have been found in rocks from the Devonian period, some 350 million years old.

Modern seeds develop from a mature **ovule**, the female gametophyte being embedded within a fleshy **nucellus**. One or more layers of the integuments, which develop into the seed coat, enclose the nucellus. The thickness and chemical composition of the **integuments** ultimately determine the nature of the **seed coat** and thus its protective quality. Some seed coats become exceedingly thick and highly impermeable to water and gases. The survival capability of such seeds is very high.

In most modern seeds, the embryo matures before dispersal, whereas ancient seeds apparently completed embryo development after dispersal. Survival ability would appear to be best with embryo development prior to dispersal, and such embryos might have a selective advantage. Even today some seeds have **after-ripening dormancy** because the embryo is not fully developed. Newly harvested sugar beet seeds exhibit this characteristic.

Storage of reserved food within the seed is critical for survival. Within the endosperm, or cotyledons, various storage molecules (starch, protein, and lipids) become available for energy during the germination process. As germination proceeds, respiration rates increase dramatically, and storage molecules are hydrolyzed to produce the glucose substrate for respiration. This food reserve is usually more than adequate and sometimes stored reserves are available weeks or months after the germination process begins. Such storage capacity is vital to survival during long periods of dormancy, perhaps hundreds of years for some species.

This remarkable adaptation—the seed—increased the chances of survival on land. It allows for continuation of the life cycle under conditions unfavorable for growth. Frozen soil, drought, extreme heat, and other environmental factors usually inhospitable to plant growth are endured in the seed stage; this resistant, reproductive structure simply goes dormant until unfavorable conditions have passed.

There are five extant groups of seed plants: the **cycads**, the **ginkgos**, the **conifers**, the **gentophytes**, and the **angiosperms**. The first four of these comprise the gymnosperms, or plants with naked seeds, the angiosperms, or flowering plants, have enclosed seeds.

Gymnosperms

The gymnosperms apparently evolved during the Paleozoic era from an intermediate group of plants that were more highly vascularized than the lower plants. These now-extinct groups were the **trimerophytes**. They had no leaves; the main axis formed lateral branches that divided several times. Some of the lateral branches terminated in sporangia, which produced a single type of spore. These **progymnosperms** eventually gave rise to plants with leaves and seeds.

The movement of land and increasing aridity presented the problem of moving male gametes to the female gamete for fertilization. Whereas ferns and lower plants facilitate the reproductive process by having motile sperm swim through water to reach the egg, most large land plants failed to develop a mechanism for ensuring that water-based reproduction will continue. Only in the cycads and in *Ginkgo* do the relic motile sperm swim to the region of fertilization. Instead, airborne pollen evolved, and pollination had to be accomplished by wind or by animals. In most gymnosperms, the immature male gametophytes, the pollen grain, is borne by the wind to the vicinity of the female gametophyte within an ovule. Following pollination, the pollen grain germinates and produces a pollen tube, which delivers the sperm directly to the archegonium.

Conifers

Although the total number of conifer species is not large, they are the largest group of living gymnosperms, comprising about 50 general and 550 species. These plants have adapted remarkably to aridity by means of sunken stomata and thick cuticles, and the needlelike and scale leaves have a much reduced surface area, as shown in Figures 14-4a and 14-4b. There is reason to believe that the evolution of conifers coincides with tremendous selection pressure during the worldwide aridity of the Permian period. During the early Tertiary period, some genera were more prominent than they are in the modern landscape. On all the northern continents, vast regions were and still are covered with conifers. The pines are unquestionably the most economically important conifers. There are about 90 species widely distributed throughout North America, Europe, and Asia.

Figure 14-4a

Conifers dominate much of the landscape in forests throughout the world. Conifer leaves consist of needles formed singly or in bundles called fascicles.

Figure 14-4b

Some conifers have scalelike leaves, however, their juvenile leaves are needles, and as they mature they turn into scales.

Reproduction

The reproduction structure in conifers is a **cone (stobilus)**, and male and female gametes are produced on separate cones. Both types of cone are found on the same plant, the male cones occurring on the lower branches and the more conspicuous female cones being borne above. It is not uncommon to find 1-, 2-, and even 3-year-old female cones on the same tree in pines, although most conifers produce mature cones in one season. Compared with that of angiosperms, the reproductive cycle in conifers is very long. The period between pollination and fertilization may be many months, whereas in angiosperms the period is a matter of hours or days. Male cones are quite small, usually no more than 1 or 2 cm in length. Each of the spirally arranged microsporophylls (scales) contains two **microsporangia**, and each microsporangia contains many **microspore mother cells**, which undergo meiosis to produce four haploid **microspores**. Each of these develops into a winged pollen grain with two **prothallium cells** (having no apparent function), a generative cell, and a **tube cell**. The mature pollen grains are shed by the male cones in great abundance in the spring, and sometimes coniferous forest are shrouded in yellow clouds of pollen, even covering the entire surface of lakes. The bladderlike air sacs increase the buoyancy of the pollen, and it can travel hundreds of miles before falling inside the scales of an ovulate cone.

The female cone is technically a modification branch with spirally arranged scales known as **seed-scales complexes**. Each scale bears two ovules on the upper surface. Each ovule consists of a nucellus surrounded by the integument, which eventually gives rise to the seed coat. A tiny opening at the base of the integument, the micropyle, is the site of penetration of the pollen tube. Within the megasporangium located between the micropyle and nucellus, a megaspore mother cell undergoes meiosis, producing four megaspores. The three nearest the micropyle disintegrate, leaving a single megaspore to produce the gametophyte containing the egg. Pollination occurs in the spring. The ovulate cones open the scales while in an upright position so that the pollen grain can fall downward into the crevice. A drop of sticky fluid at the micropyle captures the pollen grain and pulls it closer to the ovule. During this process, the scales again close, providing additional protection from the environment. When the pollen grain comes in contact with the nucellus, it forms a pollen tube that grows slowly toward the megagametophyte. At this time, the megaspore mother cell has not undergone meiosis. This process is exceedingly slow, sometimes requiring as long as 6 months for initiation and an additional 6 months for complete development. Fertilization is not accomplished until approximately 15 months after pollination (see Figure 14-5).

Figure 14-5

The life cycle of conifers is a very slow one, sometimes requiring as long as 6 months for initiation and an additional 6 months for complete development. Fertilization is not accomplished until approximately 15 months after pollination.

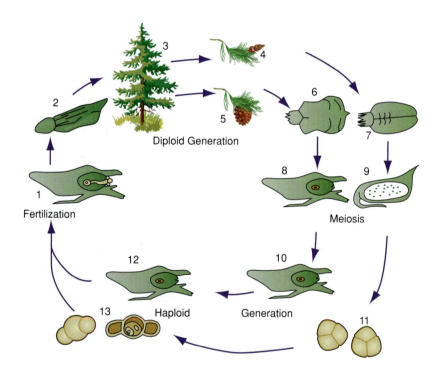

In gymnosperms, two sperm cells are eventually produced, but one disintegrates and the other fertilizes the egg. Even though more than one egg may be fertilized and more than one embryo may start to develop, a process known as **polyembryony**, all but one generally aborts, and a single embryo develops to maturity within each seed. The mature cone completes development during the fall of the second year following pollination. The mature conifer embryo consists of a root/shoot axis with several, usually eight, cotyledons or true seeds. The environmental conditions essential for germination are similar to those for the angiosperms.

Ecology and economic importance

Conifers' adaptation to aridity, particularly true of the pines, has allowed colonization in many parts of the world. Vast areas are covered by coniferous forest, particularly the temperate regions of the northern hemispheres. Secondary growth is rapid and extensive; thus, conifer wood is usually considered to be softwood, and angiosperm wood is considered to be hardwood, but the distinction is far from absolute. Many conifers produce relatively hard wood. The largest and oldest organisms in the world are conifers.

Ginkgo

One of the broad leaf deciduous street trees of many cities throughout the world is the **maidenhair tree** (*Ginkgo biloba*), as shown in Figure 14-6. Few nonbotanists recognize this tree as a gymnosperm, yet its morphological features, including naked seed, place it within this group. The unusual leaves are fan shaped and sometimes deeply lobed. *Gingko* is the only living species of a group of organisms that extended into the late Paleozoic era and were widespread in the Mesozoic area. *Gingko* is probably extinct in the wild, and all specimens now growing have been cultivated. The species has spread rapidly during recent years because it is an excellent street tree for metropolitan areas, and it is particularly resistant to air pollution. Tokyo's famous cherry trees, many of which have succumbed to industrial pollution, have been replaced with *Gingko*.

Figure 14-6

Ginkgo biloba *is exceptionally resistant to air pollution and makes an excellent street tree.*

Gingkos are dioecious trees; that is, male and female sex structures are borne on different trees. The ovules are borne in pairs on the end of short stalks, and seeds are produced in the fall. Fertilization is very much delayed and may not occur until after the ovules have been shed from the tree. The seeds produce butyric acid, which has an offensive odor, and consequently male trees are usually propagated vegetatively, and therefore are easy to perpetuate only the male plants (see Figure 14-7).

Figure 14-7

Ginkgo *or maidenhair fern tree do not have the basic characteristics of a gymnosperm, but in fact the seeds are born naked, and the fossil records place this single species as a primitive gymnosperm.*

Cycads

These are palmlike gymnosperms, very different from either the conifers or *Gingko* and native to subtropical and tropical regions. In the Mesozoic era, during the reign of the dinosaurs, cycads were dominant vegetation. There are approximately 10 genera with 100 species living at the present time. The only species native to the United States is *Zamia pumila*, which grows in southern Florida. *Cycads*, which are shown in Figure 14-8, produce a trunk with some secondary tissue, and the leaves are usually clustered at the apex to produce a palmlike effect. Pollen and seed cones are borne on separate plants, and the cone can be very large in some species.

Figure 14-8

Cycad plants are dioecious. There are only 10 genera and about 100 species living at present.
Courtesy of North Carolina Zoological Park.

Gnetophytes

This unusual group of plants consists of three genera with some 70 species. Like the angiosperms, this single group of gymnosperms conducts water via vessels. *Ephedra*, which inhabits arid regions, is a shrubby gymnosperm with extensive branching and scalelike leaves. It is the only genus native to the United States. One species *Ephedra antisyphilitica*, the so-called **Mormon's tea**, was brewed by the early settlers as a substitute for true tea, and the liquid was thought to have medical properties. *Ephedra* also produces **ephedrine**, an important **bronchodilator**. *Gnetum* is a group of trees and climbing vines found throughout tropical rain forest. They have thick, leathery leaves and are easily mistaken for typical dicots.

Welwitschia is one of nature's strangest products. It grows only in the desert regions of Africa, and most of the plant is covered with sand. The aerial parts consist of woody, straplike leaves. On old plants, the leaves are torn and tattered, the splits giving the impression of more than two leaves.

Angiosperms

Unquestionably today's dominant worldwide vegetation consists of the flowering plants. In addition to their agricultural importance, their applications are abundant in commerce and industry. Most of the plant products for consumer use come from the angiosperms. This giant group of plants, which include some 200,000 species of dicots (grouped into about 250 families) and some 70,000 species of monocots (grouped into about 60 families), is characterized by a diversity not seen in any of the other plant groups.

From the sedimentary strata of the fossil record, one can determine layer upon layer of ferns and primitive gymnosperms, the lush vegetation of the Carboniferous forest about 200 million years ago. Then suddenly, angiosperms appear in sediments from about 127 million years ago. **Paleobotanists** have generally concluded that angiosperms evolved considerably earlier than the fossil record indicates, perhaps as long as 175 to 200 millions years ago. The argument suggests that conditions that lead to selection pressures (changing and diverse climates) existed in the tropical highlands, not in the valleys and lowlands where the sedimentation would be expected to occur. It would be possible for a group of organisms to have evolved in such high elevations and have no fossil record for many millions of years.

Known distribution patterns and the fossil record suggest that angiosperms did evolve from now-extinct broad-leafed gymnosperms in the hills and uplands of Gondwana, which began breaking apart just about the same time the angiosperms appeared. The separation of South America and Africa was not complete in the tropical regions until about 90 million years ago. Presumably the early angiosperms radiated into Laurasia in the region of what is now North Africa, the Iberian Peninsula, and the Middle East. Even though the Indian subcontinent was moving rapidly northward at this time and eventually collided with Asia, there is little evidence of major angiosperms crossing this link; the Indian subcontinent had undergone such major climate changes during its move northward that most species became extinct.

The fossil record also suggests that by 75 million years ago many flowering plant families were well established, and some of these are common in today's angiosperm flora including; birch (*Betula*), alder (*Alnus*), oak (*Quercus*), elm (*Ulmus*), sycamore (*Platnus*), basswood (*Tilia*), chestnut (*Castanea*), maple (*Acer*), beech (*Fagus*), sweet gun (*Liquidamber*), hickory (*Carya*), and *Magnolia*.

A number of modern angiosperms have returned to an aquatic habitat where the environmental stresses are not serious. Whether the plant is submerged or floating, water stress in an aquatic environment is seldom a factor in growth and reproduction. Light and gas exchange can be compromised, but species that have managed to survive did so through specific anatomical and morphological adaptations. In *Lemna minor* (duckweed), for example, the tiny leaves float on the surface of the water, and roots lie just below the surface.

Angiosperm evolution, as in other organisms, is characterized by the development of new species as a world climate change. Unquestionably one of the major selective forces is the lack of a constant water supply, and many species have managed to survive because variants within the species were capable of coping with aridity. Because lack of water is such a compelling selective factor, proportionately fewer organisms manage to compete well in arid ecosystems. Species that do survive are often very different morphologically from their ancestors.

Evolution of the Flower

The shoot apex of flowering plants is a remarkable structure: It sometimes produces leaves and sometimes produces flowers, fruits, and seeds depending on a number of complex internal and external factors. A vegetative apex that produces leaves is said to be indeterminate in that new leaves arise whenever climate conditions are suitable for growth and development. Even in periods of dormancy, the vegetative apex remains intact, and primordial leaves wait for the onset of favorable conditions. This same shoot apex, cued by certain environmental and/or hormonal factors, suddenly stops producing a flower, fruit, and seeds; thus, it becomes a determinate organ. The conversion from indeterminate to determinate status is characteristic of angiosperms. Perhaps no single event of plant evolution has had such a significant impact as has the evolution of the flower. It is important to remember that floral parts have evolved from leaves, and indeed flowers are really nothing more than modified leaves. Most of what we know about evolution of flowers is based on morphological comparison of modern forms because flowers are generally too delicate to be preserved in the fossil record. A primitive flower was found in sedimentary deposits in Sweden. Apparently the progenitor of the rose, it is thought to be approximately 80 million years old. Such discoveries are rare, and the fossil record is not likely to reveal the minute, step-by-step history of floral evolution.

Recall that a flower is a determinate shoot bearing various leaflike appendages. The carpel is derived from a folded blade without a well-defined stigmatic surface to which pollen grains adhere. Instead, hairs on the margin of the blade acted as pollen traps. Gradually, with the evolution of carpels, the stigma became more specialized and relocated near the top of the structure. In early angiosperms, the ovules were probably arranged in rows near the edges of the inner surface. As evolution proceeded, the number of ovules decreased. Likewise, the primitive flower contained a number of separate carpels within the ovary, which have been fused and/or reduced in number with increasing specialization.

Stamens too, have evolved from leaves, although most modern stamens bear them little resemblance. One of the primitive flowering plants, *Magnolia*, does produce flat, broad stamens, however. It is thought that ancestral stamens were nothing more than leaf blades with sporangia near the center of the blade. As specialization took place, the blade narrowed to produce what is now the filament, and the sporangium was left at the tip of the modified leaf. Increased specialization has led to fused stamens. In some cases, they are fused to the corolla. Some very advanced species have sterile stamens that have become modified to produce nectar, although typical nectaries are not derived from stamens.

Most sepals are green, photosynthetic, and leaflike. They have apparently been derived directly from leaves with little modification. Petals are occasionally derived from sepals, but generally they are specialized from stamens and have become broadened, pigmented, and otherwise modified for pollinator attraction. Petals, like stamens, generally have a single vascular strand, whereas sepals, like leaves, have three or more vascular connections. In the more advanced families, the petals are fused into a tube, and the stamens are often fused to that tube. Likewise, the sepals may fuse into a tube in the more recently evolved groups.

The typical primitive flower was made up of many carpels, stamens, petals, and sepals (all distinctly separated), and all structures were spirally arranged on the tip of the stem. Figure 14-9 shows a modern flower that still retains those characteristics: the *Magnolia*. By comparing *Magnolia* with an advanced species such as an orchid, one can make some generalization about the trends in floral evolution:

Figure 14-9

Magnolia is one of the most primitive flowering plants. It has broad flat stamen. It is thought that these stamen were leaf blades.

1. Flowers have evolved from many indefinite parts to a few definite parts.

2. The number of kinds of parts has been reduced from four in the primitive flower to three, two, or one in modern flowers. (Actually, there were only carpels and stamens in the primitive flower, then complete flowers with petals and sepals, then flowers reduced to the three, two, or one part.)

3. The primitive position of the ovary was superior, the advanced position inferior.

4. The radial symmetry of primitive flowers has given way to bilateral symmetry in the more advanced flowers.

The most specialized dicot family is the sunflower family (*Asteraceae*), and the most specialized monocot family is the orchid family (*Orchidaceae*), as shown in Figure 14-10. Both also have the largest number of species.

Figure 14-10

The most specialized dicot family is the Asteraceae, *and the monocot family is the* Orchidaceae.
Courtesy of Monica Haddix.

Evolution of Fruit

A fruit is a mature ovary. In some cases it retains floral parts. Depending on the arrangement of the carpels, fruits can be simple, multiple, or aggregated. Modifications of floral parts have, in some cases, led to fruits that include the ovary but are largely composed of the organs. In the apple, for example, the receptacle enlarges, growing around the carpels to become the fleshy portion of the fruit. The ovary itself becomes the apple core (and is thus of secondary importance as a food source).

Many fruits have developed hard walls that protect the seeds inside. Such a protective mechanism helps to ensure survival of the species, even though it can make seed germination and establishment more difficult.

Coevolution

The evolutionary processes are normally described in terms of individuals and populations of a given species changing in adaptive (or nonadaptive) ways. The variability in sexually reproducing individuals allows natural selection to function in several ways. The selective forces, however, are usually described in general terms of the organism's "environment," which all too often brings to mind only abiotic and climate features. The complete environment acting on organisms also includes interactions with other living organisms. The term **coevolution** refers to two or more groups of organisms evolving in parallel and interdependently. Neither group "causes" the other to change; both groups develop new types independently. However, whether a new type succeeds is often a direct function of the interrelationships between the groups. Coevolution then refers to successful adaptations only.

Insect-flower coevolution

One of the most striking examples of coevolution is that of flowers and insects. Although today there are some 245,00 species of flowering plants and over 750,00 species of insects, 200 million years ago angiosperms were just beginning to evolve, and only a few different insect species since that time attests to the high development of this relationship. It is one of the most successful examples of coevolution known to scientists.

This interdependence revolves around the selective advantage it provides both groups. Flowers are more efficiently cross-pollinated by insect visitors, and the insects are provided a reliable food source. Since gymnosperms are wind-pollinated today, as they were when angiosperms evolved, undoubtedly insect pollination is one of the advantages that allowed flowering plants to become more successful than gymnosperms.

The earliest visitations were probably by beetles feeding randomly on soft plant tissues, pollination droplets of the gymnosperm ovules, and sap or resin exuded from leaves and stems (see Figure 14-11). During their random foraging, the beetles accidentally carried pollen from plant to plant. If primitive flowers formed by broad-leaved gymnosperms accidentally produced a more nutritious tissue or fluid that increased beetle visitations, pollen sticking to their bodies would be carried from flower to flower. This would have increased the effectiveness of cross-pollination over wind-pollination, and increased cross-pollination frequencies resulted in more ovules developing into seed and greater total genetic variability in successive generations of offspring.

Figure 14-11

Beetles were probably the first pollinators. However, now that is overlooked. Beetles have an acute sense of smell better than their vision. Therefore, odors that simulate their common food sources attract them.

As mentioned, most flowers are self-incompatible—their own garments cannot fertilize them. Thus, new genetic material from a different plant of the same species can constantly be introduced. Cross-pollination ensures maximum genetic variability possible through sexual reproduction.

As the number of surviving offspring and variability increase, flowers having even greater insect attraction developed, producing even for regular and frequent visitation and cross-pollination. For the insects, more plants with more flowers meant greater amounts of nutritional material and a reliable food source. These flowers were numerous enough to become primary or even exclusive food sources for the insects. This interdependence has occurred in a spiraling way: More and different types of insects, resulting in increased cross-pollination efficiency, which resulted in increased flower diversity from which insects could feed, and so on. Unlike the chicken and the egg quandary (which came first?), flowers and insects diversified, multiplied, and succeeded in parallel—they coevolved.

It is probable that several features unique to angiosperms evolved as a direct response to insect pollination. Beetles are generally not very dainty or specific feeders; thus, an ovule is as desirable a food as other floral tissue. Early flower types that lacked sufficient surrounding protective tissues for their ovules did not survive. Today, all flowering plants have **protected ovules**. Subsequent development of entire ovaries surrounded by protective tissue has resulted in **inferior ovary** position.

As insect groups other than beetles evolved, even greater flower variability and insect-flower specificity evolved. From approximately 50 million years ago on, the fossil records indicate continual diversification among insect groups such as the bees, butterflies, and moths. Correspondingly, flower size, shape, color, and organizational complexity also have undergone remarkable change. The efficiency of bisexual flowers, in which an insect is able to pick up and deliver pollen in the same visit, exemplifies continuing flower evolution. Grouping of flowers into inflorescence has also expedited pollination by insect visitors. The development of petals from stamens resulted in an additional floral structure having a phenomenal number of different shapes and specific modifications.

The larger number of flower and insect types that exist today is a result of this process of coevolution. Visitation specificity ranges from general, as in the primordial relationships, to specific one-to-one relationships. It is important to point out that, although the general trend in pollinator-flower coevolution is toward greater specificity, the resulting advantages are there only as long as both partners are successful in all other ways as well. If either member of a highly interdependent association fails, the other will fail (unless it is able to survive by previously unused mechanisms).

Dispersal into new areas is one of the critical requirements for plant success that is very closely tied to pollinator range expansion. In a sense, pollinators are the only legs plants have. Even if seed is dispersed into new areas, unless those plants are able to sexually reproduce, they are ultimately doomed to failure in that habitat. Interestingly, most successful colonizers are weedy plants found in new, disturbed, and changing habitats; these plants have unspecialized pollination systems.

Pollinator Specificity

The importance of pollinators for sexual reproduction in most flowering plants cannot be overemphasized. Because plants lack mobility, cross-pollination and the genetic variability it allows would be very limited without the activities of pollinators. The role of the "birds and the bees" is vital, since these two groups participate in some of the most reliable and species-specific pollinator-flower relationships. A brief and much generalized description of the most common pollinator organisms and the kind of flowers they visit will provide a better perspective of the proceeding discussions.

Beetles

Even though beetles were probably the earliest pollinators and their role in the successful evolution of angiosperms is important, today they are relatively overlooked as pollinators (see Figure 14-11). Generally, beetles are large, awkward, and poorly adapted as a flower pollinator. Their mouthparts are adapted for chewing; thus, they feed on edible tissues and pollen, seldom nectar. Most beetles that do visit flowers do not depend exclusively on flowers for their nourishment; rather, they primarily feed on other plant parts or dead animals, and as pollinators they are undependable. Certainly, not all beetle visitations are random; there are many examples of regular, predictable, and even highly specific visitations. There are even some beetles with mouthparts modified for nectar feeding and flowers that have developed attractants for beetle visitation.

Beetles pollinate even highly modified flowers, such as some orchids, but the typical flowers visited by beetles are far less specialized. Usually large and solitary or small and grouped into a large inflorescence, these flowers are often dull white to greenish with strong fruity or decaying proteinlike odors. Beetles' sense of smell is much more acute than is their vision, and they are attracted by odors that simulate their common food sources—fruit, carrion, and dung.

The flowers are usually open, flat or bowl shaped, and shallow, with plentiful pollen and accessible sex organs. In addition, the ovary is normally well protected, safe from the beetles' indiscriminate feeding habits and chewing mouthparts. Some well-known solitary flowers pollinated by beetles are species of the *Magnolia* genus, poppies, cactus, and lily. Inflorescence in the carrot family (umbels), dogwoods, and tropical members of the *Fagaceae* (beech family) are also beetle pollinated.

Bees

Of all the flower-visiting pollination organisms, bees are definitely the most well adapted, specific, and numerous. From small nonsocial bees to the larger social honeybee and bumblebees, these insects are experts in flower recognition, feeding, and pollination. Bees have hairy bodies, which are ideal for pollen transport, and bumblebees are known to carry as many as 15,000 grains per individual. Bees are able to readily learn shapes, patterns, and colors, and they have mouthparts modified for nectar feeding and pollen collecting. In addition, social bees have a large food demand, providing for themselves and their brood. They also have a communication system to inform others in their colony of the exact location of a bountiful food source, as shown in Figure 14-12.

An interesting phenomenon of bee vision is their range of color perception. Bees have a visible spectrum that is shifted into the ultraviolet wavelengths but out of the red range. Therefore, they are able to detect patterns produced on petals by ultraviolet-reflecting pigments but are essentially red colorblind. One of the groups of secondary plant compounds, the flavonoids, include some pigments often found in petals that reflect ultraviolet patterns. Two related species, both with the same color (to us) yellow flower, have distinct colors to a bee visitor—hence, differential visitation.

Bee and flower coevolution is specific and highly complex; as a result, some of the most unusual and highly modified flowers are bee pollinated. Typical bee-pollinated flowers are irregular, sturdy, fairly deep, and often have a "landing platform." Usually bright yellow or blue (but not red), bee-pollinated flowers commonly have color nectar guidelines running from the outer edges of the top surface of lower petals down the tube of the flower. Many also have semiclosed flower throats, which help prevent nectar thieving by smaller insect visitors too weak to force their way past the closure. Many flowers have hairy areas near the stigma that groom pollen off the bee's body to ensure pollination.

The flower tubes are often structurally designed so that only the appropriate bee species can get to the nectar—the reward for visiting that flower. In addition, the pathway to the nectar supply also requires that the visiting bee come in contact with the stigma, guaranteeing pollination. Pollen from the flower being visited is also deposited on the bee's legs, back, or underside, often by complicated modifications that break open anther sacs or force the departing bee to come into contact with the mature anther.

One of the most specialized and highly evolved bee-flower interdependencies involves a wild orchid in the genus *Ophrys*. The flowers of a given species open at a time in the spring when males of the coevolved bee species emerge. The flowers are the same size and shape as the female of the bee species (which does not emerge until later). The male bees are instinctively attracted to the flower and attempt to copulate with it. Although notably unsuccessful in this effort, the thrashing around on the flower does result in pollen deposition on the male bee's body. Repeated attempts to copulate with these flowers guarantee the next generation of bees. **Pseudocopulation**, as the bee-flower activity is termed, occurs between several specific members of the *Ophrys* genus species of bee, wasp, and even some flies.

Common to many bee flowers is a postpollination change in the appearance of the flower, probably induced chemically by the development of pollen tubes. Flowers are known to have their visual attractant markers change after pollination. The potential bee visitors then do not recognize the flower and thus pass it by. Were these bees to visit such flowers anyway, they would find no nectar and the desirability of future visits to flowers of that species would be diminished. Changes such as the fading of the nectar guidelines, wilting of the landing platform petal, general flower closure and withering, and color dullness are adaptations of some bee flowers that secure even greater visitation and pollination efficiency and success.

Butterflies and moths

Even though butterflies and moths are closely related and have similar morphology, as pollinators they are quite different. Butterflies, shown in Figure 14-13a, are active in the daytime and have good vision but a weak sense of smell. Moths, shown in Figure 14-13b, are nocturnal and have a well-developed sense of smell. Butterflies light on the flower, whereas moths hover. Both suck the thin nectar through their long, thin, hollow tongues, and neither has to provide food for developing young.

Usually yellow, blue, or in some cases red, butterfly flowers have a long, thin floral tube with a sturdy outer flower structure for adequate landing. Butterfly flowers include many members of the sunflower family (*Asteraceae*), *Lantana*, and various trumpet-shaped blossoms.

Figure 14-13a

Butterflies and moths look alike but pollinate at different times. Butterflies are active in the daytime and have good vision but a weak sense of smell.

Moths visit white, pale yellow, or pink flowers that are open at night and produce a strong, heavily sweet perfume. These flowers also have deep tubes but with open or bent-back margins that allow hovering moths to reach the nectar with their long tongues. Generally, moth flowers produce more nectar than butterfly flowers because hovering expends more energy.

The hawk moth is a competitor of hummingbirds for their food sources. Essentially as large and remarkably similar in appearance to a hummingbird, this moth has a long proboscis for sucking up the thin sugary nectar of typical bird flowers. More commonly however, moth flowers are larger than bird flowers and open only at night. *Yucca*, evening primrose (*Oenothera*), and a number of night-blooming cactus species are moth-pollinated flowers.

The *Yucca* is pollinated only by the four species of a single moth genus (*Tegeticula*), which in turn depends on the *Yucca* flower for its entire reproductive cycle. The female moth collects pollen from one flower and transports it to another. It pierces the ovary wall and lays eggs inside the ovary. Only about 20% of the developing *Yucca* seeds are destroyed by the feeding larvae, which eat their way out of the ovary when they are mature so that they may pupate on the ground below the plant. The female yucca moth ensures a food source for her larvae while pollinating the flower.

Birds

Different kinds of birds act as pollinators by feeding on nectar or on insects within the flowers. Sparrows are known to visit spring crocus and have been observed pollinating pear trees. Honeycreepers pollinate the Hawaiian lobelias, and the sunbirds of Africa and Asia are known pollinators. The most well-known bird pollinators, however, are the hummingbirds of North America. These tiny animals expend incredible amounts of energy in flight, especially in hovering while feeding. They have keen eyesight, being most responsive to reds and some yellows, but do not have a well-developed sense of smell. Their long thin beaks enable them to reach the abundant, thin, sweet nectar. Their feathers carry large amounts of pollen, picked up primarily on the front and top of their heads when they come into contact with stamens extruded beyond the floral tube (see Figure 14-14).

Typical flowers pollinated by hummingbirds are red, with a good supply of thin, sweet nectar found at the bottom of a slender floral tube. The lips of the flowers are usually curved back out of the way, and they are normally a solid color, lacking nectar guides. These flowers usually have little or no scent, which, in combination with their color and long slender tubes, usually make them unnoticeable to most insect visitors. Sugar ants are a notable exception, but they provide very little competition to hummingbirds. Striking examples of bird flowers include *Erythrina*, *Aquilegia* (columbines), orchids, *Salvia*, *Mimulus,* and *Lobelia* species that have red flowers.

Figure 14-14

Hummingbirds look for red flowers and help to pollinate flowers with a long slender floral tube. Courtesy *of Web Shots Free Photo.*

Flies

Flies display greater variation in their methods and tendencies of pollination than any other insect groups. Primitive flies parallel beetles in their lack of sophistication; highly specialized flies are comparable to bumblebees and hawk moths in complexity. Some South African flies have a 5 mm long proboscis and the ability to hover with nectar feeding. In spite of the range of variability, the best known pollinators are the carrion and dung flies. Because they lay their eggs in diseased or decaying animal flesh or on fresh dung, they are attracted by the putrid odors of decomposing protein. Flowers that are pollinated by these flies usually attract with strong odors but offer no reward to the visitor. The "carrion" plant *Stapelia* is the best known of the fly-pollinated flowering plants, carrying the attraction mimicry even a step further. *Stapelia* flowers vary in size depending on the species, but they are flat and open with the dull yellow color of decaying meat, complete with reddish streaks. Most notable, however, is the odor, which does not invite a second sniff.

Ants

Ants are exceptionally fond of sweets, from table sugar to flower nectar. They are also so small that they can raid a flower without touching anthers or stigma. In addition, their bodies are hard and not well adapted to pollen transport, and they are aggressive defenders of a newfound food source. As pollinators, therefore, ants are essentially noneffective, although there are a few isolated examples of larger ants providing nonspecialized or accidental pollination for some flowers.

Their aggressive defense of a food source often chases away other insects, including those which would actually effect pollination were they allowed to visit the blossom. Some flowering plants have actually evolved "ant guard" adaptations that deter ant visitations. One of the most successful is the ring of sticky glandular hairs on the stem immediately below the flower of *Viscoria vulgaris*. Stiff hairs projecting downward on the stem or outward in the throat of a corolla tube are other common ant guards that have evolved in some species.

Mosquitoes

Mosquitoes also are too small and ill designed for effective pollination; however, some flowers are, in fact, mosquitoes pollinated. Certain small and inconspicuous orchids are visited by both male and female mosquitoes, which feed on nectar rather than blood. The mouthparts of several mosquitoes' species are actually modified, which feed on nectar rather than bloodsucking. The food demands of mosquitoes are very small, and even though they carry pollen with them as they visit from flower to flower; they seldom need to visit from flower to flower beyond those on the same branch of one plant. As cross-pollinating agents, then, they are less effective than other insects that also visit many of the same flowers.

Bats

Like birds, bats, as shown in Figure 14-15, have several effective pollinating features. They are large, have a rough (furry) surface for holding large amounts of pollen, and can move rapidly across large distances. Most bats are insect eaters, but a number of vegetarian bat species exist. Fruit-eating bats are found worldwide, and it is hypothesized that nectar and pollen feeding developed from such lines; as it did in bird species.

Figure 14-15

Bats have several effective pollinating features. They are large and have rough surfaces for holding large amounts of pollen.

Pollinator bats are nocturnal, have an acute sense of smell over great distances, and unlike most bats, have acute vision (although color-blind). Their sonar system is less developed; therefore, flying in densely vegetated areas is difficult. Many of these bats can also hover like hummingbirds and have long, slender noses and long tongues.

As these bats fly from flower to flower, they lap up the nectar, often eat parts of the flower, and transport large amounts of pollen in their head fur. Although some species ingest pollen only accidentally while feeding on nectar, others use long tongues to lick pollen off their heads, a major portion of their diets.

It is now known that some bats actually depend exclusively on nectar and pollen for their food, and the flowers that these bats visit contain high amounts of protein in their pollen.

Although flower-visiting bats enjoy mostly tropical distribution, some migrate in the summer as far north as the southern United States and northern Mexico, feeding on *Agave* (century plant) and cactus flowers. Bat pollination is also known in Australia (on an introduced *Agave*) and in Asia as far north as the Philippines. In Africa, bat pollination does not extend north of the Sahara, and in the Pacific Islands their distribution extends south to Fiji. Certain plant distribution can be explained by knowing about bat-pollinated flower types. For example, there is a banana plant (*Musa fehi*) that is adapted to bat pollination. However, it is thought to have been introduced to Hawaii, not native, because bats are not indigenous to Hawaii.

Typical bat-pollinated flowers open only at night and only for one night. They are often drab greenish to pink-purple and sometimes white or creamy. They emit a strong fruity or fermenting odor at night and have a very large quantity of both nectar, and pollen held in larger or numerous anthers. These are generally large, sturdy, and solitary or in the inflorescence positioned outside the foliage of the plant for increased pollinator accessibility.

Other Angiosperm-Animal Coevolution

In the evolution, diversification, and adaptation of the flowering plants, other specific coevolution with various animal groups has taken place. Fruit and seed dispersal mechanisms and morphological adaptations to prevent herbivory are obvious and easily demonstrated examples. Less evident are chemical groups that are now thought to exist as a result of coevolutionary interactions. These were once regarded by plant biochemists as dead-end or nonessential secondary compounds. But many of these chemical groups are now thought to be involved in specific processes of plant protection, pollinator attraction, and feeding stimulation.

We have already mentioned certain flavonoid compounds—ultraviolet wavelength pigments in some flower petals that are visible to bee pollinators. Closely related chemically are the anthocyanins, which are visible wavelength flower pigments. The *Ophrys* flowers provoke pseudocopulation because of visible patterns produced by anthocyanin pigment.

Plant Palatability

Plant palatability is significantly influenced by secondary compounds. Both insect and vertebrate animal groups including humans are attracted or repulsed by a range of secondary compounds that affect the sense of taste. Members of the mustard family, for example, contain compounds that deter many insects and people from feeding on them, whereas other insects feed only on these plants. The acid taste and pungent odor of horseradish, cabbage, cauliflower, radish, watercress, mustard, Brussels sprouts, and rutabaga is caused by the same class of chemical compounds. Scientists have proven that this group of plants when eaten by humans can cut down the cause of certain cancers. These undoubtedly coevolved with specific insect groups, protecting the plants from excessive herbivory by most insects while providing a constant food source for insects adapted to tolerate or even be actually attracted to these compounds.

Alkaloids

A large group of loosely related, nitrogen-containing compounds are **alkaloids**, which are toxic. Toxicity, combined with an unappealing bitter taste, protects plants containing such compounds from being eaten. In the milkweed family (*Asclepiadaceae*), for example, alkaloid compounds and cardiac glycosides combine to act as severe toxins to most vertebrate animals. Foxglove (*Digitalis purpurea* in the *Scrophulariaceae*) also produces cardioactive glycosides that are used as a treatment for heart disease but in higher doses can endure a heart attack in vertebrate consumers. Certain insects, however, are unaffected by these compounds and preferentially feed as larvae or as adults on the tissue of such plants. In fact, these insects even "advertise" that their bodies contain these compounds by being brightly colored and patterned. They are recognizable to potential predators, which are predominantly vertebrates, and after one experience of acute digestive upset, vomiting, and diarrhea, the predator avoids further ingestion of such insects.

Monarch butterflies feed on milkweeds, and their distinctive visibility to birds acts as a protective device rather than as an attractant. So effective is this coloration and pattern that the viceroy butterfly has an example of mimicry, a process that has occurred repeatedly in both plant and animal groups. For many plant species that have evolved in a parallel manner, they are protected by their close resemblance.

Many of the secondary compounds that have evolved in plants as protective adaptations or as attractants produce strong effects in humans. The opium poppy, the hemp plant (*Cannabis*, or marijuana), and peyote cactus are among these plants. All contain secondary compounds, mostly alkaloids that occur in nature because of specific coevolutionary phenomena.

Biochemical Evolution

As plants evolve into different forms and develop new strategies for adaptation, they may also be selected against if they do not also develop new modes of metabolism; in some cases, the biochemistry actually changes. There is strong evidence that both plants and animals adapt to environmental stresses by synthesizing **isozymes**, or alternate forms of an enzyme, which may allow the plant to carry on photosynthesis, respiration, protein synthesis, or other roles under changed environmental conditions. Given plants appear to synthesize isozymes for water stress, salt stress, heat stress, and cold stress. Those individuals of a population that do so may survive in a changing environment, whereas the plants capable of synthesizing only the normal enzyme may fall.

Biochemical evolution also includes the alternative strategies for carbon fixation (photosynthesis). Most plants fix carbon via the C3, or Calvin, cycle. Other plants that have evolved under conditions of high temperature, high light intensity, and water stress have developed an alternative method, the C4 pathway. The C4 phenomenon is widespread among monocot and dicot families, which have evolved in the hot and drier parts of the world. Such plants have managed to survive continental movements and changes in world climates largely because they have been able to change certain aspects of their metabolism.

Ecology and Importance of Angiosperms

A unifying theme of this text is the application of plant science in a human-dominated world. Everywhere one looks, flowering plants influence our lives. Except for the coniferous forest, which provides timber for housing and other construction, angiosperms are the dominant plants for food, forage for animals, commercial fibers, industrial products, and medicines. Even most of our landscaping materials are flowering plants. There is no accurate method to assess the absolute value of angiosperms in our lives. A price tag can be placed on food or other items of trade, but higher plants essentially and complexly influence our very being. Societies that depend on firewood for survival are exquisitely aware of the importance of angiosperms.

Finally, the angiosperms have so totally invaded the terrestrial ecosystems that species are found almost everywhere. No other group of plants has been so successful in filling niches under the most extreme environmental conditions. Such plants compete most effectively for available light energy and contribute incalculably as producer organisms.

 ## Summary

1. Approximately 450 million years ago a primitive multicellular marine green algae washed up on shore and was able to survive the ultraviolet radiation of the sun because of an atmospheric

ozone layer that had been accumulating since the first cyanobacteria began photosynthesizing some 3 billion years ago.

2. In addition to the bryophytes, primitive vascular plants began evolving. Their vascular system plus protected sex cells, spores, and cuticle development allowed them to succeed and diversify rapidly on land. Three extinct primitive vascular groups gave rise to the four major divisions of living seedless vascular plants.

3. Ferns have fiddlehead coiled leaf fronds, and the reproductive structures, sori, occur on the underside of the fronds. The sporangia within each sorus produce spores that germinate and develop into a prothallus, which grows into a heart-shaped gametophyte. Ferns are most important economically as ornamentals.

4. Seed plants are even better adapted to extreme climates than are spore-bearing plants. The two seed-producing groups are the gymnosperms and angiosperms.

5. Gymnosperms include the cycads, gnetophytes (including *Ephedra*), *Gingko,* and conifers. By far the largest and most evident group is the conifers, which reproduce by having separate male and female cones. From wind pollination to fertilization takes about 15 months in pine, and with an additional year before mature seeds are ready to be released by the female cone.

6. The angiosperms are today's dominant vegetation, comprising some 235,000 species. They are thought to have evolved from a broad-leaved gymnosperm between 125 and 200 million years ago. Flowering plants occupy a wide range of ecological habitats and vary in size from a few centimeters to 100 m tall.

7. Flower parts evolved from leaves, and primitive flowers commonly have many of each floral part. More highly derived flowers have reduced numbers of floral parts, an inferior ovary, and bilateral symmetry. The *Magnoliaceae* are considered to be one of the most primitive angiosperm families, the *Orchidaceae* and the *Asteraceae* are among the most highly evolved families. Fruits have also evolved into more complex types.

8. Flowering plants and insects have coevolved, providing both groups with several advantages. Originally only beetles acted as pollinators; now wasps, bees, butterflies, moths, ants, mosquitoes, birds, and bats all act as pollinators for an incredible array of flowering plants.

9. Beetles are clumsy, general visitors; bees and wasps are highly specialized visitors attracted by complex shapes, color (except red) markings, and even ultraviolet patterns. Butterflies are active in the daytime, visiting yellow, blue, and red flowers with thin nectar and sturdy corollas. Moths are nocturnal and prefer stronger-smelling light-colored flowers. Hummingbirds are the most common bird pollinators, visiting red, tubular flowers with

a lot of thin nectar. Flies, ants, and mosquitoes effectively pollinate a wide range of flower types, including some that attract flies by their rotting meat smell. Bats visit large, nocturnal flowers with a strong perfume smell, thin nectar, and copious pollen.

10. Secondary plant compounds are thought to protect plants from herbivores and insects. Some of these compounds have strong physiological effects on humans. No other plant groups have more utility to human society than the angiosperms.

Something to Think About

1. What made the environment of the Earth allow plants to start photosynthesize?

2. Plants with vascular systems were able to survive and succeed on land because of what?

3. How many Pterophyta exist on Earth?

4. Diagram how Pterophyta reproduce.

5. Compare and contrast the spore-producing plants with the seed-producing plants.

6. Which is the largest group of plants?

7. List the difference in angiosperms and gymnosperms.

8. How long does it take for pollination and fertilization to complete?

9. Trace the evolution of angiosperms.

10. What did the flower evolve from?

11. Name the plant family that is considered to be one of the most primitive.

12. Describe flower symmetry.

13. Which two plant families are considered to be the most evolved?

14. List different pollinators.

15. Which plant group is considered to be the most utilitarian to human society?

Suggested Readings

Belly, P. R., and R. Hemsley. 2004. *Green plants: Their origin and diversity*. Cambridge, MA: Cambridge University Press.

Dowson, J., and R. Lucas. 2005. *The natural habitats, challenges, and adaptations*. Portland, OR: Timber Press.

Gaston, K. J., and J. I. Spicer. 2004. *Biodiversity: An introduction*. Oxford: Blackwell Publishing Professional.

Huston, M. 2002. *Biological diversity*. Cambridge, MA: Cambridge University Press.

Moran, R. C. 2004. *A natural history of ferns*. Portland, OR: Timber Press.

Raven, P., et al. 1999. Biology of Plants (6th ed.). New York: W. H. Freeman.

Internet

Internet sites represent a vast resource of information. The URLs for Web sites can change. Using one of the search engines on the Internet, such as Google, Yahoo!, Ask.com, or MSN Live Search, find more information by searching for these words or phrases: **after riping, dioecious, generative cell, integument, microsporangia, microspore mother cell, nucellus, ovule, polyembryony, seed coat, self-incompatible, and tube cell.**

PART 5

Plants and Society

CHAPTER 15

Putting Down Our Roots

Although much research remains to be done on the beginning of agriculture, there is already a great deal of interesting knowledge in this area. The most popular theory is that agriculture originated earlier in temperate regions than it did in the tropics. One idea to support this theory is that of need: There was no shortage of food and no need for shelter, clothing, or fire in the tropics, whereas all of these were concerns for those living in colder climates. More evidence for this theory is that the ancestors of many important domesticated crops are from the temperate regions. However, early agriculture in the tropics is difficult to document because the warm, moist conditions are not conducive to the preservation of archaeological remains; thus, we have more evidence from drier, colder regions on which to base our theories. With the diverse climates and soil types major agronomic crops developed in the United States, including: corn, wheat, barley, oats, rice, potatoes, soybeans, cotton, dry beans, peas, peanuts, and alfalfa.

After completing this chapter, you should be able to:

- Document the development of agriculture through the ages
- Compare the early humans that survived on hunting and gathering with the twenty-first century hunters and gatherers
- Name the earliest crop plants
- Discuss why certain plants became crops for humans
- Explain and define the three different types of plants
- Describe how the trade routes helped the spread of agriculture crops
- Discuss the major agronomic crops in the United States

Key Terms

slash and burn	hemp plant	millets
Yellow River	nitrophiles	maize
wadis	taro	rye
multipurpose plant	irish potato	oats
mulberry tree	tuber	sorghum
baobab tree	cassava	first agriculture
century plant	tiller	revolution

Beginnings of Agriculture

Located in the Fertile Crescent of the Middle East, formed by the Tigris and Euphrates River, the ancient city of Jarmo has been studied by archaeologists, who have documented the existence of agricultural activities there between 10,000 and 12,000 years ago. Jarmo is located in the foothills of the Zagros Mountains, which is now Iraq. Although it is hot and dry today, at the end of the last glacial period climates were cooler and wetter in that area, which made it ideal for agriculture.

Nomadic tribes migrated annually in the fall and winter out of the Zagros foothills to the nearby valleys where forage and water availability lured the animals on which these people depended for meat and hides. These valleys also provided the plants needed for food, fiber, and fuel during the winter; they were not available at the higher elevation of the surrounding mountains.

In the spring, the animals slowly moved back toward the foothills of the mountains and spent the summer in these cooler elevations, where forage was plentiful for the animals as well as for the wandering nomads. This cycle was repeated year after year, and it is probable that the routes and temporary camps used by these nomadic people were the same on each trip. Each such camp had a designated trash dump area in which seeds and roots of the gathered plants were occasionally thrown. The composting of these dump areas produced rich nutrient conditions in which many of the plants sprouted and grew in greater densities than normally found in nature. As the people returned yearly to these areas, they must have gradually realized that using these plants near the campsite was more efficient than wandering far and wide gathering from wild populations. This realization could well have slowly evolved into experimentation by planting increased numbers of selected plants near the camp. Finally, enough food could be grown in such a manner to last year-round, with only occasional hunting forays required to provide meat and hides.

Once these people limited their nomadic activities to one or two semipermanent annual camps and were increasingly dependent on cultivated plants, intensified agricultural activities developed fairly rapidly. Because of reduced hazards associated with a nomadic existence and more leisure time, increasing population size also resulted from this sedentary lifestyle. This further intensified the need for adequate food supplies from cultivation. The domestication of animals followed, until finally villages, such as Jarmo, became firmly established.

In such villages, the evolution from a nomadic existence to a stable one with a stored food supply allowed other advances. With a sedentary existence, material goods could be accumulated, and the potters, weavers, tanners, artisans, and scholars became important members of the community. Advanced civilization rapidly evolved from such beginnings.

The Tehuacán Valley

We now know that agriculture developed independently in other areas of the world at a similar time in history. In the Tehuacán Valley of Mexico, a parallel sequence of events led to an agricultural society dated by archeologists also at 10,000 to 12,000 years ago.

The Tropics

An alternate theory of how agriculture may have first developed centers on the tropical regions of the world. Since this region of the world had ample supplies of water, year-round growing seasons, and warmth, there was less need to cultivate plants. Popularly termed the "genius" theory, this idea holds that since there were abundant food resources, free time was available for other pursuits; greater effort was applied to a consideration of planning, and it follows that the idea to concentrate some of the useful plants of their areas close to the village soon developed. Because of the vegetative density of the tropics, this required clearing some land and planting seeds or underground tubers and rhizomes. Since these areas would revegetate in a normal succession of plant species, a subsequent clearing of new areas every few years was necessary. This has been called the **slash and burn** technique, and because it allows the balance of nature to be reestablished, it has been a very successful small-scale agricultural practice in the tropics for thousands of years. How many thousands are not known because of the lack of preserved archaeological remains to study and date.

The **Yellow River** of China runs through a tropical area in which agricultural activities are thought to have existed as long as 15,000 years ago. It is interesting that this area did not develop the complex societies that resulted from agricultural beginnings in the temperate areas of the Middle East and North America (Mexico).

Egypt

Because of the existence of datable archaeological evidence, the Fertile Crescent of the Middle East and the Tehuacán Valley of Mexico have long been accepted as the areas in which agriculture had its earliest beginnings. It is also estimated that agricultural activity was present in the Yellow River area of China 15,000 years ago. But none of these sites has produced evidence as old as that discovered along the Nile River in central and southern Egypt. In this area, some agricultural activity was practiced as early as 18,000 years ago. At that time, the level of the Nile River was much higher than today, and its annual floods provided water to large depressions in the adjacent sand dunes, called **wadis**. While the water was available, agricultural intensification of wild barley was practiced each year until the heat and lack of rainfall finally dried up the wadis and forced the seasonal farmers back to the more plentiful banks of the Nile. The seminomadic lifestyle apparently continued in this region for 4,000 to 6,000 years but did not result in large, permanent villages, increased population size, or a total dependence on agriculture. Thus, it seems likely that the earliest plant domestication began in areas where agriculture provided only a partial food resource for small hunting and gathering populations. These did not develop into the more complex and advanced cultures that resulted from the separate agricultural beginnings in the Fertile Crescent and the Tehuacán Valley.

Earliest Crop Plants

The period during which people first cultivated part or all of their plant food sources was almost certainly after the most recent glaciation. Cultivation arose independently in several different parts of the world. Until agriculture was used as the primary source of food supply in these early societies, the associated advance of civilization did not develop.

With some certainty, the first plants to be cultivated were those that had been gathered by these societies. We do know that every important civilization depended on cereal crops as the mainstay of their agricultural base. In addition, plants were more likely to have been cultivated if they had many uses or if they were abundant locally and easy to grow.

In the tropical regions the coconut palm (*Cocos nucifera*), as shown in Figure 15-1, was and still is a very important **multipurpose plant**.

Figure 15-1

The coconut palm tree is a multipurpose plant thought to have been one of the earliest plants used by humans. It is found worldwide in tropical climates and grows on sandy beaches, having been distributed by a buoyant fruit.

Its fruit provided plentiful, nutritious, and bacteria-free "milk" and a solid flesh that can be eaten fresh or dried into copra, from which coconut oil can be extracted. The buoyant and watertight outer husks are useful containers, and the fiber of the husk can be made into rope and matting. Palm leaves are used for roof thatching on cottages, which are often constructed using palm trunks for the main supports. In addition, sap from the stem can be fermented into an alcoholic drink or evaporated to produce sugar.

The **mulberry tree** (*Morus*) in China provided fruit for human consumption, leaves for silkworms to eat, and a beautiful yellow dye from the wood. In African savanna areas, the **baobab tree** (*Adansonia digitata*) is also a multipurpose plant. Its fruits contain seeds rich in oil, and pulp of the fruit, high in vitamin C and tartaric acid, is made into a popular drink. The leaves have medicinal value, and the trunk provides fiber from which rope can be made. Even the trunks of old baobab trees are hollowed out for water storage during dry periods. The **century plant** (*Agave*), as shown in Figure 15-2, of Mexico yields fiber from its leaves and fluids from which several alcoholic drinks, including tequila, are made. Locally the "meat" of the developing flower stalk is a nutritious food source.

Figure 15-2

The Agave, *or century plant, is still a highly used multipurpose plant in Mexico.*

One of the best known of the multipurpose plants is the **hemp plant** (*Cannabis*). Requiring high levels of nitrogen, this plant might have originally grown along with other weedy **nitrophiles**. In the nitrogen-rich rubbish piles of the nomadic camps, it was cultivated for fiber from the stems, oil from the seeds, and the medicinal properties of its leaves.

Root and Stem Crops

Because of the ease with which they may be cultivated and because of their high carbohydrate levels, plants with underground storage parts were probably early crop plants. Some of these had roots, and other stems, modified for carbohydrate storage, especially starch. Easily harvested by use of a digging stick, these edible underground parts could well have had their earliest agricultural beginnings in rubbish piles. Deliberate replanting of the leftover pieces of root or stem increased the already abundant food supply for these nomadic people. The use of such root crops in an early form of cultivation almost certainly existed in many different parts of the world.

The **taro** (*Colocasia*) from Asia and the similar tannia (*Xanthosoma*) of the West Indies are both known to have been early root crops.

They both have carbohydrate corm—swollen underground stems with stem buds that can each produce a new plant. The **Irish potato** (*Solanum tuberosum*) is a modern crop **tuber** having buds (potato "eyes") that can be planted to produce new plants. **Cassava** (*Manihot utilissima*) is another tropical plant valued for its root; tapioca is made from this plant.

Cereals

As food plants, these members of the grass family have several desirable traits. Under cultivation, they yield a large amount of grain per acre, and that grain—the single-seeded fruit—contains carbohydrates, minerals, fats, vitamins, and protein. The grains are compact and dry, which allows for long-term storage, or they can be ground into flour, which also stores well. Additionally, the grass stems (straw) can be woven or thatched into basket, bedding, and housing.

Cereal plants can be encouraged to produce lateral shoots when the upper part of the plant stem is cut off. This process is called tillering and occurs when animals are grazed on young plants. Domestication of grazing animals not only provided these societies with food and skins, but it also increased the density of grain-producing cereal stems (**tiller**) for harvest.

Cereals do well in plain areas near mountains of semiarid regions. According to archaeological records, **millets** (*Setaria*, *echinochloa*, and *Panicum* species) were among the earliest cereal crops to have been cultivated in China. However, the crops around which advanced civilizations developed in China and Southeast Asia is rice (*Oryza*), a cereal plant of the wet lowlands.

In the Middle East, wheat (*Triticum*) was cultivated in the hilly regions of the Zagros Mountains, and barley (*Hordeum*) was grown in "upper" (southern) Egypt as early as 18,000 years ago. Both were later introduced into the lower elevations of the river valleys after the development of irrigation techniques. Civilization had then spread from the mountainous highlands to the Tigris-Euphrates and Nile River valleys.

Maize (*Zea*; Indian corn) is the cereal crop of the Americas, cultivated in the Tehuacán Valley of Mexico at least 8,000 years ago. As in other areas of the world, sophisticated irrigation systems were developed that allowed cultivation to expand and yields to increase. By the time of Christ, chinampas, as seen in Figure 15-3, had been developed. These long, narrow strips of land bordered on three sides by irrigation canals yield several crops per year because there is no need to allow the land to lie fallow for part of each year. The rich muck of the canals is dredged each year and spread on the strips of land, thus, replenishing the topsoil fertility. This maintains the canals while returning the rich nutrients to the soil.

The chinampas system of irrigated agriculture was being practiced by the Mixtecs when the Aztecs conquered the region. The high productivity of this system was a major reason the Aztecs were

Figure 15-3

The Aztecs created artificial islands in shallow waters on which they could raise crops. These floating island gardens, called chinampas, can still be seen in Xochimilco, near Mexico City, where they attract many visitors.

able to dominate such a large region in such a short period of time. When the Spanish conquistadors arrived in Mexico City in 1519, they found the Aztec emperor was receiving 7,000 tons of corn, 5,000 tons of chilies, 4,000 tons of beans, 3,000 tons of cocoa, 2 million cotton cloaks, and several tons of gold, amber, and other valuables each year from his subjects. All this was possible because of advanced agricultural systems, which include the chinampas.

Other cereal crops that were cultivated include **rye** (*Secale*), **oats** (*Avena*), and **sorghum** (*Sorghum*). Rye was first developed as a secondary crop growing as a weed among the primary crop species. Just as wheat was introduced from the Mediterranean region, so was rye; and since rye does better in colder climates than does wheat, it replaced wheat as the primary cereal crop in such areas and was grown even as far north as the Arctic Circle. Oats probably developed as a major crop plant in a similar way because it is also tolerant of a wide variety of climates. Sorghum is known to have been cultivated in much of Africa, not only for its grain but also for the straw, which was used in the construction of walls and roofs of the village houses and for weaving baskets and sleeping mats. The events discussed in this chapter occurred slowly over hundreds and even several thousands of years. The advent of agriculture, therefore, was truly a slow evolution from nomadic hunting and gathering to a stable society in which the components of civilization could develop.

Because of the much longer period that humans existed as small hunting and gathering groups, maybe as long as 2 million years, the relative suddenness of the beginnings of agriculture are often popularly referred to as the **First Agriculture Revolution**. Successive advances in agriculture techniques and productions have also been termed revolutions, and in fact a couple of such advances truly were revolutionary.

RESEARCH AND INNOVATION

Until the 19th century, agriculture in the U.S. shared the history of European and colonial areas and was dependent on European sources for seed, stocks, livestock, and machinery, such as it was. That dependency, especially the difficulty in procuring suitable implements, made American farmers more innovative. They were aided by the establishment of societies that lobbied for governmental agencies of agriculture; the voluntary cooperation of farmers through associations; and the increasing use of various types of power machinery on the farm. Government policies traditionally encouraged the growth of land settlement. The Homestead Act of 1862 and the resettlement plans of the 1930s were the important legislative acts of the 19th and 20th centuries. Also, in 1862, the drive for agricultural education culminated in the passage of the Morrill Land Grant College Act.

In the 20th century steam, gasoline, diesel, and electric power came into wide use. Chemical fertilizers were manufactured in greatly increased quantities, and soil analysis was widely employed to determine the elements needed by a particular soil to maintain or restore its fertility. The loss of soil by erosion was extensively combated by the use of cover crops, contour plowing, and strip cropping.

Selective breeding produced improved strains of both farm animals and crop plants. Hybrids of desirable characteristics were developed; especially important for food production was the hybridization of corn in the 1930s. New uses for farm products, by-products, and wastes were discovered. Standards of quality, size, and packing were established for various fruits and vegetables to aid in wholesale marketing. Among the first to be standardized were apples, citrus fruits, celery, berries, and tomatoes. Improvements in storage, processing, and transportation also increased the marketability of farm products. The use of cold-storage warehouses and refrigerated railroad cars was supplemented by the introduction of refrigerated trucking, by rapid delivery by airplane, and by the quick-freeze process of preservation, in which farm produce is frozen and packaged the same day that it is picked. Freeze-drying and irradiation have also reached practical application for many perishable foods.

Scientific methods have been applied to pest control, limiting the excessive use of insecticides and fungicides and applying more varied and targeted techniques. New understanding of significant biological control measures and the emphasis on integrated pest management have made possible more effective control of certain kinds of insects.

Chemicals for weed control have become important for a number of crops, in particular cotton and corn. The increasing use of chemicals for the control of insects, diseases, and weeds has brought about additional environmental problems and regulations that make strong demands on the skill of farm operators. Now genetic engineering of pest-resistant crops helps overcome the problems encountered by the use of chemicals.

In the 1990s high-technology farming, including hybrids for wheat, rice, and other grains, better methods of soil conservation and irrigation, and the growing use of fertilizers led to increases in food production, not just in the U.S. but in much of the rest of the world. U.S. farmers, however, still have the advantage of superior private and government research facilities to produce and perfect new technologies.

U.S. Agricultural Crops

The following section covers major agronomic crops of the United States, including: corn, wheat, barley, oats, rice, potatoes, soybeans, cotton, dry beans, peas, peanuts, and alfalfa.

Corn

The botanical name for corn is *Zea mays*. Worldwide, corn is better known as maize. Several varieties are grown including dent corn, flint corn, flour corn, sweet corn, popcorn, and pod corn, as shown in Table 15-1.

Table 15-1

Botanical Varieties of Corn		
Type	**Variety**	**Features**
Dent corn	*indentata*	Primary commercial feed corn; grain yellow or white; kernels form dent on top upon drying.
Flint corn	*indurata*	Produced primarily for starch.
Flower corn	*amylacea*	Easily ground and is used for production of cornmeal or corn flour.
Sweet corn	*saccharata*	Sugary kernels when slightly immature; the horticultural vegetable crop.
Popcorn	*praecox*	Subform of flint corn with a hard starchy endosperm; moderately high moisture content; fibrous endosperm explodes when heated.
Pod corn	*tunicata*	Thought to be an ancestor or relative to corn; individual kernels enclosed in husks or pods; not produced commercially.

Successful corn production requires an understanding of the various management practices and environmental conditions affecting crop performance. Planting date, seeding rates, hybrid selection, tillage, fertilization, and pest control all influence corn yield. A crop's response to a given cultural practice is often influenced by one or more other practices. The keys to developing a successful production system include:

- To recognize and understand the types of interactions that occur among production factors, as well as various yield limiting factors.
- To develop management systems that maximize the beneficial aspect of each interaction.

Knowledge of corn growth and development is also essential to use cultural practices more efficiently to obtain higher yields and profits.

Temperature

Corn can survive brief exposures to adverse temperatures—low-end adverse temperatures being around 32°F and high-end ones being around 112°F. Growth decreases once temperatures dip to 41°F or exceed 95°F. Optimal temperatures for growth vary between day and night, as well as over the entire growing season. The optimal average temperatures for the entire crop growing season, however, range between 68°F and 73°F.

Approximately 100 to 150 GDDs (heat units) are required for corn to emerge. Improved seed vigor and seed treatments allow corn seed to survive up to three weeks before emerging if soil conditions are not excessively wet. An early morning soil temperature of 50°F at the 1/2- to 2-inch depth usually indicates that the soil is warm enough for planting. Corn germinates very slowly at soil temperatures below 50°F.

Planting hybrids of different maturities reduces damage from diseases and environmental stress at different growth stages (improving the odds of successful pollination) and spreads out harvest time and workload.

The appropriate planting depth varies with soil and weather conditions. For normal conditions, planting corn 1 1/2 to 2 inches deep provides frost protection and allows for adequate root development. Shallower planting often results in poor root development.

Insects of corn include corn earworm, European corn borer, and aphids. Insects are generally controlled by extensive use of insecticides from tassel emergence through kernel drying. Integrated pest management (IPM) using scouting, biological models of pest populations, targeted insecticide use, and narrow-target pesticides have reduced the use of insecticides, improved production profitability, and reduced environmental hazard.

Diseases affecting corn are Southern leaf blight, Northern leaf blight, and diplodia rot. Diseases are controlled primarily through selection of disease-resistant cultivars, good management techniques, and applying appropriate targeted fungicides based upon biological-meteorological models.

Fertilizer requirements vary according to soil tests. A corn crop removes nitrogen, phosphate, potassium, and various micronutrients from the soil. These must be replenished by a fertilization program.

The corn-soybean rotation is by far the most common cropping sequence used in the Midwest. This crop rotation offers several advantages over growing either crop continuously. Benefits to growing corn in rotation with soybean include more weed control options, fewer difficult weed problems, less disease and insect buildup, and less nitrogen fertilizer use.

No-till cropping systems, which leave most of the prior crop residue on the surface, are more likely to succeed on poorly drained soils if corn follows soybeans rather than corn or a small grain, such as wheat. This yield advantage to growing corn following soybean is often much more pronounced when drought occurs during the growing season. Corn is harvested in the fall with a combine. Frequently it has to be dried down before storage.

Wheat

The botanical name for wheat is *Triticum* spp; however many varieties are available.

Variety Selection

Variety selection should be based on winter hardiness, standability, disease resistance, and yield potential (see Figure 15-4). Although differences in winter hardiness exist among varieties, planting date has the greatest effect on winter survival. The yield potential of available varieties is generally in excess of 150 bushels per acre. This yield is not approached, however, primarily because of a short grain fill period caused by high air temperatures in late June. The ideal air temperature during grain fill is 68° to 76°F (20 to 24°C).

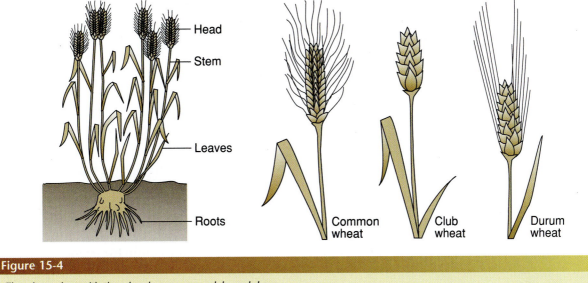

Head

Stem

Leaves

Roots

Common wheat

Club wheat

Durum wheat

Figure 15-4

The wheat plant with three heads—common, club, and durum.

Disease must be controlled if high yields are to be obtained. Both varietal resistance and fungicides are available and may be combined to provide a wide spectrum of protection. Although most available varieties have excellent standability, excessive seeding and nitrogen rates, or their combination, cause lodging, which results in reduced yield.

Because seed size varies from variety to variety and year to year, seeding rates should be based on the number of seeds per foot of row rather than pounds per acre. The ideal seeding rate is 1 million to 1.5 million seeds per acre.

Lodging Control

Lodging is a serious deterrent to high yields. Cultural practices that tend to increase grain yield also increase the likelihood of lodging (when grain falls over to the ground). Using recommended seeding rates, applying proper rates of nitrogen, and selecting lodging-resistant varieties prevents lodging in high-yield environments where yields of 100 bushels per acre are anticipated.

When lodging occurs, the severity of foliar disease increases, resulting in reduced grain yield and quality. Additional effects of lodging are reduced straw quality and slowed harvest. The prevention of lodging increases dividends through a combination of reduced input costs and improved grain and straw quality.

Disease Control

Disease is often the major factor limiting yield of wheat. Effective disease management requires knowledge of the diseases most likely to occur in a production area. Producers fine-tune their disease control strategies for those few diseases encountered each year. Correct diagnosis is the cornerstone to effective control, and producers with little experience identifying diseases should seek help from competent sources, such as a university extension or an agricultural consulting service.

A comprehensive wheat disease management program consists of the following practices:

1. Selecting varieties with resistance to the important diseases in the area. Monitoring wheat diseases aids a producer in selecting varieties with resistance to the common diseases of his or her community.

2. Planting well-cleaned, disease-free seed, treated with a fungicide that controls seeding blights, bunt, and loose smut.

3. Planting in a well-prepared seedbed.

4. Rotating crops—never plant wheat where the previous crop was wheat or spelt. A two- to three-year rotation from wheat prevents most pathogens from surviving in fields.

5. Plowing down residues from heavily diseased fields. Plowing enhances decomposition of residue and death of the disease-causing fungi.

6. Using a well-balanced fertility program based on a soil test. Apply sufficient amounts of phosphorus, nitrogen, and potassium in the fall for vigorous root and seedling growth.

7. Controlling grass weeds. Destroying volunteer wheat, quack grass, and other grass weeds in and around potential wheat fields reduces the amount of disease inoculum available to infect the crop.

8. Applying fungicides only if warranted. Scout fields from flag leaf emergence through flowering. Foliar fungicides are able to control the following diseases: powdery mildew, leaf rust, Septoria tritici leaf blotch, Septoria nodorum leaf, and glume blotch. Leaf rust and powdery mildew are the most severe. Know symptoms, severity ratings, and disease thresholds before scouting fields. Fungicide application following flowering is usually not economical.

Table 15-2 lists some of the common wheat diseases.

Seed treatments are an important part of any wheat production system. Seedborne diseases include loose smut, common bunt, and seedborne Septoria and scab. When wheat is planted into a well-prepared seedbed, with good moisture for quick emergence, controlling these diseases and establishing good stands is easy. However, seed treatment cannot compensate for planting in soil that is too wet, too dry, poorly prepared, or for planting at the wrong depth. No-till seeding increases the likelihood of several diseases developing.

Table 15-2

Some Common Wheat Diseases and Disorders

Disease or Disorder	Symptoms	Environment	Control
Head scab	Spikelets of head turn straw colored, glume edges with pink spore masses, kernels shriveled white to pink in color	Warm, wet weather during flowering period	Seed treatment for infected seed; crop rotation; plow down corn residues
Powdery mildew	Powdery white mold growth on leaf surfaces	High humidity, 60°–75°F, high nitrogen fertility and dense stands	Resistant varieties; crop rotation; delayed planting; fungicides
Leaf rust	Rusty red pustules scattered over leaf surface	Light rain, heavy dew, 60°–77°F, 6–8 hour leaf wetness for germination and infection	Resistant varieties; balanced fertility; fungicides
Septoria tritici leaf blotch	Leaf blotches with dark brown borders, gray centers speckled with black fungal bodies	Wet weather from mid-April to mid–May, 60°–68°F, rain 3–4 days a week	Seed treatment; plant less susceptible varieties; crop rotation; balanced fertility; fungicides
Septoria nodorum leaf and glume blotch	Lens-shaped chocolate brown leaf lesions with yellow borders, brown to tan blotches on upper half of glumes on heads	Wet weather from mid-May through June, 68°–80°F, rain 3–4 days each week	Seed treatment; plant less susceptible varieties; crop rotation; balanced fertility; fungicides
Tan spot	Lens-shaped, light brown leaf lesions, yellow borders	Moist, cool weather during late May and early June	Plow down infested residues; crop rotation; balanced fertility; fungicides
Cephalosporium stripe	Chlorotic and necrotic interveinal strips extending length of leaf	Cold, wet fall and winter with freezing and thawing causing root damage	Crop rotation; bury infested residues; control grassy weeds; lime soil to pH 6.0–6.5
Take all	Black, scruffy mold on lower stems and roots, early death of plants	Cool, moist soil through October–November and again in April–May	Crop rotation; control weed grasses; balanced fertility; use ammonium forms of N for spring top dress; avoid early planting
Fusarium root rot	Seedling blight (pre- and postemergence) wilted, yellow plants, roots and lower stems with whitish to pinkish mold; root rot plants have brown crowns and lower stems	Dry, cool soils, drought stress during seed filling	Seed treatments for seedling blight; delayed planting; balanced fertility; avoid planting after corn
Barley yellow dwarf	Stunted, yellow plants, leaves with yellowed or reddened leaf tips	Cool, moist seasons	Delay planting until after the Hessian fly safe date; balanced fertility
Wheat spindle streak mosaic	Discontinuous yellow streaks oriented parallel with tapered ends forming chlorotic spindle shapes	Cool, wet fall followed by cool spring weather extending through May	Resistant varieties

Fertilization

Fertilization programs for wheat include consideration of nitrogen, phosphorus, and potassium seeds. Providing adequate nitrogen for the wheat crop is an important step toward high yields. However, as the nitrogen rate increases, the potential for lodging and disease also increases. Depending on the level of soil organic matter, carryover nitrogen from previous crops and yield goal, the amount of fertilizer nitrogen required varies greatly. Small grains respond well to phosphorus fertilizer on soils testing. The small grain response to potassium application is less than that of phosphorus.

Considerations for No-Till Wheat

The successful production of no-till wheat requires the proper management of a different set of inputs than for other crops because wheat must survive the winter while maintaining vigor and fighting disease organisms. Key factors for success include having a smooth seedbed; proper seeding depth, rate, and date; the absence of carryover herbicides; and proper seed treatments and residue management.

Wheat grows well on a range of soils, but does not grow well on poorly drained soil, especially during wet periods. The major cause of stand loss is standing water and the formation of ice sheets where water accumulates. Adequate surface and subsurface drainage is absolutely necessary, and more important for wheat than for other crops. Wheat should not be no-tilled in fields that were wet at the time of the soybean harvest or where soil compaction is present. When planting, grain drills should be adjusted to penetrate crop residue and place the seed one-inch deep.

The severity of several wheat diseases increases when tillage is removed from the production system. Typically, no-till seeding reduces the amount of vegetation produced in the fall. For this reason, and to assure fall tillering, no-till seedings should be made as soon as possible after the fly-safe date. Twenty pounds of starter nitrogen should be applied preplant, along with the other recommended fertilizers, to accelerate root system development and vegetative growth.

The benefits of no-till wheat include reduced production costs and much less stand loss resulting from heaving in spring. Eliminating tillage helps retain soil moisture needed for germination and emergence, and can result in more rapid emergence and better stands when soil moisture is low at seeding.

Barley

The botanical name for barley, which is shown in Figure 15-5, is *Hordeum vulgare*. Two cultural practices are used to grow barley—winter and spring.

Barley is used mainly for livestock feed. Studies show that ton for ton ground barley may be equal to corn in feeding value for dairy cows when used as 40% to 60% of the grain mixture. It also is sold

Figure 15-5

Most of the barley grown in the United States is used for livestock feed.

to the malting industry for use in making beer and other alcoholic beverages. Barley does particularly well when the ripening season is long and cool. Although it can withstand much dry heat, it does not do well in hot humid weather because of the prevalence of diseases under such conditions. It grows better with moderate rather than excessive rainfall. Some spring barley varieties mature earlier than oat, rye, and wheat. Barley can be grown farther north and at higher altitudes than any other cereal.

Barley is one of the most dependable crops in areas where drought, frost, and salt problems occur. It is best adapted to well-drained medium and fine textured soils with a pH of 7 to 8. It generally does not produce satisfactory yields on sandy soils. Barley often lodges when grown on soils high in nitrogen, resulting in low grain yields. Nitrogen management is therefore very important. Barley has the highest tolerance for salts of any of the small grains, but the soils in north central and northeastern Minnesota generally do not have this problem.

Winter Barley

Although barley is an excellent animal feed and easily replaces corn in rations, very little winter barley is grown in colder areas because of a general lack of winter hardiness. If grown, barley should be seeded between September 15 and 30 to increase its chances of surviving the winter. Soil pH between 6.5 and 7.0 and recommended levels of phosphorous will improve winter survival.

Spring Barley

Yield performance of spring barley in some areas has been erratic because of late spring planting due to cool, wet growing conditions, which have delayed maturity. When seeding is accomplished early, fertilization is adequate, and good growing conditions occur, high yields follow.

Bowers

Bowers is an awnless six-rowed barley developed by the Michigan Agricultural Experiment Station and released in 1979. Bowers has exhibited medium-early maturity and medium height. The variety tillers well and usually has large heads. Bowers has exhibited some resistance to a number of barley diseases.

Oats

Figure 15-6 shows the oat plant. The botanical name for oats is *Avena sativa*.

When used as a companion crop, oat often is removed during early flowering for forage. Oat grain is mostly used as livestock feed with a small amount sold for use in high-protein cereals. Oat grows well on a wide range of soil types, especially if the soil is well-drained and reasonably fertile. Oat is less sensitive than wheat or barley to

Figure 15-6

Oats provide bulk and protein to the diets of animals as well as food for humans.

soil conditions, especially to acidity. Oat generally grows best on medium and fine textured soils, but it also will produce fair yields on light sandy soils if there is sufficient moisture.

The oat plant requires more water for its development than does any other small grain. It also will yield better than the other small grains with less sunshine. Generally, however, oat yields are inferior to those of barley when moisture is a limiting factor. High temperatures during flowering can increase the proportion of empty spikelets, a condition called blast. Hot dry weather during grain fill causes early ripening and reduced yield; hot humid weather during this period favors diseases.

Nitrogen management is important in obtaining the best oat yields. If too much nitrogen is present, lodging can be a problem and yields will suffer as a result.

Several varieties of spring oats with high yield potential, good test weight, and stiff medium-short straw are available. Some varieties are resistant to the common diseases of oats.

Spring oats is the first crop to be planted in the spring. The selection of fields with well-drained soil is essential to permit timely planting. Spring oats should be seeded as early in the spring as soil conditions permit, preferably between March 1 and April 15. Grain yields decrease rapidly as seeding is delayed past mid-April.

Although oats can be established using no-till seeding techniques, little if any crop residue should be present to allow the soil to warm rapidly and the seeds to emerge. Fall preparation of the seedbed eliminates the need for tillage in the spring and sometimes permits earlier planting than when tillage is needed. This technique also eliminates the soil compaction associated with soil preparation.

Soil pH should be above 6.0 unless a legume is also seeded, in which case the soil pH should be 6.5 to 7.0. Phosphate and potash can be applied in the fall or spring. Nitrogen should be applied in the spring anytime before emergence.

Rice

Rice is a member of the grass family. It is an annual monocot. The botanical name for rice is *Oryza sativa*. Common rice is lowland or paddy rice (requires continuous irrigation or flooded ponding). Upland rice is nonirrigated or not grown in flooded conditions. Wild Rice is not from the same genus and is not related to paddy or upland rice species. Figure 15-7 illustrates the rice plant.

There are 42 rice producing countries in the world, ranging from mountainous Himalayan to lowland delta areas. Rice is the staple food in Asia, Latin America, parts of Africa, and the Middle East.

Within the United States, the rice producing states include: Florida, Arkansas, Louisiana, California, Texas, Mississippi, and Missouri.

Rice thrives under the hot and humid conditions that characterize areas of the southern United States during the summer months.

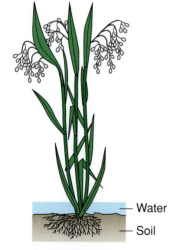

— Water

— Soil

Figure 15-7

A rice plant.

Fields that are nearly level and can be diked and flooded easily are readily adapted to rice production.

Varieties

Rice is divided into short-, medium-, and long-grain varieties, and certain varieties have distinct aromas and flavors. Rice can also be classified according to its cooking characteristics, such as stickiness. Rice is classified into three broad groups by starch and grain characteristics:

1. Waxy or glutinous rice

2. Common, translucent, or nonwaxy rice

3. Aromatic rice

Rice cultivars are characterized by maturity date or length of growing season as follows:

- Very early cultivars—90 days from germination to harvest
- Early cultivars—90 to 97 days
- Intermediate to late cultivars—98 to 105 days

Rice is further classified by cultural adaptation. Japonica is usually short-grained kernels and grown in temperate zones. Indica is usually found in tropical climates, is typically long-grained.

Rice is planted in a level, well-prepared field using a grain drill, which is similar to what is used to plant other small grains. Some rice is seeded with an airplane. Because of Florida's long growing season, rice plants can be harvested and then allowed to regrow for a second harvest or ratoon crop. Approximately half of the rice acreage regrows as a ratoon crop—regrown from young shoots or tillers.

The length of time from planting to harvest is determined by the rice variety and weather conditions. It takes an average of 120 days for the first crop to mature and another 85 days for the ratoon crop to mature.

Weed Control

Weeds are controlled by integrating cultural practices and herbicides. Field preparation includes complete weed removal through disking. Immediately after planting, an herbicide is applied that inhibits weed seed germination or kills growing weeds. As soon as the rice plants reach 5 to 6 inches (13 cm) in height, the flood is applied, which further inhibits weed growth.

Fertilizer

Soil pH needs to be 5.0 to 7.5. Fertilizer needs are determined by a soil test.

Flooding

Rice is an aquatic crop and is flooded during its growing season, this period corresponds with the rainy season in some places like south Florida. Rice fields actually serve as temporary storage for this rainfall and allow for the gradual addition of this water back into the environment.

Rice fields are flooded when the rice plant is about 6 inches high. The flood is maintained 2 to 4 inches deep until the rice grain has matured and begins to dry out. The flood is drained prior to harvest to insure firm ground for harvest equipment. About one month after the harvest operation is complete, the flood is reestablished to allow the ratoon crop to grow.

Paddy leveling and levee building are very critical to maintaining optimum and uniform paddy water depth. Soils must hold water well after flooding. Flooding is used to control some weeds and insect pests but may lead to waterborne disease spread, some water weeds, and some water insect pests. Typically, flooding begins at tillering, although flooding at planting occurs in some instances.

Pests

The fungal disease, blast, is a serious problem for rice growers. The best control available is to plant blast-tolerant varieties, although some fungicide is used in case of high disease incidence. Other diseases of rice include seedling blight, sheath blight, red rice, leaf spot, stem rot, and root rot.

When rice is grown as a rotation crop with sugarcane or vegetables, certain insects such as the rice water weevil have not become a problem. Stinkbugs are a problem and must be controlled. Fields are scouted and, when population thresholds are exceeded, control measures with approved insecticides are started. Stinkbugs pierce the immature kernels with their proboscis and, by sucking the juices, reduce the quality of the seed. Other insect pests include rice water weevil, leafhoppers, thrips, and grasshoppers.

Insect-feeding and wading birds are welcome additions to the rice fields. Blackbirds and bobolinks (small migratory sparrowlike birds) are not. Blackbirds not only pull up the new seedlings and eat the seed but also eat the mature grain. Bobolinks eat large amounts of mature grain just before harvest. Both of these birds can seriously reduce yields.

Harvesting

Once the rice is mature and the fields have been drained, the rice field is harvested with a conventional grain combine. The moisture content of the rice kernel is about 19% at harvest, much too high for quality storage. The first stop for the rice after harvest is the drying bins. There, heated air is forced through the rice until the moisture content has been reduced to 12% to 13%. The rice is then stored in silos until it is milled and sold as white rice.

Potato

The potato is a member of the *Solanaceae* or nightshade family. It is related to such vegetables as tomato, pepper, and eggplant. It is also related to important drug plants such as datura and belladonna. The potato of world commerce is a tetraploid, which means it has four sets of chromosomes in *Solanum tuberosum*. Wild potatoes are usually diploid (two sets of chromosomes). This is important for plant breeders who may want to move favorable characteristics, such as disease resistance, from wild populations into commercial cultivars. Other important species of potato are *S. andigena, S. phureja,* and *S. stenotonum.*

The potato is a herbaceous dicot that reproduces asexually from tubers, hence the species name *tuberosum*. The tubers form at the end of underground stems called stolons. The tubers are the edible portion of the plant, and botanically they are stems not roots. They are stems because they contain all the morphological features of stems. They have buds (the eyes), leaf scars (the eyebrows), and lenticels (see Figure 15-8). Lenticels are small pores that allow stems to exchange gases with the air. The tuber is a storage organ. The plant produces sugars in the leaves. These sugars are converted to starch, which is stored in the tubers. The potato then uses the starch to produce new plants the following growing season.

Potato flowers range in color from white to purple and produce small berries that are very poisonous. The berries contain viable seed, but in the past these were not used to produce new potato plants because the offspring were highly variable in such traits as yield and eating quality. The leaves and aboveground stems of the potato plant are also toxic to humans.

The potato is a very versatile crop. It is produced commercially on every continent on Earth, except Antarctica. Potatoes are produced commercially throughout western Europe, but only sporadically in their native range. In the native areas, potatoes are produced mainly for direct consumption. In the United States, potatoes are produced commercially in every state.

The potato is a cool season crop. Mean temperatures in the range of 60° to 65°F (6° to 18°C) are optimal for high yields. The formation of tubers decreases at soil temperatures above 68°F (20°C) and is almost completely inhibited above 84°F (29°C). This presents a problem in tropical areas where soil temperatures are usually in this range. A goal of plant breeders is to find potato strains adapted to tropical conditions. Soil temperature is very important in potato growth. Since tuber pieces are planted, this is one of the main determining factors in the growth pattern of the plant. In cool soils, the tubers sprout and emerge slowly. At 52°F (12°C) it takes 30 to 35 days for complete emergence. The optimum temperature for emergence is around 72°F (22°C). Higher temperatures seem to retard emergence. Soil temperature is also important in determining yield.

The temperate region of the world is ideal for potato production because the plants are started and the tops are established in spring, when temperatures are low. The warm weather of summer causes high rates of photosynthesis, which produces plenty of starch for transport to the tubers. The length of the growing season should be from 90 to 120 frost-free days. In parts of the world where there are shorter growing seasons, potatoes can be grown because the extremely long days of summer compensate.

Both photoperiod and temperature affect tuber initiation (tuberization). In general, short days initiate tubers. For long days, tuberization occurs if night temperatures are well under 68°F (20°C). Maximum tuber set occurs when nights are around 54°F (12°C). Interestingly, the temperature-sensitive part of the plant is the top and not the stolons or roots.

Low nitrogen levels in the plant favor tuber initiation. This is an important consideration when determining when to make side-dress applications of fertilizer to a potato crop. High light intensity also seems to enhance the tuberization process.

A loose-textured, well-drained soil with a pH of 5.0 to 6.5 is best, but potatoes will grow on almost any soil. Diseases can cause problems on soil that has high clay content. Generally speaking, the soil needs to be about 4 feet deep as this is the depth of rooting of the potato plant. If the soil texture is too "tight" (high in clay), the tubers that form can be misshapened. A total of 18 to 30 inches (500 to 750 mm) of water is needed throughout the entire life of the crop. This can be in the form of rainfall and/or irrigation. Potatoes respond better if the supply of water is uniform throughout the part of the season when the plant is actively growing. Potatoes require about 0.75 inches of water per week.

Potatoes usually require the application of nitrogen (N), phosphorus (P), and potassium (K), fertilizers. The normal method of fertilizer application is to apply half the needed fertilizer prior to planting and half as a side-dress application.

Potatoes are propagated vegetatively in developed countries, including the United States. Seed pieces weighing 1.5 to 2 oz (40 to 60 g) are used. Large potatoes can be cut by hand with a sharp knife, although commercial growers use automated equipment for this procedure. When the potatoes are cut, they need to be kept at 64° to

70°F (18° to 21°C) and 85% to 90% relative humidity (RH) for several days. This allows the wound to heal, and protects the tuber from soilborne decay. A good dusting with a fungicide is often substituted for this practice. Each piece of the cut tuber must contain an "eye." Since the eye of the potato is actually a bud, this is where the potato sprout will come from. The pieces are planted 2 to 4 inches (5 to 10 cm) deep in beds 30 to 36 inches (76 to 91 cm) apart with 9 to 12 inches (23 to 31 cm) between the seed pieces.

Because potatoes are propagated vegetatively, the use of disease-free planting stock is very important for good potato production. Many virus diseases are passed from insects feeding on the foliage of potato plants. The disease moves through the stem into the tuber. If infected tubers are used as seed, the new plants are also infected with the disease. Certified seed potatoes are grown in cool areas.

True Potato Seed

The benefit of using pieces of tuber as seed is that all individuals planted from that tuber have the same genetic makeup. In essence they are clones. This allows growers to know what to expect when a certain cultivar is planted as far as yield components, and other quality factors. The seed that is produced in potato berries will germinate and grow into a plant with tubers. In the past, however, the yield from these plants has been very variable. Over the last 10 to 15 years, new breeding techniques have been developed that allow for the introduction of high-yield genes into all the progeny that grow from these seeds. This has led to the production of several commercial cultivars that are grown from true botanical potato seed. These cultivars are presently used mostly by home gardeners and professional market gardeners. The development of this technology should aid in the spread of potato production into developing countries, where the nutritional benefits of potatoes are greatly needed.

Pests

The major insect pests of potato include wireworms, white grubs, Colorado potato beetle, aphids, and leaf hoppers. Wireworms and white grubs are important because they feed directly on the tubers, causing injury, which can lead to sites where diseases can enter. Aphids are important in the spread of virus diseases.

The diseases of major importance are early and late blight, fusarium and verticillium wilt, scab, bacterial ring rot, blackleg, and at least 10 viruses. More and more viruses are found every year in different parts of the world.

Weeds are a major problem in potato production. If the potato crop can be kept entirely weed free for the first six weeks after the emergence of the crop, there is very little competition from weeds. The potato plants get large enough that the ground is shaded and very little additional weed growth occurs. Research has shown that there is no benefit in terms of yield from controlling weeds longer than the first six weeks after crop emergence.

Harvesting

As the potato tubers mature, the tops of the plants begin to turn yellow. In areas of early frost in the fall, the potato vines are killed. In most of North America, chemical vine killers are used. After the vines die, tubers are allowed to cure in the ground for several days to one week. This allows the potato skin to "set" by getting a little thicker and more firm. Large commercial growers dig and handle the tubers by machine, as shown in Figure 15-9. Damage to the tubers opens infection ports through which disease can enter, especially bacterial soft rot in storage. Bruising lowers the quality and price received by the grower.

After harvest, the tubers must be hardened before they can be stored for long periods. This is usually done in a period of four to five days at 61° to 70°F (16° to 21°C) in high humidity. This is usually done by packing houses. Some growers own their own packing houses, but many are owned by brokers. Once the potatoes are cured, they can be packed for immediate consumption, or stored for use later.

For those potatoes that are going to the processor, it is not necessary to cure the tubers after harvest. In some instances, the potato vines are not killed prior to harvest.

Figure 15-9

Potato harvesting occurs in the fall after the vines have been knocked down.

Storage

Potatoes can be stored for almost a year under the proper conditions. Very cool temperatures 40° to 50°F (4° to 10°C) at 90% RH is ideal (see Figure 15-10). The potatoes must be protected from freezing, since this destroys tubers. If temperatures are too low, the starch contained in the tuber is converted to sugars. Then, when the potatoes are cooked, the sugar turns brown. This is especially bad for the processing industry.

Potato tubers must be stored in the dark to prevent greening. The green color is caused by chlorophyll, which is what plants use to trap the sun's energy. But sunlight also causes the production of the alkaloid, solanine. Green potatoes contain solanine. Solanine is a toxic substance and can cause illness in most cases, or death if too much is consumed.

Soybean

The botanical name for the soybean is *Glycine max*. It is a native of Asia where it was cultivated long before written history. The date of planting has more effect on soybean grain yield than any other production practice. Regardless of planting date, row width, or plant type, the soybean crop should develop a closed canopy (row middles filled in) prior to flowering or by the end of June, whichever comes first. An early canopy results in high yields because more sunlight is intercepted and converted into yield than when row middles do not fill in until late in the growing season.

Adequate, vigorous stands are more difficult to obtain with early planting. Seed treatments, good seed-soil contact, and reduced seeding depths help establish vigorous stands. Herbicide programs must provide weed control for a longer time until the crop is large enough to suppress weed growth through competition.

Most soybean varieties, shown in Figure 15-11, have genetic yield potentials well over 100 bu/A. A variety's adaptability to the environment and production system where it will be used sets the production system yield potential. Varieties should be selected with characteristics that will help them perform well in the cultural system and environment to be used rather than on their yield record alone. Where excessive growth and lodging are problems, varieties that are medium to short in height with good standability should be selected.

Figure 15-11

A soybean plant.

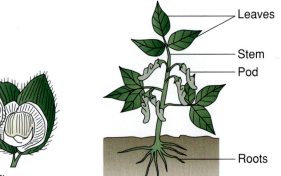

Flower

Diseases

Phytophthora root rot is a serious soybean disease everywhere soybeans are grown. Varieties are most susceptible in the seedling stage. Susceptible varieties should not be grown in poorly drained soils or on soils known to have a history of the disease. Seed varieties with good field tolerance should be treated with a fungicide that aids in the control of Phytophthora damping off, or the soil may be treated. Resistant varieties should be treated with a seed treatment fungicide to control Phytophthora damping off if they are to be planted in poorly drained soil using reduced or no tillage.

Pythium and Rhizoctonia root rots are also common, and many varieties are susceptible. Damage to plant stands is greatest on poorly drained soils and during seasons of high rainfall. Pythium is controlled by fungicide seed treatments and seedling infections of Rhizoctonia are controlled by seed treatments containing fungicide.

Sclerotinia stem rot may be severe when wet weather occurs during flowering. Some varieties are less susceptible than others, but there is no known resistance. Stem symptoms appear as water-soaked lesions followed by cottony growth and eventually large, black, irregular shaped sclerotia resembling mouse droppings. Wide rows (20 to 30 inches) and reduced seeding rates aid in control by permitting air to move through the canopy to dry plant leaves and the soil surface.

Brown stem rot can severely reduce yield. The fungus enters the plants through the roots and slowly grows upward into the xylem, where it interferes with the flow of water. The disease symptoms develop after flowering and are identified by an internal browning of the stem in August. Foliar symptoms are rare, but the leaves of infected plants may suddenly wilt and dry 20 to 30 days before maturity and remain attached to the plant.

Phomopsis seed rot can be severe when rainfall occurs intermittently during the dry down and harvest. The longer soybeans are in the field after ripening, the greater the incidence of seed rot. Harvesting soon after the soybeans mature (15% to 20% moisture) decreases the amount of seed damage. Use varieties with a range of maturities to allow for a more timely harvest. Yield and grain quality losses are greater when soybeans are not rotated with other crops.

Soybean cyst nematode (SCN) was first found in the southeastern states in the early 1950s. SCN injury is easily confused with other crop production problems, such as nutrient deficiencies, injury from herbicides, soil compaction, or other diseases.

Tillage

A desirable seedbed for soybeans should be smooth enough to permit the planting equipment to place the seed at a uniform depth. The soil particles should be fine enough to assure good seed-soil contact for rapid germination and emergence. The greater the time required for emergence, the greater the time for disease infection and loss of stand.

The freezing and thawing action on clay, silty clay, and silty clay loam soils tilled in the fall or winter to produce stale seedbeds usually have excellent seedbeds for early no-till spring planting. If these soils are tilled when too wet in the spring, a rough cloddy seedbed will result. Also, spring tillage often causes compaction of the soil below depths of 5 inches, which restricts root growth and reduces availability of water and nutrients.

Silt loam soils and those soils with less than 2% organic matter, but with good drainage, tend to have the most desirable seedbed with a late winter or spring tillage system if the soil moisture level is satisfactory for tillage. Where tillage is used, the type and amount has little effect on yield, provided it is adequate to permit the establishment of a uniform stand and Phytophthora root rot is not a problem. This is true for both heavy- and light-textured soils where organic matter contents are 2% to 6%.

Rhizobium Inoculation

If inoculation is necessary, the seed should be inoculated at the time of planting. Commercial inoculants usually contain strains of bacteria that fix nitrogen more rapidly than native strains, but they usually do not survive well from year to year.

Harvesting

Threshing consists of separating the beans from the pods (portion of the plant fruit that encases the soybean seeds). Most soybeans are harvested and threshed simultaneously by modern combines.

Drying is a postharvest phase during which the beans are rapidly dried until they reach the safe moisture level. After threshing, the moisture content of the beans is sometimes too high for good conservation (13% to 15%). The purpose of drying is to lower the moisture content in order to guarantee conditions favorable for storage or for further processing and handling of the product. Drying can be done by allowing warm, dry air to circulate around the beans. Two methods of drying are used, either natural or artificial drying.

Cotton

The botanical name for cotton is *Gossypium* spp. Of the more than 30 species, only three are of commercial importance. Cotton is a dicot, annual of almost tropical and/or subtropical origin. Its growth range is limited by length of frost-free growing season. It is classified as an agronomic, fiber crop and is grown primarily for its fiber but also its seed meal and seed oil. Figure 15-12 illustrates a mature cotton plant.

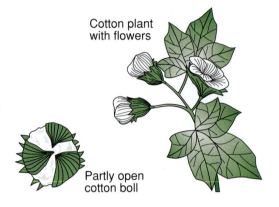

The major production areas worldwide are China and the United States. In the United States, production extends from central California to the southeastern U.S. coast (South Carolina) and south to the southern tip of Texas. Estimated annual world production of 80 million bales is produced by 70 countries. U.S. production is approximately 10 million bales per year. Each bale weighs 500 pounds.

Cotton is produced primarily for lint (fibers) from the mature cotton boll, as shown in Figure 15-13. However, the seeds left after lint removal (ginning) are used for seed oil production and as animal feed or feed supplement (cotton meal). Fibers are protuberances on epidermal cells of the ovule seed coat.

Cotton is a highly mechanized crop from planting through harvest through ginning. Originally, the crop required extensive hand labor for weed control, harvesting, and separating lint and seeds.

Cotton is a complete, perfect flower. It is self-fertile and self-pollinating with the aid of wind and insects. Cotton flowers have multiple anthers but a single, fused pistil.

Pests

Resistance to the complex association of fusarium wilt with nematodes is important in fields where the disease is known or suspected to occur. Cotton cultivars are evaluated under field conditions for their reaction to this combination of problems. The cotton root-knot nematode commonly is associated with the wilt disease, but other nematodes may also contribute to the problem. Only cultivars resistant to the complex should be planted in fields known to have nematode or fusarium wilt problems.

Successful weed control is essential for economical cotton production. Weeds compete with cotton for moisture, nutrients, and light. The greatest competition usually occurs early in the growing season. Late-season weeds, while not as competitive as early-season weeds, may interfere with insecticide applications and may cause harvesting difficulties.

Herbicides are the most effective means for controlling weeds in cotton. To be effective, however, herbicides need to be matched with the weed problem. Preplant and/or preemergence applications are important for ensuring that the cotton has the initial competitive advantage over the weeds.

Problem insects include cotton boll weevil, thrips plant bugs, pink bollworm, and budworm. Two major diseases concern cotton growers—damping off from fungi and bacterial blight.

The use of IPM technologies of scouting, introduction of biological controls (predators), targeted or narrow-spectrum pesticides, and use of biological insecticides are reducing the total chemical load and reducing production costs. There is an increasing market for "organic" cotton produced without insecticides.

Diseases are controlled through use of resistant cultivars and proper management practices; for example, sanitation, drainage, and fungicides. Seeds for planting are typically treated with fungicide to prevent seedling diseases.

Chemicals can be used just before or at planting to reduce nematode population densities. However, their use is not always justified. The grower must compare the cost of treatment with the dollar value of the anticipated yield improvement.

Crop rotation can help keep preplant nematode population densities from becoming too high and is important in managing a soil-borne fungal disease such as fusarium wilt. Grass family crops, such as small grains, corn, sorghum, millet, and forage grasses are good crops to rotate with cotton because they support few root-knot and reniform nematodes.

Planting occurs after all danger of frost is past and soil temperatures warm 68° to 77°F (20° to 25°C). Plants do well with high temperatures during growing season. This crop requires high light intensity and ample soil moisture. Cotton needs well-drained, sandy loam soil, and requires high fertilization. The major production limitation is the length of the growing season.

Crop destruction after harvest reduces the levels of pests and pathogens available to attack the following crop. Under some conditions, cotton roots can survive a long time after harvest. Fields should be tilled as soon as possible after harvest to stop reproduction of nematodes and other pests that can live in (or on) cotton roots or stubble, and to allow natural population decline to begin. Nematodes that can damage cotton can build up on weeds and other crops that may precede it, and many other crops can be affected by the nematodes that will build up on cotton. Thus, killing the roots after harvest by plowing them up and exposing them to sunlight is important to prepare for either cotton or a different crop to follow cotton.

Harvesting

Cotton is harvested with a combine and hauled to a gin. At the gin, the cotton is separated from the seed. The cotton is baled into 500 pound bales for further processing. The seed is either sold for livestock feed directly, as shown in Figure 15-14, or the oil may be extracted from it and then sold as cotton seed meal.

Figure 15-14

Whole cottonseed is used as a dairy feed.

Dry Beans

Dry beans are produced in a variety of colors and sizes. The botanical name for beans is *Phaseolus* spp. The common red kidney bean is *Phaseolus vulgaris*. Figure 15-15 illustrates the typical bean plant.

The best soil for dry bean production is a loamy soil with high organic matter and good drainage. Fine-textured soils tend to be poorly aerated and susceptible to compaction problems, while coarse-textured sandy soils tend to be droughty and susceptible to wind erosion. The best bean soils are nearly level. Steeper slopes are susceptible to water erosion.

Beans need a frost-free season of 100 to 120 days, with frequent rains or proper irrigation during the period of rapid growth and plant development.

Variety and Seed Selection

Growers select high-quality, disease-free seed. Certified seed is a dependable source of high-quality seed that has passed rigid quality standards. For colored bean types, regardless of the source, growers should have seed tested or ask their seed suppliers for results of disease tests, including common bean mosaic virus and bacterial blight. Regardless of blight test results, all bean seed should be treated with streptomycin sulfate (along with an insecticide and fungicide) to control external bacterial organisms on the seed coat surface.

Planting

Growers plant beans, if possible, following corn, or small grain seeded to a clover green manure crop, or after alfalfa. Planting beans after beans or after beets is not recommended. Producers try to choose fields for beans that are level or only slightly sloping, well-drained, medium to fine textured, with good water-holding capacity. If the field has low spots where water is likely to collect after heavy rains, use a land leveler or construct open shallow surface ditches running to an outlet to lead off excess water.

Where the soil is subject to wind erosion, some growers may seed a small grain (rye is excellent) for winter and early spring cover. They keep it mowed to a 4- to 6-inch (10 to 15 cm) height to prevent excessive moisture loss until fitting and planting. In fine-textured soils, rye can be used to dry out the ground in the spring to allow fitting under optimum moisture conditions. Growers use tillage in fitting the land for beans to avoid soil compaction.

In already compacted soils, deep tillage is recommended. This must be done when the soil moisture content is at an intermediate level.

Soil and Fertilizer

Growers should follow soil test recommendations in using fertilizers. Beans are very sensitive to fertilizer applied in contact with the seed. As a general guide, the micronutrients manganese and zinc may increase yields, particularly under high soil pH. When submitting a soil sample, these micronutrients need to be tested. Beans generally respond to nitrogen.

Planting Practices

Traditionally, the first part of June has been the preferred planting period, if soil moisture is favorable. With the development and release of full season direct-harvest types, a planting period can begin in late May, provided soil temperature [65°F (18°C) or higher] and moisture are favorable.

Seeds should be placed at a uniform depth in moist soil. Planting in dry soil or planting deep to reach moisture is not recommended. See Table 15-3 for planting rates for different dry bean classes.

Table 15-3

Suggested Planting Rates for Field Beans			
Type	Row width	Seeds/ft of row	Approx. (lbs/acre)
Black turtle	28	4–5	40
Cranberry	28–32	3–4	60
Kidney	28–32	3–4	60
Navy	28	4–5	40
Pinto	28	4	50
Yellow eye	28–32	4	60

Weed Control

Herbicide should be chosen for the type of weed problem that exists. Growers develop a program for perennial weed control that involves control in nonbean years and some cultivation to hold these weeds down in the year of growing beans.

Diseases

Several serious diseases affect dry edible beans. Some, such as bacterial blight and common bean mosaic virus, are seedborne and can be perpetuated by planting disease-infected seed. Others, such as fusarium root rot and bean rust, are not perpetuated by seed. Table 15-4 lists bean diseases and their control.

Table 15-4

Field Bean Diseases and Their Controls		
Disease	Spread	Control
Halo bacterial blight	Splashing water, insects, animals, seed	Copper sprays, disease-free seed, crop rotation, seed treatment
Common and fuscous blight	Splashing water, insects, animals, seed	Disease-free seed, crop rotation, seed treatment
Common bean mosaic virus	Aphids	Disease-free seed
Root rots	Plowing, cultivation, etc.	Tolerant varieties
Bean rust	Windblown spores	Copper and sulfur
Bean anthracnose	Infected seed	Disease-free seed
White mold	Splashing water, wind	Fungicide sprays

Other disease control measures include:

1. Planting disease-resistant varieties, when available, or seed that has been tested for seedborne diseases.

2. Plowing under all bean refuse, preferably in the fall, to reduce the disease inoculum potential from the previous season.

3. Avoiding conditions that favor fusarium root rot and other soilborne diseases. These conditions include poor soil aeration resulting from soil compaction or poor soil drainage, low soil temperatures, and planting where beans were grown the year before.

4. Avoiding working in bean fields when they are wet.

5. Being prepared to apply agricultural chemicals when halo blight is first observed in the field and to navy beans when bean rust is present three or more weeks before plants mature.

Insect Pests

Good management that yields a clean, vigorous stand of beans will assure a minimum of problems with insect control. No special equipment or operations—only those normally used in producing high yields of quality beans—are needed.

Harvesting

Beans are pulled when approximately 90% of the leaves have fallen and the stems and pods have lost all green color. Plants are pulled early in the day when the pods and stems are tough. Weather conditions at harvest time are all-important to determine the best procedure (see Figures 15-16 and 15-17).

Figure 15-16

Cutting beans is normally a two-step process. New equipment developed by Pickett Equipment cuts the beans and windrows them in one pass over the field. Courtesy of Pickett Equipment, Burley, Idaho.

Drying and Storage

If beans are harvested with more than 20% to 22% moisture, they can be stored only a few days before spoiling. At moistures of 16% to 18%, beans will store safely several months. For long-term storage, they should be below 15% moisture. Beans can be dried in most commercial grain driers.

Peas

There are many varieties of peas (*Pisum sativum*). The processing or market will determine variety grown. Peas are a cool season crop, best planted in late summer or early fall, but early spring crops can be successful where temperatures are not hot during late spring. Both bush and vine types are available for both edible pod-type and regular shelling-type peas. Bush peas can be grown in most areas, but vine types need the cooler, moister climate found along the coast. Vine peas produce more and for a longer period, but they require a trellis to climb on. Figure 15-18 shows a typical pea plant.

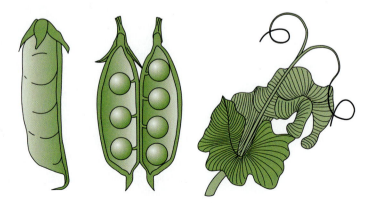

Soil

Fields should have uniform fertility, soil type, slope, and drainage to get a uniform pea crop. The best soils are silt loams, sandy loams, or clay loams. Peas need a good supply of available soil moisture, but yields may be reduced by overirrigating as well as underirrigating. Peas grown on wet soils develop shallow root systems, which cannot supply the plant's water requirements when the soil dries out later in the season. Root rot is often a problem in wet soils. Determine corrective lime and fertilizer needs by a soil test. Growers adjust pH to 6.5 or higher for maximum yields.

Soil Temperature and Planting

Good germination will occur at 39° to 57°F (4° to 14°C). The land should be plowed, harrowed and a cultipacker used lightly to ensure a firm seedbed. The land should be level in order to make harvesting more efficient. Plantings may be made as soon as the soil can be worked in the spring. Enation-resistant varieties may be planted throughout the entire planting season. Fresh market peas and edible pod peas may be scheduled on the basis of heat units and by picking requirements for given plantings. In general, April plantings will require about 70 days to harvest, May plantings about 60 days, and June plantings about 55 days.

Fertilizer

Good management practices are essential if optimum fertilizer responses are to be realized. These practices include:

- Use of recommended pea varieties
- Selection of adapted soils
- Weed control
- Disease and insect control
- Good seedbed preparation
- Proper seeding methods
- Timely harvest

Because of the influence of soil type, climatic conditions, and cultural practices, crop response from fertilizer may not always be accurately predicted. Soil test results, field experience, and knowledge of specific crop requirements help determine the nutrients needed and the rate of application. The fertilizer application for vegetable crops should insure adequate levels of all nutrients. Optimum fertilization is essential for top quality and yields. Recommended soil sampling procedures should be followed in order to estimate fertilizer needs.

Weed Control

Growers cultivate as often as necessary when weeds are small. Proper cultivation, field selection, and rotations can reduce or eliminate the need for chemical weed control.

Insect Control

Proper rotations and field selection can minimize problems with insects. Insect pests of peas are described in Table 15-5.

Table 15-5

Insect Pests of Peas

Insect	Description
Loopers (also alfalfa looper) (*Autographa californica*)	Slender, dark, olive-green worms with pale heads and three distinct dark stripes. Move in a looping manner.
Celery looper (*Anagrapha falcifera*)	Loopers pale green with no distinct marking. Adults are gray, 3/4 inches long with a white teardrop on both front wings.
Cutworms and armyworms	Large grown larvae that feed on seedlings, leaves, and pods.
Grasshoppers (several species)	May cause considerable damage in years of grasshopper abundance.

Disease Control

Proper rotations, field selection, sanitation, spacing, fertilizer, and irrigation practices can reduce the risk of many diseases. Fields can be tested for the presence of harmful nematodes. Using seed from reputable seed sources reduces risk from seedborne diseases. Some disease of peas include bacterial blight, stem rot, downy mildew, enation mosaic virus, leaf roll, powdery mildew, root rot, seed rot, and mosaic virus.

Harvesting, Handling, and Storage

The processor determines time of harvest according to a tenderometer reading, the number of other fields ready for harvest, weather, soil conditions, and the processor's need for quality. Generally, yields of shelled peas increase with increasing maturity, but quality decreases. With mobile viners, the crop is cut and swathed into windrows, threshed out by the mobile viners following swathers. Peas must be delivered to the processing plant soon after harvest, especially when the weather is hot, to avoid off-flavors. With the new pod stripping harvesters, no swathing is needed.

Green peas tend to lose part of their sugar content, on which much of their flavor depends, unless they are promptly cooled to near 32°F (0°C) and have 90% to 95% RH, after picking. Hydrocooling is the preferred method of precooling. Peas packed in baskets can be hydrocooled from 70° to 34°F (21° to 1°C) in about 12 minutes when the water temperature is 32°F (0°C). Vacuum cooling also is possible, but the peas must be prewet to obtain cooling similar to that by hydrocooling.

After precooling, peas should be packed with crushed ice (top ice) to maintain freshness and turgidity. Adequate use of top ice provides the required high humidity (95%) to prevent wilting. The ideal holding temperature is 32°F (0°C). Peas cannot be expected to keep in salable condition for more than one to two weeks even at 32°F (0°C) unless packed in crushed ice. With ice, the storage period may be extended perhaps a week. Peas keep better unshelled than shelled. Also, researchers demonstrated that the edible quality of green peas was maintained better when the peas were held in a modified atmosphere of 5% to 7% carbon dioxide at 32°F (0°C) than in air for 20 days.

Peanuts

The scientific name for the peanut is *Arachis hypogaea*. Figure 15-19 shows a typical peanut plant with the pods underground. Peanut yields and quality rise and fall each year based on weather conditions during the growing season. Moisture and temperature are the two weather factors that have the most impact on crop yields.

Figure 15-19

Peanut plant showing peanuts at the end of the pegs. Courtesy of USDA/ARS.

The germination process begins when soil temperatures are above 60°F (16°C) and viable and nondormant seeds absorb about 50% of their weight in water. For practical purposes, soil temperatures need to be above 65°F (18°C) for germination to proceed at an acceptable rate. Large-seeded Virginia-type peanuts planted under favorable moisture and temperature conditions will show beginning radicle (root) growth in about 60 hours. If conditions are ideal, sprouting of young seedlings should be visible in seven days for smaller-seeded varieties and ten days for larger-seeded varieties.

Root Growth

The peanut root grows rapidly following germination. The root tip grows downward in the soil through cell division and enlargement. By the time the main shoot breaks through the soil, the tap root will be 5 to 6 inches (13 to 15 cm) deep. Lateral roots develop from the tap root and may be 1 to 2 inches (2.5 to 5 cm) long at seedling emergence. The cotyledons supply the food for early root growth. Once the vegetative tissues are producing a food supply (photosynthate) greater than their needs, some of the food is transported to the roots.

Any condition that subjects peanut roots to stress should be avoided so that adequate root capacity exists to supply the plant with moisture and nutrients. Lack of an adequate and efficient root system may limit yields.

Vegetative Growth

Vegetative growth is slow for the first three to four weeks, as leaf tissue is limited. As leaves develop and fully expand, the capacity for photosynthate production increases and vegetative growth is more rapid. In early growth stages, cool temperatures cause slow growth. As temperatures rise, vegetative growth increases rapidly. Scientists have determined that optimum peanut growth occurs at 86°F (30°C). Excessively high [above 95°F (35°C)] or low [below 60°F (16°C)] temperatures will slow growth.

High temperatures are often accompanied by drought conditions, and vegetative stress can be observed as visible wilting of the plants. Irrigation alleviates the stress of high temperatures or drought.

Other stress factors reduce the photosynthetic efficiency of the plant and limit yields. Leaf or stem diseases, insect infestations, and chemical phytotoxicity may reduce the vegetative surface area, slow photosynthate production, and reduce plant growth.

Reproductive Growth

Peanuts are indeterminate in growth habit. An indeterminate growth habit means that vegetative and reproductive growth occurs simultaneously. The plant must produce enough food to continue vegetative growth, as well as provide food for seed development.

The first flower develops about 30 to 40 days after emergence. Daily flower production is slow for the next two weeks. By mid-July, dozens of flowers may be visible each day on each plant. Temperature [about 86°F (30°C)] and moisture must be favorable for flowering to continue at a steady rate. Any environmental stress may interrupt normal flowering.

Pollination and fertilization occur quickly in the flower. If fertilization of the ovule is successful, the peg begins to grow downward. It takes about 10 days for the peg to penetrate into the soil. A week later, the peg tip enlarges and pod and seed development begins. With favorable temperatures, nutrient, and moisture conditions, mature fruit develops in nine to ten weeks, as in the peanut flower in Figure 15-20.

Figure 15-20

Pollinating a peanut flower. Courtesy *of Texas Agricultural Extension Service.*

Weather or biotic stresses may interrupt normal flowering and fruit development. High or low temperatures and big differences between day and night temperatures may stop flowering and fruit development. Drought conditions will slow growth. Diseases, insects, or weeds that effectively reduce leaf or root surface area may result in slower maturation.

Variety Selection

Yield and quality are two major factors that influence variety selection. Growers with significant disease or insect history may need to choose a variety with disease or insect resistance. Planting varieties with different genetic pedigrees reduces the risk of crop failure because of adverse weather or unexpected disease and insect epidemics.

Selecting and Managing Soil

Peanuts are best adapted to well-drained, light-colored, sandy loam soils. These soils are loose, friable, and easily tilled with a moderately deep rooting zone for easy penetration by air, water, and roots. A balanced supply of nutrients is needed, as peanuts do not usually respond to direct fertilization.

Soil pH should be in the range of 5.8 to 6.2. Peanuts grown in favorable soil conditions are healthier and more able to withstand climatic and biotic stresses.

On most peanut farms, it is not possible to plant peanuts every year in the most suitable soil types. A grower plans the long-range use of fields considering rotations and disease, insects and weed problems, along with all crops grown on the farm. Maximum income normally results when the best balance between all crop factors is found.

Crop Rotation

A long crop rotation program is essential for efficient peanut production. The peanut plant responds to both the harmful and beneficial effects of other crops grown in the fields. Research shows that long rotations are best for maintaining peanut yields and quality. A three-year rotation with two years of grass-type crops has been effective in reducing nematode and soilborne disease problems and permits better control of many weeds. These crops respond to heavy fertilization but leave adequate residual nutrients for healthy peanut growth.

Growers with a short rotation program can expect a long-term buildup of diseases such as southern stem rot, black root rot, and sclerotinia blight.

Fertilizing

Peanuts respond better to residual soil fertility than to direct fertilizer applications. For this reason, the fertilization practices for the crop immediately preceding peanuts are extremely important. Grass-type crops generally respond well to direct application of fertilizer. Growers can fertilize these crops for maximum yields and, at the same time, build residual fertility for the following crop of peanuts. Peanuts have a deep root system and are able to utilize soil nutrients that reach below the more shallow root zone of the grass-type crops.

The peanut crop is usually produced without applying any fertilizer materials during the production year. If peanut fields need fertilizers, they should be broadcast before land preparation. Fertilizing peanuts requires an understanding of the growth characteristics and nutrient needs of the plant and the ability of the soil to provide these needs.

Growers should inoculate their peanut seed or fields to ensure that adequate levels of rhizobia are present in each field. Commercial inoculants are available that can be added to the seed or put into the furrow with the seed at planting. In-furrow inoculants are available in either granular or liquid form.

Perhaps the most critical element in the production of large-seeded Virginia-type peanuts is calcium. Lack of calcium uptake by peanuts results in "pops" and darkened plumules in the seed. Seeds with dark plumules usually fail to germinate.

Calcium must be available for both vegetative growth and pod growth. Calcium moves upward in the peanut plant but does not move downward. Thus, calcium does not move through the peg to the pod and developing kernel. The peg and developing pod absorb calcium directly, so it must be readily available in the soil solution.

Adequate soil calcium is usually available for good plant growth, but not for pod development or good quality peanuts. It is important to provide calcium in the fruiting zone through gypsum applications.

Each pod must absorb adequate calcium to develop normally. Gypsum is available in three forms—finely ground, granular, and phosphogypsum. Several additional by-product gypsums are now on the market. The by-product materials vary in elemental calcium content. Studies show that all forms of gypsum are effective in supplying needed calcium when used at rates that provide equivalent calcium levels in the fruiting zone.

Land Preparation

Land preparation begins with the disposal or management of the crop residue from the previous crop. In order to promote decomposition during the winter, crop litter should be chopped or shredded and disked lightly. Seeding a cover crop can help reduce soil erosion during the winter months. Applications of lime, phosphorus, and potassium can be applied at this time if needed.

Planting

As a general rule, peanuts should be planted as soon as the risk of a killing frost is over. Varieties require from 150 to 165 days to reach full maturity. Early plantings usually give higher yields, more mature pods, and permit earlier harvesting.

Peanuts should not be planted until the soil temperature at a 4-inch (10 cm) depth is 65°F (18°C) or above for three days when measured at noon. The soil should be moist enough for rapid water absorption by the seed. The planter should firm the seedbed so there is good soil-to-seed contact.

Determining Maturity

The best tasting peanut is a mature peanut. Growers must harvest and deliver mature peanuts to the market. Maturity affects flavor, grade, milling quality, and shelf life. Not only do mature peanuts have the quality characteristics that consumers desire, they are worth more to the producer. However, the indeterminate fruiting pattern of peanuts makes it difficult to determine when optimum maturity occurs. The fruiting pattern may vary considerably from year to year, mostly because of different weather conditions.

Heat units or growing degree days have been evaluated as a means of determining maturity. Research shows that 2,600 growing degree days are needed for the earliest varieties to mature.

Disease

A long rotation is the single most powerful disease management tool available to peanut growers. Resistant varieties are usually the least expensive procedure. Proper rotation and variety selection delays or prevents the onset of most serious disease problems. Once diseases become established, expensive chemicals must often be used.

The first step in controlling peanut diseases, which occur in spite of using long rotations and resistant plants, is to correctly identify diseases and how abundant they are. This can be accomplished by scouting the fields each week to monitor disease levels.

The second step is to determine the best control methods for the identified problems. Cultural and chemical controls are usually used in combination for maximum benefit. Cultural control methods such as rotation, resistant varieties, and crop residue destruction have the general effect of reducing the number of disease-causing organisms (pathogens) such as nematodes and fungi.

Pesticides are only useful if cultural practices have not sufficiently reduced pathogen levels below a certain threshold. Thresholds are levels of disease or weather favorable for disease, which represent an economic threat to the crop. Diseases of the peanut plant include: peanut leaf spot, web blotch, and pepper spot.

Alfalfa

Unlike red or white clover, established alfalfa (*Medicago sativa*) is productive during midsummer except during extreme drought. Alfalfa is a tap-rooted crop and can last five years and longer under proper management. Whether grazed or fed as hay, alfalfa is excellent forage for cattle and horses. Figure 15-21 shows an alfalfa plant just before being cut.

Figure 15-21

An alfalfa plant.

Once established, good management practices are necessary to ensure high yields and stand persistence. These practices include timely cutting at the proper growth stage; control of insects, diseases, and weeds; and replacement of nutrients removed in the forage. Alfalfa has superior forage quality when managed properly. The major problems are getting a stand and keeping it productive.

Site Selection and Soil Fertility

Alfalfa is best adapted to deep, fertile, well-drained soils with a salt pH of 6.0 to 6.5, but it can be grown with conservative management on more marginal soils. On sites that have more moderate drainage, growers should also seed a grass, such as orchard grass or bromegrass, with alfalfa to reduce winter heaving of the alfalfa. The grass acts as a mulch during winter to reduce variations in soil temperature, which cause repeated freezing and thawing. Grasses also help prevent weed invasion by filling in spaces between alfalfa crowns.

Alfalfa requires high levels of fertility for establishment, especially phosphorus. Soil should be tested 6 to 12 months ahead of planting to determine proper amounts of fertilizer and agricultural lime for successful establishment. Soil salt pH should be 6.0 or above, which allows for good nodulation by the plant use. Growers should disk or plow down any needed limestone 6 to 12 months before seeding to give time for it to react in the soil and raise the salt pH.

In no-till seedings, growers should apply needed lime a year in advance, because it cannot be incorporated. After two years of production, the producer should take another soil sample to determine if the soil needs additional limestone or fertilizer. Top-dressing limestone or fertilizer helps maintain production potential and ensures stand longevity.

Adequate available phosphorus is a key to establishing a vigorous stand of alfalfa. Phosphorus stimulates root growth for summer drought resistance, winter survival, and quick spring growth.

Nitrogen and potash are not as important as phosphorus for alfalfa establishment, but they are needed in small amounts. Soil test recommendations normally are used to guide the application of fertilizers.

Variety Selection

Several varieties of alfalfa are available, but a limited number are adapted to certain areas. There is no single "best" variety for a particular location. The most recommended varieties are those that are consistently high yielding, moderately winter hardy, and have moderate or higher resistance to bacterial wilt, phytophthora root rot, and anthracnose.

Establishment

Alfalfa may be frost-seeded, broadcast, no-tilled, or drilled into a prepared seedbed. Growers frost-seed in January or February to allow freezing and thawing to work the seed into the soil. Planting into killed vegetation using no-till techniques or into a prepared seedbed involves less risk of failure and produces denser, more uniform stands than frost-seeding.

Whether planting no-till or into a prepared seedbed, growers should place seed no more than 1/4 inch deep for maximum emergence. With a prepared seedbed, the soil should be very firm to ensure good soil to seed contact. When broadcasting, growers should firm the field with a cultipacker or roller before and after planting. Drills that are capable of precise seed depth control and have press wheels to firm the seedbed are also excellent.

In dry years, getting a firm seedbed is critical for seedling survival. In dry years, seedlings germinate and then die in a loose seedbed because water does not move up to the upper soil layer where the young roots are.

Companion Crop

Alfalfa is often fall-seeded with small grains such as wheat, oats, and barley. Otherwise, alfalfa is broadcast into these crops during winter. The companion crop prevents excessive soil erosion, decreases weed problems, protects young alfalfa seedlings, and provides some early spring forage before the alfalfa becomes productive. Although beneficial, the small grain companion crops also compete for light, water, and soil nutrients. The companion crop should be harvested for hay or silage no later than the boot stage to minimize competition. Alfalfa often provides one hay cutting in late August to early September when seeded with a companion crop.

Seeding Rates and Mixtures

When seeded alone, growers use 15 pounds (7 kg) per acre of certified seed, which is about the equivalent of 13 pounds (6 kg) per acre of pure, live seed (PLS). When seeded with a grass, 10 pounds (4.5 kg) per acre of bulk alfalfa seed (equal to 8 pounds per acre PLS) is sufficient. Seeding rates for grasses in an alfalfa grass mixture are:

- Bromegrass—10 pounds (4.5 kg) bulk (8 pounds PLS) per acre
- Orchard grass—6 pounds (3 kg) bulk (4 pounds PLS) per acre
- Tall fescue—10 pounds (4.5 kg) bulk (8 pounds PLS) per acre
- Reed canary grass—6 pounds (3 kg) bulk (4 pounds PLS) per acre

Seeding a cool-season grass with alfalfa decreases the potential for heaving, reduces weed competition, lessens damage to soil structure by grazing animals, and reduces bloat potential when grazed. The grass will decrease forage quality but will be a major component in the first cutting only.

Growers make decisions about whether to include a grass based on the intended market or use of the alfalfa and on the winter-heaving potential of the site. If intended for dairy use or sale to a cubing plant, growers seed pure alfalfa. For grazing, beef, or horses, growers use an alfalfa-grass mixture. On sites that have a high clay content subject to heaving, alfalfa-grass mixtures are recommended.

Maintaining Alfalfa Stands

Proper management can allow growers to maintain a productive stand of alfalfa for five or more years. An annual fertility program and proper harvesting management are major factors determining stand productivity and longevity. Insects, diseases, and weeds are problems that can reduce yields and length of stand (see Figure 15-22).

Figure 15-22

A good stand of alfalfa can last five years or longer under good management.

Most alfalfa seedings initially have 15 or more plants per square foot. As the stand ages, some plants die and remaining plants spread to occupy the space. Alfalfa-grass mixtures can maintain productivity with only two alfalfa plants per square foot.

Annual Fertilization

Annual applications of phosphorus, potash, boron, and sometimes lime are necessary to maintain vigorous, productive stands. To avoid nutritional deficiencies, producers should apply fertilizer each year according to soil tests. Phosphorus fertilization of established stands keeps plants vigorous so that high yields can be maintained over time.

Potash application improves winter survival of plants and lengthens the productive life of the stand. Alfalfa stands with fewer than three plants per square foot cannot maintain high yields and are often subject to increased weed invasion.

Annual fertilizer recommendations vary according to phosphorus and potassium levels in the soil, but will be close to 15 pounds (7 kg) phosphate and 55 pounds (25 kg) potash per ton of expected yield. Growers can apply fertilizer at any time. A single application following the first cutting or a split application following the first and third cuttings are both good options. Split applications are useful for irrigated alfalfa, for high-yielding alfalfa stands, or when applying high rates of potash.

Growers should include boron in the top-dress fertilizer at a rate of 1 pound of boron per acre per year. Boron is toxic to seedlings, therefore, it should not be applied at seeding.

Producers should soil test every two to three years to make sure that soil salt pH, phosphorus, and potassium levels are adequate. Also, where needed, growers should top-dress additional lime, as needed, to keep the pH above 6.0.

Harvest Management

Stage of maturity at harvest determines hay quality and affects stand life. Forage quality (protein, energy value) declines rapidly as the plant begins to flower. Figure 15-23 illustrates what happens to protein content as the plant reaches different stages of flowering.

Figure 15-23

Graph of protein content versus flower stage of alfalfa.

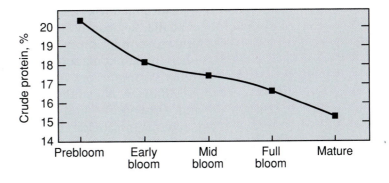

For spring-seeded established stands in the seeding year, growers take the first harvest at the mid- to full-bloom stage. Following harvests are made as flowers begin to appear.

For established stands, growers take the first and second cuttings when the plants are just beginning to bloom. For persistence of the stand, the grower may make one or two more harvests. Alfalfa should not be cut or grazed from mid-September to the first of November. This allows the plant to store root reserves for the winter. After November 1, growers can take or graze a fifth cutting if the soil is well drained or a grass is used to help prevent winter heaving. With a four-cut system, a properly fertilized stand can last six or more years.

Harvesting alfalfa in the bud stage produces three to five cuttings of high-quality hay, such as that shown in Figure 15-24. This practice, however, reduces stand life to three or four years.

Figure 15-24

Not all alfalfa is harvested and baled, some is chopped for dairy feed.

Alfalfa can be grazed without a loss of stand using small pasture units and high stocking rates. Producers should use enough animals to remove most top growth in less than 6 to 10 days. Animals are turned onto the alfalfa when alfalfa is in the bud stage. After grazing, alfalfa needs to regrow for 30 to 35 days. To reduce the chance of bloating producers, use poloxalene (bloat-inhibitor) blocks. Hungry animals should never be turned onto lush alfalfa pastures.

Pests

The alfalfa weevil and potato leafhopper are the two major pests of alfalfa. Regular monitoring of alfalfa fields is the best way to prevent economic injury from insects. Growers should spray or cut when insect populations reach economic thresholds, not after insect injury symptoms are apparent.

Alfalfa weevil adults lay eggs in the older alfalfa stems in late fall and early spring, and the larva damage mainly the first cutting. Chemical insecticides can be used when 25% of the tips are skeletonized and if there are three or more larvae per stem. Instead of spraying, some growers will cut the alfalfa when it is in the bloom stage, scouting regrowth for signs of damage.

The immature or nymph stage of potato leafhoppers stunts plants and yellows leaves. It also lowers yield and protein content by sucking juices from young upper stems. Leafhopper numbers can be large enough to warrant treatment before significant leaf yellowing occurs. Population thresholds for chemical control vary with plant height.

Weed Control

Weed control in alfalfa begins with establishing a uniform dense stand of alfalfa or alfalfa/grass. Experts recommend a preplant incorporated herbicide for conventional spring seedings of pure alfalfa. If growers want a grass in spring-seeded alfalfa, they plant the alfalfa alone using a preplant herbicide. The grass is drilled into the alfalfa stand the following spring. Numerous diskings during mid- to late summer give adequate control for late summer seedings.

Control of weeds after alfalfa emergence depends on the individual weeds, the stage of growth of the alfalfa, and whether there is a grass with the alfalfa. Several herbicides control weeds in pure stands of alfalfa, but only a few are available for use in alfalfa/grass stands.

Diseases

Alfalfa subject to several diseases, including phytophthora root rot, bacterial wilt, anthracnose and sclerotinia root, and crown rot. The best control is prevention. Growers should choose a variety with a high level of resistance to phytophthora, bacterial wilt, and anthracnose. There is no varietal resistance to sclerotinia.

Sclerotinia is particularly damaging to fall-seeded alfalfa stands south of the Missouri River. The disease has killed seedling stands, and older stands are also subject to damage. Cultural controls include deep tillage of alfalfa residue to bury the inoculum that is formed in the spring on infected alfalfa. A less practical control is to maintain a three- to four-year interval between forage legumes in a rotation. Red and white or ladino clovers are also hosts of sclerotinia and can be a source of inoculum for subsequent alfalfa crops.

Summary

1. Agriculture is documented for the Fertile Crescent region, of what is now the Middle East, to between 10,000 to 15,000 years ago. At a similar period, agriculture was also developed in the Tehuacán Valley of Mexico. One theory explaining the development of domesticated plants is the "need" theory for temperate regions such as these two.

2. Another hypothesis for the origins of agriculture is the "genius theory" for tropical areas. Domestication of plants is believed to have occurred near the mouth of the Yellow River of China approximately 15,000 years ago. Agricultural practices are now documented as far back as 18,000 years ago for middle Egypt. It is thought that era moved with the flood periods of the Nile River.

3. The earliest crop plants were probably multipurpose plants such as palms, mulberry, and hemp. Root crops are undoubtedly another group of early domesticated plants because of their ease of propagation. "Root crops" also included underground stems (tubers) such as potato and the corms of taro and tannia.

4. The food basis of all advanced civilization is a cereal crop such as rice, wheat, corn, barley, or sorghum. Irrigation systems developed in many early agriculture areas, the chinampas of Mexico being among the best known.

5. Increased travel ultimately resulted in crop introductions from one part of the world to another.

6. Major agronomic crops such as corn, wheat, barley, oats, rice, potatoes, soybeans, cotton, dry beans, peas, peanuts, and alfalfa have some very specific growing requirements.

7. Depending on cultural practices, final product, pest problems, and other environmental factors, growers must select the appropriate variety or cultivar.

8. Based on the variety or cultivar selection, growers prepare the seedbed, the fertilization, irrigation, and pest management to maximize the yield from the crop.

9. Much of successful production involves proper timing; for example, timing of planting, timing of irrigation, timing of herbicides and pesticides, and timing of harvesting.

10. Crop rotation is often helpful in controlling disease, insect pests, and nematodes.

11. Integrated pest management (IPM) at some level is always an option.

Something to Think About

1. When did hunting and gathering agriculture start?

2. Name the countries that make up the Fertile Crescent.

3. Where in the Americas was agriculture evolving at the same time it was in the Fertile Crescent?

4. What was the hypothesis for the origin of agriculture evolving in the tropical areas?

5. How far back in history are agricultural practices documented?

6. List the earliest agriculture crop plants.

7. What are the basic food crops of all advanced civilizations?

8. Name the agriculture area best known for its irrigation system.

9. Trace the migration of agriculture crops to other parts of the world.

10. Describe the production of four agronomic crops in the United States.

Suggested Readings

Ciolek, M. T. 1999. *Old World Trade Routes (OWTRAD) Project. Canberra-Asian Pacific Research*. Retrieved from http://www.ciolek.com/owtrad.html, June 19, 2008.

Connors, J. J., and S. Cordell, Eds. 1995. *Soil fertility manual*. Tucson, AZ: Potash & Phosphate Institute.

Garofalo, M. 2003. *The History of Gardening: A Timeline to the Twentieth Century, Fields of Knowledge*. Retrieved from http://www.gardendigest.com/timegl.htm, June 19, 2008.

Hamrick, D. 2003. *Ball Redbook, Vol. 2: Crop production*. 17th Ed. Batavia, IL: Ball Publishing.

Janick, J., and J. E. Simon, Eds. 1990. *Advances in new crops*. Portland, Oregon: Timber Press.

Janick, J. and J. E. Simon, Eds. 1993. *New crops*. New York: John Wiley and Sons.

Janzen, D., and R. Goodland. 1990. *Race to save the tropics: Ecology and economics for a sustainable future*. Washington, DC: Island Press.

Jones, J. B. Jr. 2002. Agronomic handbook: Management of crops, soils, and their fertility. Boca Raton, FL: CRC Press.

Mazoyer, M., and L. Roudart. 2006. *The history of world agriculture: From the Neolithic age to current crisis*. New York: Monthly Review Press.

McMahon, M., A. M. Kofranek, and V. E. Rubatzky. 2006. Hartmanns Plant science, 4th ed., Growth, development, and utilization of cultivated plants. Englewood Cliffs, NJ: Prentice Hall Career & Technology.

Simpson, B. B., and M. C. Ogorzaly. 2000. Economic botany: plants in our world. 3rd Ed. New York, NY. McGraw-Hill, Inc.

United States Department of Agriculture. 1953. *Plant diseases, the yearbook of agriculture*. Washington, DC: United States Department of Agriculture.

United States Department of Agriculture. 1961. *Seeds: The yearbook of agriculture*. Washington, DC: United States Department of Agriculture.

United States Department of Agriculture. 1992. *Economic comparisons of biological and chemical pest control methods in agriculture: An annotated bibliography*. Beltsville, MD: USDA.

Vorst, J. J. 1993. *Crop production* (3rd ed.). Champaign, IL: Stipes Publishing Company.

Internet

Internet sites represent a vast resource of information. The URLs for Web sites can change. Using one of the search engines on the Internet, such as Google, Yahoo!, Ask.com, and MSN Live Search, find more information by searching for these words or phrases: **agronomic crop, baobab tree, barley, cassava, century plant, Chinampas, coconut palm, cotton, Fertile Crescent, Agriculture Revolution, hemp, Irish potatoes, Jarmo, maize, millet, multipurpose plant, Nitrophiles, small grains, slash and burn, tannia, taro, wadis, wheat, and Yellow River.**

CHAPTER 16

Vegetables

A vegetable is the edible part of a plant that grows and dies in a single growing season. Many people enjoy a home garden, and it is part of a healthy lifestyle. These vegetables can be used fresh or stored for later production. Many vegetables are grown on a commercial, large-scale basis.

After completing this chapter, you should be able to:

- Name 10 different vegetable crops
- Describe three types of vegetable production
- Identify the basic principles of vegetable production
- Discuss the growing requirements of five vegetables
- Name five herbs and describe their growing conditions and preservation
- Describe cultural practices used in vegetable production

Key Terms

olericulture	transplantability	hydrocooling
market gardening	mulching	relative humidity (RH)
truck cropping	solarization	vapor concentration
seedbed	precooling	

Scope of Vegetable Production

The study of vegetable production is **olericulture**. The production of vegetables can be classified into three categories:

1. Home gardening

2. Market gardening

3. Truck cropping

Home gardening usually involves vegetable production for one family with little or no selling of the produce. **Market gardening** refers to growing a wide variety of vegetables for local or roadside markets. **Truck cropping** refers to commercial, large-scale production of selected vegetable crops for wholesale markets and shipping. Table 16-1 lists the annual production of commercial vegetables in the United States that are tracked by the USDA. Head lettuce, onions, watermelon, tomatoes, and cabbage are the top five commercially grown vegetables.

Potatoes are a common vegetable in most home and market gardens. Potatoes are a major agronomic crop. They are grown on a large scale in several states and their production is covered in Chapter 15.

Categorizing Vegetables

Vegetables can be categorized in the following three ways:

1. Botanical classification

2. Edible parts

3. Growing season

Botanical Classification

All vegetables belong to the angiosperm division of the plant kingdom. Next, vegetables are classified as either monocotyledons (one seed leaf) or dicotyledons (two seed leaves), as shown in Figure 16-1. From this point, vegetables are grouped by family genus, species, and variety (see Table 16-2).

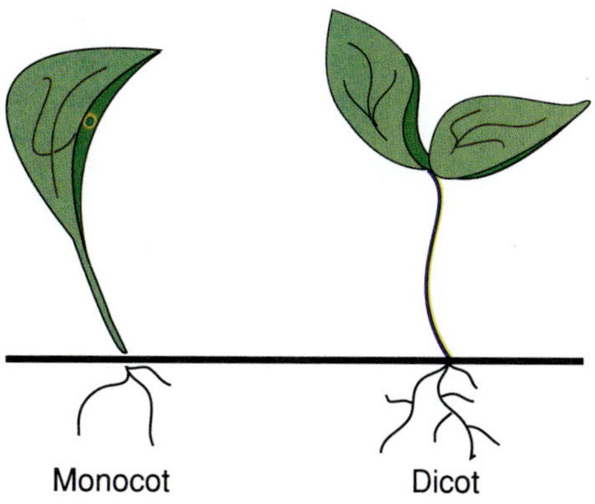

Table 16-1

Production of Commercially Produced Vegetables in the United States

Vegetable	Annual production (1,000 CWT)[1]
Head lettuce	61,137
Onions	42,016
Watermelon	28,856
Tomatoes	27,497
Cabbage	20,684
Celery	17,820
Carrots	16,243
Leaf/Romaine	15,605
Sweet corn	15,522
Cantaloupe	13,656
Bell peppers	7,389
Cucumbers	6,255
Broccoli	5,846
Cauliflower	3,921
Snap beans	3,317
Honeydews	3,124
Garlic	2,006
Spinach	1,142
Asparagus	1,069
Escarole/Endive	954
Artichokes	789
Brussels sprouts	693
Eggplant	651

[1]Based on a 25-year average of data from the National Agricultural Statistics Service, USDA.

Figure 16-1

Monocots have one seed leaf, dicots have two.

Monocot   Dicot

Table 16-2

Vegetable Varieties, Families, and Characteristics[1]

Common name	Variety[2]	Plant family[3]	Transplantability[4]	Pounds per 100 feet	Days to harvest[5]	Comments
Warm season vegetables						
Bush, beans	Snap: Bush Blue Lake, Contender, Roma, Harvester, Provider, Cherokee Wax, BushBaby, Tendercrop. Shell: Horticultural, Pinto, Red Kidney	Leguminosae (Pea)	III	45	50–60	Fertilizer at 1/4 rate used for other vegetables. Seed inoculation not essential for most soils. Flowers self-pollinate. Use shell beans green or dry.
Beans, pole	Dade, McCaslan, Kentucky Wonder 191, Blue Lake	Leguminosae (Pea)	III	80	55–70	See Beans, bush. Support vines. May be grown with corn for vine support.
Beans, lima	Fordhook 242, Henderson, Jackson Wonder, Dixie Butterpea, Florida Butter (Pole), Sieva (Pole)	Leguminosae (Pea)	III	80	65–75	See Beans, bush. Provide trellis support for pole varieties. Control stinkbugs, which injure seeds in pods.
Cantaloupes	Smith's Perfect, Ambrosia, Edisto 47, Planters Jumbo, Summet, Super Market, Primo, Luscious Plus	Cucurbitaceae (Pea)	III	150	75–90	Bees needed for pollination. Mulch to reduce fruit rots and salmonella. Harvest at full-slip stage.
Corn, sweet	Silver Queen, Gold Cup, Guardian, Bonanza, Florida Staysweet, How Sweet It Is, Supersweet	Gramineae (Grass)	III	150	60–95	Separate super-sweets (last three varieties) from standard varieties by time and distance. Sucker removal not beneficial. Plant in 2–3 row blocks.
Cucumbers	Slicers: Poinsett, Ashley, Dasher, Sweet Success, Pot Luck, Slice Nice. Picklers: Galaxy, SMR 18, Explorer	Cucurbitaceae (Melon)	III	100	50–65	Bees required for pollination. Many new hybrids are gynoecious (female flowering). Monoecious varieties have M/F flowers. For greenhouse use parthenocarpic type.
Eggplant	Florida Market, Black Beauty, Dusky, Long Tom, Ichian, Tycoon, Dourga	Solanaceae (Nightshade)	I	200	90–110	Stake eggplants. Harvest into summer. Require warm weather. Dourga is white.
Okra	Clemson Spineless, Perkins, Dwarf Green, Emerald, Blondy	Malvaceae (Mallow)	III	70	50–75	Produces well in warm seasons. Okra is highly susceptible to root-knot nematodes.
Peas, southern	Blackeye, Mississippi Silver, Texas Cream 40, Snapea, Sadandy, Purplehull, ZipperCream	Leguminosae (Pea)	III	80	60–90	See Beans, bush. The cowpea curculio is a common pest. Tiny white grub infests seeds in pods. Good summer cover crop.
Peppers	Sweet: Early Calwonder, Yolo Wonder, Big Bertha, Sweet Banana, Jupiter. Hot: Hungarian Wax, Jalapeno, Habanero	Solanaceae (Nightshade)	I	50	80–100	Mulching especially beneficial. Continue care of peppers well into summer. Mosaic virus a common disease pest. Most small-fruited varieties are attractive but hot.

Vegetable	Varieties	Family (common name)		Seeds	Days	Comments
Potatoes, sweet	Porto Rico, Georgia Red, Jewel, Centennial, Coastal Sweet, Boniato, Sumor, Beauregard, Vardaman	Convolvulaceae (Morning Glory)	I	300	120–140	Sweet potato weevils are a serious problem. Vardaman bush type for small gardens.
Pumpkin	Big Max, Funny Face, Connecticut Field, Spirit, Calabaza, Cushaw	Cucurbitaceae (Melon)	III	300	90–120	Bees required for pollination. Foliage diseases and fruit rot are common.
Squash	Summer: Early Prolific Straightneck, Dixie, Summer Crookneck, Cocozelle, Gold Bar, Zucchini, Peter Pan, Sunburst, Scallopini, Sundrops. Winter: Sweet Mama, Table Queen, Butternut, Spaghetti	Cucurbitaceae (Melon)	III	150	40–55	Summer types usually grow on a bush while winter squash have vining habit. Both male and female flowers on same plant. Common fruit rot/drop caused by fungus and incomplete pollination. Bees required. Crossing occurs but results not seen unless seeds are saved. Winter types store longest.
Tomatoes	Large fruit: Floradel, Solar Set, Manalucie, Better Boy, Celebrity, Bragger, Walter, Sun Coast, Floramerica, Flora-Dade, Duke. Small Fruit: Florida Basket, Micro Tom, Patio, Cherry, Sweet 100, Chelsea	Solanaceae (Nightshade)	I	200	90–110	Staking, mulching beneficial. Flowers self-pollinated. May drop if temperatures too high or low, or if nitrogen fertilization excessive. Some serious problems are blossom-end rot, wilts, whitefly, and leafminers. Better Boy appears resistant to root-knot.
Watermelon	Large: Charleston Gray, Jubilee, Crimson Sweet, Dixielee. Small: Sugar Baby, Minilee, Mickylee. Seedless: Fummy	Cucurbitaceae (Melon)	III	400	85–95	Cue to space requirement, not suited for most gardens. Suggest small ice-box types. Plant fusarium wilt-resistant varieties. Bees require pollination.
Cool season vegetables						
Beets	Early Wonder, Detroit Dark Red, Cylindra, Red Ace, Little Ball	Chenopodiaceae (Goosefoot)	I	75	50–65	Beets require sample moisture at seeding or poor emergence results. Leaves edible.
Broccoli	Early Green Sprouting, Waltham 29, Atlantic, Green Comet, Green Duke	Cruciferae (Mustard)	I	50	75–90	Harvest small multiple side shoots that develop after main central head is cut.
Cabbage	Gourmet, Marion Market, King Cole, Market Prize, Red Acre, Chieftan Savoy, Rio Verde, Bravo	Cruciferae (Mustard)	I	125	90–110	Buy clean plants to avoid cabbage black rot, a common bacterial disease that causes yellow patches on leaf margins. Watch for loopers.
Carrots	Imperator, Thumbelina, Nantes, Gold Pak, Waltham Hicolor, Orlando Gold	Umbelliferae (Parsley)	II	100	65–80	Grow carrots on a raised bed for best results. Sow seeds shallow and thin to proper stand.
Cauliflower	Snowball Strains, Snowdrift, Imperial 10-6, Snow Crown, White Rock	Cruciferae (Mustard)	I	80	75–90	Tie leaves around flower head at 2–3 inch diameter stage to prevent

(Continues)

Table 16-2 (*Continued*)

Common name	Variety[2]	Plant family[3]	Transplantability[4]	Pounds per 100 feet	Days to harvest[5]	Comments
		Cool season vegetables				
						discoloration. For green heads, grow broccoflower.
Celery	Utah Strains, Florida Strains, Summer Pascal	*Umbelliferae* (Parsley)	II	150	115–125	Celery requires very high soil moisture during seeding/seedling stage.
Chinese cabbage	Michihili, Wong Bok, Bok Coy, Napa	*Cruciferae* (Mustard)	I	100	70–90	Bok Choy is open-leaf type, while Michihili and Napa form round heads.
Collards	Georgia, Mates, Blue Max, Hicrop Hybrid	*Cruciferae* (Mustard)	I	150	70–80	Tolerates more heat than most other crucifers. Harvest lower leaves. Kale may also be grown.
Endive/ Escarole	Florida Deep Heart, Full Heart, Ruffec	*Compositae* (Sunflower)	I	75	80–95	Excellent ingredient in tossed salads. Well adapted to cooler months.
Kohlrabi	Early White Vienna, Grand Duke, Purple Vienna	*Cruciferae* (Mustard)	I	100	70–80	Both red and green varieties are easily grown. Use fresh or cooked. Leaves edible.
Lettuce	Crisp: Minetto, Ithaca, Fulton, Floricrisp. Butterhead: Bibb, White Boston, Tom Thumb. Leaf: Prize Head, Red Sails, Salad Bowl. Romaine: Parris Island Cox, Valmaine, Floricos	*Compositae* (Sunflower)	I	75	50–90	Grow crisp head type in coolest part of the season for firmer heads. Sow seeds very shallow, as they need light for germination. Intercrop lettuce with long-season vegetables.
Mustard	Southern Giant Curled, Florida Broad Leaf,	*Cruciferae* (Mustard)	II	100	40–60	Consider planting in a wide-row Tendergreen system. Broadleaf type requires more space. Cooked as "greens."
Onions	Bulbing: Excel, Texas Grano, Granex, White Granex, Tropicana Red. Bunching: White Portugal, Evergreen, Beltsville Bunching, Perfecto Blanco. Multipliers: Shallots	*Amaryllidaceae* (Lily)	I	100	120–160	Plant short-day bulbing varieties. For bunching onions, insert sets upright for straight stems. For multipliers, divide and reset. Bulbing onions may be seeded in the fall, then transplanted in early spring (Jan.–Feb.). Granex used for Vidalia and St. Augustine Sweets.
Parsley	Moss Curled, Perfection, Italian	*Umbelliferae* (Parsley)	II	40	70–90	Grow parsley root similarly (Hamburg type). Curley and plain types do well.
Peas, English	Wando, Green Arrow, Laxton's Progress, Sugar Snap, Oregon Sugar	*Leguminosae* (Pea)	III	40	50–70	Edible podded type are Oregon (flat) and Sugar Snap (round)—be sure to trellis.

Vegetable	Varieties	Family	Category		Days	Comments
Potatoes	Sebago, Red Pontiac, Atlantic, Red LaSoda, LaRouge, Superior	*Solanaceae* (Nightshade)	II	150	85–110	Plant 2-oz. seed pieces with eyes. Do not use table stock for seed. Remove tops two weeks before digging to toughen skin. Varieties planted by seeds produce less than from tubers.
Radish	Cherry Belle, Comet, Early Scarlet Glove, White Icicle, Sparkler, Red Prince, Champion, Snowbelle	*Cruciferae* (Mustard)	III	40	20–30	Intercrop summer type with slow growing vegetables to save space.
Spinach	Virginia Savoy, Melody, Bloomsdale Longstanding, Tyee, Olympia	*Chenopodiaceae* (Goosefoot)	II	40	45–60	Grow during coolest months.
Turnips	Root/Tops: Purple-Top White Globe, Just Rite. Tops: All Top	*Cruciferae* (Mustard)	III	150	40–60	Grow for roots and tops. Broad-cast seed in wide-row system or single file.

[1] *University of Florida Cooperative Extension Service.*

[2] *Other varieties may produce well also. Suggestions are based on availability, performance, and pest resistance.*

[3] *To practice crop rotation, group family members; avoid planting family members following each other.*

[4] *Transplantability categories: I, easily survives transplanting; II, survives with care; III, use seeds or containerized transplants only.*

[5] *Days from seeding to harvest, values in parentheses are days from transplanting to first harvest.*

Edible Parts

Vegetables are divided into three groups based on the part of the plant that is eaten:

- Leaves, flower parts, or stems
- Fruits or seeds
- Underground parts

For example, asparagus, chard, celery, lettuce, and cabbage represent edible leaves, flower parts, and stems. Sweet corn, okra, peas, peppers, tomatoes, and melons represent the vegetables of edible fruits or seeds. Onions, garlic, beets, radish, potatoes, and turnips are vegetables categorized by their edible underground parts. Some vegetables may fit in more than one category. Turnips and beets are both edible underground vegetables, but the leaves of both may also be eaten as a vegetable. Table 16-2 indicates the use of some vegetables.

Growing Season

Vegetables are either warm season or cool season according to their growing season. Table 16-2 shows which vegetables are warm season, and which are cool season.

Steps in Vegetable Production

Planning goes before planting. Regardless of the size of the vegetable production enterprise, the individual must decide where to plant, and what and how much to plant.

Site

Gardens will need to be located for convenience on a site close to a source of water with at least 8 to 10 hours of direct sunlight. Where possible, the site should allow rotation for weed and other pest control.

Soil type at the site is important. Most vegetables do best in well-drained, loamy soil, and a pH of 5.8 to 7 (see Table 16-3).

Size

The size of the vegetable growing area will depend on the ground available and the use of the produce. What to plant depends too on the use and the market for the produce. How much to plant depends on the crop or crops selected and the amount of available space.

What to Plant

What to plant is determined by the type of vegetable enterprise. For example, a home garden and market garden will have a variety of vegetables determined by the likes of a family or the demands of a roadside or farmer's market. A large-scale, commercial operation will concentrate on large areas of single vegetable crops.

Table 16-3

Crop	Fertilizer requirements[1]			pH
	Nitrogen	Phosphorus	Potash	
Asparagus	4	3	4	6–7
Bean, bush	1	2	2	6–7.5
lima	1	2	2	5.5–6.5
Beet, early	4	4	4	5.8–7
late	3	4	3	5.8–7
Broccoli	3	3	3	6–7
Cabbage, early	4	4	4	6–7
late	3	3	3	6–7
Carrot, early	3	3	3	5.5–6.5
late	2	2	2	5.5–6.5
Cauliflower, early	4	4	4	6–7
late	3	3	4	6–7
Corn, early	4	3	3	6–7
late	3	2	2	6–7
Cucumber	3	3	3	6–8
Eggplant	3	3	3	6–7
Lettuce, head	4	4	4	6–7
leaf	3	4	4	6–7
Onion	3	3	3	6–7
Parsley	3	3	3	5–7
Parsnip	2	2	2	6–8
Pea	2	3	3	6–8
Potato	4	4	4	4.8–6.5
Radish	3	4	4	6–8
Spinach	4	4	4	6.5–7
Squash, summer	3	3	3	6–8
winter	2	2	2	6–8
Tomato	2	3	3	6–7
Turnip	1	3	2	6–8

Nutrient Requirements and pH for Vegetables

1 = Light, 2 = Moderate, 3 = Heavy, 4 = Extra heavy.

The days to harvest influences the type of vegetable grown. Some areas may not provide enough growing days to produce a crop (Table 16-2). Also, if the crop is not all consumed or sold and fresh, some plans for storage and preservation should be made. Diseases, insect pests, and crop rotation may further influence the vegetable crops grown.

Site Preparation

Site preparation includes primarily the soil. This includes the addition of organic matter, lime, or fertilizer. Plowing, tilling, or spading is the first step in preparing a seedbed. The soil is worked into a fine, firm **seedbed** for planting time. To maintain the physical structure of the soil, ground should never be plowed when it is too wet. Plowing should be done at least three weeks before planting.

Many soils benefit from applications of various forms of organic matter such as animal manure, rotted leaves, compost, and cover crops. These are applied by thoroughly mixing liberal amounts of organic matter in the soil well in advance of planting, preferably at least a month before seeding. For example, in some areas, 25 to 100 pounds of compost or animal manure can be added per 100 sq. ft. When inorganic fertilizers are not used, well-composted organic matter may be applied at planting time. Due to inconsistent levels of nutrients in compost, accompanying applications of balanced inorganic fertilizer may be beneficial. Organic amendments low in nitrogen, such as composted yard trash, must be accompanied by fertilizer to avoid plant stunting.

Another type of organic matter used is green manure in the form of a cover crop. Off-season planting and plow down of green-manure crops includes legumes, ryegrass, lupine, and hairy vetch.

A pH test of the soil helps determine if liming is necessary for the type of crop being grown. If a soil test indicates a pH of 6.0 or less, lime should be applied (see Table 16-4). Lime needs are best met two to three months before the crop is to be planted. The lime must be thoroughly mixed into the soil to a depth of 6 to 8 inches and then watered to promote the chemical reaction.

Unless very large quantities of organic fertilizer materials are applied, commercial fertilizer is usually needed for gardens. A soil test also helps determine the amount and type of fertilizer that should be added to the vegetable crop. Vegetable crops have different fertilizer needs, as shown in Table 16-3, and proper fertilization will increase yields.

Planting Vegetables

Vegetables all start as seeds. Some seeds can be sown where they will grow, while others need to be started in a special environment such as a cold frame, hotbed, or greenhouse and then transplanted. Table 16-2 indicates the **transplantability** for many vegetable crops.

Depending on the size of the vegetable enterprise, the type of seed and the quantity of seed, vegetable seeds are planted by the following:

- Hand in rows or hills
- Broadcasting by hand or machine
- Seeders powered by hand or tractor

If vegetables are to be transplanted as seedlings as the tomatoes in Figure 16-2, they can be:

- Hand set
- Hand-machine set
- Machine set

Seeding rate, row and plant spacing, and seeding depth are listed in Table 16-4.

Figure 16-2

Transplanting tomatoes to the field for commercial production.

Table 16-4

Planting Guide for Vegetables				
Crop	Seeds/plant per 100 ft	Spacing (in) rows	plants	Seed depth (in)
Warm season vegetables				
Beans, bush	1 lb	18–30	2–3	1–2
Beans, pole	1/2 lb	40–48	3–6	1–2
Beans, lima	2 lb	24–36	3–4	1–2
Warm season vegetables				
Cantaloupes	1/2 oz	60–72	24–36	1–2
Corn, sweet	2 oz	24–36	12–18	1–2
Cucumbers	1/2 oz	36–60	12–24	1–2
Eggplant	50 plts/1 pkt	36–42	24–36	1/2
Okra	1 oz	24–40	6–12	1–2
Peas, southern	1/2 oz	30–36	2–3	1–2
Peppers	100 plts/1 pkt	20–36	12–24	1/2
Potatoes, sweet	100 plts	48–54	12–24	—
Pumpkin	1 oz	60–84	36–60	1–2
Squash, summer	1 1/2 oz	36–48	24–36	1–2
Squash, winter	1 oz	60–90	36–48	1–2
Tomatoes, stake	70 plts/1 pkt	36–48	18–24	1/2
Tomatoes, ground	35 plts/1 pkt	40–60	36–40	1/2
Watermelon, large	1/8 oz	84–104	48–60	1–2
Watermelon, small	1/8 oz	48–60	15–30	1–2
Watermelon, seedless	70 plts	48–60	15–30	1–2
Cool season vegetables				
Beets	1 oz	14–24	3–5	1/2–1
Broccoli	100 plts/1/8 oz	30–36	12–18	1/2–1
Brussels sprouts	100 plts/1/8 oz	30–36	18	1/2–1
Cabbage	100 plts/1/8 oz	24–36	12–24	1/2–1
Carrots	1/8 oz	16–24	1–3	1/2
Cauliflower	55 plts/1/8 oz	24–30	18–24	1/2–1
Celery	150 plts/1/8 oz	24–36	6–10	1/4–1/2
Chinese cabbage	125 plts/1/8 oz	24–36	12–24	1/4–3/4

(Continues)

Table 16-4 (*Continued*)

Crop	Seeds/plant per 100 ft	Spacing (in) rows	plants	Seed depth (in)
Cool season vegetables				
Collards	100 plts/1/8 oz	24–30	10–18	1/2–1
Endive/Escarole	100 plts	18–24	8–12	1/2
Kale	100 plts/1/8 oz	24–30	12–18	1/2–1
Kohlrabi	1/8 oz	24–30	3–5	1/2–1
Leek	1/2 oz	12–24	2–4	1/2
Lettuce: crisp, butterhead, leaf, and romaine	100 plts	12–24	8–12	1/2
Mustard	1/4 oz	14–24	1–5	1/2–1
Onions, bulbing	300 plts/sets 1–1 1/2 oz	12–24	4–6	1/2–1
Onions, bunching	800 plts/sets 1–1 1/2 oz	12–24	1–2	2–3
Onions, multipliers	800 plts/sets 1–1 1/2 oz	18–24	6–8	1/2–3/4
Parsley	1/4 oz	12–20	8–12	1/4
Peas, English	1 lb	24–36	2–3	1–2
Potatoes	15 lbs	36–42	8–12	3–4
Radish	1 oz	12–18	1–2	3/4
Spinach	1 oz	14–18	3–5	3/4
Strawberry	100 plts	36–40	10–14	—
Turnips	1/4 oz	12–20	4–6	1/2–1

Source: Florida Cooperative Extension Service.

THE GREEN REVOLUTION

The term *Green Revolution* is used to describe the transformation of agriculture in many developing nations that led to significant increases in agricultural production between the 1940s and 1960s. This transformation occurred as the result of programs of agricultural research, extension, and infrastructural development largely funded by the Rockefeller Foundation, the Ford Foundation, and national governments. The Green Revolution was based on the application of science and technology.

Scientific research processes led to new varieties of rice and wheat, and in a few short years led to tremendous increases in yield. The Green Revolution began when an interdisciplinary team of researchers in Mexico, headed by geneticist Dr. Norman Borlaug, introduced several new varieties of wheat that far outproduced any of the existing varieties, provided that water and fertilizer were present in abundance. In parts of the world where these inputs are available, increase in productivity has been astounding. The program began in 1944, and within 20 years Mexico changed from a wheat importer to a wheat exporter. Since the 1950s, wheat production in Mexico has more than quadrupled. India and Pakistan, among wheat growing nations, made similar gains. This dramatic change in worldwide productivity primarily included cereal crops.

Despite the success associated with the Green Revolution, critics have pointed out that the gains were made only with tremendous energy inputs.

The new varieties fail to perform in regions where irrigation or fertilizers are not available. Some individuals contend that only large corporate farms will be able to afford the productivity, and small landowners will be forced out of business.

Overall the Green Revolution has had major social and ecological impacts, which have drawn intense praise and equally intense criticism. Dr. Borlaug was awarded the Nobel Peace Prize in 1970 in recognition of his contributions to world peace through increasing food supply.

Agricultural scientists continue to recognize the need for improved production of all crop plants. Plant biotechnologists are working on many of the food crops to make crop plants that are resistant to insects, herbicides, and salt spray. This work continues into the twenty-first century. Currently, about 70% of all crops grown for consumption have been genetically transformed. This is the future of all crop plants. Many questions have been answered, but there are many more out there that need to be answered.

The approach since the beginning of mechanized agriculture had been to select desirable crops and then tailor the environment accordingly. As long as water, fertilizer, pesticides, shading, supplemental light, enriched CO_2, heating, cooling, and other factors were inexpensive, the genetic constraints have been the upper limits of crop production. Through genetic engineering, the genetic constraints are being overcome and many new plant varieties that are resistant to many of the production problems are available for farmers to use.

Cultural Practices

Common cultural practices include weed control, irrigation, and pest and disease control.

Weed Control

The primary purpose of cultivation is to control weeds. Weeds are easier to control when they are small. Depending on the size of the garden, weed control is best accomplished by hand pulling, hoeing, mechanical cultivation, or mulching. Chemical herbicides may be used on larger operations.

Mulching involves the artificial modification of the surface of the soil with straw, leaves, paper, polyethylene film, or a special layer of soil. The mulch is spread around the plants and between the rows. Mulching helps control weeds, regulates soil temperatures, conserves soil moisture, and provides clean vegetables. Often, plastic film is used as a mulch with cantaloupes, cucumbers, eggplants, summer squash, and tomatoes.

Commercial vegetable operations plan for irrigation to ensure a crop. Rainfall is seldom sufficient or uniform for a high-yielding vegetable crop. Depending on their location, some gardeners may get by with rainfall alone, but the dry regions of the West need irrigation. The type of irrigation will depend on the location. Vegetable operations use sprinkler systems, drip systems, surface irrigation, subirrigation (water permeates soil from below), and trickle irrigation.

Pest and Disease Control

Vegetable producers scout the crop twice weekly for insect damage and spray with the appropriate pesticide when needed. Soil-inhabiting insects, including mole crickets, wireworms, cutworms, and ants can be controlled with a broadcast preplant application of pesticides.

Pesticides should be applied strictly according to manufacturers' precautions and recommendations. Use of pesticides should be limited to only as necessary to control insects and diseases. During the harvesting season, application should be stopped. Finally, pesticides need to be applied in the early evening to avoid killing bees and reducing pollination.

Some soils contain nematodes, microscopic worms that can seriously reduce growth and yield of most vegetables by feeding in or on their roots. Nematode damage is less likely in soils with high levels of organic matter and where crops are rotated so that the same members of the same family are not planted repeatedly in the same soil. Excessive nematode populations may be reduced temporarily by soil **solarization**. To solarize soil, vegetation is removed, soil is broken up, and then wetted to activate that nematode population. Next, the soil is covered with a sturdy clear plastic film during the warmest six weeks of summer. High temperatures (above 130°F) must be maintained during this time for best results.

Diseases can be controlled by exclusion—the purchase of only disease-free plants. Also, plants and growing areas should be examined for common symptoms of diseases to avoid gross movement of infested soil.

Eradication is another tool for disease control. Certain soilborne diseases such as damp-off, root and stem rots, and wilts are especially troublesome on old crop sites. Site and crop rotation can slow or prevent the incidence of certain soilborne diseases. Growers avoid growing vegetables of the same family repeatedly in one area, and they watch for early disease symptoms (see Figure 16-3).

Choosing resistant varieties prevents diseases. Growers choose adapted varieties with resistance or tolerance to the diseases common in their area.

Planting fungicide-treated seed prevents disease. Untreated seeds can be dusted with a captan or thiram fungicide. Many common diseases can be controlled with either chlorothalonil, maneb, or mancozeb fungicide. Powdery mildews can be controlled with triadimefon, sulfur, or benomyl, and rusts with sulfur or ziram. Bacterial spots are controlled with basic copper sulfate plus maneb, or mancozeb.

Figure 16-3

An adult Colorado potato beetle devouring a potato leaf. Courtesy of USDA/ARS.

🌿 Harvesting

Vegetables should be harvested when they are at the peak of maturity. At this point, the produce has reached its optimal size, the flavor is fully developed, texture is just right, it keeps best, and it produces a quality processed product. The best harvesting time varies with each vegetable crop. Some vegetables hold their quality for only a few days, while others can hold their quality over a period of several weeks (see Figure 16-4). Table 16-5 provides guidelines for harvesting some vegetable crops.

Figure 16-4

For optimum flavor and nutrition, timing in harvesting garden produce is critical. Courtesy of USDA/ARS.

When harvesting produce from some vegetable plants like peas, beans, and cucumbers, growers take care not to damage plants. Injured plants may be killed and stop producing fruit. Also, growers should never harvest vegetables when foliage is wet, as this practice may spread plant diseases.

Depending on the type of operation, crop harvesting is done either by hand or by mechanical harvesters. Mechanical harvesters for larger commercial operations are designed to prevent injury to the crop.

Table 16-5

Some Guidelines for Harvesting Vegetables	
Vegetable	**When to Harvest**
Asparagus	Not until third year after planting when spears are 6 to 10 inches above ground while head is still tight. Harvest only 6 to 8 weeks to allow for sufficient top growth. The "fern," which then develops, should be left to grow for the rest of the summer.
Beets	When 1 1/4 to 2 inches in diameter.
Broccoli	Before the individual flowers in the head begin to open their yellow flowers. The dark green head should be tight and flat topped. Side heads will develop after central head is removed, but will be smaller at maturity.
Cabbage	When heads are solid and before they split. Splitting can be prevented by cutting or breaking off roots on one side with a spade after a rain.
Carrots	When 1 to 1 1/2 inches in diameter.
Cauliflower	Before heads are ricey, discolored, or blemished. Tie outer leaves together above the head when curds are 2 to 3 inches in diameter; heads will be ready about 12 days after tying.
Vegetable	**When to Harvest**
Cucumbers	When fruits are slender and dark green before color becomes lighter. Harvest daily at season's peak. If large cucumbers are allowed to develop and ripen, production will be reduced. Keep large fruit from forming; otherwise fewer cukes will be formed. For pickles, harvest when fruits have reached the desired size. Pick with a short piece of stem on each fruit.
Eggplant	When fruits are half grown, before color becomes dull.
Kohlrabi	When balls are 2 to 3 inches in diameter.
Lima beans	When the seeds are green and tender, just before they reach full size and plumpness, and when pods first reach the stage when they open easily.
Muskmelons	When stem easily slips from the fruit, leaving a clean scar.
Onions	For fresh table use, when they are about 1 inch in diameter. For boiling, select when bulbs are about 1 1/2 inches in diameter. For storage, when tops fall over, shrivel at the neck of the bulb, and turn brown. Allow to mature fully, but harvest before heavy frost.
Parsnips	Delay harvest until after a sharp frost. Roots may be left safely in ground over winter and used the following spring before growth starts. They are not poisonous if left in ground over winter.
Peas	When pods are firm and well filled, but before the seeds reach their fullest size.
Peppers	When fruits are solid and have almost reached full size. For red peppers, allow fruits to become uniformly red.
Potatoes	When tubers are large enough. Tubers continue to grow until vines die. Skin on unripe tubers is thin and easily rubs off. Such tubers will not store well. For storage, potatoes should be mature and vines dead.

(Continues)

Table 16-5 *(Continued)*	
Vegetable	**When to Harvest**
Pumpkins/Squash	Summer squash are harvested in early immature stage squash when skin is soft and before seeds develop. Winter squash and pumpkin should be well matured on the vine. Skin should be hard and not easily punctured by the thumbnail. Cut fruit off vine with a portion of stem attached. Harvest before heavy frost.
Snap beans	Before pods are full size and while seeds are about 1/4 developed, or 2 to 3 weeks after first bloom.
Sweet corn	When kernels are fully filled out and in the milk stage as determined by the thumbnail test. Use before the kernels get doughy. Silks should be dry and brown, and tips of ears filled tight. Generally, corn is ready at 19 or 20 days after silking, unless weather is cool.
Tomatoes	When fruits are a uniform red, but before they become soft. High-quality fruit can be obtained by harvesting at any time after pink color is evident and leaving such fruits to sit indoors for a few days.
Turnips	When 2 to 3 inches in diameter. Larger roots are coarse, textured, and bitter.
Watermelon	When the underside of the fruit turns yellow or when snapping the melon with the finger produces a dull, muffled sound instead of a metallic ring. Also, curly tendrils on fruit stem probably will be turning brown.

Storing

The most effective method ever devised to increase the storability of fresh horticultural commodities is cooling. As a general rule of thumb, every 18-degree decrease in temperature increases storability two- to threefold. Reducing the temperature of a tomato, for instance, from a field temperature of 100° to 50°F, will improve storability eight- to twentyfold. This shows the importance of avoiding even short holding periods at field temperatures. In this example of tomatoes, those held one to two hours at field temperatures before cooling to 50°F would store approximately one day less than those cooled immediately to 50°F. The remarkable effect of temperature on vegetable storage is not only due to its influence on the living tissue of the commodity being stored, but also on decay organisms. The most crucial temperature range for decay control seems to be 32° to 40°F. At these temperatures, many decay organisms become inactive, or nearly so. Optimal cooling strategies require rapid cooling to the lowest temperature the commodity can withstand without inducing chilling or freezing injury. The process of rapidly removing heat from vegetables before shipping is called **precooling**.

Hydrocooling is one method of precooling. In this method, vegetables are immersed in cold water long enough to lower the temperature to the desired level.

In addition to the effect of temperature, storage humidity also has a marked influence on storability. The rate of water loss is proportional to the gradient in the concentration of water between the commodity and its environment. For a given temperature, water loss is proportional to the difference between the **relative humidity (RH)** of the interior of the commodity and the relative humidity of the environment. The internal

RH is usually considered to be near 100%, thus, the gradient is usually 100 minus the RH of the air. Knowing this, one can calculate that a storage RH of 80% will cause a commodity to lose water five times faster than a storage RH of 96%, a typical storage humidity. For tomatoes at 70°F, this would amount to about 0.5% weight loss per day.

Interestingly, the amount of water vapor the atmosphere can hold declines with temperature, so the gradient in water **vapor concentration** (the gradient equals the water vapor concentration inside the product minus that of the surrounding air) for a commodity at 80% RH would be smaller if it were held at lower temperatures. In fact, the gradient at 80% RH is roughly four times smaller at 50°F than at 80°F.

The combined effects of temperature and RH can be used to describe a common, but undesirable, field phenomenon that occurs when vegetable temperatures are higher than surrounding air temperatures. In this situation, the driving force behind water loss, the water vapor gradient, is higher for the warmer vegetables, and they lose moisture more rapidly than cooler vegetables even though both are exposed to air of the same humidity. For example, if the air temperature is 80°F and the commodity temperature is 100°F, the water vapor concentration in the product would be twice that if it were at 80°F, leading to markedly greater moisture loss. If the RH of the 80°F air were at 50%, the 100°F vegetables would lose moisture three times faster than those at 80°F. Even more to the point, the rate of water loss for the 100°F product would be approximately 30 to 50 times faster than the product stored properly at 50°F and 90% to 95% RH.

To minimize water loss and the resulting decline in yield, temperature should be reduced to the minimum for the commodity, and the RH at that temperature needs to be elevated to the maximum for the commodity. Table 16-6 provides the storage temperature and relative humidity for some common vegetables. Even when stored at the proper temperature and relative humidity, storage time for vegetables varies.

Table 16-6

Storage Conditions and Life for Common Vegetables			
Vegetable	Temperature (°F)	Relative humidity (%)	Storage life
Peas	32	95–98	1–2 weeks
Peppers	45–55	90–95	2–3 weeks
Potatoes	38–40	85–95	4–5 months
Summer squash	41–50	95	1–2 weeks
Sweet corn	32	95–98	5–8 days
Tomatoes	46–50	90–95	4–7 days
Turnip	32	95	4–5 months

Herbs

Besides vegetables, many people consider growing some of the herbs on a large scale, since a market for these seems to be developing. Herbs,

such as the spearmint in Figure 16-5 and the chives in Figure 16-6, are grown for seasoning, aroma, or medicine. Table 16-7 provides propagation, growing conditions, and preservation details for a number of common herbs.

Figure 16-6

Chives plant (Allium schoenopras).

Figure 16-5

Spearmint plant (Mentha *spp.*).

Table 16-7

Common Herbs				
Name/Latin name	Type of plant	Method of propagation	Growing conditions	Preserving
Angelica *Angelica archangelica*	Biennial or perennial	Seed	Moist, rich, soil; partial shade	Crystallize stems in sugar syrup
Anise *Pimpinella Anisum*	Annual	Seed, self sows	Full sun, rich soil	Store dried seed in closed container
Balm *Melissa officinalis*	Perennial	Seed or divide root clumps	Full sun or partial shade	Dry or freeze leaves
Basil *Ocimum Basilicum*	Annual	Seed	Full sun, rich soil	Dry or freeze leaves
Bay, sweet *Laurus nobilis*	Shrub	Semihard cuttings in fall	Sun or partial shade, well drained soil	Dry
Borage *Borago officinalis*	Annual	Seed, self sows	Full sun, well drained soil	Possibly dried, but difficult
Burnet *Poterium Sanguisorba*	Perennial	Seed	Full sun, well-drained, light soil	None
Caraway *Carum carvi*	Biennial	Seed	Full sun	Dry seed heads

(Continues)

Table 16-7 (*Continued*)

Name/Latin name	Type of plant	Method of propagation	Growing conditions	Preserving
Chervil *Anthriscus cerfolium*	Annual	Seed, self sows	Partial shade, moist soil, well drained	Use fresh leaves or dried
Chives *Allium Schoenoprasum*	Perennial	Seeds or divide clumps	Full sun or some shade, rich soil	Dry or freeze leaves, seeds fresh
Cicely, sweet *Myrrhis odorata*	Perennial	Seeds or divided roots	Shade or partial shade, moist soil	Dry or freeze leaves
Coriander *Coriandrum sativum*	Annual	Seed	Full sun	Dry seeds
Costmary *Chrysanthemum balsamita*	Perennial	Divided roots	Full sun or partial shade, rich soil	Dry or freeze leaves
Dill *Anethum graveolens*	Annual	Seed	Full sun, moist soil, well drained	Dry leaves and seeds
Fennel *Foeniculum vulgare*	Perennial	Seed	Full sun	Dry leaves and seeds
Garlic *Allium sativum*	Perennial	Plant from separated cloves	Full sun, light soil	Dry bulbs
Horseradish *Armoracia rusticana*	Perennial	Root cuttings	Full sun or partial shade, moist soil	Store main root in sand in cool, dark, dry place
Hyssop *Hyssopus officinalis*	Perennial	Seed, self sows or divided roots	Full sun, light, well-drained soil	Dry or freeze leaves
Lovage *Levisticum officinale*	Perennial	Seeds or root division	Full sun or partial shade, rich, moist soil	Dry leaves
Marjoram *Origanum Majorana*	Perennial	Seeds or divided plants, or tip cuttings	Full sun, sheltered, rich soil	Dry leaves
Mint *Mentha* spp.	Perennial	Cuttings, root division	Partial shade, rich, moist soil	Dry or freeze leaves
Oregano *Origanum vulgare*	Perennial	Seed, cuttings, root division	Full sun	Dry leaves
Parsley *Petroselinum crispum*	Biennial	Seed	Sun or partial shade, rich, moist soil	Freeze
Rosemary *Rosmarinus officinalis*	Perennial	Seed, cuttings, layerings	Full sun or partial shade, light, well-drained soil	Dry or freeze leaves
Sage *Salvia officinalis*	Perennial	Seed, cuttings, division	Full sun, sandy soil	Dry leaves
Savory *Satureja hortensis*	Annual	Seed	Full sun	Dry leaves
Sesame *Sesamum indicum*	Annual	Seed	Full sun	Dry seeds
Shallot *Allium cepa*	Perennial	Plant from separated cloves	Full sun	Use fresh or dry bulbs
Sorrel *Rumex* species	Perennial	Seeds or root division	Full sun or partial shade, moist soil	Dry or freeze leaves

(Continues)

Table 16-7 (*Continued*)

Name/Latin name	Type of plant	Method of propagation	Growing conditions	Preserving
Tarragon *Artemisia dracunculus*	Perennial	Root division, cuttings	Full sun	Dry or freeze leaves
Thyme *Thymus vulgaris*	Perennial	Seed, cuttings, root division	Full sun, sandy soil	Dry leaves

Summary

1. Vegetables are either monocotyledons or dicotyledons.

2. The parts of vegetables that are eaten vary from the seeds or fruits, to the leaves, flower parts, or stems to some underground part.

3. Vegetable production enterprises vary in size from the home gardener to truck cropping.

4. Site selection is important for a successful garden. Factors to consider in site selection include water supply, other vegetation, soil type, and size.

5. Soil preparation for a garden means plowing or spading and can often mean adding organic matter, lime, and fertilizer.

6. Depending on the type of garden crop, seeds can be planted directly into the garden or the plants may be started in a greenhouse and transplanted.

7. Once the vegetables are planted, the crop requires cultivation, weed control, irrigation, mulching, and pest control for a successful crop.

8. For the best quality, vegetables should be harvested at the peak of maturity. To maintain quality after harvest, vegetables need to be stored under the proper conditions.

9. Vegetables vary in the length of time they can be stored. Precooling a vegetable at harvest increases the success of storage.

Something to Think About

1. When is the best time to transplant vegetable plants?

2. What is the scientific name for the mustard family of vegetables?

3. Describe the process for the rapid removal of heat from a vegetable crop before storage or shipment.

4. Describe factors to consider for the proper storage of vegetables.

5. Discuss how the pH of the soil can influence the type of vegetables grown.

6. What factors should be considered when selecting a site for vegetable production?

7. Describe the edible parts of plants often called vegetables. Give examples of each.

8. How should the soil be prepared and maintained for growing vegetables?

9. Discuss the care of the vegetable crop once it is planted.

Suggested Readings

Acquaah, G. 2004. Horticulture: Principles and Practices. 3rd Ed. Englewood Cliffs, NJ: Prentice Hall.

Huang, K. C. 1993. *The pharmacology of Chinese herbs*. Boca Raton: FL, CRC Press.

Phillips, E., and C. C. Burrell, Eds. 2004. *Rodale's illustrated encyclopedia of herbs*. Emmaus, PA: Rodale Press.

Newcomb, D., and K. Newcomb. 1989. *The complete vegetable gardener's sourcebook*. New York: Prentice Hall Press.

Preece, J. E., and P. E. Read. 2005. The Biology of Horticulture: An Introductory Textbook. 2nd Ed. Hoboken, NJ. Wiley Publishing.

Readers Digest. 2003. *Illustrated guide to gardening*. Pleasantville, NY: The Readers Digest Association.

Reiley, H. E. 2006. *Introductory horticulture*. Albany, NY: Delmar Publishers.

Wijesekera, R. O. B. 1991. *The medicinal plant industry*. Boca Raton, FL: CRC Press.

Internet

Internet sites represent a vast resource of information. The URLs for Web sites can change. Using one of the search engines on the Internet, such as Google, Yahoo!, Ask.com, or MSN Live Search, find more information by searching for these words or phrases: **hydrocooling, market gardening, home gardening, olericulture, truck cropping, vapor concentration, solarization, vegetable production, and food preservation.**

CHAPTER 17

Small Fruits

Small fruits include grapes, blueberries, strawberries, and the brambles. Brambles are red, black, and purple raspberries, and the erect and trailing blackberries. Also included as brambles are loganberry, boysenberry, dewberry, and tayberry. Small fruits are grown in a variety of locations and climates, depending on the type. Many are able to grow in cooler climates.

After completing this chapter, you should be able to:

- Describe a fertilizer program for a vineyard
- List the characteristics of ripe grapes
- Name the procedures for weed control in vineyards
- Explain how to select a site for a vineyard
- Describe a site and its preparation for strawberry production
- Explain the cultural practices of weed control, mulching, and fertilizing
- List eight steps for better strawberry production
- Diagram four planting systems for strawberries
- List the three main soil requirements for highbush blueberry production
- Identify the two types of commercially grown blueberries
- Outline a fertilizer and pruning program for bramble fruits
- List five key points to consider when selecting a site for brambles
- Identify seven measures for disease prevention in raspberry planting
- Describe currant and gooseberry production

Key Terms

trellis	canes	hill system
training	runners	brambles
pruning	heeling in	

🌼 Grapes

Grapes are a popular homegrown fruit as well as an important commercial crop. Grapes have many uses and are very nutritious. They are the best known vine fruit. They are consumed fresh, as juices and wines, as raisins, jam and jelly, and as frozen products. Grapes are native to the United States. Many of the domesticated varieties have native wild grape ancestry in at least one parent. American fine wine grapes are native to central Asia. Grapes are easily grown commercially in most areas of the United States and have a wide range of flavors. They do not grow well in arid sections and in areas having extremely high temperatures and very humid conditions.

Basically, three kinds of grapes are grown in the United States.

1. European or Old World grapes (*Vitis vinifera*). These make up about 95% of the grapes grown in the world and are used for table, raisin, and wine grapes.

2. North American grapes (*Vitis labrusca* or *Vitis rotundifolia*). These are hardy, native, disease-free grapes. Well-known examples include the Concord, Delaware, and Niagara.

3. Hybrids. These grapes are crosses between the European grape varieties and the native American species. They are selected for the quality of fruit from their European ancestors and for their disease and insect tolerance from their native American ancestors.

Grapes, shown in Figure 17-1, require 3 to 4 years for the vines to mature before producing fruit. They also require a growing season of at least 140 frost-free days.

Figure 17-1

ARS-developed Crimson Seedless grapevines. Courtesy of USDA/ARS.

Site

A standard variety like Concord needs 160 to 165 days. Variety selection will depend on the growing season. The varieties listed in the variety section are early to midseason varieties. The vines are healthier if good weather occurs after the crop is produced so the vines recover from producing the crop and have time to get ready for winter.

The soil should be well drained, 30 inches deep, and have a pH from 5 to 8. An excessively rich soil, high in organic matter, produces heavy, late maturing crops with a low sugar content. Light soils produce light yields of early maturing fruit with high sugar contents but vine growth is weak. There should be a minimum of cold–warm–cold cycles during the winter. Exposure to alternating cold and warm temperatures causes winter injury. The site should allow for good air drainage to reduce the amount of frost injury.

Varieties

When selecting varieties, growers consider the number of days required to mature the crop and the cold temperature hardiness. Varieties listed next mature fairly early and are hardy. Other varieties should be considered only if they mature in a time comparable to the growing season for a specific area. Other characteristics would be the type and uses of the fruit.

- Concord—160 to 165 days, a blue variety, hardy, fruit good for juice, jelly, jam, wine
- Niagara—150 to 155 days, a white grape, somewhat less hardy than Concord
- Delaware—150 to 155 days, a pink red grape, somewhat less hardy than Concord, low vigor
- Fredonia—146 to 151 days, a blue variety, less productive than Concord
- Golden Muscat—a white variety, vigorous, productive
- Moore Early—146 to 151 days, a blue variety, fruit quality not as good as Concord, the fruit sometimes cracks badly
- Beta—very early, very hardy, productive, a blue grape, berries with high sugar and acid
- Seyval—Early midseason, a white grape, moderately vigorous, productive

Table Varieties

Fresh table grapes come in three basic colors: green (sometimes called white), red, and blue-black. More than 50 kinds of table grapes are currently in production, but the following list describes 17 major varieties. Each variety possesses a distinct color, taste, texture, and history.

Table grapes are often covered with natural bloom, which is a delicate white substance common on many soft fruits. The bloom protects the grape from moisture loss and decay.

The Greens

- Perlette Seedless—The first grape of the season, the Perlette, is light in color, almost frosty green with a translucent cast; the berries are nearly round. *Perlette* means "little pearl" in French.
- Thompson Seedless—Almost everyone is familiar with this grape's light green color, oblong berries, and sweet, juicy flavor. The variety may have originated in southern Iran.
- Sugraone—The Sugraone berry is bright green and elongated. The fruit offers a light, sweet flavor and a distinctive crunch.
- Calmeria—This grape carries the nickname "lady fingers" so called for its elongated, light-green and delicately sculpted berries. A winter treat, this seeded grape has a mild, sweet flavor with an unforgettable tang.

The Reds

- Flame Seedless—The result of a cross between Thompson Seedless, Cardinal, and several other varieties, the Flame Seedless is a round, crunchy, sweet grape with a deep-red color.
- Crimson Seedless—This blush-red variety has firm, crisp berries with a sweetly tart, almost spicy, flavor.
- Premium Red—This red seedless variety has medium-sized berries with a sweet flavor.
- Emperatriz—These red, medium-sized berries show slight yellow hues throughout their firm skins. The Emperatriz is sweet and fruity.
- Rouge—These seeded, dark red berries are large and oval with a firmly crisp, thick-skinned texture and a mildly sweet, earthy flavor.
- Ruby Seedless—Grown commercially in the San Joaquin Valley since 1968, the Ruby Seedless is a deep-red, tender-skinned grape.
- Emperor—Large, deep-red clusters and a lasting flavor characterize this seeded variety that was first planted in California in 1863. In East Coast cities, where European traditions remain strong, the Emperor is very popular.
- Red Globe—The large, remarkable clusters of the Red Globe contain plum-sized seeded berries. The Red Globe is popular for both eating and decorating during the holiday season.
- Christmas Rose—Another relative newcomer, this light-red seeded variety ripens through December. Developed from four older varieties, the berries are large with a tart-sweet flavor.

The Blue-Blacks

- Exotic—Born in 1947 in Fresno, California, Exotic berries are plump and juicy and grow in long, beautiful clusters. A cross between the red Flame Tokay and the Ribier, this seeded grape is crisp and mild in flavor.
- Fantasy Seedless—These blue-black sweet berries are oval, thin-skinned, and firm. Fantasy's conical clusters have medium-sized berries with pale green flesh and a mellow flavor.

- Ribier—This dark blue-black seeded grape crossed the Channel from Orleans, France, in 1860 to become an English "hothouse" variety. The skins are firm and the taste is mild.
- Niabell—This Concord-type variety features thick-skinned, round berries ranging in color from purple to black with an earthy, rich flavor.

Planting

Growers plant grape vines in the spring. Nurseries offer either one-year- or two-year-old plants. The two-year-old plants cost more than one-year-old plants. At planting time, the dead or broken roots are cut off and the tops are pruned back to two or three buds. The grower should set vines at 8 foot spacings in rows 8 to 9 feet apart and plant the vines 2 inches deeper than they were growing in the nursery.

PLANT BREEDERS PRODUCE NEW BERRY VARIETIES

Producing better varieties of plants is an on-going challenge for plant breeders. Thanks to the work of scientists at the Fruit Laboratory in Beltsville, Maryland, growers will have four new selections of commercially available blueberries to choose from. The Little Giant variety is bred for the cooler climates of Washington, Oregon, Michigan, New Jersey, and North Carolina. It offers an alternative variety for frozen and processing markets, and can be planted with other northern highbush blueberry varieties for cross pollination.

Pearl River, Magnolia, and Jubilee are new varieties and more suited to the warmer climates in the Gulf Coast and southeastern United States. They can be interplanted with other southern highbush blueberry varieties to ensure fruit set, early ripening, and maximum yield. Each of the four new varieties is productive and disease resistant.

Also, two new June-bearing strawberries have been introduced by the Agricultural Research Service (ARS) plant breeders for the Middle Atlantic and adjacent regions. Primetime is a midseason berry. It bears fancy, good-quality, large fruit. Latestar is a late-season variety. It produces large, attractive fruit. Both varieties are recommended for shipping and local markets and resist multiple fungal diseases. They produce well on either light or heavy soils, and in matted rows or in hill culture. Plants are available in season at nurseries.

Propagation

Grapes are propagated commercially by hardwood cuttings. If a disease-resistant rootstock is needed for certain varieties, plants are grafted or budded to a disease-resistant seedling rootstock.

Trellis Construction

Only one type of **trellis** and training method will be discussed. The trellis is the two-wire trellis. Growers should use durable posts about 3 inches in diameter. About 2 1/2 feet of the 8-foot-long post should be buried in the ground. Posts are set at 16 foot intervals along the row. End posts may need to be longer to give good support and strength. If the end posts are weak or poorly braced, the trellis will be weak and sag.

Wires fasten to the post so they slide through the fastening device. This allows the tension on the wires to be increased to keep them taut. The bottom wire should be 3 feet from the ground, the top wire about 5 feet. Number 10 galvanized wire is suitable for use on grape trellises.

Training

The **training** system described is called the four-arm Kniffin. Plants trained to the four-arm Kniffin system will consist of a vertical stem that reaches to the top wire plus an upper and lower branch on each side of the trunk. These branches are tied to the wires. During the first year, the strongest shoot is trained to grow upward and is fastened to the first wire. During the second growing season, it reaches and is tied to the top wire. During the winter, cut off the portion of the trunk that extends above the wire. During the second growing season, the vine will have sent out side branches near the bottom wire. One branch on each side of the trunk should be trained to grow on the bottom wire. During the second winter, these side branches should be pruned back to five buds each at the time the excess trunk is cut off. During the third growing season, the upper branches form.

Pruning

Fruit is only produced on the previous year's wood. Vines allowed to grow for many years without pruning accumulate much old, unproductive wood. The object of **pruning** is to remove all the old wood and leave four **canes** that will produce next year's crop. The four canes are the four arms or branches described in the training section. In addition, two to four other canes, called renewal spurs, are left to produce the fruiting canes for the following year. Renewal spurs only have two to three buds and like the four fruiting canes originate from the trunk.

Grape pruning regulates the health and vigor of the vines. This is done by pruning the vines so the fruit load is about what the vines are able to mature. The system is based on the vigor of the vines in the previous growing season.

Grapes are pruned during the dormant season in late winter or early spring. Buds that have begun to swell are easily knocked off during the pruning operation. When pruning, growers select the four canes that will be the fruiting wood for the coming growing season. They also select the canes to serve as the renewal spurs. The renewal spurs should originate on the trunk near the point where a wire crosses. This positions canes properly. Once the renewal spurs and fruiting wood have been selected, all other growth is removed. The wood left should be dark brown and slightly larger than a pencil.

Weed Control and Fertilizing

Weeds are controlled by cultivating no deeper than 3 to 4 inches. The amount of fertilizer needed varies with the soil. Soil testing and foliar analysis will help monitor plant needs. Growers avoid giving too much nitrogen.

Harvesting

Grapes should be left on the vine until fully ripe because ripening stops once the grapes are harvested. Color is not a good indicator of ripeness as it changes two to three weeks before the grapes are ripe. The stem of ripe grape clusters will be brownish and wrinkled. Ripe grapes are easily pulled from the cluster.

Pest Control

Grapes are sprayed for insects and fungal diseases. Growers should consult a local county agricultural agent or locate the recent USDA bulletin on spraying grapes. Major diseases that require control are black rot, powdery mildew, and downy mildew. The major insect pests include the flea beetle, leaf hopper, berry moth, and Japanese beetle.

Strawberry

The strawberry is moderately easy to grow when compared to other types of fruits. A strawberry planting lasts about 3 years before it needs replanting.

Site

A good site slopes about 2 feet in 100 feet. This allows good drainage of both water and air. Good air drainage allows the cold air to flow off reducing susceptibility to frost injury. Plants in low areas may be covered by water during winter thaws. Then, when the temperature drops after the thaw, the freezing water may kill the plants.

The best growing conditions are full sun and sandy to gravely loam with a pH of 6.0 to 6.5 and a good supply of organic matter. Good drainage is essential for best growth. Strawberry **runners** root better on light soils. Everbearing varieties may give the best results on rich soil. Growers avoid sites infested with nutsedge, quackgrass, or persistent problem weeds and spray problem weeds with herbicides before planting the strawberries. If the strawberries follow sod, there could be problems with white grubs or wireworms.

Flower Bud Formation

June bearers form flower buds in the short days of late summer and fall. This is why the strawberry patch should not be neglected after the berries have been picked. Poor care late in the season leads to a poor crop the following year. Low temperatures are needed for the flower buds to complete their development. Everbearers form flower buds in the longest days of summer and flower and fruit in summer and fall. The longer days of summer trigger runner formation (see Figure 17-2).

Figure 17-2

Parts of the strawberry.

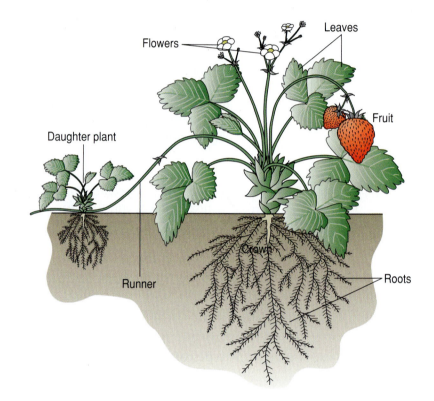

Varieties

Many varieties of strawberries are available. Some varieties are listed. Everbearing varieties may not be as productive or have as high a quality as June bearing varieties. Some varieties include—

- Raritan—Midseason, good flavor and yields but susceptible to stele and wilt diseases.
- Delite—Late, resistant to stele and wilt but forms too dense a matted row.
- Redchief—Midseason, resistant to stele and wilt but berries are hard to cap.
- Holiday—Midseason, large, firm berries but plants are not very disease resistant.

- Guardian—Midseason, resistant to stele and wilt but berries are light fleshed, rough and green tipped.
- Earliglow—Early, resistant to stele and wilt.
- Midway—Midseason variety.
- Marlate—Late, high-quality berries but low yields; berries have light flesh, hard to cap.
- Gem—Everbearer, hardy and productive but a poor runner producer, berries are soft, acid.
- Ozark Beauty—Everbearer, best everbearer for north central United States.
- Scarlet—Midseason variety for home gardens.

Planting

Strawberries fit into a crop rotation system when grown in the vegetable garden. Strawberries should not follow strawberries, tomatoes, peppers, or eggplants. These crops are all susceptible to verticillium wilt. Where wilt has been a problem, growers should use only disease-resistant varieties. The soil should be tested before planting. The best yields of strawberries, as seen in Figure 17-3, are obtained from new plants set new each year. Everbearers give the best crop the year they are planted. June bearers give the best crop the year after they are planted.

Figure 17-3

Strawberries in the field and ready to harvest. Courtesy of USDA/ARS.

Growers purchase virus-free plants. Healthy strawberry plants have medium to large crowns and large root systems consisting of light colored roots. Plants with black roots are old and should not be planted.

Strawberries can be planted as soon as the soil can be worked in the spring. When the plants arrive, they should be unpacked and either planted or heeled in. **Heeling in** is the temporary planting of plants in a trench. Plants should not be left heeled in for longer than 2 to 3 weeks.

Strawberry plant crowns should be at soil level. If set too deep, the crown rots. The roots should not be allowed to dry out while the plants are waiting to be planted. Roots are spread out like a fan when planting then firm down the soil around them. The spacing depends on the training system used. Early spring planting promotes the formation of highly productive runner plants.

The most common system is the matted row because it is easy to establish. The plants are set at 2 feet in rows 3 to 4 feet apart. Plants should not be allowed to form too many runner plants. An overcrowded matted row produces small berries. The runner plants should be 4 to 6 inches apart with the matted row 15 to 18 inches wide. The finished rows should be about 18 inches apart. This system is used most with June bearers but it is the least productive training systems for strawberries.

In the hedgerow system, plants are set 15 to 18 inches apart in rows spaced 24 to 30 inches apart. Only two runners are allowed to grow, one from each side of the plant. The runners are trained to grow in line with the row, not out to the sides. In the double hedgerow system, each plant is allowed to form four runners that are placed diagonally from the row. Looking down on a double hedgerow system row it would look like a row of X's. The mother plant would be at the center of each X, and the runners would form the four arms. In either hedgerow system, the runner plants end up being about 1 foot apart. This system produces high-quality fruit but considerable time is spent getting it established.

To establish the spaced bed system, plants are set 24 to 36 inches apart in rows that are 42 to 48 inches apart. The runners are placed by hand to create a finished row that is 15 to 24 inches wide. Plants are spaced 8 inches apart, and after the bed is filled, all other runners are removed. A great deal of labor is involved in this system. Once established the planting must be gone over to remove surplus runners. The spaced bed system produces larger berries and higher yield than the matted row.

The **hill system** is used for everbearers. Plants are set 12 to 15 inches apart. Because all runners are removed, three rows can grow together to form one large row. The three rows are spaced 1 foot apart. Growers stagger the plants so they do not line up across the row. Space the triple rows 2 feet apart. All the runners are removed before they are 2 weeks old. This system produces the maximum size berries.

Fertilization

Before planting, build up soil organic matter. Well-rotted manure is good for strawberries. Ten days after planting, growers fertilize with 12-12-12 at the rate of 2 to 3 pounds per 100 feet of row. The application is repeated in four to six weeks. Fertilizer should not be more than 4 inches away from the crown. Fertilizer can be broadcast over a row if the foliage is dry. Applying too much nitrogen causes the plants to make excessive leaf growth and causes soft berries that rot easily. Too much nitrogen will also delay ripening.

Bearing beds should not be fertilized in the spring. Growers wait until after the berries are harvested.

Removing Flower Stalks on New Plants

The flower stalks should be removed from new plantings. Stalks are removed as soon as they appear until the first of July. Allowing the plants to set fruit reduces the amount of runner formation. This reduces the number of plants thus reducing the yield the following year. Varieties that form many runners may only need the flower stalks removed until the plants are established. Blossoms are taken off everbearing varieties for the first 60 to 80 days, and then the plants can be allowed to set fruit. There is no benefit gained by allowing the fruit to form and then not picking it.

Weed Control

In home plantings, the primary weed control methods are shallow cultivation and hoeing. Problem perennial weeds are killed with herbicides before planting the strawberries. Mulching will help control weeds. Chemicals may control some weeds in strawberry plantings.

Mulching strawberry plantings is strongly recommended. The mulch provides winter protection, eliminates dirty berries, delays blooming, reduces weed growth, conserves moisture, and decreases fruit rot. Mulch for winter protection is applied when the temperature is consistently about 20°F, normally in November. Plants may be damaged if the mulch is applied too early or too late. If applied to early, a number of warm days may injure the plants. If applied too late, some crown injury may occur due to cold temperatures. Crown injury occurs at temperatures below 20°F. The mulch should be put on before the ground freezes. A 2- to 3-inch layer of straw or hay may be used for mulch material. A 1-inch layer of sawdust may be used. The mulch should be left on as long as possible to delay bloom in the spring. Check the plants often. When the leaves begin to grow yellowish green, growers remove the mulch. The plants will grow through a thin layer of mulch.

Watering

Strawberries need generous watering during and immediately following bloom. If the plants get too dry while developing fruit, the fruits will be smaller than they should be. Daytime waterings should be about 1 to 1 1/2 inches of water per week. This includes any rain that occurs during the week.

Frost Protection

Sprinkling can be used for frost protection. As water freezes on the plants it gives up heat. This heat can prevent plant injury even though they will be covered with ice. Sprinkling provides protection down to 22°F. Sprinklers are started as soon as the temperature at plant level reaches 32°F. The sprinklers will have to run as long as there is ice on the plants. Once all the ice has melted, the sprinklers may be turned off. Another method of frost protection is covering the plants with the recently removed winter protective mulch.

Harvesting

The first ripe berries appear about 30 days after the first blossoms open. Hot weather hastens ripening, shortens the harvest season, and makes frequent picking necessary. In moderate weather harvesting, every other day should be sufficient. In hot weather, it may be necessary to harvest every day. Growers pick the berries in the morning when they are cool, and harvest only the ripe berries at a single picking. Berries to be frozen should be left on the plants until fully mature. When harvesting, diseased, rotted, or injured berries should be picked and thrown away to help control rots.

Renovating the Bed

A strawberry bed may last only two to three years. The best year will be the first year after planting, but by the third year the yield will be down by about two-thirds. A strawberry patch should be renovated only if the plants are vigorous and healthy. The first step, once harvest is done, is to mow off the foliage about 1/2 inch above the crown. The plant crowns are injured if the mower setting is too low. Then the rows are narrowed down 10 to 12 inches and thinned, leaving only the most vigorous looking plants.

For renovation to succeed, it must be done right after harvest. After renovation, manage the bed just as though it were a new planting. If the mowing and renovation is delayed too long, yields the next year will be reduced. The yearly application of fertilizer should be put on right after renovation.

Blueberry

The lowbush blueberry is the low growing type that grows wild. The highbush blueberry is the type usually planted in commercial plantings. This information applies to the highbush blueberry. Blueberries, shown in Figure 17-4, do not need cross-pollination in order to produce a crop. Some varieties are: Bluecrop, Bluejay, Rubel, and Jersey. Grow varieties that mature at different times to prolong the harvest.

Site

Highbush blueberry needs an average growing season of 160 days and is badly injured by temperatures of 20° to 25°F.

A loose soil is best. A mixture of sand and peat gives excellent results. Heavier textured soils are suitable if acidic and high in organic

Figure 17-4

Blueberry plant with fruit. Courtesy *of USDA/ARS.*

matter. Peat soils may stimulate late growth that fails to harden before winter or such soils are often in frost pockets. Blueberries must have an acid soil with a pH between 4.0 and 5.1. The pH can be lowered with applications of sulfur. If the pH is very high, it may not be practical to try to lower the pH. A soil with a constant moisture supply is best. The water table should be within 14 to 22 inches of the soil surface most of the time, but good surface drainage is needed.

Good air drainage reduces frost injury and helps control diseases. Problem weeds should be controlled before the blueberries are planted. Unless a home gardener has almost ideal conditions for blueberries, it may be wise to grow some other fruit.

Planting

The best planting stock is two or three years old. The three-year-old stock usually costs more. The best planting time is early spring. Set the plants 4 feet apart in rows 10 feet apart. The plants should be set about 2 inches deeper than they were growing in the nursery. Growers can mix a shovelful of acid peat into each hole at planting time if the soil is sandy and low in organic matter.

Cultivation

Blueberries are shallow rooted so cultivation should be no deeper than 2 to 3 inches. A cover crop may be sown after harvest.

Mulching

Most organic materials may be used, but they should be allowed to weather before being applied to the blueberries. The mulch should be about 6 to 8 inches deep. Until the mulch has decomposed, the amount of nitrogen should be doubled. Leguminous mulches such as clover or soybean vines may be harmful.

Fertilizer

Growers avoid using nitrate forms of nitrogen and chloride forms of potassium. The nitrate and chloride may be toxic to blueberries. Growers use a blueberry fertilizer such as 16-8-8-4. The 4 refers to magnesium. New plants should be fertilized four weeks after planting. Established blueberries are fertilized in April before growth starts. The fertilizer should be spread evenly and kept off wet plants to prevent injury. A very sandy soil may need a second fertilization. A foliar analysis may be helpful in identifying specific nutritional needs.

Pruning

Fruit is produced on the previous season's wood. New plantings need no pruning until about the third year. Growers prune during the dormant season to remove the small twiggy growth near the base of the plant. After the third year, growers remove dead or injured branches, fruiting branches close to the ground, spindly bushy twigs on mature branches, and older stems of low vigor. Heavier pruning yields larger berries but fewer of them. Old black canes are removed at ground level.

Watering

Blueberries need 1 to 2 inches of water at 10-day intervals during dry weather.

Harvesting

Blueberries ripen over a period of several weeks. Three to five pickings are needed to harvest all the berries. Pick only the ripe berries. A reddish tinge means the berry is not yet ripe.

Pest Control

Diseases and pests of blueberries include mummy berry, other fungal diseases, scale insects, plum curculio, fruit worms, leaf hoppers, and the blueberry maggot. Occurrence of these depends on location. These can be controlled with insecticides and fungicides.

Raspberry

Red raspberries are considered a bramble. Other **brambles** include the loganberry, boysenberry, dewberry, and tayberry. Once established, a red raspberry planting should produce over a number of years, as shown in Figure 17-5. The most productive years are usually the third through the sixth.

Figure 17-5

Red raspberry plant with fruit.

Varieties

Some varieties include: Latham, Heritage, Fall Gold, Taylor, September, Amber, Canby, Fall Red, Golden Queen, Hilton, Sentry, and Newburgh. Sodus and Clyde are varieties of purple raspberries. Black raspberry varieties include: Logan (New Logan), Cumberland, Allen, Bristol, and Huron.

Site

A site with some slope to aid water and air drainage is preferred. Protection from wind will help. The junction of the cane and crown is quite weak so strong winds may cause the cane to fall over, especially if the cane is carrying a heavy fruit load. If protection cannot be provided, a trellis will support the plants.

When possible, red raspberries should be 300 feet away from other cultivated or wild raspberries to help control virus disease. Raspberries should not follow tomatoes, eggplants, potatoes, peppers, and other raspberries. These crops are susceptible to verticillium wilt, a soilborne disease that builds up if susceptible crops are grown. Black and purple raspberries are most susceptible.

The best soil is a well-drained, slightly acid loam or clay loam although raspberries are fairly tolerant. Growers avoid sites with a subsoil that prevents good drainage or good root penetration.

Herbicides control problem perennial weeds before the raspberries are planted.

Planting

Red raspberries can be planted in early spring as soon as the soil can be worked. When the plants arrive from the nursery they need to be kept cool or, if they cannot be planted right away, they should be heeled in or stored. If stored, the storage area should have a temperature of about 35°F.

The plant spacing depends on which of the several possible training systems is used. During planting, the plants should not dry out and plant them about 1 inch deeper than they were growing at in the nursery. The portion of the stem that was below ground is a different color. The hole should be big enough to allow the roots to spread out normally. The soil is firmed around the roots and the tops are cut back to about 6 inches. Cutting back may be done before or after planting. There can be a heavy loss of newly planted raspberry plants, but suckers produced by the surviving plants can be used to fill in gaps.

Training Systems

The hedgerow is the most common system for growing red raspberries. The plants are set 2 1/2 to 3 feet apart in rows from 6 to 10 feet apart. Where space is limited, use closer spacings. Suckers will fill in the row, but do not allow rows to get wider than 1 to 1 1/2 feet. Wider rows are more difficult to spray and harvest. A two-wire trellis, with one wire running down each side of the row can be used on windy sites. A single wire, running down the middle of the row and to which plants are tied, may also be used. Everbearing varieties especially need support.

To establish the hill system, plants are set 5 to 6 feet apart in rows 5 to 6 feet apart. A stake is driven into the ground next to each plant. As the plant produces suckers, five to eight healthy, vigorous suckers spaced at intervals around the stake are kept. The canes produced by the suckers are tied to the stake with two or three ties during the spring pruning. The tied suckers form a roughly triangular outline. The canes are cut back to the height of the stake in the spring.

The linear system is established just as is the hedgerow system. The only difference is that all the suckers produced by the plants are removed. The new fruiting canes come from the plant crown and are not suckers. The suckers come up at a distance from the plants. The crown shoots all arise from the plant crowns.

Fertilizing

Growers fertilize 10 to 14 days after planting. Fertilizer should be 3 to 4 inches from the shoots and canes. The second year fertilization can be increased. After the last cultivation, a cover crop may be sown. Any annual crop, such as oats or Sudan grass, which dies during the winter, may be used.

Watering

Raspberries use much water, especially when fruiting. They need about 1 inch of water per week and perhaps more during hot windy weather. A lack of water is a serious problem during the time from just before fruiting through the fruiting period. Watering is most critical from the time the fruit begin to show color until picking has been completed. A good water supply in late summer enhances cane vigor and enhances productivity in the following year. Water raspberries during the day.

Weed Control

Cultivation controls weeds but should not be deeper than 3 to 4 inches to avoid injuring raspberry roots. Growers cultivate until harvest, then once or twice after harvest. Herbicides labeled for use on raspberries may also be used.

Pruning

Red raspberries are pruned when dormant and after canes have fruited. The canes are biennial so a cane emerges and grows during one year then bears a crop of berries and dies the following year. The exception would be everbearers. Remove canes that have fruited right after harvest. The early removal of these canes may help control pest problems and maximize the water and nutrients available to new canes. Everbearers produce a crop during the late summer and fall of the same year. The same canes have a crop the following spring but it is not as large as the fall crop. The cane dies after the spring crop is harvested and can be removed. This gives growers an opportunity to do different types of pruning. If the large, fall crop is enough to satisfy family needs, the canes can be cut to the ground during the winter or early spring. This effectively eliminates the spring crop. It also eliminates doing both an after harvest and a dormant pruning. Only one pruning will be needed. If the second crop is wanted, prune off any winter killed cane tips during the dormant season. Remove canes completely when they have finished bearing the spring crop.

Other types of red raspberries need a dormant pruning to remove weak or damaged canes. In the linear or hill systems thin the canes to six to eight per hill. In the hedgerow, the canes should be spaced 8 inches apart. In the hill and linear systems, shorten the canes to about 5 1/2 feet. In the hedgerow system, shorten the canes to 4 feet. If the canes are shorter than these heights, take off only the portion that has been winter injured. Dormant pruning is done before the buds swell in the spring. If pruned too early, winter kill may reduce the height further. Dormant pruning reduces the number of suckers to keep the rows from becoming jungles.

Pest Control

Insects are not usually as destructive to raspberries and other bramble fruits as are diseases. Control of disease introduction is most important, but some spraying may be necessary depending on location. Growers should consult a local county extension agent or university for details on spraying and controlling diseases in their location.

🌀 Blackberry

The two types of blackberries are erect and trailing. The trailing blackberry is also called dewberry. These are usually tied to trellises and ripen earlier than the erect types.

Varieties

Trailing varieties may not be as hardy as the erect types in some climates. Lucretia is a trailing blackberry variety. Some erect blackberry varieties include: Alfred, Baily, Darrow, and Hedrick (see Figure 17-6).

Figure 17-6

Drawing of blackberry plant (Rubus fruticosa) *with fruits.*
Artist: A. A. Newton; Courtesy of USDA/ARS.

Site

A good moisture supply is needed, especially when the fruits are ripening. If the site is low or poorly drained, winter hardiness of the plants is reduced.

Planting

The erect types are propagated by suckers or root cuttings with root cuttings preferred. Trailing types may be propagated from root cuttings or tip layers. Thornless varieties only come from tip layers.

Plants that cannot be planted immediately can be heeled in. If the plants are dry, soak them in water for several hours prior to planting. Before or just after planting, cut the tops back to 6 inches. The planting depth should be the same as it was in the nursery. Plant erect types 5 feet apart in rows 8 feet apart with trailing varieties 4 to 6 feet apart in rows 8 feet apart. Vigorous trailing varieties are spaced at 8- to 12-foot spacings in rows 10 feet apart.

Training

Erect blackberries are most easily trained to a single wire suspended 30 inches from the ground. As canes grow, they are tied to the wire when they cross it. The trailing types are trained to a two wire trellis. The first wire is 3 feet from the ground, and the second wire is 5 feet from the ground. The canes are tied horizontally along the wire or the canes are fanned out then tied to a wire where they cross. Sometimes the canes fruiting this year are tied on one wire and canes fruiting the following year are tied on the other wire.

Pruning

The cane tips of erect blackberries are pinched out in summer when the canes are 30 to 36 inches tall. The pinching stimulates the formation of laterals. In winter, cut the laterals back to 12 inches. Fruiting canes are removed once the crop has been harvested. When the fruiting canes are being removed, thin out the new canes. Leave three to four new canes on each plant or five to six canes per linear foot or row. Remove all suckers that appear between the rows.

Pruning trailing types is not as complicated. When the fruit have been harvested, remove the canes that produced the crop. At the same time, thin the new canes. Leave 8 to 12 canes or if the variety is semitrailing leave 4 to 8 canes.

Fertilizing

Blackberries are fertilized when they blossom. Growers use 5-10-5 fertilizer at 5 to 10 pounds per 50 feet of row.

Weed Control

Cultivation is the primary means of weed control. Growers cultivate only 2 to 3 inches deep near the row, and stop cultivating about a month before freezing weather arrives.

Currant and Gooseberry

Currants and gooseberries cannot be planted in some areas unless a permit is obtained because black currants can be a host to white pine blister rust. A planting of currants or gooseberries should last for 10 to 12 years.

Varieties

Some red currant varieties include: Red Lake, Wilder, Cascade, Prince Albert, and Viking. White grape is a white currant variety.

Gooseberry Varieties

Two types of gooseberries are available—American and European. The European varieties have larger and more flavorful fruits. The American varieties are healthier and hardier. American gooseberry varieties are Downing, Houghton, and Poorman. European gooseberry varieties include: Fredonia, Chautauqua, and Industry.

Site

Currants and gooseberries prefer cool moist growing conditions. The soil should be well drained and high in organic matter avoiding light sandy soil. A dry soil may cause premature leaf drop on gooseberries causing the fruit to sunscald due to lack of shading. Gooseberries like a partially shaded growing area. Select a site with good air circulation and avoid frost pockets.

Planting

Growers control any problem perennial weeds before planting. The soil should be worked the fall before planting and if available work in well-rotted manure in the fall or early spring prior to planting. Fall planting right after the plants go dormant is best. Spring planting will have to be very early, as currants and gooseberries are quick to begin growth. Use one- or two-year-old plants and space them 4 to 5 feet apart in rows 8 to 11 feet apart. Remove any broken or injured roots or branches. Prune the top to within 6 to 10 inches of the ground. Set the plants so the lowest branch is just below the soil surface. Make sure the soil is firmed down around the roots.

Cultivation and Mulching

Cultivation should be shallow and continued until the harvest is completed. Mulches may be used as a substitute for cultivation, however, a 6-inch layer of mulch can attract mice. Any natural material may be used as a mulch.

Fertilizer

The best fertilizer for currants and gooseberries is manure. If applied annually, the plants will be productive. On infertile soil, use a fertilizer with a 1:1:1 ratio such as 12-12-12. Fertilize in the fall after growth stops or in the spring before growth starts. If fresh sawdust or straw are used as a mulch, double the fertilizer rates the year the mulch is applied.

Pruning

Growers prune when the plants are dormant in late winter or early spring. At the end of the first season, remove all but six or eight of the most vigorous shoots. At the end of the second season, leave four or five one-year shoots and three to four of the two-year shoots. At the end of the third season, leave three to four first-, second-, and third-year shoots. Each plant should have a total of nine to twelve canes.

Prune older plants so they have six to ten fruiting canes and three to four replacement canes. Leave only enough new canes to replace the older canes that are removed. Wood older than 3 years produces inferior fruit. Remove all branches that lie on the ground. The center of the bush should be fairly open.

Harvesting

Harvesting may last over a period of time as long as a month.

National Grower/Horticultural Organizations

The following organizations address growers' concerns nationwide through dialogue, annual meetings, and publications.

- North American Bramble Growers Association. c/o Harry J. Swartz, Department of Horticulture, University of Maryland, College Park, MD 20742.
- North American Strawberry Growers Association. P.O. Box 1245, Tarpon Springs, FL 34688.

Summary

1. Grapes, blueberries, strawberries, and brambles are considered small fruits.

2. Brambles include red, black, and purple raspberries, and the erect and trailing blackberries. Also included as brambles are loganberry, boysenberry, dewberry, and tayberry.

3. Small fruits are grown in a variety of locations and climates, depending on the type. Many are able to grow in cooler climates.

4. Small fruits require a wide variety of cultural practices, depending on type, use, and location.

5. Small fruits are propagated by asexual methods.

6. After harvesting, which is often by hand, the small fruits may be eaten fresh, preserved, or further processed.

Something to Think About

1. How often should red raspberries be picked?

2. Describe two trellis types used in grape production.

3. What is the best test for ripeness in table grapes?

4. What type of use can be made of the new day neutral everbearing strawberries?

5. Why should strawberries not be planted in soil recently planted with tomatoes, potatoes, peppers, or eggplants?

6. What is the best pH range for the cultivation of blueberries?

7. How can root rot of blueberries be prevented?

8. Why are blueberries pruned?

9. List the three main soil requirements for highbush blueberry production.

10. Identify the two types of commercially grown blueberries.

11. List five key points to consider when selecting a site for brambles.

12. Describe the characteristics of ripe grapes.

13. What is the best site for a vineyard?

14. Discuss eight steps for better strawberry yields.

15. Diagram four planting systems for strawberries.

16. Name seven points to consider when selecting a strawberry variety.

17. Outline a fertilizer and pruning program for bramble fruits.

18. Identify seven measures for disease prevention in bramble fruit planting.

Suggested Readings

Adams, C. R., and M. P. Early. 2004. Principles of Horticulture. 4th Ed. Boston, MA: Butterworth-Heinemann.

Eck, P. 1988. *Blueberry science.* New Brunswick, NJ: Rutgers University Press.

Ellis, M. A., R. H. Converse, R. N. Williams, and B. Williamson, Eds. 1991. *Compendium of raspberry and blackberry diseases and insects.* St Paul, MN: American Phytopathological Society.

Jennings, D. L. 1988. *Raspberries and blackberries: Their breeding diseases and growth.* San Diego, CA: Academic Press Limited.

Maas, J. L., Ed. 1998. *Compendium of strawberry diseases.* St Paul, MN: American Phytopathological Society.

Pearson, R. C., and A. C. Goheen. 1988. *Compendium of grape diseases*. St Paul, MN: American Phytopathological Society.

Poincelot, R. P. 2004. *Sustainable Horticulture: Today and Tomorrow*. Englewood Cliffs, NJ: Prentice Hall.

Pritts, M., and A. Dale. 1989 *Day-neutral strawberry production guide*. (Inf. Bull. 215). Ithaca, NY: Cornell Cooperative Extension.

Reader's Digest. 2003. *Illustrated guide to gardening*. Pleasantville, NY: The Reader's Digest Association.

Rieger, M. 2006. *Introduction to Fruit Crops*. Binghamton, NY: Food Products Press.

Internet

Internet sites represent a vast resource of information. The URLs for Web sites can change. Using one of the search engines on the Internet, such as Google, Yahoo!, Ask.com, or MSN Live Search, find more information by searching for these words or phrases: **grapes, strawberries, blueberries, brambles, raspberries, blackberries, small fruits, loganberry, boysenberry, dewberry, tayberry, and vineyards.**

CHAPTER 18

Fruit and Nut Production

Fruit operations provide high-quality and bountiful varieties of fruits and nuts. Homegrown fruits can be enjoyed at the peak of ripeness. Supermarket fruits aim for this quality. Commercial fruit and nut enterprises supply the fruits and nuts for the supermarkets nationwide.

After completing this chapter, you should be able to:

- Identify the benefits of fruit or nut production as a personal enterprise
- Name fruit and nut crops
- Describe how to plan and prepare a site for fruit or nut production
- Identify how to plant fruit and nut trees
- Describe appropriate cultural practices
- List some procedures for harvesting and storing a fruit or nut crop
- Discuss how to prune and why prune

Key Terms

pomologist	semi-dwarf	rootstock
drupe	dwarf	vegetative
pome		

🌿 Fruit and Nut Production Business

Today, the production of fruit and nut crops is more a business than a way of life. Production requires the grower to be a good financial manager, as well as a scientist and farmer. The person who is a fruit grower or fruit scientists is called a **pomologist**. People who work in the industry must be able to propagate fruit and nut trees, as well as plant and transplant, prune, thin, train, and fertilize them.

The production, harvest, and marketing of fruits and nuts is a large industry. The United States accounts for a sizable portion of the combined world crops of apples, pears, peaches, plums, prunes, oranges, grapefruits, limes, lemons, and other citrus fruits. Table 18-1 lists the production of tree fruits and nuts in the United States. Also, Figure 18-1 shows citrus production in the United States for the most recent 10 years.

Table 18-1	
Noncitrus Fruit and Nut Production	
Crop	**Average production (1,000 ton)[1]**
Apples	5,420
Apricots	98
Cherries	336
Figs	50
Nectarines	220
Olives	109
Peaches	1,148
Pears	924
Plums & Prunes	853
Tree Nuts	
Almonds	436
Hazelnuts	26
Pecans	116
Walnuts	224

[1]*Production of three recent years.*
Courtesy of National Agricultural Statistics Service, USDA.

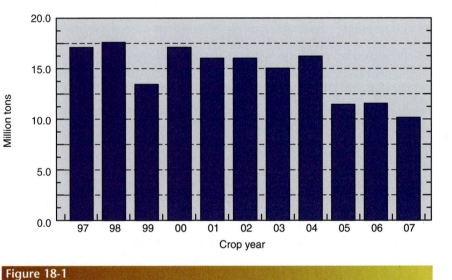

Figure 18-1

Citrus historical trends.

🌿 Types of Fruits and Nuts

The tree fruits and nuts include many types and varieties. Temperate fruit trees and their common and scientific names are listed in Table 18-2. Subtropical and tropical fruit trees are listed in Table 18-3.

Besides their family and scientific name, fruit trees can be identified as drupe or pome. A **drupe** is a fruit with a large hard seed called a stone. **Pome** fruits have a core and embedded seeds.

Fruit trees can also be grouped according to their growth habit— standard, semi-dwarf, and dwarf. A standard tree has its original rootstock and grows to normal size (Figure 18-2). For example, a standard apple tree can grow to be 30 feet high. **Semi-dwarf** and **dwarf** trees are standard varieties of trees grafted onto dwarfing rootstocks. These

Table 18-2

Temperate Fruit Trees		
Group	**Common name**	**Scientific name**
Pome fruits	Apple	*Malus domestica*
	Pear	*Pyrus communis*
Drupe or stone fruits	Peach	*Prunus persica*
	Nectarine	*Prunus persica*
	Plum	*Prunus* spp.
	Cherry	*Prunus* spp.
	Apricot	*Prunus armeniaca*

Table 18-3

Subtropical and Tropical Fruit Trees		
Group (family)	**Common name**	**Scientific name**
Citrus	Orange	*Citrus sinensis*
	Grapefruit	*Citrus paradisi*
	Lemon	*Citrus limon*
	Tangerine	*Citrus reticulata*
	Limes	*Citrus aurantifolia*
	Tangelos	*C. reticulata* × *C. paradisi*
Lauraceae	Avocado	*Persea americana*
Oleacea	Olive	*Olea europaea*
Palmae	Date	*Phoenix dactylifera*
Moraceae	Fig	*Ficus carica*
Musaceae	Banana	*Musa* spp.
Caricaceae	Papaya	*Carica papaya*
Anacardiaceae	Mango	*Mangifera indica*

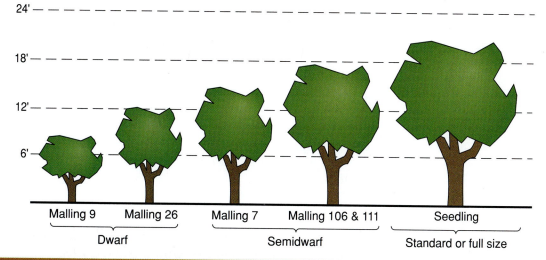

Figure 18-2

Rootstocks control sizes of fruit trees.

rootstocks cause the tree to produce less annual growth, and the size of the tree produced is called a dwarf or semi-dwarf (Figure 18-2). Semi-dwarf trees reach about 10 to 15 feet in height while dwarf trees only reach 4 to 10 feet.

Table 18-4 identifies the common nut trees, and Table 18-1 indicates the average production of some of these nuts in the United States.

Table 18-4

| | Nut Trees | |
Family	Common name	Scientific name
Rosaceae	Almond	*Prunus amygdalus*
Juglandaceae	English Walnut	*Juglans regia*
	Pecan	*Carya illinoensis*
Betulaceae	Filbert	*Corylus avellana*
Proteaceae	Macadamia	*Macadamia ternifolia*

Selecting Fruit or Nut Trees

Besides being grown commercially, fruit and nut trees may be grown in home gardens or as a part of a landscaping plan. Growers need to consider the climate, rootstock selection, frost susceptibility, fertility, pollination, and growth patterns and fruiting dates.

Climate

Some fruit and nut trees are more suited for one type of climate than another. For example, the citrus fruits grow only in the tropical and subtropical climates such as Florida, Texas, and California (Table 18-3). Growers should seek advice on fruit and nut trees that will grow in their areas.

Rootstock

Rootstock selection is important since many of the fruit and nut trees are propagated by grafting. Disease-free, hardy rootstock should be sought for fruit and nut trees. Nurseries will provide this information about their trees.

Frost Susceptibility

Locations for fruit and nut trees should be protected from frost. If an area is susceptible to late frosts, growers should select late blooming varieties. Planting trees on hillsides, near ponds, or near cities will help prevent frost damage. Commercial growers use irrigation sprinklers, heaters, and fans to help reduce frost damage.

Fertility

Producing high-quality fruits and nuts depends on the fertility of the soil. Generally, a pH of 5.5 to 6.0 is required. Fruit crops need to be fertilized annually, but too much fertilizer can damage the roots and create other problems. Depending on the type of fruit or nut tree grown, growers should seek expert advice for their location.

Pollination

Most fruit crops require pollination. It is necessary for the fruit to set or form. Depending on the type of fruit or nut tree grown, the pollination requirements can vary. Sometimes trees need to have several varieties that bloom at the same time for adequate cross-pollination. Other trees require a tree just for its pollen. Table 18-5 indicates some of the fruit and nut trees that need a pollinizer. Bees also help with pollination.

Table 18-5

Fruit and Nut Tree Production Guidelines			
Crop	Space per trees (ft)[1]	Pollinizer tree needed	Approx. years to bearing	Sprays usually required to control
Apples	5–40	Sometimes	2–10	Codling moth,[2] scab
Apricots	15–25	No	6–7	Brown-rot bacterial canker
Butternut	30–40	Yes	3–5	None
Cherries, sour	14–20	No	3–5	Fruit fly[2]
Cherries, sweet	20–35	Yes	6–7	Fruit fly,[2] bacterial canker
Chestnut	20–40	Yes	5–7	None
Figs	12–20	No	5–6	None
Filberts	15–20	Yes	5–6	Filbert moth,[2] bacterial blight
Hickory	20–40	Yes	10–14	None
Papaw	15–20	Yes	10–14	None
Peaches, nectarines	12–15	No	4–5	Leaf curl, borers, coryneum blight, brown rot
Pears	10–20	Yes	5–7	Fire blight, scab, codling moth[2]
Persimmons	15–20	Yes	8–10	None
Plums/Prunes	10–20	Some varieties	3–5	Crown borers, brown rot
Walnuts, black	30–40	No	10–12	Husk rot[2]
Walnuts, English	40–50	No	10–12	Husk fly,[2] blight

[1]The vigor of the variety and the rootstock, and the amount of pruning, also determine space requirements.
[2]Insect, if uncontrolled, causes wormy fruit or nuts.
Courtesy of Oregon State University Extension.

Growth Patterns and Fruiting Dates

As Tables 18-6 and 18-7 indicate, growing fruit or nut trees is a long-term commitment. Some trees can take anywhere from 2 to 12 years to bear sufficient fruit or nuts, but they will continue to bear for years.

Table 18-6

Starting and Bearing Time for Fruit Trees		
Tree	Years to start	Years bear
Apple	3–10	50–100
Apricot	3–4	10–25
Cherry, sour	3–4	15–25
Cherry, sweet	4–5	25–75
Peach	2–3	5–15
Pear	3–5	25–75
Plum	3–4	10–20
Quince	3–4	10–20

Table 18-7

Some Fruit and Nut Varieties and Approximate Time of Maturity		
Variety	Approx time of maturity	Comments
Apples		
Lodi	July 15–30	Yellow, won't keep
Earlygold	Aug. 1–15	Yellow, crisp
Stark Summer Treat	Aug. 1–15	Red, good flavor
Summer Red	Aug. 1–15	Red, good flavor
Gravenstein	Aug. 15–30	Pollinated by Lodi, not hardy, best sauce apple
Jonamac	Sept. 1–10	Red, McIntosh-like
Elstar	Sept. 10–20	Tart, good flavor, cool climate
Gala	Sept. 15–25	Sweet, good flavor, heat-tolerant
Jonagold	Sept. 15–30	Big, good flavor, cool climate, need pollinizer
Spartan	Sept. 20–30	Red, productive
Delicious	Sept. 25–Oct. 5	Standard red, scabs badly
Golden Delicious	Oct. 1–10	Yellow, flavorful, very productive
Empire	Sept. 20–30	Small, red flavorful
Braeburn	Oct. 5–15	Flavorful, stores well, productive
Fuji	Oct. 10–25	Sweet, flavorful, stores well
Granny Smith	Oct. 15–30	Tart, stores well
Newtown Pippin	Oct. 10–20	Green, tree vigorous, slow to produce
Apples, scab-resistant varieties		
Redfree	Aug. 5–15	Small, red, mild
Chehalis	Aug. 15–25	Yellow, big, long picking season
Prima	Sept. 1–10	Big, red, pits
Nova Easygro	Sept. 10–20	Good flavor
Liberty	Sept. 20–30	Best flavor, red
Jonafree	Sept. 25–Oct. 5	Medium, good flavor
Apricots		
Puget Gold	July	—
Rival	July	Mild flavor
Royal (Blenheim)	July	Self-fruitful
Moongold	July	Cold-hardy, pollinized by Sungold
Sungold	July	Pollinized by Moongold, hardy

(Continues)

Table 18-7 (*Continued*)

Variety	Approx time of maturity	Comments
Apricots		
Chinese	July	Resist frost
Cherries, sour varieties		
Montmorency	July	Michigan strain best
North Star	July	Dwarf variety
Cherries, sweet varieties		
Royal Ann	Mid	White, pollinized by Corum
Bing	Mid	Black, pollinized by Van, Corum
Lambert	Late	Black, pollinized by Van, Corum
Van	Early	Black, pollinized by Bing, Lambert
Sam	Mid	Black, pollinized by Lambert
Bada	Mid	White, pollinized by Royal Ann, Bing, Lambert, semi-dwarf
Stella	Mid	Self-fruitful, black
Compact Stella	Mid	Smaller than Stella
Chestnuts		
Revival	Sept.	Pollinized by Carolina
Carolina	Sept.	Pollinized by Revival
Layeroka	Sept.	Reliable producer
Chinese seedling	Sept.	Pollinizer for Layeroka
Figs		
Brown Turkey	Aug.	Large, brown
Desert King	Aug.	Green, large, sweet
Lattarula	Aug.	Green, golden inside
Filberts (some regions too cold)		
Barcelona	Oct.	Standard variety, pollinized by Davianna
Davianna	Oct.	Light producer, pollinized by Barcelona
Nectarines (fuzzless peaches)		
Stark Red Gold	Aug.	Southern and northeastern Oregon only
Harko	Aug.	Better fruit set
Genetic dwarfs	Aug.	Grown in pots, take inside for winter
Peaches		
Veteran	Aug. 20–25	Regular bearer
Red Haven	Aug. 5–10	Most popular, clingstone until fully ripe
July Elberta	Aug. 1–20	Old favorite
Early Elberta	Aug. 24–28	Old favorite
Rochester	Aug. 24–30	Old favorite
Reliance	Aug. 5–10	Resistant to cold
Frost	Aug.	Resists leaf curl
Genetic dwarfs	Summer	Very small trees, grow in pots, indoors in winter
Pears, European varieties		
Bartlett	Aug. 15–30	Pollinized by Anjou, Fall Butter
Anjou	Sept. 5–20	Pollinized by Bartlett, needs 45–60 days of cold storage before ripening
Bosc	Sept. 10–30	Pollinized by Comice
Cascade	Sept. 10–30	Pollinized by Bosc, needs 45–60 days of cold storage before ripening

(Continues)

Table 18-7 (*Continued*)

Variety	Approx time of maturity	Comments
Pears, European varieties		
Seckel	Aug. 30–Sept. 10	Pollinized by Anjou, Bosc, Comice
Pears, red varieties		
Red Bartlett	Aug. 15–30	Pollinized by Anjou, Fall Butter
Reimer Red	Sept.	Pollinized by Bartlett
Red Anjou	Sept.	Pollinized by Bartlett
Starkrimson	Aug. 1–15	Pollinized by Bartlett
Pears, Oriental varieties		
Chojuro	Sept.	Pollinized by Nijisseiki, Shinseiki
Nijisseiki (20th Century)	Sept.	Pollinized by Chojuro, Shinseiki
Shinseiki	Aug.	Pollinized by Nijisseiki, Chojuro
Kikusui	Aug.	Pollinized by Chojuro, Nijisseiki
Persimmons		
Fuyu	Nov.	Seedless Japanese
Garrettson	Nov.	American, small
Early Golden	Nov.	American, small
Plums, cold-resistant varieties		
Mount Royal	Sept.	Self-fruitful
Superior	Sept.	Pollinized by Pipestone
Ember	Oct.	Pollinized by Superior
Plums, European varieties (prunes when dehydrated)		
Italian	Sept. 10–30	Tart, "purple plum"
Brooks	Sept. 20–30	Bears regularly, large
Parsons	Sept. 1–15	Pollinized by Stanley, sweet
President Plum	Sept. 20–30	Pollinized by Stanley
Moyer Perfecto	Oct. 1	Best dried, sweet
Stanley	Sept. 1–15	Bears but brown rots
Plums, Oriental varieties		
Early Golden	July	Apricot-like flavor
Red Heart	Sept.	Pollinized by Shiro
Shiro	Aug.	Pollinized by Red Heart
Burbank	Aug.	Pollinized by Elephant Heart
Walnuts, black varieties		
Thomas	Oct.	Seedlings inferior
Ohio	Oct.	—
Myers	Oct.	—
Walnuts, English varieties		
Franquette	Late Oct.	Standard variety, limited hardiness
Spurgeon	Late Oct.	Late bloomer, hardy
Chambers #9	Late Oct.	Heavy producer, moderately hardy

Courtesy of Oregon State University Extension Service.

Most fruit trees go through a vegetative and productive period of growth. During the **vegetative** period, the tree grows vigorously and rapidly. In some trees, this may be 4 to 5 feet in a single season. When the tree enters the productive period, growth slows and the fruit buds become larger and plumper (see Figure 18-3).

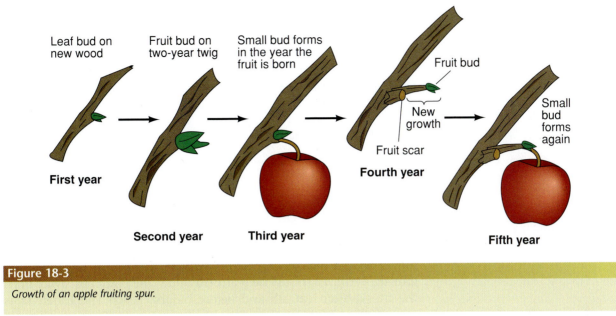

First year — Leaf bud on new wood

Second year — Fruit bud on two-year twig

Third year — Small bud forms in the year the fruit is born

Fourth year — Fruit bud / New growth / Fruit scar

Fifth year — Small bud forms again

Figure 18-3

Growth of an apple fruiting spur.

Besides growth patterns, different varieties of the same fruit mature at different times of the year. Selection should be based on this and the variety that will satisfy market demands (refer back to Table 18-5).

Site and Soil Preparation

The site should allow for maximum exposure to the sunlight, since most trees require full sunlight for maximum production. Land with a slight slope and good air circulation helps prevent frost damage. Soil at the site needs to be well drained and of medium texture. Before planting, the soil should be sampled and tested for fertilizer requirements and pH. Also, the soil needs to be deeply plowed so the root systems can penetrate into the soil.

Planting

Trees from nurseries can be planted in the spring or the fall. Both seasons have their advantages. Each type of fruit or nut tree will have its own planting distance. Refer back to Table 18-6 for the suggested space per tree needed for some of the common fruits and nuts.

Growers need to follow the guidelines for the actual planting. Each fruit or nut tree will have its own specifications. In general, the hole should be dug to be 1 1/2 to 2 feet wide and about 1 1/2 feet deep. The trees are planted so that the uppermost root is no more than 2 inches below the ground level. With dwarf trees, the graft union should be 2 to 3 inches above ground level. The roots need to be spread out in the hole and dead parts trimmed. Newly planted trees are lost because of the following:

- Roots suffocate by too deep planting
- Water standing in the hole

- Late planting leading to top growing before the roots
- Drought from lack of irrigation or weed competition
- Fertilizer placed in the hole

Mulching newly planted trees with several inches of sawdust, bark, gravel, or plastic will help establishment and early growth. Fertilizers and herbicides should not be applied the first year. Pruning the top immediately after planting to restore the normal ratio of roots to top will help trees become established.

Cultural Practices

For successful production, trees need fertilizer, irrigation, pruning, and disease and pest control.

Fertilization

After the first season, some trees will need nitrogen (N) to hasten growth. Fertilizer should be scattered under the branches, away from the trunk, after leaf fall and before bloom. Peach and filbert trees require more fertilizer than other fruits and nuts. Trees in grass sod will require much more N than where the ground is clean cultivated.

Irrigation

Depending on the type of tree and the location, trees will need irrigation, even in wet climates. How the trees are irrigated will depend on the area. Some trees may be flood irrigated and others drip or sprinkler irrigated.

Pruning

Pruning and training of young trees establishes a strong framework of branches that will support fruit. Each tree type has a specific way that it should be pruned. For example, peach trees are pruned for an open center and V-shaped pattern, as shown in Figure 18-4. Apple trees are pruned into a Christmas-tree scaffold, as shown in Figure 18-5.

Figure 18-4

A peach tree pruned to the open center of a V-shaped pattern.

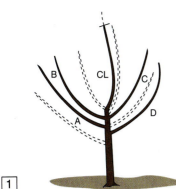

1

Sketch of a young apple tree after one year's growth. All limbs with broken lines should be removed. The central leader (CL) should be tipped if it is more than two feet long.

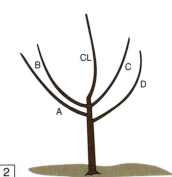

2

The same tree as in 1, after pruning.

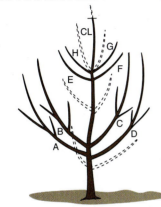

3

Sketch of how the tree in 2 may look after the second year. Branches with broken lines should be removed. The central leader (CL) should be tipped if it grew more than two feet.

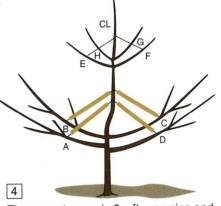

4

The same tree as in 3, after pruning and spreading of branches. All limbs in the first tier of branches (A, B, C, & D) have been spread with wooden spreaders with a sharp-pointed nail in each end, to illustrate the beginning of the Christmas-tree shape. Limbs E and F only, in the second tier of branches, have been spread with wire with sharpened ends.

Figure 18-5

An apple tree showing the cuts to be made for proper pruning.

Disease and Pest Control

Control of diseases and pests is major concern of fruit and nut growers. Since the crops are so diverse, so are the control measures. Refer back to Table 18-5 for a list of some of the pests that need to be controlled in some fruits and nuts. Proper pesticides must be selected and used correctly for the fruit or nut tree being sprayed. Timely and thorough spraying is required to control diseases and insects. Some growers will spray only if excessive damage appears imminent, and insect predators can help keep populations under control.

URBAN AGRICULTURE MOVEMENT

Half of the world's population lives in cities. By 2015 about 26 cities in the world are expected to have a population of 10 million or more. To feed a city of this size – at least 6000 tons of food must be imported each day. Recognizing these facts and the importance of food, an urban agriculture movement is gaining momentum.

Simply, urban Agriculture is the practice of cultivating, processing and distributing food in, or around a town or city thereby contributing to food security and food safety for those living in the town or city. Since urban agriculture promotes energy-saving, local food production, it is considered a "sustainable practice." Credible agencies recognize the urban agriculture movement.

The Food and Agriculture Organization of the United Nations (FAO; http://www.fao.org) defines urban agriculture as an industry "that produces, processes and markets food and fuel, in response to the daily demand of consumers within a town, city, or metropolis, on land and water dispersed throughout the urban area, applying intensive production methods, using and reusing natural resources and urban wastes to yield a diversity of crops and livestock." (Smit, J., A. Ratta, and J. Nasr, 1996, *Urban Agriculture: Food, Jobs, and Sustainable Cities.* United Nations Development Programme (UNDP), New York, NY)

The Council on Agriculture, Science and Technology, (CAST; http://www.cast-science.org/) describes urban agriculture as "a complex system encompassing a spectrum of interests, from a traditional core of activities associated with the production, processing, marketing, distribution, and consumption, to a multiplicity of other benefits and services that are less widely acknowledged and documented. These include recreation and leisure; economic vitality and business entrepreneurship, individual health and well-being; community health and well being; landscape beautification; and environmental restoration and remediation." (Butler, L. and D.M. Moronek (eds.) 2002. *Urban and Agriculture Communities: Opportunities for Common Ground.* Council for Agricultural Science and Technology (CAST). Ames, IA)

With new thinking and new designs for productive urban farms, the idea of locating agriculture in the city takes on many characteristics and offers some exciting opportunities for the future.

Harvesting and Storage

Apples are mature when they easily separate from the tree when twisted upward and when they taste good. They should be picked before the core gets areas with a glassy appearance (water core).

Sweet cherries, apricots, figs, plums, prunes, and peaches taste ripe when ready for picking. Ripening will continue after harvest. For canning or drying, they should be left on the tree until completely ripe. Sour cherries are ripe when they come off the tree easily without the stems.

European pears should be picked when still green but when they easily separate from the tree. Most varieties other than the

Bartlett require a month or more of cold storage before they will ripen properly. Oriental pears can be picked when they are sweet and juicy.

Figs are ripe when they are very soft and droop on their stems. Persimmons ripen late in the fall when they become soft and lose astringency.

Nuts fall to the ground when they are mature. They can be gathered after falling to the ground or they may be shaken to the ground or into a harvester. Nuts are dried for storage, and dried nuts can be frozen.

Fruit needs to be stored cool (near 32°F) but not frozen. A good storage room is insulated against heat and freezing. Humidity in the storage room should be moderate.

National Grower/Horticultural Organizations

The following organizations address growers' concerns nationwide through dialogue, annual meetings, and publications.

- American Pomological Society. 103 Tyson Building., University Park, PA 16802.
- American Society for Horticultural Science. 701 N. St. Asaph Street, Alexandria, VA 22314-1998.
- International Apple Institute. P.O. Box 1137, McLean, VA 22101, (703) 442-8850.
- International Dwarf Fruit Tree Association. 14 S. Main Street, Middleburg, PA 17842.

Summary

1. Fruit and nut trees are grown on a commercial scale or by home gardeners.

2. Trees are adapted to a variety of climatic conditions, but all trees are susceptible to frost damage.

3. Trees can be standard size, semi-dwarf, or dwarf.

4. Fruit and nut trees can be found for temperate climates and subtropical or tropical climates.

5. Depending on the type of tree selected, a variety of cultural practices are necessary.

6. Proper site selection, soil type, and fertility need to be considered when planting trees.

7. Many trees require pruning.

8. Trees pollinate by wind or insects, and some growers bring in bees to increase pollination.

9. Harvest and storage depend on the crop. With proper storage conditions, the fruit or nut can be kept fresh longer.

Something to Think About

1. What type of sunlight requirements do most fruits require?
2. How deep should the hole be to plant a fruit tree from a nursery?
3. Name five fruit trees and five nut trees.
4. List four drupe fruits.
5. Identify two temperate fruit trees and two subtropical fruit trees.
6. Describe the storage of apples.
7. Why are fruit trees pruned?
8. Compare the advantages of planting in the spring to planting in the fall.
9. How are fruits and nuts harvested?
10. Compare standard, semi-dwarf, and dwarf trees.

Suggested Readings

Barritt, B. H. 1991. *Intensive orchard management.* Yakima, WA: Good Fruit Grower.

Childers, N. F., and W. B. Sherman, Eds. 1988. *The peach: World cultivars to marketing.* Gainesville, FL: Horticultural Publications.

Galletta, G. J., and D. G. Himelrick. 1990. *Small fruit crop management.* Englewood Cliffs, NJ: Prentice Hall.

Janick, J., R. W. Schery, F. W. Woods, and V. W. Ruttan. 1974. *Plant science. An introduction to world crops* (2nd ed.). San Francisco, CA: W. H. Freeman and Company.

Litz, R. E. 2005. Biotechnology of fruit and nut crops. Cambridge, MA. CABI Publishing.

McMahon, M., A. M. Kofranek, and V. E. Rubatzky. 2006. Hartmann's Plant science, 4th Ed., Growth, Development, and Utilization of Cultivated Plants. Englewood Cliffs, NJ: Prentice Hall Career & Technology.

Rom, R. C., and R. F. Carlson. 1987. *Rootstocks for fruit crops.* New York: John Wiley and Sons.

Westwood, M. N. 1993. *Temperate-zone pomology: Physiology and culture 3rd Ed.* Timber Press, Inc.

Internet

Internet sites represent a vast resource of information. The URLs for Web sites can change. Using one of the search engines on the Internet, such as Google, Yahoo!, Ask.com, and MSN Live Search, find more information by searching for these words or phrases: **fruit production, fruits and nuts, nut production, fruit pests, fruit diseases, apples, apricots, cherries, figs, olives, pears, peaches, plums, prunes, almonds, hazelnuts, pecans, walnuts, and orchards.**

CHAPTER 19

Flowers and Foliage

People grow a variety of flowers and foliage as a business to sell to other people for various occasions and to decorate their surroundings. To commercially produce flowers and foliage requires knowledge of the specific cultural practices for each type of plant whether it is grown outdoors or in a greenhouse. These plants can be perennials, biennials, or annuals.

After completing this chapter, you should be able to:

- Name four types of flowers or foliage plants that people purchase at different times of the year
- Describe the time range from planting flowers until the flowers produce income
- Identify the pH range for perennial plants
- Describe watering precautions for some flowers and foliage
- Discuss planting of perennials and annuals
- Explain the winter care of perennials
- Name two methods of asexually propagating perennials
- Discuss considerations when selecting seed
- Describe an indoor starting media
- Explain how to transplant flowers or foliage outdoors
- Identify four types of bulbs
- Discuss how bulbs can be forced
- Identify the light requirements of flowering houseplants
- Name four common indoor flowering plants
- Describe the greenhouse environment
- List five pests of bulbs

Key Terms

deadheading	media	rootbound
drifts	vermiculite	cold frames
bare root	dusting	
staking	forcing	

🌿 Flower Business

People buy flowers and foliage for various reasons and at different times of the year as Table 19-1 shows. Successful growers recognize the specific times certain flowers and foliage are purchased and how long these take to produce so they can have product ready when the customer is ready to buy. Table 19-2 shows the average time from plant to sales for different flower and foliage crops.

Table 19-1

Flowers and Foliage Use and Peak Seasons for Sales		
Category	**Crops**	**Peak seasons**
Personal and environmental adornment	Cut flowers and cut foliage or greens	Christmas, Valentine's Day, Easter, Mother's Day, and Memorial Day
Personal environment, intimate	Flowering potted plants	Christmas, Easter, and Mother's Day
	Tropical foliage plants	Holidays and January
	Hanging plants	Holidays, spring, and fall
	Landscape plants, outdoors	Spring, fall
	Patio container plants	Spring, fall
Personal environment, background	Bedding plants	Spring, fall
	Woody ornamentals	Spring, fall
	Herbaceous perennials	Fall, spring
Sustenance	Vegetable transplants	Spring, fall

Table 19-2

Time from Planting of Flowers to First Income	
Crop	**Time to income**
Cut flowers	3–4 months
Potted flowering plants	3–4 months
Potted foliage plants (6 inch)	4–6 months
Woody plants in containers	3–18 months
Cut foliage	1–2 years

Flowers and foliage are adapted to a variety of growing conditions and climates as Table 19-3 shows. These conditions need to be considered unless a greenhouse is used.

Table 19-3

Temperature Requirements and Crop Examples	
Crop requirements	**Crop examples**
Cool night temperatures	Cut flowers like carnations, snapdragons, stock heather; flowering potted plants like cyclamen, fuchsia, tuberous begonia
Medium night temperatures	Cut flowers like roses
Warm night temperatures	Foliage plants
Cold winters	Deciduous shrubs
Warm summers	Ornamental plants like junipers

🌸 Value of Floriculture

Not counting the value of personal flower gardens, floriculture in the United States has a total value of about $3 billion annually. Bedding plants, pulled flowers, foliage, cut flowers, and cut greens all contribute to the value of floriculture. And as Table 19-4 shows, some states lead in floriculture. Figure 19-1 illustrates the trends in floriculture production over several years.

Table 19-4

Floriculture Crops– Top Five States by Value of Sales, 2006 for Operations with $100,000+ Sales							
Commodity	Rank	Value ($1,000)	1	2	3	4	5
		15 states	CA	FL	MI	TX	NC
Total Value Wholesale		3,834,912	1,007,463	786,614	363,158	256,468	190,334
			26.3%	20.5%	9.5%	6.7%	5.0%
		15 states	CA	MI	TX	FL	NC
Annual Bedding/ Garden Plants	1	1,281,113	231,199	192,861	156,058	118,776	108,712
		33.4%	18.0%	15.1%	12.2%	9.3%	8.5%
		15 states	CA	FL	NY	PA	NC
Potted Flowering Plants	2	619,925	205,990	79,791	50,288	39,162	38,356
		16.2%	33.2%	12.9%	8.1%	6.3%	6.2%
		15 states	FL	CA	TX	HI	IL
Foliage Plants for Indoor or Patio Use	3	542,533	366,414	97,893	19,301	15,254	7,600
		14.2%	67.5%	18.0%	3.6%	2.8%	1.4%
		15 states	CA	FL	WA	HI	OR
Cut Flowers & Cut Cultivated Greens	4	520,725	338,666	96,302	20,088	18,051	15,306
		13.6%	65.0%	18.5%	3.9%	3.55%	2.9%
		15 states	CA	SC	MI	NJ	OH
Potted Herbaceous Perennial Plants	5	507,346	70,186	67,484	45,970	45,834	39,394
		13.2%	13.8%	13.3%	9.1%	9.0%	7.8%
		15 states	FL	MI	CA	WA	PA
Propagative Floriculture Materials	5	363,270	95,489	81,587	63,529	26,017	20,673
		9.5%	26.3%	22.5%	17.5%	7.2%	5.75%

Courtesy of USDA-NASS, July 2007.

🌸 Flowering Herbaceous Perennials

Perennial plants live for many years after reaching maturity, producing flowers and seeds each year. Perennials are classified as herbaceous if the top dies back to the ground each winter and new stems grow from the roots each spring. They are classified as woody if the top persists, as in shrubs or trees. Most garden flowers are herbaceous perennials meaning the tops of the plants, the leaves, stems, and flowers die back to the ground each fall with the first frost or freeze (see Table 19-5). Every spring, new plant tops arise from the roots, which persist through the winter. Any plant that lives through the winter is said to be hardy.

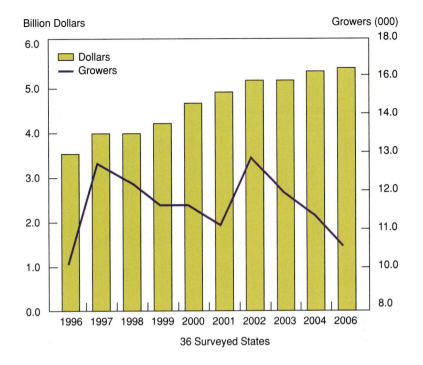

Figure 19-1

Trends in floriculture production.
Courtesy of NASS/USDA.

There are advantages to perennials, the most obvious being that they do not have to be set out, like annuals every year. Some perennials, such as delphiniums, have to be replaced every few years. Another advantage is that with careful planning a perennial flower bed will change colors, as one type of plant finishes and another variety begins to bloom. Also, since perennials have a limited blooming period of about two to three weeks, **deadheading**, or removal of old blooms, is not as frequently necessary to keep them blooming. However they do require pruning and maintenance to keep them attractive, as the rose in Figure 19-2. Their relatively short bloom period is a disadvantage, but by combining them with annuals, a continuous colorful show can be provided. Most require transplanting every three years.

Growing Perennials

While perennials do not require replanting each year, as do annuals, they still require care. For best results, initial planning, proper soil preparation, and occasional maintenance are necessary. With proper attention to these details, a perennial garden can provide color and interest in the landscape throughout the growing season.

Selection

The site will influence what species of perennials can be grown. Most flowering perennials prefer six to eight hours of sun per day, making a southeast exposure ideal. Areas of shade will reduce the numbers of species that can be grown. Exposure to wind will vary depending

Table 19-5

Various Flowering Perennials

Name	When to plant seed	Exposure	Germination time (days)	Spacing	Height	Best use	Color	Remarks
Achillea mille folium (yarrow)	Early spring or late fall	Sun	7–14	36"	24"	Borders, cut flowers	Yellow, white, red, pink	Seed is small. Water with a mist. Easy to grow.
Alyssum saxatile (golddust)	Early spring	Sun	21–28	24"	9"–12"	Rock garden, edging, cut flowers	Yellow	Blooms early spring. Good in dry and sandy soils.
Anchusa italica (Alkanet)	Spring to September	Partial shade	21–28	24"	48"–60"	Borders, background, cut flowers	Blue	Blooms June or July. Refrigerate seed 72 hours before sowing.
Anemone pulsatilla (windflower)	Early spring or late all for tuberous	Sun	4	35"–42"	12"	Borders, rock garden, potted plant, cut flowers	Blue, rose, scarlet	Blooms May and June. Is not hardy north of Washington, DC.
Anthemis tinctoria (golden Daisy)	Late spring outdoors	Sun	21–28	24"	24"	Borders, cut flowers	Yellow	Blooms midsummer to frost. Prefers dry or sandy soil.
Arabis alpine (rock cress)	Spring to September	Light shade	5	12"	8"–12"	Edging, rock garden	White	Blooms early spring.
Armeria maritima alpine (sea pink)	Spring to September	Sun	10	12"	18"–24"	Rock garden, edging, borders, cut flowers	Pink	Blooms May and June. Plant in dry sandy soil. Shade until plants are well established.
Aster alpinus (hardy aster)	Early spring	Sun	14–21	36"	12"–60"	Rock garden, borders, cut flowers	White	Blooms June.
Astilbe arendsii (false spirea) 'Europa' 'Fanal' 'Deutchland' 'Superba'	Early spring	Sun	14–21	24"	12"–36"	Borders	Pink, red, white	Blooms July and August. Gives masses of color.
Begonia evansiana (hardy begonia)	Summer in shady, moist spot	Shade	12	9"–12"	12"	Flower bed	Yellow, pink, white	Blooms late in summer. Can be propagated from bulblets in leaf axils.
Bergenia purpurascens (bergamot)	Late winter	Light shade	10	18"	2'–3'	Medicinal	Pink, red	Hummingbirds love it.

Courtesy of USDA.

Figure 19-2

A red rose is a symbol of love and a beautiful addition to any landscape. Courtesy of USDA/ARS.

on the site, and thought should be given to wind protection, particularly if growing taller perennials such as delphinium or lilies.

Planning

The perennial garden is important to insure continued bloom and desired combinations of color, texture, and height. Growers should draw a plan of the garden to scale using graph paper. Most perennials have a limited period of bloom, and growers choose plants to give a succession of flowering. This requires selecting a range of perennials, which have flowering periods that collectively cover the whole growing season. The blooming period of a particular species can usually be classified as spring, early summer, midsummer, or late summer/fall. A properly planned perennial garden will include plants from each of these flowering groups to create a season-long succession of bloom throughout the garden.

Perennials look best when planted in **drifts** of several plants, rather than as single plants or in rows. A rough guide for spacing would be 6 to 12 inches between dwarf plants, 12 to 18 inches between medium-sized plants, and 18 to 36 inches between tall perennials. Plant taller species toward the back of a flower border.

Soil Preparation

This is probably the most important factor in determining the success of a perennial planting. Soil with good water drainage is necessary. It is particularly important that the soil not stay excessively moist during the winter dormant period. Incorporating organic matter such as compost or peat moss helps improve soil drainage.

Spading or rototilling the soil to a minimum depth of 8 to 10 inches is also important. Soil preparations should be done in the fall or even one year ahead to remove all weeds that may germinate. With poor soil conditions, raised beds containing improved soil are better. Most perennials grow best in soil with a pH range of 6.5 to 7.0. A soil test can be made to determine the soil pH. Make any needed adjustments before planting.

Application of total vegetation killers can be added to beds before planting also. These chemicals control quack grass, thistle, and other perennial grasses and broadleaf weeds well. Read and follow all label directions.

Planting

Perennials are generally planted in the spring (April to May). Plants bought from mail-order nurseries are often shipped **bare root**, not potted. Growers should plant these as soon as possible after receiving. Container-grown (potted) perennials can be planted throughout the growing season, but spring is generally preferred. The proper time to plant will depend on how these perennials were produced by the grower.

Container-grown plants that have been exposed to outside temperatures throughout the winter can be planted as soon as the soil can be worked, about the same time as trees and shrubs. Perennials produced in a greenhouse during winter should not be planted until after danger of frost (below 32°F) is past, much like annual bedding plants and vegetable transplants.

If planting in loose soil, plant crowns may end up higher than planned. If soil does settle after planting, add mulch or additional soil to cover exposed crowns. Growers need to get rid of weeds before planting. Once the plants are in, the weeds are harder to remove.

Perennials can be planted in spring or fall, anytime the soil is workable and plants are available. Dormant plants are set out in early spring, four to six weeks before the last frost. Actively growing plants can be planted spring through fall. Perennials do best if planted before October 1 because roots can establish themselves before the ground freezes. Use stakes to mark where the plants will be set. A hole should be dug large enough to provide space for the roots. If the weather does not allow for immediate planting, growers need to keep plants in a cool, dark spot and make sure the peat moss around the roots does not dry out.

Watering

Although water requirements of perennials can vary greatly from species to species, most require supplemental watering until well established. One inch of water per week is a general rule. Once established, many species require watering only during prolonged dry periods. Selection of species adapted to drier climates will help reduce the need for supplemental watering.

Occasional overhead watering can harm perennials by causing disease and resulting in a shallow root system less able to withstand periodic dryness. Less frequent, more thorough watering applied directly to the soil is better.

Fertilization

Fertility of the soil can be improved before planting with the incorporation of a complete fertilizer, such as 4-12-4 or 5-10-5 at a rate of 2 to 3 pounds/100 square feet. Avoid turf fertilizers high in nitrogen as these may promote excessive foliage production at the expense of a strong root system and good flower production. With proper soil preparation and improvement of soil fertility before planting, most perennials require little additional fertilization. Application of a "starter" fertilizer when perennials are first planted may aid in more rapid establishment of a good root system. For established plants, a little bone meal or super phosphate (0-19-0) worked into the soil around the plant in the spring can be beneficial.

Maintenance

Once established, most perennials require only routine maintenance. Taller species may require **staking**, particularly in windy areas. Staking is best done when plants are first sending up new growth, before they have become top heavy. Pinching back new growth may produce bushier plants that are less likely to require staking.

A mulch applied to the soil will help suppress weeds while conserving moisture. Do not place a mulch too close to the crown as this may hold excess moisture and result in disease problems.

Insects and diseases are generally not severe threats to most perennials, particularly if locally adapted species are selected and proper care is given to the plants.

Winter Protection

Perennials damaged or killed during the winter are not usually injured directly by cold temperatures, but indirectly by frost heaving. Frost heaving occurs when the soil alternately freezes and thaws, resulting in damage to the dormant crown and root system. This action can be reduced by a winter mulch, which helps prevent rapidly fluctuating soil temperatures. A mulch about 3 inches thick is best. Evergreen boughs, clean straw, or other loose, coarse materials are good mulches. Materials such as tree leaves or grass clippings may compact too much around the plant, inhibiting water drainage and promoting disease development. Apply winter mulches after the ground freezes, usually in late November. Unless the perennials are evergreen, cut back the dead foliage to about six inches before applying the mulch, and discard the dead foliage and other debris. In the spring, mulch is gradually removed as new growth develops.

Dividing and Transplanting

Most perennials can be divided. Periodic division is needed to maintain vigor and maximum flower production. It is usually done every three to four years. Some perennials should never be divided.

The time of year when perennials are divided is a major factor in determining the success of this procedure. Species that bloom from midsummer to the fall are best divided in the early spring, before much new growth has begun. Perennials that bloom in the spring or early summer should be divided in the fall, or after the foliage dies. Exceptions are iris and daylilies, which are divided immediately after flowering.

To divide a perennial, the plant is removed from the soil by digging around and under the entire plant and lifting it carefully from the soil to avoid as much root damage as possible. All adhering soil is dislodged by hand or with a gentle stream of water from a hose. Growers should remove and discard diseased parts and cut back the top of the plant (stems, shoots, leaves) to about 6 inches. Divisions are usually taken from the outer perimeter of the plant, as this younger area tends to produce more vigorous growth. The plant can be divided by carefully breaking it apart by hand or by cutting with a heavy, sharp knife. Divide the plant in such a way that each new division has three to five "eyes"—buds that will produce new shoots.

Growers should replant the new divisions as soon as possible. They also should rework the soil if necessary to improve drainage and structure. A winter mulch is needed for divisions that are replanted in late summer or fall to help prevent frost heaving.

ETHYLENE INJURY IN FLORICULTURE

Ethylene (C_2H_4) is considered to be a plant hormone, a growth regulator, and a potentially harmful pollutant of ornamental crops. It has sometimes been called the death hormone, because it promotes the aging and ripening of many fruits and flowers. It is a simple organic substance that is active at very low concentrations. Ethylene and related substances are produced when almost any material is incompletely burned. It also evolves naturally from plant materials that are aging, ripening, or rotting. Many ripening fruits and vegetables generate ethylene as do certain microorganisms.

Ethylene toxicity and damage is of particular importance in the shipping and handling of floral produces. Ethylene gas is often added to banana ripening rooms at food wholesalers. This can cause problems for floral products if they are handled and distributed from the same building.

Symptoms of excess ethylene include malformed leaves and flowers; thickened stems and leaves; abortion of leaves and flowers; abscission of leaves and flowers; excessive branching and shortened stems. Hastened senescence (aging) of cut flowers occurs when they are exposed to ethylene. Flower corms or bulbs should never be stored with apples, because serious injury occurs. Fusarium spp. disease organisms give off ethylene and can cause flower abortion of tulips and delayed bud activity in roses. Many flowers such as carnations will begin to age in a matter of hours, and snapdragon and geranium florets will prematurely shatter after exposure to only twenty parts per billion (ppb). Some plants are far more sensitive to ethylene than others. Orchids, carnations, gypsophila, delphinium, and antirrhinums are extremely sensitive, whereas roses and chrysanthemums appear to be more tolerant.

Fumes from welding, auto exhaust, and local trash burning can cause extensive damage to greenhouse crops. Natural and propane gases are normally free of impurities and burn very cleanly. However, near perfect or complete combustion must occur when gas burners are used in or near greenhouses. Burners must have proper exhaust venting with no leaks, and they must have adequate air (oxygen) intake for proper combustion.

Not all harmful forms of ethylene are from external sources. At certain times, endogenous ethylene products, that is, ethylene produced within the plants themselves, can contribute to premature senescence of flowers. This can happen whenever plants or cut flowers are mishandled or stored improperly.

Finally, ethylene is not all bad for floriculture. It is used to induce flowering in bromeliads, and has been used successfully as a growth regulator and a chemical pinching agent in some floriculture crops.

Diseases and Insects

Perennials in general are healthy plants. Producers should select resistant varieties. They plant perennials in conditions of light, wind, spacing, and soil textures that are suited to them. After planting, the spent flowers, dead leaves, and other plant litter should be removed as these serve as a source of reinfestation. Growers need to know the major insect and disease pests (if any) of each specific plant type grown so that problems can be correctly diagnosed and treated as they arise.

Propagation

Plants can be propagated from tip or root cuttings.

Flowering Annuals

Annual flowers live only one growing season during which time they grow, flower, and produce seed, thereby completing their life cycle. Annuals must be set out or seeded every year since they do not persist. Some varieties will self-sow or naturally reseed themselves. This may be undesirable in most flowers because the parents of this seed are unknown and hybrid characteristics will be lost. Plants will scatter everywhere instead of their designated place. Examples are alyssum, petunia, and impatiens. Some perennials, plants that live from year to year, are classed with annuals because they are not winter hardy and must be set out every year, such as begonias and snapdragons. Annuals have many positive features. They are versatile, sturdy, and relatively cheap. Plant breeders have produced many new and improved varieties. Annuals are easy to grow, produce instant color, and most important, they bloom for most of the growing season.

There are a few disadvantages to annuals. They must be set out as plants or sowed from seed every year, which involves some effort and expense. Old flower heads should be removed on a weekly basis to insure continuous bloom. If they are not removed, the plants will produce seed, complete their life cycle, and die. Many annuals begin to look worn out by late summer and need to be cut back for regrowth or replaced.

Annuals offer the gardener a chance to experiment with color (as in the petunias in Figure 19-3), height, texture, and form. If a mistake is made, it is only for one growing season.

Annuals are useful for the following:

- Filling in spaces until permanent plants are installed
- Extending perennial beds
- Filling in holes where an earlier perennial is gone or the next one has yet to bloom
- Covering areas where spring bulbs have bloomed and died back
- Filling planters, window boxes, and hanging baskets
- Planting along fences or walks
- Creating seasonal color

Figure 19-3

Petunias provide a variety of color for landscapes.

Site Selection and Preparation

Different annuals perform well in full sun, light shade, or heavy shade. Light, soil characteristics, and topography should be considered. The slope of the site will affect temperature and drainage. Also, the texture, fertility, and pH of the soil will influence the plant's performance.

Preparing the soil in the fall is the best time. Proper preparation of soil will increase success in growing annuals. Growers first have the soil tested and adjust the pH if needed. Check and adjust drainage. If drainage is poor, growers may plan to plant in raised beds. The next step is to dig the bed. Often growers will add 4 to 6 inches of organic matter to heavy clay to improve soil texture. They dig to a depth of 12 or 18 inches and leave until fall or early spring. In spring, fertilizer is added, the area is spaded again, and the surface is raked smooth.

Seed Selection

To get a good start toward raising vigorous plants, growers always buy good viable seed packaged for the current year. Seed saved from previous years usually loses its vigor. It tends to germinate slowly and erratically and produces poor seedlings. Seed needs to be kept dry and cool until planted. If seed must be stored, it should be placed in an airtight container, refrigerated, and stored with a material that will absorb excess moisture. Growers should buy hybrid varieties. Plants from hybrid seed are more uniform in size and more vigorous than plants of open pollinated varieties. They usually produce more flowers with better substance.

Starting Plants Indoors

The best **media** for starting seeds is loose, well drained, fine textured, low in nutrients, and free of disease causing fungi, bacteria, and unwanted seeds. Many commercial products meet these requirements.

Clean containers are filled about two-thirds full with potting medium. The medium is leveled and moistened evenly throughout. It should be damp but not soggy. A furrow is made 1/4 inch deep. Large seeds are sown directly into the bottom of the furrow. Before sowing small seed, growers fill the furrow with **vermiculite**, and then sow small seed on the surface of the vermiculite. Seed may be sown in flats following seed package directions or directly in individual peat pots or pellets, two seeds to the pot.

After seed is sown, all furrows are covered with a thin layer of vermiculite, then watered with a fine mist. A sheet of plastic may be placed over seeded containers and then they are set in an area away from sunlight where the temperature is between 60° and 75°F. Bottom heat is helpful.

As soon as seeds have germinated, the plastic sheeting is removed and seedlings are placed in the light. If natural light is poor, fluorescent tubes can be used. Seedlings should be placed close to the tubes. After plastic is removed from the container, the new plants need watering and fertilizing, since most planting material contains little or no plant food. Growers use a mild fertilizer solution after plants have been watered.

When seedlings develop two true leaves, plants are thinned in individual pots to one seedling per pot. Those in flats, such as the seedlings in Figure 19-4, are transplanted to other flats, and spaced 1 1/2 inches apart, or to individual pots.

Figure 19-4

Seedlings ready to be transplanted.

Planting Times

As a general rule, growers delay sowing seeds of warm weather annuals outdoors or setting out started plants until after the last frost date. Seeds of warm weather annuals will not germinate well in soils below 60°F. If soil is too cold when seed is sown, seeds will remain dormant until soil warms and may rot instead of germinating. Some cold-loving annuals should be sown in late fall or very early spring.

Sowing Seed Outdoors

To seed annuals successfully, growers sow seed in vermiculite filled furrows. Annuals seeded in the garden frequently fail to germinate properly because the soil hardens on the surface keeping the water out. The furrows in soil are about 1/2 inch deep. If soil is dry, producers water the furrow, and then fill it with fine vermiculite and sprinkle with water. Make another shallow furrow in the vermiculite and sow the seed in this furrow. Seed should be sown at the rate recommended on the package. Mulch can be used until the plants are receiving enough sunlight.

Setting Out Transplants

By setting out started plants in the garden, producers can have a display of flowers several weeks earlier than if sown by seeds of the plants. This is especially useful for annuals that germinate slowly or need several months to bloom. Started plants can be purchased or produced. Buy only healthy plants free of pests and diseases.

Before setting out transplants, growers harden them off by setting the plants outside during the day. After the last frost date, annual plants may be set out. A hole is dug for each plant large enough for the root system to fit comfortably. Plants are lifted out from the flat with a block of soil surrounding their roots.

Setting Plants

If plants are in fiber pots, growers remove the paper from the outside of the root mass and set the plant in a prepared planting hole. When setting out plants in peat pots, growers set the entire pot in the hole but remove the upper edges of the pot so that all of the peat pot is covered when soil is firmed around the transplant. If a lip of the peat pot is exposed above the soil level, it may produce a wick effect, pulling water away from the plant and into the air. After setting the plants, growers water them and provide protection against excessive sun, wind, or cold if needed while the plants are getting settled in their new locations. Inverted pots, newspaper tunnels, or cloaks can be used.

Thinning

When most outdoor-grown annuals develop the first pair of true leaves, they should be thinned to the recommended spacing. This spacing allows plants enough light, water, nutrients, and space for them to develop fully above and below the ground. If they have been seeded in vermiculite filled furrows, excess seedlings can be transplanted to another spot without injury. Zinnias are an exception to this rule of thinning. In many varieties of zinnias, flowers will appear with a large nearly naked corolla and few colorful petals. This phenomenon is sometimes referred to as Mexican hats. To avoid such plants, growers sow two or three seeds at each planned location. Then they wait until the plants bloom for the first time, and remove the plants with this undesirable characteristic. Next, the remaining plants are thinned to the recommended 8- to 12-inch spacing.

Watering

Growers do not rely on summer rainfall to keep flower beds watered. They plan to irrigate them from the beginning. The entire bed needs to be moistened thoroughly, but not watered so heavily that the soil becomes soggy. After watering, soil is allowed to dry moderately before watering again. Drip systems are good for watering.

Sprinklers are not as effective as soaker hoses. Water from sprinklers wets the flowers and foliage making them susceptible to diseases. Structure of the soil may be destroyed by impact of water drops falling on its surface. The soil may puddle or crust, preventing free entry of water and air.

Mulching

Mulches help keep the soil surface from crusting and aid in preventing growth of weeds. Organic mulches can add humus to the soil. Grass clippings make a good mulch for annuals, if they do not mat. Sheet plastics also may be spread over the soil surface to retard evaporation of water and to prevent growth of weeds.

Cultivation

After plants are set out or thinned, cultivate only to break the crusts on the surface of the soil. When the plants begin to grow, stop cultivating and pull weeds by hand. As annual plants grow, feeder roots spread between the plants. Cultivation may injure these roots. In addition, cultivation stirs the soil and uncovers weed seeds that germinate.

Deadheading (Removing Old Flowers)

To maintain vigorous growth of plants, and assure neatness, growers remove spent flowers and seed pods. This step is particularly desirable if growing ageratum, calendula, cosmos, marigold, pansy, scabiosa, or zinnia.

Staking

Tall growing annuals like larkspur or tall varieties of marigold or cosmos need support to protect them from bad weather. Tall plants are supported by stakes of wood, bamboo, or reed large enough to hold the plants upright but not large enough to show. Stakes should be about 6 inches shorter than the mature plant so the blossom can be seen. Staking begins when plants are about one-third their mature size. Stakes are placed close to the plant but not so close as to damage the root system. The stems of the plants are secured to stakes in several places with paper-covered wire or other materials that will not cut into the stem. Plants with delicate stems can be supported by a framework of stakes and strings in crisscrossing patterns.

Fertilizing

When preparing beds for annuals, fertilizer should be added according to recommendations given by soil sample analysis, or by seeing plants that have grown on the site. Lime may also be needed if the soil test results indicate. Dolomitic limestone should be used rather than hydrated lime. Ideally, lime should be added in the fall so it will have time to change the pH. Fertilizer should be added in the spring so it will not leach out before plants can benefit from it.

Additional fertilizers may be needed after annuals have germinated and started to grow. This is especially true if organic mulches are added because microorganisms decomposing the mulch take up available nitrogen. A fertilizer high in nitrogen should be used in these situations. Work the fertilizer in the soil around the plants being careful not to touch the stems. Fertilizers should be applied to damp soil.

Bulbs

People use the term *bulbs* to refer to corms, tubers, and rhizomes. While all these structures contain an embryonic plant and stored plant food, they are all different in appearance and their method of propagation.

Bulbs are also used in wooded areas with evergreen ground covers, in rock gardens, with evergreen shrubs to add color, and as cut flowers.

Site Selection and Soil

Bulbs grow well in well-drained loam. To improve soil texture, organic matter can be added to the soil in the form of compost, bark, and manure. Neutral soil (pH 7) is best for bulbs. Limestone can be added if the pH is below 6.

Planting Bulbs

Growers plant bulbs with a bulb planter, nursery spade, or hand trowel. Bulbs such as crocus, narcissus, and hyacinth are planted in the fall, while dahlia, amaryllis, gladiolus, and similar bulbs are planted in the spring.

Each type of bulb has a recommended planting depth and spacing. As a general rule, bulbs should be planted the same distance apart as their planting depth (see Figure 19-5).

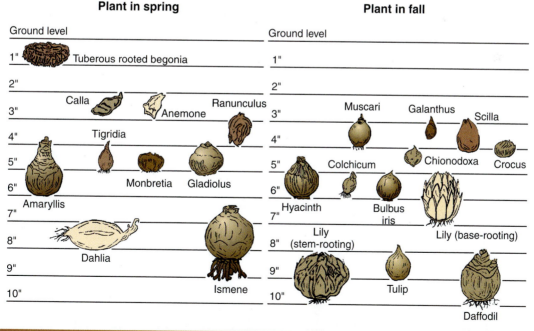

Figure 19-5

Planting depth of bulbs. Courtesy of Brooklyn Botanic Garden.

Fertilizing

Bulbs should be fertilized by adding a small amount of fertilizer (5-10-5) to the bottom of the bed and covering it with soil before planting the bulb. Just before growth starts each spring, bulbs should receive a light application of a complete fertilizer, for example 5 pounds of 5-10-5 for every 100 square feet of bed area.

Care After Flowering

Bulbs flowering in the spring should be dug up after the foliage turns yellow and dies. Some bulbs such as tulips and hyacinths need to be dug every year and replanted for high-quality flowers. Other bulbs such as daffodil, crocus, lily, and colchicum need to be dug up only every three to five years, thinned, and replanted. Digging the bulbs after all the foliage dies and no green remains in the foliage ensures that the bulb completed its growth cycle and all the food is stored in the bulb.

Dusting the bulbs with a pesticide prevents insects and rodents from attacking bulbs during storage. Any bulbs showing signs of disease or damage are removed before storage. Bulbs are stored in peat moss or sawdust.

Bulb Pests and Diseases

Good cultural practices can control many of the pests. This includes such practices as weeding and keeping the bed free of trash so as not to provide a home for insects and disease organisms. Chemical control is also effective and necessary. Growers should seek expert advice before applying chemicals to bulbs. Table 19-6 lists the most common types of bulbs, pests, damage and control measures.

Table 19-6

Common Bulb Pests and Diseases and Their Control		
Host, pest, disease	Damage	Control
Amaryllis		
Spotted cutworm	Feeds on flowers at night.	Scatter cutworm bait or spray with Sevin.
Bulb mites	Rotting bulbs. See *Hyacinth.*	Discard soft bulbs.
Narcissus bulb fly	Decaying bulbs. See *Narcissus.*	Discard soft bulbs.
Leaf scorch, Red blotch	Reddish spots on flowers, leaves, bulb scales; stalks deformed.	Discard bulbs or remove diseased leaves. Avoid heavy watering.
Gladiolus		
Thrips	Leaves silvered, flowers streaked, deformed.	Spray with lindane in spring. Dust corms before storing.
Botrytis and other flower blights	Flowers, leaves, stalks spotted, then blighted.	Spray with zineb (Dithane Z-78 or Parzate).
Corm rots, scab	Lesions on corms, spots on leaves.	Dust with Arasan before planting.
Yellows (due to a soil fungus)	Plants infected through roots, turn yellow and wilt.	Choose resistant varieties.

(Continues)

Table 19-6 (*Continued*)

Host, pest, disease	Damage	Control
Hyacinth		
Bulb mites	Minute; less than 1/25 inch, white mites in rotting bulbs.	Discard infested bulbs.
Aphids, several species	Leaves are curled; virus diseases may be transmitted.	Spray with malathion, rotenone, or nicotine.
Bulb nematode	Dark rings in bulbs.	Discard.
Soft rot	Vile-smelling bacterial disease; often after mites.	Discard.
Iris (bulbous)		
Tulip bulb aphid	See *Tulip.*	See *Tulip.*
Gladiolus, iris thrips	Leaves russeted or flecked, flowers speckled or distorted.	Spray or dust with malathion or lindane.
Leaf spot	Light brown foliage spots with reddish borders.	Spray with zineb or bordeaux mixture; clean up old leaves.
Lily		
Aphids (lily, bean, melon, peach, other species)	Curl leaves, transmit mosaic and other virus diseases.	Spray with malathion, being sure to cover underside of leaves.
Botrytis blight	Oval tan spots on leaves, which turn black, droop.	Spray with bordeaux mixture.
Mosaic and other virus diseases	Plants mottled, stunted.	Rogue infected plants. Start lilies from seed in isolated portion of garden.
Narcissus		
Narcissus bulb fly	Fly resembling bumblebee lays eggs on leaves near ground in early summer. Larva, fat, yellow maggot 1/2 to 3/4 inch long, tunnels in rotting bulb.	Sprinkle naphthalene flakes around plants to prevent egg-laying. Before planting, dust trench with 5% chlordane and dust over bulbs after setting.
Bulb nematode	Dark rings in bulb.	Discard bulbs. Commercial growers treat with hot water, adding formalin to prevent rot.
Basal rot	Chocolate-colored dry rot at base of bulbs.	Inspect bulbs before planting.
Smoulder (botrytis rot)	Rots the foliage and flowers in cold, wet seasons; leaves stick together when they emerge and infected bulbs rot in storage	Remove diseased plants. Put new bulbs in new location. Applications of fungicides such as chlorothalonil, thiophanate-methyl, iprodione, and mancozeb when new growth emerges in the spring.
Scorch	Affected plant parts are often bent or deformed at the point of infection; brown spots or blotches with yellow borders develop.	Minimize moisture on the leaves and flower stalks by careful watering; provide good ventilation and plenty of light; discard heavily infected bulbs.
Tulip		
Tulip bulb aphid	Powdery white or grayish aphids common on stored bulbs.	Dust with 1% lindane before storing.
Green peach, tulip leaf, and other aphids	Transmit viruses to growing plants.	Spray or dust with malathion or lindane.

(Continues)

Table 19-6 (*Continued*)		
Host, pest, disease	Damage	Control
Tulip		
Botrytis blight, fire	Plants stunted, buds blasted, white patches on leaves, dark spots on white petals, white spots on colored petals, gray mold, general blighting. Small, shiny black sclerotia formed on petals, foliage rotting into soil and on bulbs.	Discard all infected bulbs. Plant new tulips in new location. Spray with ferbam or zineb, starting early spring. Remove flowers as they fade, remove all tops as they turn yellow.
Cucumber mosaic	Yellow streaking or flecking of foliage.	Do not grow near cucurbits or gladiolus.
Lily mottle viruses	Cause broken flower colors, mottled foliage, in tulips.	Do not plant near lilies. Control aphids.

Courtesy of the Brooklyn Botanic Garden.

Forcing Bulbs

To enjoy the color and fragrance of flowering bulbs in the winter, bulbs can be forced. **Forcing** bulbs artificially breaks dormancy so they will flower when brought into a warm room. High-quality bulbs can be forced by potting them and storing them at 40° to 50°F for 10 to 12 weeks. Next, the potted bulbs are brought into a cool partially lit room. Bulbs then grow and bloom within 5 weeks after being taken from the cool storage. The process is challenging and provides plants for another market.

Flowering Houseplants

Indoor plants have added color and variety to many places of work and play for hundreds of years. To grow indoor plants successfully there are several factors to consider. These considerations are light, temperature, water/humidity, and general care. Of these factors, light and temperature are the most critical.

Light

Light intensity and duration influence the growth and development of indoor plants. Too little light causes some indoor plants to grow tall and leggy and to lose leaves because the long, thin, weak stems cannot support the plant due to limited photosynthesis.

Five different categories describe the light requirements of indoor plants:

1. Full sun: at least 5 hours of direct sunlight per day

2. Some direct sun: brightly lit but less than 5 hours of direct sunlight

3. Bright indirect light: considerable light but no direct sunlight

4. Partial shade: indirect light of various intensities and durations

5. Shade: poorly lit and away from windows

Temperature

Indoor plants survive best at constant temperatures. For optimum indoor growth, temperature should remain in a range from 60° to 68°F. Temperature in an area also interacts with light humidity and air circulation. Plants have various temperature requirements and should be placed in areas that match their requirements.

Water/Humidity

Water is essential. But water for indoor plants is not as much as many individuals assume. Many indoor plants die from overwatering. Like the requirements for light and temperature, each plant varies in its requirement for water. Without knowing the specific requirements, indoor plants should be watered when they are a little on the dry side.

Indoor plants are watered from the top, by soaking the pot in water, or by placing water in a second container at around the bottom of the pot. The water used should not be too hot or too cold. Besides water, plants need moisture in the air (relative humidity). Finally, good drainage is essential for potted indoor plants.

General Care

Under general care, plants need fertilizing on a regular basis for good health. A balanced fertilizer such as 5-5-5 should be applied at regular intervals. There are two types of fertilizers, slow release and soluble. With slow-release the plant absorbs the fertilizer as needed, which eliminates the possibility of overfeeding or fertilizer burn. The soluble fertilizer must be dissolved in water before applied. Be sure to read manufacturer's directions before applying as these fertilizers may also be concentrates. Do not fertilize when the plant is dormant or very dry. Flowering plants need more fertilizer. Groom a plant by dusting, collecting dead leaves, and loosening dirt. Misting can be done with mild dish washing soap and water to control insects.

After time, the root systems of potted plants become restricted by the pot or other container. This creates a condition called **root-bound**, and the plant must be repotted to stay healthy. In general, repotted plants are moved to a pot no more than 2 inches wider than the original. Flowering plants should be repotted after the flowers have faded. Repotting is also a time to check the health of the root system and remove any damaged or unhealthy roots.

The method chosen for propagation depends on the type of plant— herbaceous or woody, flowering or foliar. Both sexual and asexual methods of propagation are used. Chapters 4 and 5 cover the methods of plant propagation. Common asexual methods used with indoor plants include: leaf and stem cuttings, removal of plantlets, and air layering.

Common Flowering Indoor Plants

A few common indoor flowering plants are African violets, fuchsias, gardenias, geraniums, and impatiens. Some that grow outdoors as well as indoors are wax begonias, ageratums, verbenas, and petunias.

Commercial Greenhouse Environment

Three purposes of a greenhouse include:

1. Provide a controlled environment for plants grown on a large scale.

2. Grow plants in areas where outdoor growth during winter seasons is not possible.

3. Extend the growing season for plants that would normally go dormant (see Figure 19-6).

Figure 19-6

A large commercial greenhouse provides the optimum growing conditions.

Basically, greenhouses are structures used to start and to grow plants year-round. It should be located to receive the maximum amount of sunlight. Further, greenhouses control temperature, moisture, ventilation and climate.

For optimal growth of plants, the temperature in greenhouses is controlled and monitored. Great variations in temperature cause fast vegetative growth or the lack of plant growth.

Plant growth in a greenhouse is also dependent on moisture (humidity). Moisture aids in helping plants maintain their shape and nutrient transport. Watering plants controls growth and helps maintain the humidity. The amount of water needed by plants depends on the type of plants and the conditions outside the greenhouse (see Figure 19-7).

Ventilation or the movement and exchange of air is important for optimum growth. Ventilation helps ensure the proper temperature and humidity in the greenhouse.

When considering a greenhouse, the climate of the area must also be considered. The climate will directly influence the structure, the heating and the cooling systems. Potential greenhouse growers should seek the best advice possible before construction or purchase of a greenhouse.

Greenhouse Features

Greenhouses can be constructed of several different types of materials, including: aluminum, iron, steel, concrete blocks, or wood. Materials used to cover the green house can be glass, soft plastic (polyethylene, vinyl or polyvinyl fluoride), fiberglass, shade fabric, acrylic rigid panels, or polycarbonate rigid panels. Some shapes for greenhouses are detached A-frame truss, quonset-style, and ridge and furrow. There are also **cold frames**, hotbeds, and lath houses. Types of greenhouse structures used for floriculture in recent years include: glass, fiberglass and other rigid plastic, plastic film, shade and temporary cover, and open ground (see Figure 19-8).

Figure 19-8

A greenhouse on open ground.

Summary

1. Flowering plants and foliage plants are grown for personal enjoyment and decoration.

2. These plants may be grown outdoors or indoors, and people grow their own or purchase these plants from nurseries and greenhouses.

3. Perennials have the advantage of regrowing year to year. While perennials do not require planting each year, they still require care.

4. Perennials require winter protection.

5. When gardeners or growers understand the features and characteristics of each type of flowering plant or foliage, they can combine plantings to give the best display of color.

6. Soil preparation, watering, and proper fertilization are necessary for flowering and foliage plants.

7. For the best results from annuals, growers should select hybrid seed. Some annuals do best when thinned after planting.

8. Watering methods need to be checked since some plants become susceptible to disease if sprinklers are used.

9. Bulbs are another way to produce flowering plants. Some bulbs are forced cold storage.

10. While many flowering and foliage plants are grown outdoors, many are produced in greenhouses. Here, the lighting, temperature, and watering can be closely controlled and different types of plants produced on a year-round basis. These are sold for resale and for home use indoors and outdoors.

Something to Think About

1. Name four common indoor flowering plants.
2. List four pests of bulbs.
3. What is the best pH range for flowering perennials?
4. How should bulbs be protected from insects, diseases, and rodents?
5. What is transplanting?
6. How are bulbs forced?
7. Why do people purchase flowering and foliage plants?
8. Describe a greenhouse environment.
9. Identify the features of a starting media.
10. Describe the time from planting to the time of sale for different types of plants.

Suggested Readings

Boodley, J. and S. E. Newman. 2008. The commercial greenhouse, 3rd Ed. Albany, NY: Cengage Delmar Publishers.

Dole, J. M. and J. L. Gibson. 2006. Cutting propagation: A guide to propagating and producing floriculture crops. Batavia, IL: Ball Publishing.

Griner, C. 2000. Floriculture design & merchandising. Albany, NY: Cengage Delmar Publishers.

Hamrick, D. Ed. 2003. Ball RedBook: Crop production (volume 2) 17th Ed. Batavia, IL: Ball Publishing.

Hamrick, D. Ed. 2003. Ball RedBook: Greenhouses & equipment (volume 2) 17th Ed. Batavia, IL: Ball Publishing.

Reader's Digest. 2003. *Illustrated guide to gardening*. Pleasantville, NY: The Reader's Digest Association, Inc.

Internet

Internet sites represent a vast resource of information. The URLs for Web sites can change. Using one of the search engines on the Internet, such as Google, Yahoo!, Ask.com, and MSN Live Search, find more information by searching for these words or phrases: **greenhouse production, floriculture, greenhouse construction, flower bulbs, annual flowers, potted plants, perennial flowers, indoor plants, and houseplants.**

CHAPTER 20

Forage Grasses and Sod

Forage grasses capture the energy of the sun and then they are harvested directly by livestock or by mechanical methods and fed to livestock later. Forage grasses include Kentucky bluegrass, Canada bluegrass, orchard grass, perennial ryegrass, reed canary grass, smooth bromegrass, tall fescue, timothy, prairie grass, garrison grass, and bermudagrass.

Interest in commercial sod production has increased due to increased demand for an instant turf by many building contractors and their customers. Sod production involves growing a solid stand of desirable grass species and then harvesting it intact with a thin layer of soil and roots attached to it. Most sod operators also ship the product to market and many offer custom installation.

After completing this chapter, you should be able to:

- Discuss the cultural practices for growing forage grasses
- Give the common and scientific name for five forage grasses
- Name six insect pests and tell which crops they affect
- Identify four criteria for selecting forage grasses
- List warm-season and cool-season forage grasses
- Explain the importance of seed quality
- Describe the soil preparation for a growing sod
- Identify the best soil types for growing sod
- Discuss the water requirements of sod
- Name four turfgrasses
- List four factors to consider when selecting a grass species
- Describe the fertilizer needs of growing turfgrass
- List the micronutrients that may need to be considered for grass
- Identify three types of mowers used on turfgrass
- Describe the harvesting of sod
- List three factors to consider before installing sod
- Explain the purpose of fumigation
- Name three insect pests of turfgrass
- List three diseases of sod
- Describe nematode damage and control
- Identify six weed problems of sod production

Key Terms

palatability	sod-forming	percent purity
bunch grass	endophyte	sod

Key Terms, *continued*

hardpan
planed
preplant fumigation

muck soils
effluent
soil probe

installation
roll

Forage Grasses

This discussion of forage grasses includes Kentucky bluegrass, Canada bluegrass, orchard grass, ryegrass, reed canary grass, smooth bromegrass, tall fescue, timothy, prairie grass, garrison grass, and bermudagrass. Management of all the forage grasses is similar and are covered in a separate section. Tables 20-1 and 20-2 summarize some of the features of the forage grasses.

Table 20-1

Agronomic Adaptation and Characteristics of Perennial Forage Grasses							
Forage species	Minimum adequate drainage[1]	Tolerance to pH < 6.0	Minimum adequate soil fertility	Drought tolerance	Persistence	Seedling aggressiveness	Growth habit
Kentucky bluegrass	SPD	Medium	Medium	Low	High	Low	Dense sod
Orchard grass	SPD	Medium	Medium	Medium	Medium	High	Bunch
Perennial ryegrass	SPD	Medium	Medium to high	Low	Low	Very high	Bunch
Reed canary grass	VPD	High	Medium to high	High	High	Low	Open sod
Smooth bromegrass	MWD	Medium	High	High	High	Medium	Open sod
Tall fescue	SPD	High	Medium	Medium	High	High	Variable
Timothy	MWD	Medium	Medium	Low	High	Low	Bunch
Garrison grass	VPD	Medium	Medium	High	High	Low	Open sod
Switchgrass	SPD	High	Low to medium	Excellent	High	Very low	Bunch
Big bluestem	MWD	High	Low to medium	Excellent	High	Very low	Bunch
Indiangrass	MWD	High	Low to medium	Excellent	High	Very low	Bunch
Eastern gamagrass	PD	High	Medium to high	Good	High	Very low	Bunch

[1]*Minimum drainage required for acceptable growth: WD = well-drained; MWD = moderately well-drained; SPD = somewhat poorly drained; PD = poorly drained; VPD = very poorly drained.*

Table 20-2

Suitability of Perennial Forage Grass Species to Different Types of Management and Growth Characteristics

Species	Frequent, close grazing	Rotational grazing	Stored feed	Periods of primary production	Relative maturity[4]
Kentucky bluegrass	HS[1]	HS	NR[2]	Early spring and late fall	Early
Orchard grass	NR	HS	HS	Early spring, summer, fall	Early–medium
Perennial ryegrass	NR	HS	HS	Early spring and late fall	Early–medium
Reed canary grass	NR	HS	HS	Early spring, summer, fall	Medium–late
Smooth bromegrass	NR	S[3]	HS	Spring, summer, fall	Medium–late
Tall fescue	NR	HS	HS	Early spring, summer, fall	Medium–late
Timothy	NR	S	NR	Late spring and fall	Late
Garrison grass	NR	S	NR	Spring, early summer	Medium–late
Switchgrass	NR	HS	HS	Summer	Very late
Big bluestem	NR	HS	HS	Summer	Very late
Indiangrass	NR	HS	HS	Summer	Very late
Eastern gamagrass	NR	HS	S	Summer	Very late

1. *Highly suitable.*
2. *Not recommended.*
3. *Suitable.*
4. *Relative time of flower or seedhead appearance in the spring; depends on species and variety. Warm-season grasses mature in midsummer, exact time varies by species.*

Kentucky Bluegrass

Kentucky bluegrass (*Poa pratensis*) is one of the most predominant pasture grasses. It is adapted to well-drained loams and heavier soil types. Bluegrass will be dominant in pastures only if the soil has a salt pH of 5.3 or higher and at least a medium level of phosphorus. Although the grass often becomes dormant during dry weather, it will survive severe droughts. Figure 20-1 shows a Kentucky bluegrass plant.

Kentucky bluegrass is a rhizomatous plant that produces a dense sod. Optimum temperatures for growth range from 60° to 90°F (16° to 32°C). Serious injury will occur if the grass is subjected to continuous soil and air temperatures of 100°F (38°C) and above. Bluegrass is relatively unproductive in midsummer, but yields can be increased and sustained by favorable moisture and nitrogen fertilization.

At comparable growth stages, bluegrass contains more energy per pound of dry matter than smooth bromegrass. It is also extremely palatable during periods of rapid growth. Because of its prostrate growth habit, in early clipping studies it did not compare in yields with taller growing species such as bromegrass, orchard grass, or reed canary grass. However, later studies with grazing animals, similar to those in Figure 20-2, found it to be nearly equal with these grasses in productivity. The carrying capacity of these fields can be increased by weed control, addition of legumes, and fertilization.

Pastures of bluegrass overgrown with weeds generally results from overgrazing and under fertilization. Bluegrass should not be grazed lower than 2 or 3 inches (5 to 8 cm). Overgrazing opens the sod to weeds and reduces the vigor and growth rate of bluegrass. Overgrazing causes very poor root and rhizome development.

Pasture weeds are annuals, biennials, and perennials. Annual species such as ragweed, fleabane, and sunflower originate from seed each year and die after the seed has matured in the summer or fall. Perennials originate from well-established root systems that survive for many years. Seeds of perennials also produce new infestations.

Ironweed, goldenrod, dock, thoroughwort, and vervain are examples of this most difficult group to eradicate.

Repeated mowing or clipping may gradually reduce a stand of weeds. Mowing must be timed properly for good results. The early bud stage, when root reserves are low and weeds are under stress, is most favorable for effective control of biennials or perennials. Weeds mowed at this time have a slow recovery rate. To prevent seed production, annual weeds should be mowed before seed heads mature. The optimum stage for mowing annual and perennial weeds usually does not occur at the same time.

In many cases, spraying with an herbicide is the most effective way to control broadleaf weeds growing in bluegrass pastures. The best time to spray for weed control is when the weeds are growing rapidly and are in the vegetative stage. Using an herbicide and mowing or clipping is an excellent way to control weeds while maintaining lespedeza in a bluegrass sod.

Legumes are often established in existing bluegrass fields to improve production and strengthen the production curve during the summer months. Bluegrass with a legume will produce nearly as much beef as orchard grass, timothy, or fescue with the same legume.

Birdsfoot trefoil, ladino clover, lespedeza, red clover, or alfalfa may be seeded in bluegrass sods. Trefoil is probably the best suited companion crop with bluegrass because of its heavy summer production when bluegrass is semidormant. Trefoil-bluegrass pastures can be expected to produce twice as much as straight bluegrass fertilized with 60 pounds (27 kg) of nitrogen. Production with trefoil will also be much more uniform throughout the growing season than on heavily fertilized straight bluegrass pastures.

To add legumes to existing bluegrass fields, producers till the grass sod during late fall, winter, or very early spring enough to kill approximately one-half the sod. Band-seed the trefoil or other legumes using a high analysis phosphate fertilizer such as 0-45-0.

In many locations, the legume seeding should be done during January or February if it is broadcast. If it is band-seeded with a drill, it may be done as late as March. Seedings made after April 1 must compete with the grass.

Pure bluegrass sods should be fertilized annually with at least a 60-20-20 fertilizer. In fields with medium to high potash levels, the potash may be omitted. If legumes are to be maintained in a stand, the nitrogen should be omitted and a higher level of potash and phosphate used. For example, a 0-30-60 fertilizer is well suited to bluegrass-lespedeza or trefoil pastures.

Although Kentucky bluegrass usually appears spontaneously, it is occasionally seeded as a pasture grass. When bluegrass is seeded for this purpose, common or commercial seed is used. It is usually sown in a mixture with other grass and legume seeds. Usually 2 to 4 pounds (1 to 1.8 kg) per acre of bluegrass seed are added to a normal forage mixture. When seeded alone, use 10 pounds (4.5 kg) per acre of certified quality seed. Seeding in the fall before October 15 is best.

Canada Bluegrass

Canada bluegrass (*Poa compressa*) is also found in many Kentucky bluegrass fields in Missouri. This is usually a signal that the field is lacking in lime or phosphate or both because Canada bluegrass is more tolerant of acid, low phosphate soils than Kentucky bluegrass.

Canada bluegrass has a bluer foliage than Kentucky bluegrass, matures later, is less productive and makes very slow or little recovery after it is grazed. It is an inferior species, and fields infested with it should have a soil test taken and the soil nutrient deficiencies corrected.

Orchard Grass

Orchard grass is a versatile perennial bunch-type grass (no rhizomes) that establishes rapidly and is suitable for hay, silage, or pasture. It has rapid regrowth, produces well under intensive cutting or grazing, and attains more summer growth than most of the other cool-season grasses. Orchard grass tolerates drought better than several other grasses (refer back to Table 20-1). Orchard grass grows best in deep, well-drained, loamy soils. Its flooding tolerance is fair in the summer but poor in the winter. Orchard grass is especially well suited for mixtures with tall legumes, such as alfalfa and red clover. The rapid decline in **palatability** and quality with maturity is a limitation with this grass. Timely harvest management is essential for obtaining good quality forage.

The botanical name for orchard grass is *Dactylis glomerata*, and Figure 20-3 shows the orchard grass plant. Improved varieties of orchard grass, with high yield potential and improved resistance to leaf diseases, are available. When seeding orchard grass-legume mixtures, select varieties that mature at about the same time. The later-maturing varieties are best suited for growing with alfalfa because they match the maturity development of alfalfa and are easier to manage for timely harvest to obtain good quality forage.

Figure 20-3

Orchard grass is known for its rapid early spring growth.

Perennial Ryegrass

Perennial ryegrass is a **bunch grass** suitable for hay, silage, or pasture. Perennial ryegrass produces excellent quality and palatable forage, is a vigorous establisher, has a long growing season, and is high yielding under good fertility when moisture is not lacking (see Tables 20-1 and 20-2). Because it is less winter hardy than other grasses, perennial ryegrass is best seeded in combination with other grasses and legumes. Perennial ryegrass can be grown on occasionally wet soils. It is less competitive with legumes than orchard grass, and is usually later to mature than orchard grass. Like orchard grass, perennial ryegrass can withstand frequent cutting or grazing. It is more difficult to cut with a sickle bar mower and is slower to dry than other grasses.

The botanical name for perennial ryegrass is *Lolium perenne*. Figure 20-4 illustrates a ryegrass plant.

Figure 20-4

Perennial ryegrass.

Reed Canary Grass

Reed canary grass (*Phalaris arundinacea*), as shown in Figure 20-5, is a tall, leafy, coarse, high-yielding perennial grass, tolerant of a wide range of soil and climatic conditions (see Table 20-1). It can be used for hay, silage, and pasture. It has a reputation for poor palatability and low forage quality. This reputation was warranted in the past, but new varieties are available that make this forage an acceptable animal feed, even for lactating dairy cows. Reed canary grass grows well in very poorly drained soils, but is also productive on well-drained upland soils. It is winter hardy, drought tolerant (deep-rooted), resistant to leaf diseases, persistent, responds to high fertility, and tolerates spring flooding, low pH, and frequent cutting or grazing. Reed canary grass forms a dense sod. Limitations of this grass include: slow establishment, expensive seed, and forage quality declines rapidly after heading.

Only low-alkaloid varieties (Palaton, Venture, Rival) are recommended if the crop is to be used as an animal feed. These varieties are palatable and equal in quality to other cool-season grasses when harvested at similar stages of maturity. Common reed canary grass seed should be considered to contain high levels of alkaloids, and is undesirable for animal feed (see Table 20-2).

Smooth Bromegrass

Smooth bromegrass is a leafy, **sod-forming** perennial grass best suited for hay, silage, and early spring pasture. It spreads by underground

Figure 20-5

Reed canary grass grows in wet areas too wet for the production of other grasses.

rhizomes and through seed dispersal (see Tables 20-1 and 20-2). Smooth bromegrass is best adapted to well-drained silt-loam or clay-loam soils. It is a good companion with cool-season legumes. Smooth bromegrass matures somewhat later than orchard grass in the spring and makes less summer growth than orchard grass. It is very winter hardy and, because of its deep root system, will survive periods of drought. Smooth bromegrass produces excellent quality forage, especially if harvested in the early heading stage. It is adversely affected by cutting or grazing when the stems are elongating rapidly (jointing stage). It should be harvested for hay in the early heading stage for best recovery growth.

The botanical name for smooth bromegrass is *Bromus inermis* and Figure 20-6 shows the plant. Improved high-yielding and persistent varieties are available. Some varieties are more resistant to brown leaf spot, which may occur on smooth bromegrass. These improved varieties start growing earlier in the spring and stay green longer than common bromegrass, which has uncertain genetic makeup.

Tall Fescue

Tall fescue is a deep-rooted, long-lived, sod-forming grass that spreads by short rhizomes. It is suitable for hay, silage, or pasture for beef cattle and sheep (see Tables 20-1 and 20-2). Tall fescue is the best cool-season grass for stockpiled pasture or field-stored hay for winter feeding. It is widely adapted, and persists on acidic, wet soils of shale

Figure 20-6

Bromegrass is grown throughout the northern half of the United States.

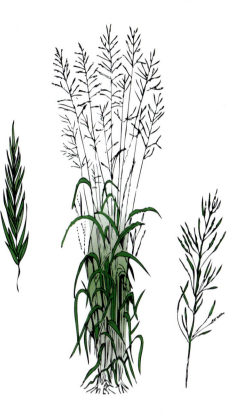

origin. Tall fescue is drought resistant and survives under low fertility conditions and abusive management. It is ideal for waterways, ditch and pond banks, farm lots, and lanes. It is the best grass for areas of heavy livestock and machinery traffic. Figure 20-7 illustrates a tall fescue plant.

Many older tall fescue pastures contain a fungus (**endophyte**) growing in the plant. The fungal endophyte is associated with poor palatability in the summer and poor animal performance. Several health problems may develop in animals grazing endophyte-infected tall fescue, especially breeding animals. Deep-rooted legumes should be included with tall fescue if it is to be used in the summer. Legumes improve animal performance, increase forage production during the summer, and dilute the toxic effect of the endophyte when it is present.

The botanical name for tall fescue is *Festuca arundinacea*. Newer endophyte-free varieties or varieties with very low endophyte levels (less than 5%) are recommended if stands are to be used for animal feed. Kentucky-31 is the most widely grown variety, but most seed sources of this variety are highly infected with the endophyte fungus, and should not be planted. When buying seed, growers should make sure the tag states that the seed is endophyte-free or has a very low percentage of infected seed. Because endophyte-free varieties are less stress tolerant than endophyte-infected varieties, they should be managed more carefully.

Figure 20-7

Fescue is a widely adapted grass, but palatability is a problem causing poor animal performance unless endophyte-free varieties are used.

Timothy

Timothy (*Phleum pratense*) is a hardy perennial bunchgrass that grows best in cool climates. Its shallow root system makes it unsuitable for droughty soils. It produces most of its annual yield in the first crop. Summer regrowth is often limited because of timothy's intolerance to hot and dry conditions. Timothy is used primarily for hay and is especially popular for horses. It requires fairly well-drained soils (see Tables 20-1 and 20-2). Timothy is less competitive with legumes than most other cool-season grasses and is adversely affected by harvesting or grazing during the jointing stage (stem elongation). Because it is easily weakened by frequent cutting, a sufficient recovery period is necessary for accumulation of energy reserves for regrowth. Figure 20-8 shows the timothy plant.

Prairie Grass

Prairie grass ('Matua') is a tall-growing perennial bunchgrass introduced from New Zealand and the only variety of prairie grass currently sold in the United States. It is adapted to well-drained soils with medium to high fertility and pH of 6 or higher. It is a type of bromegrass, but differs from smooth bromegrass in that it does not spread by rhizomes, and it produces seedheads throughout the growing season. This grass is

Figure 20-8

Timothy grass grows best when temperatures are between 65° and 72°F.

not as winter hardy as the other cool-season grasses. It winterkills easily in colder states.

Although its forage quality compares well with that of other cool-season grasses, prairie grass may contain lower levels of trace elements. Prairie grass grows well during the fall and can be managed to provide early spring grazing. It should be harvested or grazed on a monthly schedule during the fall. Stockpiling growth dramatically reduces winter survival of this grass.

Garrison Grass

Garrison grass, also called creeping meadow foxtail, is a vigorous, sod-forming, cool-season grass. It is especially well suited for low meadows and pastures, and survives extended periods of standing or running water. Garrison grass is adapted to a broad range of soils, but sufficient moisture is required for good growth. It is very cold hardy, and has good drought tolerance. Specially designed grassland drills are the best method of seeding this grass because the seed is light and fuzzy. A cultipacker seeder can be used in the fully opened position.

Perennial Warm-Season Grasses

The native, perennial, warm-season grasses have potential to produce good hay and pasture growth during the warm and dry mid-summer months. These grasses initiate growth in late April or early May, and produce 65% to 75% of their growth from mid-June to mid-August. Warm-season grasses produce well on soils with low moisture holding capacity, low pH, and low phosphorous levels. Warm-season grasses complement cool-season grasses by providing forage when the cool-season grasses are less productive. The management requirements for warm-season grasses are different than those normally followed for cool-season grasses particularly in establishment procedure, fertilization, and grazing management.

Switchgrass, Indian grass, and big bluestem are winter hardy and grow in many areas. They do best on deep, fertile, well-drained soils with good water-holding capacity. Gamagrass and Caucasian bluestem also have potential in some areas (see Tables 20-1 and 20-2). Detailed information on these species and their management is available at County Extension offices.

Bermudagrass, (*Cynodon dactylon*) is grown extensively throughout all the tropical and subtropical areas of the world and in the southern areas of the United States. Bermudagrass is an aggressive plant that spreads rapidly by seeds, stolons, and rhizomes. It is used for hay crops, pasture, and lawns and turf in warmer areas. Bermudagrass grows best on fertile clay soils in full sun.

Improved hybrid cultivars are more productive, more frost tolerant, and more drought resistant than common bermudagrass. Figure 20-9 shows the bermudagrass plant.

Managing Forages

All forage crops respond dramatically to good management practices. Higher yields and improved persistence result from paying attention to the basics of good forage management as outlined next.

Seeding Year Management

Establishing a good stand is critical for profitable forage production. Begin by selecting species adapted to soils where they will be grown. Crop rotation is an important management tool for improving forage productivity. Crop rotation reduces disease and insect problems. Disease pathogens accumulate and can cause stand establishment failures when seeding into a field that was not rotated out of a forage for at least one year. Rotating to another crop for at least a year is the preferred method.

Growers ensure fields are free of any herbicide carryover that can harm forage seedlings. Current labels on herbicides provide more information on crop rotation restrictions. Many producers are successfully using reduced tillage and no-tillage seeding methods (see Figure 20-10).

Fertilization and Liming

Proper soil pH and fertility are essential for optimum economic forage production. Take a soil test to determine soil pH and nutrient status at least six months before seeding. This allows time to correct

Figure 20-10

Fine grass and legume seeds must be slightly covered and left in a compact seedbed. The cultipacker presses the seeds into the seedbed.

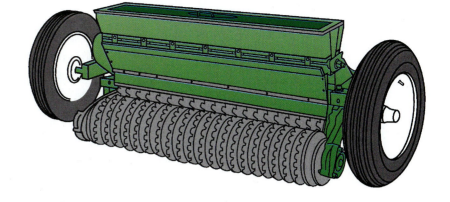

deficiencies in the topsoil zone. The topsoil in fields with acidic subsoils should be maintained at higher levels of pH than fields with neutral or alkaline subsoils. Producers maintain topsoil zone pH at the following levels:

- pH 6.8 for forage legumes on mineral soils with subsoil pH less than 6.0
- pH 6.5 for forage grasses on mineral soils with subsoil pH less than 6.0
- pH 6.5 for alfalfa on mineral soils with subsoil pH greater than 6.0
- pH 6.0 for other forages on mineral soils with subsoil pH greater than 6.0
- pH 6.0 for Kentucky bluegrass pastures

Soil pH should be corrected by application of lime when topsoil pH falls 0.2 to 0.3 pH units below the recommended levels. With conventional tillage plantings, lime should be incorporated and mixed well in the plow layer at least six months before seeding. If more than 4 tons per acre are required, half the amount should be plowed down and the other half incorporated lightly into the top 2 inches. If low rates of lime are recommended or if a split application is not possible, the lime should be worked into the surface rather than plowed down. This assures a proper pH in the surface soil where seedling roots develop and where nodulation begins in legumes.

Corrective applications of phosphorus and potassium should be applied prior to seeding regardless of the seeding method used. Fertilizer recommendations for forage legumes and tall grasses are found in Extension bulletins and through soil tests. Starter applications of 0-40-40 may be applied at seeding. Seedling vigor of cool-season forage grasses is enhanced on many soils by nitrogen applied at seeding time. Growers may apply 10 to 20 pounds (4.5 to 9.0 kg) of nitrogen per acre when seeding grass-legume mixtures, and 30 lbs/A (14 kg/A) when seeding pure grass stands. Starter nitrogen applications of 10 lbs/A (4.5 kg/A) may be beneficial with pure legume seedings, especially under cool conditions and on soils low in nitrogen. Manure applications incorporated ahead of seeding may also be beneficial.

Seed Quality

High-quality seed of adapted species and varieties should be used. Seed lots should be free of weed seed and other crop seed, and contain only minimal amounts of inert matter. Certified seed is the best assurance of securing high-quality seed of the variety of choice, as shown on the label in Figure 20-11. Purchased seed accounts for only 20% or less of the total cost of stand establishment. Buying cheap seed and seed of older varieties is false economy. The producer should always compare seed price on the basis of cost per pound of pure live seed, calculated as follows:

- **Percent purity** = 100 percent − (percentage inert matter + percentage other crop seed + percentage weed seed)
- Percent pure live seed (PLS) = percentage germination × percentage purity
- Pounds of PLS = pounds of bulk seed × percentage PLS

Figure 20-11

A grass seed label gives important information.

```
MEADOW PASTURE MIX
LOT #J574  NET 50#
PURITY  GERM
22.00   80    BROME
21.00   88    ORCHARD
09.75   90    TIMOTHY
10.25   88    FESCUE
18.50   90    PERENNIAL RYE
06.00   90    BLUE GRASS
05.95   90    CLOVER

3/96    TESTED
0.75    CROP
5.05    INERT
0.75    WEED

GLOBE SEED AND FEED CO., INC.
TWIN FALLS, ID 83301
```

Weed Management in Forages

The best way to minimize weed problems in forages is to establish and maintain a vigorous stand of the forage crop. Most of the weed control in healthy forage crops is from competition provided by the forage crop. Growers fertilize properly, follow good cutting management practices, use disease-resistant varieties, control problem insects, and use herbicides selectively and judiciously to keep the forage stand productive and competitive with weeds. Weeds reduce forage yields, but more importantly they often reduce forage quality and palatability. Because the feed value of weeds differs among species, the relative importance of controlling them varies by species. Perennial and biennial weeds should be controlled with herbicides in the previous crop or prior to planting a forage crop.

Herbicides are usually most needed during establishment, especially in pure legume seedings. Grassy weeds are usually more competitive against legumes than broadleaf weeds. To ensure good stand establishment, growers should provide good weed control during the first 60 days after planting. Preplant or postemergence herbicides are available. With a postemergence herbicide program, the grower can take the "wait and see" approach to weed control, and base herbicide use on need. Because timing of application is critical with a postemergence program, growers should check the labels for application details and restrictions.

Late summer seedings are usually less subject to serious infestations of annual grasses and most summer annual broadleaf weeds. Preplant herbicides are not recommended. Winter annual weeds may germinate in late summer seedings, and severe infestations can cause stand establishment problems. Herbicides are available for control of these weeds, and fall applications are more effective than spring applications.

No-Till and Minimum-Till Seeding

Many producers are successfully adopting minimum and no-tillage practices for establishing forage crops. Advantages include soil conservation, reduced moisture losses, lower fuel and labor requirements, and seeding on a firm seedbed. All forage species can be seeded no-till. No-till forage seedings are most successful on silt loam soils with good drainage. Consistent results are more difficult on clay soils or poorly drained soils. Weed control and/or sod suppression is absolutely critical for successful no-till forage establishment, because most forage crops are not competitive in the seedling stage.

Fertilization

Growers soil test well ahead of seeding to determine lime and fertilizer needs. For no-till seedings, growers should take soil samples to a 2-inch depth to determine pH and lime needs, and to a normal 8-inch depth to determine phosphorus and potassium needs. If possible, producers should make corrective applications of lime, phosphorus, and potassium earlier in the crop rotation when tillage can be used to incorporate and thoroughly mix these nutrients throughout the soil. When this is not feasible, lime, phosphorus, and potassium applications should be made at least eight months or more ahead of seeding to obtain the desired soil test levels in the upper rooting zone.

Lime and phosphorus move slowly through the soil profile. Once soil pH, phosphorus, and potassium are at optimum levels, surface applications of lime and fertilizers maintain these levels. Attempts to establish productive forages often fail where pH, phosphorus, or potassium soil test values are below recommended levels, even when corrective applications of these nutrients are surface applied or partially incorporated just before seeding.

Seeding Equipment and Methods

Because most forage crops have small seeds, careful attention must be given to seed placement. A relatively level seedbed improves seed placement. A light disking may be necessary before attempting to seed. Seeds are planted shallow [1/4 to 1/2 inch (6 mm to 12.7 mm), in most cases], in firm contact with the soil (see Figure 20-12).

Crop residue must be managed to obtain good seed-soil contact. Chisel plowing or disking usually chops residue finely enough for conventional drills to be effective. When residue levels are greater than 35%, no-till drills are recommended.

Establishment

When following corn, growers should plant as soon as the soil surface is dry enough for good soil flow around the drill openers and good closure of the furrow. Perennial weeds should be eradicated in the previous corn crop. If perennial weeds are still present, growers should apply Roundup before seeding. If any grassy weeds or winter annual broadleaf weeds are present in the field, growers can use Gramoxone Extra or Roundup before seeding.

Most drills can handle corn grain residue, but removal of some of the residue (e.g., for bedding) often increases the uniformity of stand establishment. Most drills do not perform as well when corn stalks are chopped and left on the soil surface.

No-till seeding in August following small grain harvest permits the producer to optimize small grain and straw yields without also being concerned with trying to establish an interseeded forage crop in the spring. No-till seeding of the forage crop helps conserve valuable moisture.

Weeds should be effectively controlled in the small grain crop, and herbicides applied at the proper stage of small grain development. Growers should remove straw after small grain harvest. It is not necessary to clip and remove stubble. It may be removed if additional straw is desired. Stubble should not be clipped and left in the field, as it may interfere with good seed-soil contact when seeding forages.

Insect Control

Insects can be a serious problem in no-till seedings, especially those seeded into old sods. Slugs can be especially troublesome where excessive residue is present from heavy rates of manure applied in previous years. Chemical control measures for slugs are limited to a metaldehyde bait.

The Sod Farm

Ideally, a site to be chosen for a **sod** farm should be based on several criteria: location (distance) in relation to targeted market, accessibility to major roads and highways, available water quantity and quality, soil type, land costs, and preparation requirements.

In order to reduce shipping costs and because sod is a perishable product, a sod farm should be as near to an urban area as is practical. Sod that is stacked on pallets should be unstacked and laid within 72 hours after harvest, preferably within 24 hours. This is especially critical during summer months. Refrigerated trucks have been used to prevent sod deterioration when high-quality sod is transported over long distances. Sod on pallets waiting to be loaded or unstacked should be kept as cool as possible. Placing pallets in a shaded environment such as under trees or under shade cloth prolongs the sod's life.

Production

Production practices are divided into several areas: establishment, primary cultural practices, pest management, and harvesting. Establishment involves land preparation, soil improvement, irrigation installation, and turf planting.

Land Preparation and Establishment

Prior to planting, the new turfgrass site should be prepared to correct any present problems and to avoid harvesting difficulties. Preparation includes land clearing, removal of trash, land leveling, tilling, installation of drainage and irrigation systems, roadway and building site selection, soil fumigation, and land rolling. The cutter blade on the sod harvester rides on a roller, allowing the unit to bridge the little hills, valleys, and holes in the field. If the surface irregularities left by poor soil preparation are too severe, the blade will not uniformly cut the sod. Then the yield will be reduced. Proper soil preparation also eliminates layers or hard pans, provides better air and water movement, and enhances deep rooting.

Growers soil test the area under consideration to determine lime and fertilizer nutrient requirements. Apply and incorporate these amendments prior to turf establishment.

Where **hardpan** is a problem, the land is subsoiled to break up any hardpans and then plowed with either a moldboard or chisel plow to a depth of 10 inches. This practice of breaking the subsurface hardpan should not be followed if subsurface irrigation is being used. Subsoiling is followed with soil incorporation of preplant fertilizer or liming material. The seedbed is firmed with a cultipacker roller. The surface must be as smooth and uniform as possible so maintenance and harvesting problems are minimized. After cultipacking, the use of a laser plane for land leveling is suggested. The field should be **planed** (leveled or smoothed) in several directions to eliminate as many surface irregular spots as possible. After planing, dry soil is considered too fluffy if footprints are more than 1 inch deep. In this case, the field should be firmed by rolling it.

Preplant fumigation is strongly recommended where previous weed, disease, and nematode problems existed. Major weeds in sod production include common bermudagrass, nutsedge, torpedograss, sprangletop, and crabgrass.

Sod is grown in several general soil groups. These include clay, sands, and muck soils. The agricultural suitability of these soils is determined by their ratio of sand, silt, clay, and organic matter fractions.

Clay soils do not drain well and stay wet for extended periods. Precious harvest days may be lost due to the wet ground. Also, due to these soils holding so much water and their high bulk densities, clay soils are heavy to haul. Loam soils, in general, have good moisture-holding capacity, drain well, are easy to work, and are relatively light in weight for transport. These contain approximately 40% sand, 40% silt, and 20% clay. Next to muck soils, loam soils are most desirable as growing media. Ideally, these soils should have at least 5% organic matter and 15% or less clay. Sandy loams are desirable because of good drainage. Traffic and harvest operations may be performed sooner after water application.

Muck soils are found in old bogs, river deltas, and lake beds. They contain high organic matter and have good water holding capacity. Nitrogen is also readily available through mineralization of organic matter. Muck soils are typically low in potassium, phosphorous, and various minor elements. Length of sod production on muck soil is usually shorter and production costs are less. Muck soils have less bulk density versus sandy or clay soils, so they weigh less on a unit basis and are cheaper to transport. Muck soils are the most desirable for sod production.

Sod production is not recommended for deep, pure sandy soil, for example sand-dune-type sand, due to the difficulty of maintaining adequate soil moisture and nutrient levels. Furthermore, such soils typically have high levels of nematodes, which adversely affect soil quality and handling.

Proper soil water management is an important key to successful (and profitable) sod production. Poorly drained fields are unsuitable for competitive sod production. These fields often remain saturated, thus unworkable, for extended periods following substantial rainfall. Fields that are poorly drained need to be designed so that individual beds are crowned before planting. Lateral drain lines or ditches also need to be installed to intercept this surface drainage and to lower the water table to manageable levels.

Irrigation

Irrigation is required for quality sod production. Ample water of good quality should be a priority during the planning stage. Water sources include wells, sinkholes, ponds, streams, and canals, as well as **effluent** sources from nearby municipalities and industrial sites. Effluent or grey water can be an excellent and inexpensive source of irrigation. These water sources may fluctuate widely in pH, salt, and nutrient levels. Many municipalities also require a contract stating that the grower must accept a certain number of gallons per given time whether irrigation is needed or not. These are issues that should be addressed early in the planning stage if effluent water is to be used. Irrigation systems normally involve center pivots, lateral pivots, walking or traveling guns, or subirrigation (raised water tables).

A **soil probe** is a very useful tool in irrigation management. The depth the soil is dry or wet can easily be measured with this and irrigation scheduling adjusted accordingly. Tensiometers are soil moisture sensing devices that measure the suction created by drying soil. If used correctly, the data gathered from these instruments' gauges can be used to determine irrigation scheduling. After the grass is planted, irrigation becomes the most important single factor for successful stolon establishment, as shown in Figure 20-13. It is critical not to plant more area than can easily be irrigated at one time.

Figure 20-13

Deep watering promotes deep, healthy root growth. Shallow watering promotes shallow rooting and leaves the grass susceptible to injury by drought.

Deep watering

Shallow watering

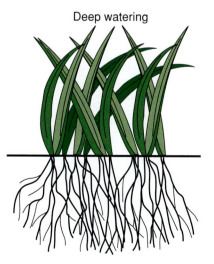
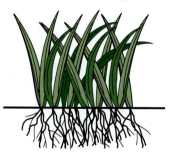

Turf and Selection and Planting

Determining which turfgrass is best for a particular situation is based on several factors. Table 20-3 lists the major warm- and cool-season grasses. In sandy soil, a deep-rooted grass is necessary. If properly maintained, bahiagrass and St. Augustinegrass provide deep rooting and increased drought resistance. If the purchaser is willing to allot more time, energy, and economic resources to turf maintenance, a finer-textured species is suggested. Included is one of the bermudagrass or zoysiagrass cultivars. In addition, centipedegrass is available for those regions with heavier, acidic soils and for those persons with less resources and time available for upkeep. Other considerations for selecting a grass species include insect and disease resistance, nematode susceptibility, seedhead/shoot growth rate, and frost and shade tolerance (see Figure 20-14).

Table 20-3

Warm and Cool Season Turfgrasses in the United States	
Cool season	**Warm season**
Colonial bentgrass (*Agrostis tenuis*)	Bermudagrass (*Cynodon dactylon* L)
Creeping bentgrass (*Agrostis palustris*)	Zoysiagrass (*Zoysia japonica* Steud)
Kentucky bluegrass (*Poa pratensis*)	Buffalograss (*Buchloe dactyloides*)
Tall fescue (*Festuca arundinacea*)	St. Augustinegrass (*Stenotaphrum secundatum*)
Fine fescue (various *Festuca* species)	
Red fescue (*Festuca rubra* L)	
Perennial ryegrass (*Lolium perenne*)	
Crested wheatgrass (*Agropyron cristatum*)	

NATIONAL TURFGRASS EVALUATION PROGRAM

Various turfgrass research programs contribute to the development of a nationwide database of unbiased information on cultivar performance. These programs participate in the National Turfgrass Evaluation Program (NTEP), a not-for-profit cooperative effort of the United States Department of Agriculture (USDA) and the Turfgrass Federation, Inc. Varieties of turf are evaluated for spring green-up, density, drought tolerance, disease or weed activity, color, and overall quality. Some of the trials include:

- Tall fescues
- Zoysiagrasses
- Bermudagrasses
- Buffalograsses

NTEP links the private and public sectors of the industry through the common goals of improving grasses, developing new cultivars, and establishing uniform evaluation standards. NTEP trials, which usually last five years, are replicated at many locations throughout the country and include most of the available varieties of each of the subject species.

Results from research facilities nationwide are collected, analyzed, and disseminated by NTEP annually. Seed companies and plant breeders use the NTEP information to determine grass adaptation and quality ratings.

Grasses in the trials are mowed weekly during the growing season, fertilized, and irrigated according to proper management for the area. No secondary management practices are used. Quality ratings are taken monthly and reported to NTEP yearly.

Other turfgrass research areas include the following:

- Resource efficiency of water, nutrition, pest management, energy, and labor
- Input in such areas as lawns, golf courses, parks, and grounds
- Environmental enhancement of urban and suburban areas using turfgrass
- Turfgrass persistence and performance with increased traffic and use on sports facilities

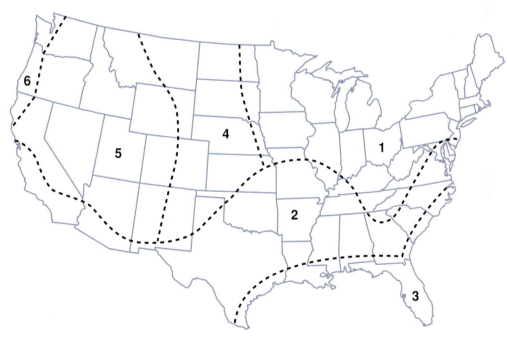

Climatic regions in which the following grasses are suitable for lawns:

1. Kentucky bluegrass, red fescue, and colonial bentgrass. Tall fescue, Bermuda, and zoysia grasses in the southern part.

2. Bermuda and zoysia grasses. Centipede, carpet, and saint augustine grasses in the southern part; tall fescue and Kentucky bluegrass in some northern areas.

3. Saint augustine, Bermuda, zoysia, carpet, and Bahia grasses.

4. Nonirrigated areas: crested wheat, buffalo, and blue grama grasses. Irrigated areas: Kentucky bluegrass and red fescue.

5. Nonirrigated areas: crested wheatgrass. Irrigated areas: Kentucky bluegrass and red fescue.

6. Colonial bentgrass, Kentucky bluegrass, and red fescue.

Figure 20-14

Regions of grass adaptations.

Fertilization

Proper fertilizing for sod production normally reflects the need for grass regrowth following establishment or cutting of the prior crop. Nitrogen is the most important nutrient regulating this regrowth. Generally, higher rates and frequencies of nitrogen application reduce the production time for a crop. Excessive nitrogen rates force excessive topgrowth at the expense of the roots, thus, reducing the "liftability" of the sod. Economics also dictate, to an extent, the amount and frequency of nitrogen use.

A balance needs to be maintained between all major and minor elements since the unavailability of any nutrient may weaken or delay the production process. Sod managers should test all fields before planting and yearly thereafter to regulate pH and nutrient levels and needs of the particular grass being grown.

Soils in some areas naturally provide adequate phosphorous and soil pH levels. Growers apply phosphorous and liming material (if necessary) prior to planting. The optimum soil pH for St. Augustinegrass, bermudagrass, and zoysiagrass is approximately 6.0 to 6.5. Centipedegrass and bahiagrass have an optimum soil pH of 5.0 to 5.5.

Following the first mowing, growers apply additional fertilizer. Once the sod has covered, fertilizer scheduling is largely dictated by economics. Obviously, if sod orders are strong, the grass needs to be aggressively fertilized to minimize production time. If sales are slow, sod should be fertilized less to save on fertilizer and maintenance costs such as mowing and watering.

Bermudagrass and zoysiagrass respond exceptionally well to ample fertilization. Quickest turnaround of these grasses occurs with monthly nitrogen application at the equivalent of 50 lbs N/acre per application. This schedule should continue unless cold weather halts growth or economics dictate otherwise. A 2:1 or 1:1 ratio of nitrogen to potassium fertilizer should be used with each application to encourage strong rooting. Phosphorus should be applied as suggested by a yearly soil test, see the fertilizer recommendations in Table 20-4.

Table 20-4

	Fertility Requirements and Some Recommended Fertilizers for Bermudagrass and Zoysiagrass	
Time to apply	**Nitrogen applied per 1,000 square feet**	**Acceptable fertilizer per 1,000 square feet of area**
April or May	1 lb.	8 lbs. 12-4-8 **or** 10 lbs. 10-6-4 **or** 5 lbs. 20-10-10
June	1 lb.	8 lbs. 12-4-8 **or** 10 lbs. 10-6-4 **or** 5 lbs. 20-10-10
July	1 lb.	8 lbs. 12-4-8 **or** 10 lbs. 10-6-4 **or** 5 lbs. 20-10-10
August	1 lb.	8 lbs. 12-4-8 **or** 10 lbs. 10-6-4 **or** 5 lbs. 20-10-10

St. Augustinegrass is normally fertilized every six to eight weeks during the growing season. As with bermudagrass and zoysiagrass,

St. Augustinegrass should be fertilized with a 2:1 or 1:1 nitrogen to potassium ratio fertilizer and phosphorus added as suggested by a yearly soil test. If overfertilized in summer with quickly available nitrogen sources, St. Augustinegrass becomes more susceptible to chinch bug infestation and grey leaf-spot disease. These problems can be minimized by using slow- (or controlled) release nitrogen sources and supplemental iron applications. These are discussed later.

Bahiagrass and centipedegrass are fertilized less than the other sod-grown grasses. Bahiagrass is fertilized yearly with 100 to 200 pounds of total nitrogen per acre. Again, economics and desired sod turnaround time dictate which rate range is used. Two applications per year are made if 100 pounds of nitrogen are used, with these being equally divided between early spring (April–May) and summer (July–August).

Centipedegrass has a very specific fertilization schedule. If over-fertilized long-term with nitrogen, centipedegrass will develop thatch, decreased winter survival and reduced rooting. The end result, referred to as "centipedegrass decline," is characterized as death or extremely weak spots roughly 2 to 20 feet in diameter that develop as the grass resumes growth in spring. Normally, centipedegrass decline does not develop until several years after establishment. So, sod managers should fertilize centipedegrass similarly to St. Augustinegrass for one year after establishment.

If the grass is not harvested within 18 months after establishment, then the fertility rate needs to be reduced to minimize the occurrence of centipedegrass decline. Established centipedegrass should be fertilized only two to three times yearly with 23 to 45 pounds of actual nitrogen per acre. In some areas, additional potassium should be considered in early fall to encourage proper rooting prior to winter. Supplemental iron or manganese application may be needed if unacceptable leaf chlorosis forms.

Some soils are low in micronutrients. If recommended by soil testing, at least two applications of micronutrients are suggested per year, but more may be required. Several iron products are used. The least expensive and most commonly used source is ferrous sulfate. Ferrous sulfate contains 21% iron and is quick-acting, but color enhancement lasts only three to four weeks. Chelated iron products are more expensive but have been formulated to hold their greening effect for a longer period of time.

Iron should be sprayed on most turfgrasses to enhance color, especially near harvesting time. These are often injected into the irrigation system but may also be applied in a dry or spray solution form. Application of 20 to 40 pounds of elemental iron (e.g., 100 to 200 lbs of ferrous sulfate) may be timed approximately one to two weeks prior to harvesting to enhance color. To prevent burn, irrigations must be applied immediately after iron application during periods of high temperature.

Liquid fertilizers are often used by injecting them into the irrigation system. Ammonium nitrate is the primary nitrogen source used for this. The major problems with using fertilizer in irrigation systems involve difficulties in maintaining uniform distribution and concerns with possible fertilizer leaching.

Mowing

After irrigation as the first priority, mowing is perhaps the second most important turfgrass cultural practice for sod producers. Mowing helps control turfgrass growth and many undesirable weeds that are intolerant to close mowing. Sod fields require a mowing schedule similar to a well-maintained home lawn.

Three basic mower types include reel, rotary, and flail. A reel mower is most desirable because highest possible mowing quality is achieved due to a cleaner cut. Rollers on a reel-type mower also help smooth the sod field for easier, more uniform harvesting. Reel mowers should always be used the last four or five mowings before harvest. This produces the finest cut available, and maximizes sod quality. Rotary mowers are acceptable for St. Augustinegrass, centipedegrass, and bahiagrass production if blades are properly sharpened and balanced. Flail mowers are widely used in bahiagrass production until sod has a uniform dense stand, and then growers switch to a reel or rotary mower.

Regardless of the mower type, the blades should be well maintained and sharpened. Dull blades reduce turf quality by leaving grass tips shredded and bruised. Shredded tips dry easily, leaving brown tissue that grows slowly, especially in hot weather. Mowers are big, heavy pieces of equipment. Ruts, which cause harvest losses, may develop if these machines are used when soils are too wet.

New sod fields are generally mowed once every one to two weeks until complete coverage is obtained, depending on grass growth and weed encroachment. Mowing frequency will vary for established sod, depending on the fertility level, season of year, species, and seedhead production. Figure 20-15 shows the mowing height.

Figure 20-15

No more than one-third of the top growth should be removed per mowing.

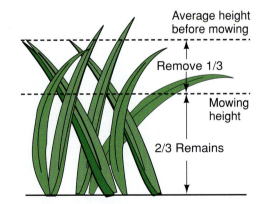

Average height before mowing

Remove 1/3

Mowing height

2/3 Remains

Growers establish a mowing frequency to ensure no more than one-third of the leaf area is removed at any one mowing. See the recommended mowing heights for different turfgrasses in Table 20-5. Maintaining this schedule will allow for clipping return to the field for nutrient recycling. An example of proper mowing frequency is a grass that is normally mowed at a height of 1 inch. In order not to remove more than one third of the leaf area, it should be mowed before exceeding 1½; inches. If that growth occurs in 3 days, then the field should be mowed every 3 days; if the growth requires 2 weeks, then that should be the mowing frequency. Established bermudagrass and zoysiagrass sod fields typically are mowed every 3 days, while centipedegrass, St. Augustinegrass, and bahiagrass are mowed once every 7 to 10 days.

Table 20-5

	Recommended Mowing Heights for Different Turfgrasses		
Species	Mowing height range (inches)	Species	Mowing height range (inches)
Bahiagrass	2–4	Colonial bentgrass	0.5–1.0
Bermudagrass		Fine fescue	1.5–2.5
Common	0.5–1.5	Kentucky bluegrass	1.5–2.5
Hybrids	0.25–1	Perennial ryegrass	1.5–2.5
Carpetgrass	1–2	Tall fescue	1.5–3.0
Centipedegrass	1–2	Crested wheatgrass	1.5–2.5
St. Augustinegrass	1.5–3.0	Buffalograss	0.7–2.0
Zoysiagrass	0.5–2.0	Blue grama	2.0–2.5
Creeping bentgrass	0.2–0.5		

Grass clippings may or may not be picked up. If removed, sweepers and vacuums are used. The purpose of removing clippings is to prevent them from filtering down into the turf stand and turning brown. When the sod is delivered, the presence of these brown clippings may cause the sod to appear to have less density than it really has. Clipping disposal is a major problem. With restrictions on burning, dumping in landfills, and problems with odor, disposal is a problem to many producers. If clippings are removed, then the removal should begin during the first or second months before harvest. This timing will help prevent the browning effect clippings may impose and prevent having disposal problems throughout the entire growing life cycle.

Installation

Proper soil preparation and turf maintenance procedures must be followed to ensure the survival and desirable aesthetics of sod. General sod is installed in the spring and summer. Year-round **installation** is possible in some southern areas if fall and winter temperatures remain conducive for turf growth. The following steps are suggested for laying sod:

1. Soil test to determine nutrient deficiencies.

2. Apply recommended nutrients, especially phosphorous and potassium, plus other soil amendments and incorporate these by tilling 6 to 8 inches deep.

3. Allow soil to settle by irrigating or rolling. Rake or harrow the site to establish a smooth and level final grade. The finish grade should be about 1 inch below walks and drives.

4. Prior to sodding, irrigate the soil to cool the surface and provide initial moisture to roots. If this is not performed, the sod roots will be subjected to initial heat and water stress damage resulting in lower sod survival.

5. When laying the sod, the first strip should be laid along a straight edge. For better knitting, stagger each piece of sod, similar to a bricklayer's running pattern, so that none of the joints are in a line. Each piece should be fitted against others as tightly as possible. Fill in gaps with clean soil to reduce weed encroachment.

6. Smooth the surface and encourage rooting by rolling.

7. Irrigate heavily to ensure good root zone moisture. This is especially necessary when laying a different sod-grown soil type over another. Provide good moisture for at least two weeks following planting. Gradually decrease the frequency between irrigations to an "as needed" basis.

8. **Roll** and/or topdress with clean root-zone soil to help smooth the sod surface.

9. Fertilize with nitrogen approximately 7 to 14 days following planting. Irrigate immediately after application.

10. Mow after the grass reaches the proper height.

After the grass is installed, problems that arise may or may not be the grower's responsibility. Sod with dead edges indicates that it became too dry before installation. Weeds within sod are probably brought in and are the responsibility of the grower. Weeds that grow between the seams of installed sod are due to faulty installation.

Improper irrigation after the installation of sod is always a problem. Enough water should be applied to thoroughly wet the root zone underneath the soil surface. Other problems often develop from spills of bleach, gasoline, and other commercial chemicals, which may complicate the ability to determine the cause of damage.

Pest Management

Preplant fumigation with materials such as methyl bromide, dazomet (Basamid), or metam-sodium (Vapam) may be required when sod farms are established on land previously used for row crop farming. Fumigating will reduce perennial weed species such as bermudagrass, nutsedge, torpedograss, and sprangletop. Soil sterilization will also reduce nematode populations that are difficult to control once the grass is established. The sod field must be fumigated at least every five years to help control weeds, nematodes, and other pests.

Methyl bromide is expensive due to the plastic cover required to ensure activity and may only be applied by a certified applicator. This material provides better pest control, and the treated area can be planted within 48 hours after the cover is removed.

Metam-sodium or dazomet do not require a cover, but a certain amount of efficacy is sacrificed. If a cover is not used, metam-sodium, once applied, requires incorporation into the soil. Incorporation is achieved by rolling, irrigation, and/or tilling the material to the depth of desired control (usually 6 to 8 inches). Poor performance will result if this incorporation is not performed. A minimum waiting period of 14 to 21 days is required before planting in metam-sodium- or dazonet-treated soil.

Weed Control

If preplant fumigation is not feasible, the use of a nonselective herbicide such as glyphosate is required on weed-infested fields. Weed-infested sod will reduce the salability of the product. Three applications of glyphosate spaced four to six weeks apart are necessary for postemergence control of perennial weeds such as bermudagrass or torpedograss. These should begin in spring after temperatures are consistently warm and weeds are actively growing. If spray applications cannot be made prior to field establishment, spot treatments of competitive weeds such as bermudagrass will be required thereafter.

Weeds can be introduced into a field in many ways. Irrigation water from open canals, ditches, or ponds often contains weeds. Soil introduced during soil preparation, such as a landplane pulling untreated soil into a field, leaves weeds. Birds, wind, soil erosion, and man also deposit weed seeds. Good housekeeping by keeping ditches and fence rows clean and by washing equipment before entering a weed-free field does benefit the sod producer.

Once the grass is established, weed management involves proper mowing, cultural practices to promote turf competition, and use of herbicides. Many upright growing broadleaf weeds can be controlled effectively through the use of continuous mowing. These include ragweed, pigweed, cocklebur, and morning glory. Growers should mow these prior to seedhead emergence to help prevent reinfestation from seed.

Grassy weeds, which are a problem in sod production, include annual bluegrass, crabgrass, goosegrass, vaseygrass, signalgrass, sprangletop, torpedograss, and bermudagrass. Broadleaf weeds include purslane, betony, pusley, pennywort (dollarweed), oxalis, and spurge. Purple, yellow, annual, globe, cylindrical, and Texas nutsedges are also weed problems.

Herbicide recommendations are updated constantly. Growers should refer to publications providing the latest recommendations. Immature weeds (seedlings) are most susceptible to herbicides, and certain turf varieties can be damaged when air temperatures exceed 80° to 85°F at the time of herbicide application. The turf should not be under moisture or mowing (scalping) stress when treated with an herbicide.

Insect Control

Insect pests are generally grouped into three categories: shoot feeding, root feeding, and burrowing. Southern chinch bugs, spittlebugs, grass scales, and bermudagrass mites suck plant juices.

Chinch bug damage is normally associated with St. Augustinegrass. Damage is apparent as yellowish to brown patches in turf and appears sooner on turf under moisture and/or heat stress. The cultivars 'Floralawn,' 'FX-10,' and 'Floratam,' provide some degree of resistance to chinch bugs.

Insect shoot feeders, which eat grass leaves, include sod webworms and armyworms. Armyworms feed during the day, while sod webworms feed at night. Injured grass has notches chewed in leaves, and grass has an uneven appearance.

Root-feeding and burrowing insects include mole crickets, white grubs, and billbugs. Mole crickets injure the turf through their extensive tunneling, which loosens soil, allowing desiccation to quickly occur. Mole crickets may be flushed out by applying water with 2 teaspoons of household soap per gallon per 2 square feet on fresh tunnels. If present, crickets will surface and die within several minutes. White grubs and billbugs are root feeders and are typically C-shaped. Grub damage is erratic with patches of turf first showing decline and then yellowing. Under severe infestation, sod may actually be removed by hand. Monitoring these insect populations involves cutting three sides of a sod piece and laying this back. If there is an average of three or more grubs per square foot, an insecticide is needed.

Other insect pests that disrupt the sod surface or are a nuisance to man include ants, fleas, and ticks, as shown in Table 20-6.

Table 20-6

Common Turf Problems		
Turf insects	**Turf diseases**	**Other problems**
Ants	Anthracnose	Dogs
Armyworms	Brown patch	Gophers
Billbugs	Copper spot	Ground squirrels
Cinch bugs	Dollar spot	Mice
Cutworms	Fairy ring	Moles
Grubs	Fusarium blight	Human vandalism
Leaf hoppers	Leaf spots	Vehicles and equipment
Mites	Net blotch	
Mole crickets	Nematodes	
Periodical cicadas	Powdery mildew	
Scale	Pythium blight	
Sod webworm	Red thread	
Weevils	Rots	
Wireworms	Rusts	
	Slime molds	
	Smuts	

Disease Control

Disease development requires three simultaneous conditions:

1. A virulent pathogen

2. A susceptible turfgrass

3. Favorable environmental conditions

Environmental conditions that favor incidence of most turf diseases include periods of high humidity, rain, heavy dews or fogs, and warm temperatures (but not always). Turf that is fast growing and succulent from nitrogen overfertilization is typically more susceptible to disease and other pest invasion. Ideally, irrigate early in the day to minimize the time in which turfgrass remains moist.

If a disease problem is suspected, growers should prepare a sample for laboratory diagnosis. For these situations, do the following:

1. Sample the affected area before fungicide application

2. Sample from marginal turf areas between diseased and healthy turf

3. Cut a 3- to 4-inch plug from each area with symptoms

4. Place these in paper bags or cardboard boxes and do not add water

5. Submit the sample to the nearest County Extension Office

6. Remember to complete a specimen data form with each sample

The major diseases that occur on sod-grown grasses are dollar spot on bermudagrass and bahiagrass, and grey-leaf spot on St. Augustinegrass. Dollar spot disease forms brown patches approximately the size of a silver dollar. On bahiagrass, dollar spot disease is generally more localized on individual leaves. Normally, dollar spot disease can be eliminated by a light nitrogen fertilization to encourage turf plants to outgrow the disease symptom.

Grey-leaf spot disease of St. Augustinegrass normally occurs during hot, humid weather. The use of excessive quick-release nitrogen or the use of atrazine or simazine during these conditions encourages this disease. If fertilized during the summer, use lower quick-release nitrogen rates or use a slow-release nitrogen source on St. Augustinegrass. Foliar applied iron also promotes desirable turf color without over-stimulating disease occurrence.

Sometimes Pythium and brown patch disease affect St. Augustinegrass. Both diseases reduce rooting and turf appearance. Pythium normally occurs in poorly drained areas where water stands. Brown patch also occurs in wet areas and is most pronounced in spring and fall months when grass growth is slow.

Nematode Damage

Nematodes are small, microscopic worms that normally feed on or in plant roots. If populations become severe, plants wilt under moderate moisture stress, are slow to recover after rain or irrigation, and gradually decline or "melt out." Weeds that commonly become a problem in nematode infested areas include spotted spurge and Florida pusley. Turf roots often become stubby, shortened, and turn black. Due to extensive root damage, plants are not able to withstand stresses such as drought, insect, or disease invasion. Sampling of soils for a nematode assay is the only sure way to determine if they are in high enough populations to cause damage. Soil samples are similar to those in the disease control section. They should be submitted to a reputable nematode lab.

Control begins with those management practices that favor good turf growth. These include proper watering, fertilization, and mowing practices. Few nematicides are available. Proper turf management is becoming increasingly important to mask nematode presence.

Harvesting

Turfgrass is harvested when sod has developed enough strength to remain intact with minimum soil adhering when cut. Time required to produce a marketable sod from initial establishment depends on turfgrass species, soil type, and growing conditions. Time typically required between harvests for most turf sod is listed as actual growing months in Table 20-7.

Table 20-7

Time in Growing Months from Planting to Harvest for Some Grasses		
	Growing month	
Cultivar	**Initial establishment**	**After harvest**
Common centipedegrass	18	6–12
Centennial centipedegrass	18	9–15
Tifgreen Bermudagrass	6–12	3–6
Tifway Bermudagrass	6–12	4–8
Emerald zoysiagrass	12–24	13–20
Matrella zoysiagrass	12–24	15–20
Meyer zoysiagrass	12–24	11–18
St. Augustinegrass	10–18	10–18
Bahiagrass	12–24	12–24

Several weeks prior to harvest, the turf should be conditioned in order to enhance its color. Suggested practices include mowing only with a reel mower, applying iron within two weeks of harvest, and applying no chemicals during the week prior to harvest. Using a sweeper or vacuum to remove mowing clippings the last three to four weeks leading up to harvest also improves the turf's appearance.

Sod must never be cut when under moisture stress. The cutter blade bounces out of the ground, the sod has little strength, and turf is under stress by the time the owner receives it.

Mechanical sod cutters harvest strips 12 to 16 inches wide and 2 to 3 feet long, as shown in Figure 20-16. Growers with less than 100 acres commonly use a small, hand-operated, walk-behind unit that has a 150 to 200 sq. yd. cutting capacity per hour. Larger growers usually use tractor-mounted and/or self-propelled harvesters capable of cutting 600 to 800 sq. yd. per hour. Sod is stacked on wooden pallets either in rolls or as flat slabs. The amount of sod harvested can be doubled if sod is rolled instead of stacked as flat slabs. Rolled, harvested sod must also be more mature. Approximately 400 to 500 sq. ft. of sod is stacked per pallet with a forklift required for placing pallets on transport trucks. A tractor-trailer load typically consists of 10,000 sq. ft. of sod. Forklifts that are rear-mounted on tractor-trailers provide a quick and easy method for unloading.

Recently, improvements allow larger rolls to be harvested. These big rolls of sod are typically cut as a continuous roll 42 inches wide and up to 100 feet long. This allows up to 24 100-foot rolls to be hauled on a semitrailer totaling 8,400 sq. ft. of grass. The roll lies like a carpet and generally is more stable and requires less water for establishment compared to traditional slab sod since fewer cut edges are exposed. Currently, the big rolls are being used for stabilization of roadsides and landfills. Less labor is involved in installing the big rolls on large area jobs but is more cumbersome on smaller sod installation jobs such as lawns.

Thickness of soil removed during harvesting varies with turfgrass species. Removing the least amount of soil is the objective of an efficient sod harvest. Soil conservation must be a priority in order to ensure long-term productivity of the soil. Ideally, 1/4 to 1/2 inch of root zone should be removed when sod is cut. Sod that is thin-cut is easier to handle, less expensive to transport, and knits in more quickly than thicker-cut sod, but is more susceptible to drought injury.

Figure 20-16

Harvesting sod for placement in a new location. Courtesy of University of Wisconsin Extension.

Growers harvest up to 40,000 sq. ft. per acre per cutting. Normal yields are generally between 28,000 and 38,000 sq. ft. per acre. A 2-inch ribbon of grass is typically left between harvested strips for reestablishment from stolons. Bermudagrass producers often clean-cut a field because bermudagrass reestablishes from rhizomes, as well as from stolons. Centipedegrass and St. Augustinegrass must re-cover the ground with stolons from ribbons left between harvested strips. Once harvesting has been performed, these strips should be lightly incorporated into the soil by rototilling and rolled to smooth the soil surface. If this is not done, the remaining strips will provide a bumpy surface for mowing, fertilizing, and harvesting equipment. If practical, harvest the second crop at 90 degrees to the first to minimize this uneven surface. For bahiagrass fields, ribbons may or may not be left. In either case, the fields are usually reseeded to hasten recovery.

Separating turfgrass cultivar areas in the field is necessary to prevent contamination from adjacent areas. Normally, this is achieved by carefully planning, before establishment, with the use of service road or drainage ditches between cultivars. If these barriers are not used, a minimum of 8 feet of tilled or bare soil must be maintained between grasses. A nonselective herbicide such as glyphosate may be used to maintain bare soil.

Marketing

Wholesale buyers for most sod producers consist of landscape maintenance/contractors, garden centers, building contractors, homeowners, and golf course/athletic field superintendents. Growers with small acreage and/or limited tractor-trailer shipping capabilities generally sell to homeowners and lawn care professionals.

Shipping costs generally limit the competitive range for most producers. Delivery charges are typically determined per load, per loaded mile, or per square yard. The weight of sod grown on mineral soils is about 5 pounds per square foot. Sod grown on muck soil is generally less expensive to produce and lighter in weight, so it can be transported over longer distances still at a competitive price.

Delivery means for growers will differ. For large producers, usually an 18-wheel, tractor-trailer rig is preferred. Many job sites do not have unloading facilities, so rear-mounted portable forklifts are brought along with the sod. Smaller producers or smaller loads will best be served by appropriately sized trucks.

Sod pallets used normally are 48 inches square and are built from inexpensive lumber. Locating and maintaining adequate pallets can be a problem for the manager.

Costs and Returns

Costs and returns vary considerably with location, equipment, and labor available, and with management practices. Generally, prices for sod increase as the farm size decreases. Capital investments for sod farms include land, buildings, and equipment. Variable costs include labor, fuel, fertilizer, pesticides, repairs, and parts. Fixed costs include insurance, taxes, depreciation, land charge, management charges, and others.

 **Summary**

1. Forage grasses capture the energy of the sun and then they are harvested directly by livestock or by mechanical methods and fed to livestock later.

Internet

Internet sites represent a vast resource of information. The URLs for Web sites can change. Using one of the search engines on the Internet, such as Google, Yahoo!, Ask.com, and MSN Live Search, find more information by searching for these words or phrases: **forage grasses, companion crop, percent live seed, seed quality, turfgrass, sod farms, preplant fumigation, turfgrass mowers, grass species, soil, soil probe, warm season grass, cool season grass, bermudagrass, zoysiagrass, St. Augustinegrass, centipedegrass, bahiagrass, and turfgrass installation.**

Plants of Medicine, Culture, and Industry

CHAPTER 21

Human society has had, from the beginning, a strong dependence on the botanical world, especially for food and wood products. If primitive humans found plants that eased pain, soothed stomach distress or provided relief of any kind, it was undoubtedly an accidental discovery. Disease per se was not understood at the time. Physical ailments were mystical and therefore were attributed to "bad spirits" or as punishment from displeased supernatural powers. Once plants that were effective in relieving human suffering were identified, knowledge of their healing properties was passed on through generation by word of mouth and later in written form.

After completing this chapter, you should be able to:

- Discuss the history of herbalism and medicinal plants
- Name places where medicine men are used as doctors today
- List documented plants that have medicinal compounds
- Name some of the curative powers of plants
- Discuss the psychoactive plants
- Discover the types of compounds in plants that are poisonous
- Describe plants that are toxic to internal organs
- Identify plants that have industrial importance

Key Terms

doctrine of signatures
quinine
pure opium
laudanum
codeine
morphine
heroin
analgesic
hyoscyamine
scopolamine
atropine
tubocurarine
curare
reserpine
ephedrine
cardiac glycosides
edema

anthraquinone
  glycosides
purgative
oxytocic agent
stimulants
hallucinogens
depressants
pseudohallucinogens
nutmeg
sedatives
stimulant
cocaine
hashish
ergot
d-lysergic acid
  diethylamide (LSD)
fly agaric

muscarine
psilocin
psilocybin
peyote
mescaline
scopolamine
oxalates
resins
phytotoxins
somatic nervous system
autonomic nervous
  system
central nervous system
locoweed
cicutoxin
wolfbane
aconitine

Key Terms, *continued*

taxine
nicotine
cis-polyisoprene

rubber
vulcanization
waxes

carnauba wax
candelilla

History of Herbalism

The early history of plant science was essentially the early history of medicinal plants, and the use of plants for treating ailments was the realm of the botanist. In the first century AD, Dioscorides compiled detailed botanicals and medicinal information on thousands of plants in his *De Materia Medica*. This book was to remain the authoritative reference on medical plants for over 15,000 years. In the fifteenth to seventeenth centuries, botany became more of a science, and the use of medicinal plants was finally a mixture of medical knowledge and ritual, not exclusively a magical practice. Herbalists were a combination of botanist, psychiatrist, and faith healer, but their knowledge of the effect plants had on the body was considerable. As more was learned about the true medicinal value of plants, less credence was given to those having only mystical powers. This was the beginning of disenchantment with much of "traditional" medicine. Early in the twentieth century, only a few medicinal plants of exceptional value and proven effectiveness continued to be recognized. However, midway in the twentieth century, the "baby boomers" started to show signs of aging, plus a continued pursuit for organic and natural foods. The medicinal effect of these natural herbs had an appeal. Most of these homeopathic herbs had not been clinically tested; little basic research has been conducted on most of the herbs used for medicinal purposes. There are approximately five to eight medicinal herbs that have been subjected to this needed research.

Modern medicine is predominately composed of scientifically developed synthetic drugs, although about 40% of all prescribed drugs are natural substances or only semisynthetic. Medical schools generally offer only a single course (pharmacognosy) dealing with medicines directly from plant, animal, and mineral origins. This has been the case for only a little over a hundred years, throughout history, and even in many cultures today, the physician has been a person with excellent knowledge of botanical knowledge.

The medicine men of the American Indians, the witch doctors and herbalists of Africa and South America, and the shamans of Asia,

576

as shown in Figure 21-1, trained practitioners who function not only in the realm of mystical and occult faith healing. Many years of training and apprenticeship are required before an individual is recognized as a medicine man or herbalist. Navajo medicine men, for example, learn about the biology and uses of over 200 different plants during their training. Their African and South American counterparts have comparable knowledge of the medicinal plants found in their parts of the world. The Miskitos of eastern Nicaragua, although occupying a very small geographical area, have medicinal uses for almost 100 different plants. Much as modern medicine often must rely on trained counselors, psychiatrists, and psychologists to treat symptoms involving the mind and the emotions of the patient, the herbalists and witch doctors have always considered treatment of their spirit an integral part of their craft. The modern Navajo Indians often blend the two worlds of medicine. The physician deals with most of their physical needs, including surgery. While the medicine man treats imbalances in body functions with his herbal teas and poultices and tends to the health of the mind and spirit.

Figure 21-1

The medicine men of the American Indians, the witch doctors and herbalists of Africa and South America, and the shamans of Asia are all trained practitioners who function in the realm of mystical and faith healing.

Nowhere has a balance of traditional medicine and modern medicine been achieved with greater success or overall acceptance than in the People's Republic of China. Few cultivars have a more thorough knowledge of herbal remedies. Along with acupuncture, this knowledge forms a solid base for their total health care. Both the traditionalists and the modern medical people accept the strength of each other's expertise, knowledgeably blending them into an approach to treatment of the whole entity—the human body, spirit, and mind. Holistic medicine is gaining in acceptance by many Western physicians, and the World Health Organization (WHO) now sponsors a program to promote and encourage traditional medicine throughout the developing countries of the world. Biomedical scientists and other health experts are showing increased interest in the plants used by health experts. The plants used by reputable herbalists and biological conservationists are receiving increased support and acceptance of their efforts to prevent plant extinction, especially in the tropics. It is estimated that 20,000 species of flowering plants are threatened because of clearing the forest for agriculture and development. It is not known how many plants may exist with as yet undiscovered potential for food, industrial, or medical use. It will never be known if these plants become extinct. The renewed interest in medicinal plants is therefore a timely and desirable activity while both the plants and the lore about their medicinal use still exist.

Medicinal Plants

A degree of difficulty exists in categorizing some plants as medicinal plants, psychoactive, and poisonous dosages. For other plants, the distinction is clear. It should be noted, however, that essentially any chemical substance added to the body could be toxic or harmful in excessive amounts.

Mandrake (*Mandragora officinarum*)

Although mandrake was written about more extensively than any of the early medicinal plants, its use was surrounded in myth and superstition. Appreciation of this plant goes back to several hundred years BC, and its use in ancient Rome is documented. The active compounds are several alkaloids, primarily hyoscyamine. They may be extracted in great quantities from the root by soaking or boiling in wine. This extract was the first effective anesthetic, sedative, and pain reliever.

Because mandrake root was so potent and because of the lack of understanding of its properties, mandrake was then considered a mystical plant. A mandrake root was a good luck charm, endowing its prowess and great wealth. A popular belief of this time was the **doctrine of signatures**, which stated that the Creator placed certain items on Earth for humans and identified their intended use by their shape. A kidney-shaped leaf was considered to be useful in treating the kidney; walnuts, with their furrowed ridges, were believed to be beneficial to the health of the brain or head; and liverworts were used to treat the liver. The shape of mandragora frequently resembled the human body, with a thick, fleshy, two-branched or forked taproot and a short stem branching into the leafy part of the plant. It did not take too much imagination to conceive of this as a human form—thus, mandragora was believed to benefit the entire body.

Biblical scholars think mandrake might have been offered to Christ on the cross to ease his pain. Shakespeare detailed important considerations in collecting the root: It must be accomplished without listening to its death shriek as it is pulled from the ground, lest the collector die of fright from the horror of the sound. Lucrezia

Borgia (most ruthless of Italian Renaissance noblewomen) used mandrake as a poison. Even in the nineteenth and twentieth centuries, extract from mandragora root have been mixed with morphine to produce a twilight sleep to ease the pain of childbirth.

Duboisia (*Solanaceae*)

In its native Australia, this plant was used by aborigines during the same period of time that mandrake was used in Europe, the Middle East, and North Africa. Also containing hyoscyamine, the leaves were used to relieve pain. The hunters ran their prey into exhaustion, which often took several days. Once the kill was made, the hunters would chew *Duboisia* leaves to ease the fatigue, hunger, and thirst from their exertions. They also used it to stupefy the emu and to poison fish.

Cinchona (*Rubiaceae*)

Native to tropical South America, this tree contains the alkaloids **quinine** and quinidine in its bark. Prepared commercially as early as 1823, quinine was used for over a hundred years to treat hundreds of thousands of malaria victims. After 1920, a synthetic chloroquine has been preferred for the treatment and suppression of malaria.

Cinchona

Cinchona was named by Linnaeus for the Countess of Chinchone, wife of the Viceroy of Peru. The countess was cured of a malarial fever in about 1638 by a tribal witch doctor (who the Viceroy subsequently appointed as the royal physician). Following a visit to Peru, King Charles II of England and several members of the French and Spanish royal families contracted the dreaded yellow fever (malaria) and sent word to Peru. The viceroy, the countess, the witch doctor, and a support entourage set sail to Europe, taking with them an ample supply of the bark. After successful treatment of King Charles and the others, the drug became widely acclaimed in Europe. Drawing the attention of Linnaeus, who named the plant and described it based on the witch doctor's description and bark samples. Later, Linnaeus sent students, to Peru, to collect botanical specimens at the invitation of the countess.

Opium Poppy (*Papaver somniferum,* Papaveraceae)

Opium poppy, a native Middle Eastern plant, contains over 25 different alkaloids, including several that are among the most important pain reliever in human history. The immature capsules yield a milky sap when cut. This sap dries to a gummy brown residue of **pure opium**. Known to the Sumerians as early as 4000 BC, opium, as shown in Figure 21-2, was used medicinally in ancient Greece, Rome, and China. During the Middle Ages, **laudanum** (opium dissolved in wine) was widely used to relive pain and produce a euphoric state.

Figure 21-2

Opium poppy's (Papaver somniferum) *distinctive fruit.*

From opium several powerful alkaloids can be extracted, the most important being **codeine**, **morphine**, and **heroin**. All these compounds are **analgesic** (pain relieving), affecting the central nervous system. Codeine is the mildest of these, producing almost no euphoria but effectively relieving minor pain and functioning as a cough suppressant.

Morphine depresses the cerebral cortex, reducing brain arousal to pain and causing a euphoric feeling, thus eliminating anxiety. Morphine also depresses the respiratory and cough centers of the brain and impairs digestive action. Prolonged use of morphine results in physical tolerance, so gradually increasing dosages are necessary to achieve the same pain relief. Prolonged use also results in addiction, which at one time was common in wounded soldiers. As with any addictive narcotic compound, discontinued use results in physical painful withdrawal symptoms.

Heroin, a product manufactured from opium, is so powerful in its effects on the body and so dangerously addictive, it is not used medicinally even by prescription in the United States. It is available only through illegal means, so the price is high and the quality and purity are erratic. As with morphine, the body builds a tolerance to heroin and requires increased dosages to produce the same effect. Advanced addiction requires such large quantities that overdoses frequently occur. Thus, from a single plant species, *Papaver somniferum*, some of the most widely used pain relievers and most dangerous addictive drugs originate.

OPIUM WARS

By the nineteenth century, "medical" opium use was widespread in Europe and North America, producing thousands of opium addicts, many who were prominent and wealthy. Opium use, however, was far more common and serious in China than anywhere else. There, it was smoked in specially constructed pipes, addicting millions by the 1830s. Although opium was banned in China by the last Ming emperor (1628–1644), British and American clipper ships nevertheless smuggled tons of opium from India into China for almost two centuries.

Beginning with territorial conquest in India in 1757, the British East India Company pursued a monopoly on opium production and export in India.

Due to the growing British demand for Chinese tea, and the Chinese refusal to accept payment other than silver bullion, the British sought to substitute another commodity for which China was not self-sufficient to alleviate the silver drain—a burden on the British economy. Opium was successfully used by the British traders to replace silver in exchange for Chinese tea for a period of decades. The product was sold by the chest in auctions in Calcutta and then smuggled into China. The East India Company used the profit to purchase teas that were in high demand in Britain. This illegal trade became one of the world's most valuable single commodity trades.

Many Chinese became addicted to opium, wreaking havoc among much of China's population. In response, the Imperial Qing dynasty halted the import of opium, demanding silver be traded instead. Lin Tse-hsü became the imperial commissioner at Canton. His purpose of seeking this appointment was to cut off the opium trade at its source by rooting out corrupt officials and cracking down on British trade in the drug. He took over in March 1839 and within two months he had taken action against Chinese merchants and Western traders and shut down all the traffic in opium. He destroyed all the existing stores of opium. Also Lin was absolutely invulnerable to bribery and corruption.

This response led to the Opium Wars, since the British were not willing to replace the cheap opium with costly silver. The First Opium War won by Britain led to seizing Hong Kong (returned to China in 2002) and to what the Chinese term the "century of shame." The Second Opium War occurred 10 years later, followed finally by the Boxer Rebellion, during which the Chinese evicted all foreigners. Seldom has a single plant affected so many or resulted in such hostility.

Belladonna (*Atropa belladona*, Solanaceae)

Like *Mandragora* and a number of other plants in this family, belladonna, shown in Figure 21-3, contains several powerful alkaloids, including **hyoscyamine**, **scopolamine**, and **atropine**. These alkaloids produce flushed skin and dilated pupils and with lethal doses, delirium and respiratory failure. Atropine specifically blocks parasympathetic nervous system effects and is used as an antidote to nerve gas poising. Spanish women used it in eye drops to dilate the pupils, thus achieving a large dark-eyed look; hence the name *belladonna*, Spanish for "beautiful women." Ophthalmologists take advantage of its pupil-dilation properties to facilitate examination of the interior of the eye.

Curare, (*Chondodendron tomentosum*, Menispermacear)

Chondodendron tomentosum is a native of South America and produces the alkaloid **tubocurarine**, the principal ingredient in **curare**. Curare interferes with nerve impulses to the skeletal (voluntary) muscles,

Figure 21-3

Belladonna (Atropa belladonna) contains several powerful alkaloids including hyoscyamine, scopolamine, and atropine.

producing a reversible paralysis. This drug has been used as a muscle antispasmodic in the treatment of rabies and in spastic conditions. It can also induce muscle relaxation during surgery without anesthesia, and helps control convulsions caused by poisons such as strychnine.

Rauwolfia (*Apocynaceae*)

This plant is native to Africa and Asia, where extracts have been used by the shamans (herbalists) to treat nervous hypertension. Rauwolfia is responsible for initiating serious consideration of plants long used by herbalists. Some 50 alkaloids are present in the several species of this genus, but **reserpine** is the medical agent responsible for lowering blood pressure. It acts by blocking impulses in the sympathetic nervous system, thus lowering blood pressure by relaxing the vessels. It also has a sedative effect on a part of the brain that controls tension. Its use in patients with mental disorders related to hypertension as well as those with high blood pressure makes this a valuable medical plant.

Ephedra (*Ephedraceae*)

Species of ephedra (see Figure 21-4) (a gymnosperm) from India and China contain an alkaloid amine, **ephedrine**, that acts as a bronchodilator. Its mode of action is to relax the smooth muscles of the bronchial tubes, increasing their diameter and improving passage of air through them. Patients with bronchial asthma, chronic bronchitis, and emphysema are treated with this drug.

Figure 21-4

Ephedra (a gymnosperm) from India and China contains an alkaloid amine, ephedrine, that acts as a bronchodilator. Ephedra has been removed from the market in the United States by the Food and Drug Administration.

Several species of ephedra also occur in western North America, but they have only small traces of ephedrine, so their effect as a bronchodilator is minimal. A tea is made from these plants, however, that is an effective diuretic. Ephedra and ephedrine are both used as dietary supplements for weight loss and enhancement of physical activities. December 29, 2003, the FDA banned the use of ephedra in any products, because of the risk of heart attacks caused by taking products that contain ephedra.

The FDA have not evaluated data concerning the safety or purported benefits of most dietary supplement products. The law provides that the manufacturer must make sure that label information is truthful and not misleading. The manufacturer is also responsible for making sure that the dietary ingredients in the supplement are safe. Manufacturers and distributors do not need to register with the FDA or get FDA approval before producing or selling dietary supplements. However, the FDA has proposed rules for dietary supplements containing ephedrine alkaloids. These proposed rules would require warning statements about the amount of alkaloid in a serving of the supplement on the label of all dietary supplements containing ephedrine alkaloids. It should also give the amount of the supplement that is safe to take.

Foxglove (*Digitalis purpurea*, Scrophulariaceae)

Native to Europe and shown in Figure 21-5, this exceptionally valuable medical plant contains several **cardiac glycosides**, including digitoxin. Prior to the discovery of this drug, thousands of people died each year from congestive heart failure, the inability of the heart to pump blood at a sufficient pressure. This resulted in **edema**, or fluid buildup in tissues throughout the body, often causing such a swollen and distorted body that movement was difficult. The ailment was known commonly as dropsy. Treatment with digitalis slows and strengthens the heartbeat, which increases the volume of blood being circulated through the body. The filtering activity of the kidneys and the blood is more effective in removing wasters and preventing fluid buildups. It is estimated that, in the United States alone, over 5 million heart disease patients daily use digitoxin, or one of the other cardiac glycosides found in this plant.

Senna (*Cassia senna*, Fabaceae), Cascara (*Rhamnus cathartica*, Rhamnaceae)

These and several other species in these two genera contain **anthraquinone glycosides**, which produce compounds having a strong **purgative** action. Constipation can be treated with these plants because they enhance peristaltic action of the colon. Several *Cassia* species have been credited periodically with special powers—as aphrodisiacs, venereal disease cures, and effective poultices for wounds and bleeding.

Figure 21-5

Foxglove (Digitalis purpurea) *contains cardiac glycosides, including digitoxin, used in treating congestive heart failure.*

Ergot (*Claviceps purpurea*)

This fungus infects wheat, rye, and other cereals, destroying the seed and producing a black spore—containing sac in its place. These spores contain two alkaloids. One of these, ergonovine, has been an effective **oxytocic agent**, stimulating uterine contraction and speeding up labor. Ergonovine is more effective in this capacity than the animal hormone oxytocin, and it also reduces postpartum bleeding, has low toxicity, and acts rapidly. Ergot also has an interesting history as a hallucinogen.

Herbal Remedies

Besides the well-known and effective medicinal plants just discussed, there are hundreds of undiscovered and poorly understood herbal remedies. Fortunately, few of these plants have toxic or addictive dangers.

Table 21-1 lists a small percentage of the plants reported as having some medicinal value in the United States alone. There are many popular books detailing the correct collection, extraction, and use of these and many more herbal remedies; however, great care should always be taken when collecting your own. An intelligent combination of home remedies and health tonic with modern medical care is becoming a popular and widely accepted health care plan in the United States. Certainly, there is room for more scientific and biomedical study of many of these herbal cures and the plants used in their preparation.

Table 21-1

Selected Herbal Remedy Plants of the United States		
Plant	**Plant part used**	**Purported value**
Aloe (*Aloe vera*)	Leaves, mucilaginous juice	A tropical pain reliever, promoting healing of burns and cuts
Alum (*Hechera* spp.)	Root, chopped, 1 teaspoon boiled for 20 minutes	An astringent for treatment of stomach flu, ulcers, sore throat (gargle), and cuts and abrasions (promotes clotting)
Asparagus (*Asparagus officinalis*)	Root chopped and steeped in boiling water to make a tea	A diuretic and laxative for treatment of gout and related joint inflammations; helps prevent kidney stones
Birch (*Betula*)	Bark	An analgesic for headaches and arthritis; methyl salicylate active ingredient
Catnip (*Nepeta cataria*)	Entire plant	A mild tranquilizer, treatment for colic or teething in young children
Comfrey (*Symphytum officinale*)	Root extraction in water	Mucilage acts as a demulcent and tonic for the respiratory tract; also used as a stomach tonic and to sooth diarrhea and dysentery; may be toxic
Dandelion (*Taraxacum officinalis*)	Leaves and flowers, steeped in boiling water to make a tea	Diuretic; treatment for kidney inflammations; as an aid to liver and spleen functions
Dogbane (*Apocynum androsaemi-folium*)	Roots	Cardiac stimulant may be toxic
Gentian (*Gentian* spp.)	Roots; chopped and steeped in hot water for 30 minutes	Considered an excellent stomach tonic for indigestion, heartburn, etc.
Mullein (*Verbascum thapsus*)	Large, basal leaves; washed dried, chopped, steeped in water	Throat, chest, and lung maladies such as coughing, asthma, and respiratory tract infections
Oak (*Quercus* spp.)	Bark or small twigs; chopped and steeped in water to make a tea	As astringent for gum inflammations, sore throat gargle, intestinal tonic
Old man's beard (*Clematis* spp.)	Branches; chopped and steeped in boiling water to make tea	Treatment of headaches, especially migraine; dilates veins but has a vasconstricting effect on the brain lining
Plantain (*Plantago major*)	Fresh leaves	Poultice; treatment for insect bites and skin abrasions; also useful for mild intestinal inflammations and hemorrhoids
Pleurisy root (*Asclepias tuberosa*)	Poultices; chopped and boiled in water	Increases perspiration and bronchial dilation; treatment for pleurisy edema; may be toxic
Willow (*Salix* spp.)	Twigs or bark; boiled in water for 2 hours	General analgesic action; reduces inflammation of joints and membranes, eases headaches, lowers fevers, and reduces neuralgia; antiseptic wash or poultice for skin abrasions, eczema, or infected wounds

As the baby boomers starting to show signs of aging , an increase in the use of many herbal dietary supplements escalated. What is a *dietary supplement*? The most logical definition would be something that supplies one or more essential nutrients missing from the diet. However, the Dietary Supplement Health and Education Act of 1994 (DSHEA) defines a dietary supplement as any product (except tobacco) that contains one of the following: (1) a vitamin, (2) a mineral, (3) an herb or botanical, (4) an amino acid, (5) a dietary substance "for use to supplement the diet by increasing total intake," or (6) any concentrate, metabolite, constant, extracts, or combinations of any of the aforementioned ingredients. Herbs, of course, are not consumed for a nutritional purpose and often are marketed with therapeutic claims. The supplement industry, which lobbied vigorously for passage of the act, included them in the definition to weaken the FDA's ability to regulate their marketing. Since the DSHEA's passage, hormones have also been marketed as dietary supplements.

Psychoactive Plants

Psychoactive plants contain compounds that act on the central nervous system to produce a mind-altering state, visions, distortions of the sense, and changes in psychomotor ability. There are three categories of psychoactive compounds: **stimulants, hallucinogens,** and **depressants.** It is not uncommon, however, for all psychoactive compounds to be considered hallucinogenic because of the mind alteration produced. The drug culture of the 1960s and 1970s dubbed these experiences "trips" and called some of these compounds "mind expansion" or "psychedelic" drugs. Modern Western culture developed new terms, new uses, and some new synthetic drugs, but psychoactive compounds have been in use for thousands of years. Hallucinogens have been a part of religious activities of many cultures, especially the more primitive ones, for a long time. Social use of such drugs has also been a regular part of many of the more advanced societies throughout history. Drug use has been acceptable in ceremonial and ritualistic context; however, it has varied from widely acceptable to forbidden and illegal in social settings.

Many of these compounds come from plants. Some of those considered to be hallucinogenic have had medical applications in lighter dosages. Mandrake and belladonna in Solanaceae fall into the medicinal, hallucinogenic, or poisonous category, depending on dosage. Others, while having no medicinal value per se, have been used to treat mental disorders and as pain relievers by altering the patient's level of consciousness. Still others are not hallucinogenic but poisonous. These are sometimes called **pseudohallucinogens** because they produce a delirium while acting on the system. The delirium makes the user very sick, and the after effect is painful. **Nutmeg** is a pseudohallucinogen, attractive for its availability and nominal cost. Several tablespoons, if it can be kept down, produces a painfully nauseating delirium from which the user comes down "hard" with a severe and extended pain toxic recovery.

Sedatives and Stimulants

Although categorized by many as hallucinogenic, opium and opiate derivatives are not; they are **sedatives**, often producing a relaxed euphoric state but no true consciousness alteration. Reserpine, the indole alkaloid form of *Rauwolfia*, is also a depressant and lowers blood pressure. Alcoholic beverages have depressant action reducing both mental and physical performance levels. Since alcoholic beverages all come from the fermentation process using plants as the sugar source, alcohol can be considered a plant product.

A drug that falls into the **stimulant** category is **cocaine**. Extracted from the leaves of *Erythroxylon coca*, as shown in Figure 21-6, a plant native to the eastern slopes of the Andes Mountains of South America, cocaine produces a feeling of euphoria, a lessening of fatigue, an absence of hunger, and increased energy. Cocaine does not produce a hallucinogenic response, is not normally physiologically addictive, and the user does not require increasing quantities due to body tolerance. It is, however, psychologically addictive, and strong doses can produce feelings of paranoia and nervous insomnia. It also damages the mucosal tissues of the nose and throat, sometimes permanently destroying these sensitive areas.

Indians in the Andes Mountains chew the leaves, which enables incredible feats of endurance without food or rest. Coca chewing is common among the people of this region, as much a cultural phenomenon as an aid to extensive manual labor. It is reported that some of the early popularity of Coca-Cola in the 1870s was because a small amount of cocaine was included in the recipe. Cocaine was banned as a component of the drink in the 1900s.

Hallucinogens

Hallucinogenic compounds distort the senses, especially visual, and produce a departure from reality. The plants that contain these compounds are all angiosperms, predominately dicots. There are no known hallucinogenic gymnosperms, ferns, bryophytes, or algae, but the fungi include a number of hallucinogenic taxa. Depending on the organism and the part that contains the active compound, one can eat the fresh or dried plant, chew the appropriate plant parts, concoct a beverage, inhale a powder, smoke the leaves, and even smear an oil-based ointment on the body.

Marijuana (Cannabis indica; C. sativa)

More than three species of *Cannabis*, as shown in Figure 21-7, are recognized, depending on the legalistic or scientific (often mixed) point of view. *Cannabis sativa* is the best known species although *C. indica* contains a higher concentration of the active compound. *C. sativa* is an erect species with many fiber cells in the stem, a commercial source for

Figure 21-6

Coco leaves (Erythroxylon coca) *contain cocaine. Courtesy of Food and Drug Administration.*

Figure 21-7

Marijuana (Cannabis) *has the typical palmately compound leaves with serrated margined leaflets. It has become the number one cash crop in several southern states. Courtesy of North Carolina Department of Agriculture.*

making hemp rope, paper, and canvas. *C. indica* is a low growing, highly branched species containing high levels of trans-tetrahydrocannabinolic acid, which converts on heating to tetrahydrocannabional (THC), the compound that produces the hallucinogenic effects. THC is found in greatest concentration in glandular hairs on the leaves, stems, and especially on the small bracts just below a pistillate flowering inflorescence. When collected from the glandular hairs separately, it is much more potent and is called **hashish**.

The physiological effects of THC on the body include an increased pulse rate, reddening of the eyes and possibly a reduction in the internal fluid pressure of the eyeball. Marijuana is credited with enhancing visual perception, reducing muscular response time and coordination, altering time perception, increasing one's inner awareness, removing tension, and increasing sex drive.

THC is not an alkaloid and is not considered toxic or physiologically addictive. Its classification as a narcotic, therefore, is a legal category and not a chemical group. Psychological dependence, however, is not uncommon.

Arguments abound today: Is marijuana a dangerous drug or a mildly hallucinogenic compound less harmful than tobacco or alcohol? *Cannabis* is probably the most widely used and highly valued multipurpose plant in history. It has served as a source for fiber, as a medicine, and as a mild hallucinogen worldwide since it was introduced into North America in 1607. Only illegal as a crop plant since the late 1940s, over 63,000 tons of hemp (*C. sativa*) were raised in 1943 to provide the Navy with rope during the war. George Washington also raised hemp on his Virginia farm. There is still the push to legalize marijuana for it medicinal properties. Research has shown that marijuana helps to control some of the side effects of chemotherapy. It has been legalized for this purpose only in a very few states.

Ergot

Ergot itself is not the hallucinogen but is the spore-producing reproductive body of a fungus, *Claviceps purpurea,* that affects rye, wheat, and other cereal crops. Within the sacs are millions of tiny black spores containing several alkaloids, including ergonovine. Chemically known as lysergic acid, it is very similar structurally to a synthetic compound, **d-lysergic acid diethylamide (LSD)**. Ergonovine produces smooth muscle and blood constriction, which causes a burning sensation. Eating infected grain produces the disease that is known as St. Anthony's fire, which could result in gangrene of the ears, nose, and other body extremities from lack of blood flow through the constricted vessels. This disease was thought to be caused by bad grain (the ergot sacs were mistaken for diseased grain), thus, indirectly an affliction from the devil. Victims would visit monasteries to repent and have the evil forces driven out. When fed with uninfected grain and given adequate fresh water and rest, most people would recover, but sometimes with the tip of their nose or ear or even a hand or foot missing.

In addition to the burning sensation and danger of gangrene, the victims of St. Anthony's fire had hallucinations, an effect of the poisoning and the lysergic acid. LSD, on the other hand, is a much more powerful synthetic relative and a popular modern hallucinogen.

Fly Agaric

The **fly agaric** mushroom is another hallucinogenic fungus (see Figure 21-8). Scientifically known as *Amanita muscaria*, these attractive mushrooms are large, bright yellow to orange-red with small white speckles on top. Their common name comes from the dead flies that are often found on the ground around them—in addition to being hallucinogenic, they are poisonous. Only the ones growing in the Old World are psychoactive, containing the alkaloid **muscarine**. This is the only known hallucinogenic compound that passes through the kidneys unaltered, and some cultures have been known to reuse the compound several times by drinking urine.

Figure 21-8

A beautiful but very poisonous mushroom with hallucinogenic properties.

In larger doses, muscarine produces blurred vision, sweating, lowered blood pressure, slow heartbeat, stomach pain, and breathing difficulty. These are the opposite symptoms of atropine poisoning, and fly agaric is used as an antidote to atropine overdoses.

Sacred Mushrooms

Several genera of mushrooms containing the alkaloid **psilocin** and **psilocybin** were used to produce hallucinogenic visions. The "mushroom cults" of modern times have most commonly used *Psilocybe mexicana* although *Conocybe* and *Stropheria* also contain significant levels of these alkaloids. These cults were originally a blend of Christian and pagan rituals, but today they are almost nonexistent found only in a few villages in the northeastern mountains of Oaxaca, Mexico.

During the Mayan period in Mexico, mushroom use was common but officially available only to the priest. In contrast, almost anyone could use sacred mushrooms during the Aztec era. Some authorities believed that widespread mushroom usage was associated with the Aztec practice of offering human sacrifices by the hundreds to the gods. Current descendants, the Mazatec Indians of Oaxaca, believed that the mushrooms may be eaten only at night, and the activity is now a closed family affair.

Peyote

Figure 21-9 shows *Lophophora williamsii*, which is a small, rounded, spineless cactus that grows flush with the ground, and is found in the southwestern United States but grows primarily in Mexico. The hallucinogenic compound found in the **peyote** is the alkaloid **mescaline**. The Aztecs used peyote as did the Comanches under the leadership of Chief Quanah Parker, who in 1875 was the last Indian chief to surrender his people to confinement on a reservation. It is now most closely associated with the religious rituals of North American Indians belonging to the Native American Church. Known for its nauseatingly bitter taste, peyote can be eaten fresh or the dehydrated buttons can be chewed to achieve the desired effects. A drink can also be made from the buttons, but it is reportedly less effective in producing the vivid hallucinations. Bad trips are possible, as are spontaneous flashback recurrences.

Figure 21-9

Peyote (Lophophora williamsii) is a spineless cactus that grows almost flush with the ground.

Solanaceae: Henbane, Belladonna, Mandrake, and Thorn Apple

Henbane (*Hyoscyamus niger*), belladonna or deadly nightshade (*Atropa belladonna*), Mandrake (*Mandragora officinarum*), and thorn apple or jimsonweed (*Datura* spp.) are all members of the same plant family. They all contain alkaloids including **scopolamine**, hyoscyamine, and atropine (hyoscyamine in the fresh plant changes to atropine when the plant is dried). These compounds have been used medically, hallucinogenically, and as toxins throughout history.

In mild doses, a feeling of timelessness and high spirits accompanies visual hallucinations and the sensation of flight. These plants were once a common ingredient in "witch's brew." When painted on the body with a straw broom, the alkaloids were absorbed through the skin and produced these reactions. The image of witches flying on broomsticks is a carryover from this ritual.

Other Hallucinogens

Other hallucinogenic plants include the South American *Virols* of the nutmeg family, which is administered to others by blowing a powder extracted from resin in the bark up the nose through a hollow snuff tube. Another South American plant, *Banisteriopsis,* yields an intoxicating beverage that produces a wide range of hallucinogenic responses. The Aztecs also used morning glories (*Ipomoea violacea* and *Rivea corymbosa*), shown in Figure 21-10, which contain alkaloids similar to lysergic acid and its relatives.

Figure 21-10

Morning glory (Ipomoea) *was used by the Aztecs as a hallucinogen.*

Today's ornamental varieties often contain these compounds and are known by common names such as "pearly gates" and "heavenly blues." North American Indians used mescal beans (*Sophora secundiflora*) before they began to use peyote. Peyote was preferred because of the high toxicity of mescal beans. Interestingly, these two plants were often found in identical habitats, and the Indians believed that the easily visible mescal bean plants were provided to lead them to the peyote cactus growing near them. They wore necklaces made of mescal beans during their peyote rituals.

Poisonous Plants

Of the approximately 500,000 plant species in the world, less than 1% of them are thought to be toxic to humans. In spite of this statistic, poison control centers in the United States find that plants consistently rank in the top three as sources of possible poisoning; for children under 5, plants are the number one danger. This is a result of the significant increases in numbers of houseplants in homes, offices, and malls.

Historically, poisonous plants have been less a problem of accidental consumption than a purposefully ingested agent for medical or hallucinogenic purposes. The chemistry of plant toxicity is much better understood today; hence, the reduced number of accidental poisonings resulting from incorrect dosage types.

Classes of Poisonous Compounds

Plant-derived medicines include a variety of chemical compounds. Some of these can be poisonous if not taken as directed. Following is a list of compounds and the plants they are derived from.

Internal Organ Poisons

These act primarily on the kidneys, liver, and stomach and include several distinct chemical types.

1. Alkaloids contain nitrogen and usually taste bitter. They are commonly found in several plant families, including the Solanaceae, Fabaceae, Papaveraceae, Rubiaceae, Apocynaceae, and Ranunculaceae.

2. Glycosides

   a. Cyanogenic glycosides included hydrocyanic acid (HCH), found in peach and cherry pits.
   b. Cardiac glycosides are the compounds found in *Digitalis*. They affect the heart.

3. Oxalates (oxalic acid) are found in rhubarb leaves and several other plants and produce severe stomach pain.

4. Resins are found in milkweed, laurels, and water hemlock and are especially potent toxins.

5. Phytotoxins are proteins, such as the active ingredients of castor bean plants.

Nerve Poisons

These are alkaloids: The toxin acts on different parts of the nervous system.

1. Somatic nervous system toxins act on the striated skeletal or conscious control muscles. Curare is an example in this category.

2. Autonomic nervous system toxins affect the smooth muscles of the heart. An example is alkaloids of the Solanaceae.

3. Central nervous system toxins, the opiate group, act on the brain and spinal cord.

Irritants

These compounds burn the skin, eyes, and throat, usually causing swelling, redness, welts, and even weeping lesions. Many are resins.

Mineral Poisons

Some plants accumulate unusually high concentrations of specific minerals found in the soil or in the air. **Locoweed** (*Astragalus mollissimus*) is a selenium accumulator. Sufficient ingestion by cattle or horses can cause "blind staggers" and even death. Lethal levels are difficult to consume, however, with horses requiring ingestion of 500 pounds of plants over a six-week period and cattle needing well over 2,000 pounds over a two- to three-month period. Some roadside weeds accumulate lead from exhaust fumes and can produce lead poisoning in animals that feed on them.

Allergens

This is not actually a category of "poisonous" compounds because, unlike the proceeding, allergens produce a toxic response only to those people (or other animals) who have species sensitivity. The true poisons produce a response in all individuals. Poison ivy (*Rhus toxicodendron*), shown in Figure 21-11, is easily the most common allergen. Over 100 million people in the United States alone are sensitive to it.

Figure 21-11

Poison ivy (Rhus toxicodendron) *is probably the most widespread and serious plant allergen.*

Deadly Plants

The degree of toxicity among poisonous plants varies with the nature of the active compound and with its concentration in the plant. Some plants are only mildly poisonous and others deadly. Table 21-2 includes some of the plants commonly reported to the Poison Control Center in the United States. This list includes most of the highly dangerous plants that one might encounter in the home; several of the most lethal wild and domesticated plants are also included in the following discussion.

Table 21-2

Common Poisonous House and Yard Plants

Common name	Scientific name (family)	Poisonous parts	Toxic compound	Symptoms
House plants				
Castor beans Rosary pea	*Ricinus communis* (Euphorbiaceae)	Seeds must be chewed	Ricin and lectin	Nausea, muscle spasms, purgation convulsions, and death; lethal dose 8 seeds
Mistletoe	*Phoradendron* (Loranthaceae)	All parts; especially berries	Toxic amines; protein	Stomach upset and in severe cases death from inability to breathe
Dumb cane	*Dieffenbachia* (Araceae)	All parts	Calcium oxalate and proteolytic enzymes; asparagine	Burning sensation in mouth and swelling of mouth and throat tissue could lead to asphyxiation in small children
Caladium	*Caladium* (Araceae)	All parts including tuber	Calcium oxalate and/or irritant juices; asparagine	Burning sensation in mouth and swelling of mouth and throat tissue could lead to asphyxiation in small children
Elephant's ear	*Colocasia antiquorum* (Araceae)	All parts	Calcium oxalate and/or irritant juices; asparagine	Burning sensation in mouth and swelling of mouth and throat tissue could lead to asphyxiation in small children
Philodendrons	*Philodendron* and *Monstera* (Araceae)	All parts	Calcium oxalate and/or irritant juices; asparagine	Burning sensation in mouth and swelling of mouth and throat tissue could lead to asphyxiation in small children
Crotons	*Croton* (Euphorbiaceae)	Seeds, leaves and stems	Croton oil	Powerful purgative action that can cause death on ingestion of small amounts
Crown of thorns	*Euphorbia milii* (Euphorbiaceae)	All parts	Milky sap contains acrid irritant	Can cause severe irritation if ingested in large quantities as complex esters
Yard and garden plants				
Daffodil	*Narcissus* (Liliaceae)	Bulbs	Calcium oxalate crystals	Nausea, vomiting, diarrhea; if eaten in quantity, could be fatal
Autumn crocus	*Colchicum autumnale* (Liliaceae)	Bulbs, flowers and seeds	Colchicine and other alkaloids	Burning in the throat and stomach vomiting; weak, quick pulse; kidney failure; respiratory failure
Star of Bethlehem	*Ornithogalum umbellatum* (Liliaceae)	All parts including bulbs	Cardiac glycosides	Nausea, gastroenteritis
Pokeweed, poke berries	*Phytolacca Americana* (Phytolaccaceae)	Roots, shoots, berries	Triterpenesaponin, lectin, mitogen	Severe gastrointestinal disturbance weakened pulse and respiration; can be fatal
Iris	*Iris* spp. (Iridaceae)	Leaves and rhizomes	A glycoside	Gastrointestinal disturbance, sometimes a burning in the mouth
Yew (English and Japanese	*Taxus* spp. (Taxaceae)	Bark, leaves, and seeds	Alkaloid, taxine	Fatal; taxine readily absorbed in intestines; death is sudden
Oleander	*Nerium oleander* (Apocynaceae)	All parts	Cardiac glycosides	Severe vomiting, bloody diarrhea, irregular heartbeat, drowsiness, unconsciousness, respiratory paralysis, and death

(Continues)

Table 21-2 (*Continued*)

Common name	Scientific name (family)	Poisonous parts	Toxic compound	Symptoms
Yard and garden plants				
Daphne	*Daphne mezereum* (Thymelaeaceae)	Berries	Coumarin glycoside, and a diterpene, mezereon	Fatal! a few berries can kill a child; burning of the throat and stomach, internal bleeding, coma, and death
Rhododendron laurel, azalea	*Rhododendron* spp. (Ericaceae)	All parts	Resin	Vomiting, weakness, and in extreme cases paralysis and death
Buckeye, horse chestnut	*Aesculus hippocas-tanum* (Hippoca-stanaceae)	Leaves, flowers, young sprouts, seeds, pods	Lactone glycoside, esculin	Vomiting, diarrhea, lack of coordination, paralysis, (used extracts of buckeye for fish poison)
Wisteria	*Wisteria* spp. (Fabaceae)	Seeds, pods	Lectin, mitogens	Mild to severe digestive upset
Jimsonweed, thorn apple	*Datura stramonium* (Solanaceae)	All parts especially the seeds	Scopolamine hyoscyamine, alkaloids	Flushed skin, dilated pupils, dry mouth, delirium, and death due to respiratory failure
Woody nightshade, bitter sweet	*Solanum nigrum* (Solanaceae)	Leaves, roots, berries	Solanine alkamine aglycones	Burning in the throat, nausea, dizziness, dilated pupils, weak convulsions; can be fatal
Irish potato	*Solanum tuberosum* (Solanaceae)	When tubers are exposed to light, the green tissue under the skin	Solanine alkamine aglycones	Burning in the throat, nausea, dizziness, dilated pupils, weak convulsions; can be fatal
Jerusalem cherry	*Solanum pseudocarp* (Solanaceae)	Berries	Solanine alkamine aglycones	Burning in the throat, nausea, dizziness, dilated pupils, weak convulsions; can be fatal
Elderberry	*Sambucus* spp. (Caprifoliaceae)	Stem, leaves, unripe berries	Alkaloids, cyanogenic glycosides	Burning in the mouth, throat, and stomach
Holly	*Ilex* spp. (Aquifoliaceae)	Berries	Toxic compound unknown	Not fatal, but purgative and emetic and most dangerous to children
English ivy	*Hedera helix* (Araliaceae)	Leaves, berries	Triterpene sapogenin, hederagenin	Vomiting, diarrhea, and nervous depression; can be serious in small children
Foxglove	*Digitalis purpurea* (Scrophulariaceae)	Leaves, seeds, flowers	Digitoxin, cardiac glycosides	Large amount can cause dangerously irregular heartbeat, mental confusion, and digestive upset; usually fatal
Lily of the valley	*Corvallaria majalis* (Liliaceae)	Roots, leaves, flowers and fruits	Cardiac glycoside, convalla toxin	Large amounts cause dangerously irregular heartbeat, mental confusion, and digestive upset; usually fatal
Monkshood	*Aconitum* spp. (Ranunculaceae)	All parts especially roots and leaves	Aconitine (alkaloid)	Numbness, paralysis of upper then lower extremities; death by respiratory paralysis; very poisonous
Rhubarb	*Rheum rhaponticum* (Polygonaceae)	Leaves (not petioles, which are edible)	Oxalates, anthraquinone glycosides	Stomach pain, nausea, vomiting and in serious cases convulsions, internal bleeding, coma, and death; very poisonous

Hemlock (Conium maculatum, Umbelliferae)

A herbaceous plant with leaves typical of the carrot family, hemlock, shown in Figure 21-12, contains the toxin coniine. It is not the same as the woody hemlock tree. This particular species is recognizable by the reddish purple spots on its stem.

Figure 21-12

Hemlock is the type of poison said to have killed Socrates.

Socrates was poisoned with hemlock in 399 BC for crimes against the state, "neglecting the gods who the city worships" and "corrupting the youth of the city" with his ideas and unwillingness to support the politicians in power. There is evidence that the officials of Athens did not want him to die; they were afraid he would have more support as a martyr than alive. Socrates was given repeated chances to plead guilty and receive a light sentence, but he refused because he had done nothing wrong. He was even given a chance to escape, which he refused. Finally, the sentence was carried out and he was given a cup of hemlock juice to drink, followed by a laudanum of wine, and probably opium to ease the pain of his death.

Socrates insisted on having the stages of the hemlock poisoning accurately observed and recorded. He walked around until his legs began to grow heavy; then he laid down and had someone feel his feet and legs and reported the progression of numbness and cold as it approached his heart. He died of respiratory failure.

Water Hemlock (Cicuta spp., Umbelliferae)

Also a herbaceous plant, water hemlock is found growing in wet habitats such as shallow streams and along the edges of ponds and lakes. Its leaves are not fernlike, as are hemlock leaves; rather, they superficially resemble marijuana leaves. The hollow stems contain a yellow sap, which has the deadly **cicutoxin** in it. Especially concentrated where the stem and root join, this toxin is even more poisonous than hemlock juice. Death is a painful process involving several convulsions and respiratory failure.

Monkshood (Aconitum napellus, Ranunculaceae)

Also known as **wolfbane**, this contains the alkaloid **aconitine**, which is a very powerful poison (see Figure 21-13). Used as an arrow poison by early Stone Age cultures, this toxin is very fast acting; a feature that once also made it a favorite poison.

Figure 21-13

Monkshood (Aconitum napellus) contains a powerful toxin credited for causing the death of the Roman Emperor Claudius.

One of the most famous poisonings accredited to aconitine, is that of the Roman Emperor Claudius, who died suddenly in 54 AD. Historians first blamed Julia Agrippina, Claudius's empress, of doing away with him so that her son by a previous marriage, Nero, could rule. It was later decided that Locusta was the poisoner, working with Stertinus Xenophon, Claudius's personal physician. Locusta also poisoned Britannicus, Claudius's son, so that Nero could rule alone rather than be co-emperor with Britannicus.

Some reports have Claudius poisoned by death angel mushrooms, but the better documented story is that he was fed only small doses of these mushrooms to evoke immediate systems. A feather, previously dipped in aconitine, was used by the physician to tickle his throat to induce vomiting and save Claudius from the attempted mushroom poisoning. That he died anyway was probably accredited to heart failure from the close call, and murder was not officially suspected.

Monkshood is also credited by historians as being the "murder weapon" in the death of Pope Adrian VI and in an unsuccessful attempt on the prophet Mohammed's life. In spite of its well-known toxicity, this plant is attractive and is grown in gardens as an ornamental, especially in Scandinavian countries.

Yew (Taxus spp., Taxaceae)

This plant, shown in Figure 21-14, was prized for the elasticity of the wood and particularly for its value in making the finest of bows. Robin Hood used a yew wood bow because of its qualities. The alkaloid **taxine** is very poisonous and is found in highest quantities in Japanese and English yew (*T. cuspidata* and *T. baccata*). The word *toxon* means "bow," and toxin means "poisonous." The genus name *Taxus* is thought to be derived from some combinations of these words.

Figure 21-14

Yew (Taxus *spp.*) *is very poisonous but also has excellent wood for bow making.*

Oleander (Nerium oleander, *Apocynaceae*)

Oleander, shown in Figure 21-15, is an extremely poisonous plant. Reported deaths include that of a child who chewed on a single leaf. Unconfirmed reports attest to the danger of eating a hot dog cooked on an oleander stick. Many people value oleander as an attractive landscaping plant without being aware of its danger (or perhaps in spite of it).

Figure 21-15

Oleander (Nerium oleander), *a popular yard ornamental, is reportedly a very poisonous plant.*

Jimsonweed (Datura stramonium, *Solanaceae*)

Jimsonweed, shown in Figure 21-16, along with many other members of this family, is highly toxic. Also known as thorn apple because of the spiny fruit and moonflower because of its night blooming white flower, the name jimsonweed stems from what occurred in 1676. During Bacon's Rebellion, a group of solders on a march near Jamestown, Virginia, added some *Datura* leaves to stew. The result was a stupefied regiment that had to be replaced. The "Jamestown Weed" in time became known as jimsonweed.

Figure 21-16

Jimsonweed (Datura stramonium) *along with many other members of the Solanaceae family are highly toxic.*

Tobacco (Nicotiana tabacum, *Solanaceae*)

Although tobacco, shown in Figure 21-17, is normally listed as a poisonous plant, the presence of the alkaloid **nicotine** in the leaves qualify it for inclusion in this section. A drop of pure nicotine is toxic enough to kill a medium-size dog in only seconds, and nicotine sulfate is one of the most effective insecticides.

Figure 21-17

Tobacco has been cultivated for centuries, but the nicotine is a health hazard.

In addition to its toxicity, nicotine is a stimulant and a physiologically and psychologically addictive drug, with physical withdrawal symptoms. In most countries, however, it is legal and is not classified as a narcotic; all other addictive alkaloids are so classified in the United States. Finally, the toxic and addictive drug is known to be a significant contributing factor in a number of respiratory ailments and in throat and lung cancer.

Industrial Plants

Plants provide us with fibers for making cloth, rope, paper, etc. Numerous dyes obtained from plants color our fabrics. Many plants have oil-rich seeds, and these oils can be extracted when they have a variety of uses. Many of them are edible, and they can also be used as lubricants, fuel, for lighting, in paints and varnishes, as a wood preservative, and waterproofing.

Latex

The story of natural rubber is one of the most interesting in the history of plant use. The latex sap of many plants contains small amounts of **cis-polyisoprene** or natural **rubber**. Only one plant has been a significant source for this material; most of the others simply do not contain enough to be commercially exploitable. A tropical tree native to the Amazon Basin, *Hevea brasiliensis*, shown in Figure 21-18, is still the only commercial source for natural rubber.

Figure 21-18

The only plant that produces natural rubber, Hevea brasiliensis. *Courtesy of Lucky Oliver.*

First known in eighteenth century Europe as a bouncy plaything of the Mexican Aztecs and the Amazonian tribes, it received its English name from Joseph Priestly who discovered that this toy would rub out pencil marks. In 1823, a Scotsman named Macintosh discovered that rubber was soluble in naphtha. When such a solution was used to impregnate cloth and the naphtha evaporated away, the result was a waterproof material suitable for protection from the rain. Even today, raincoats are still called mackintoshes in the British Isles.

In 1839, Charles Goodyear accidentally heated raw rubber with some sulfur. The result was a tougher material that would not become tacky in hot weather or stiff and brittle in cold. The process of vulcanization made rubber a much more important product. Vulcanized rubber wears longer, is more resilient, and can be molded into any shape, including that of a tire. Sadly, Goodyear did not establish a patent on this vulcanization process, and it was stolen and patented in France. Charles Goodyear died a pauper after spending the rest of his life trying to be legally credited with the discovery of the **vulcanization** process.

By the 1800s, raw rubber was in great demand, but Brazil controlled essentially all supplies and forbade the export of rubber tree seeds. After two tries, seeds were finally smuggled out of Brazil for the Royal Botanical Garden of England, and seedlings were successfully transported to the British protectorate Ceylon (Sir Lanka) Asian *Hevea* plantations have taken over as the primary supplies of rubber for the past century. Brazil never became sophisticated enough industrially to maintain early dominance.

Natural rubber shortages during World War II prompted the development of synthetic rubber, but it does not have all the properties of natural rubber. Many rubber products, including radial tires, required natural rubber. Since many of these items are being sold in increasing amounts, it is feasible that another rubber shortage could occur unless a new source can be developed for commercial rubber production.

Oils and Waxes

Oils contained in fruits, seeds, and other plant parts fall into several categories based on their chemical composition and use. Cooking oils, soaps, plastic, paints, linoleum, lubricants, and printing inks are a few examples of the range of plant oil users. Tung, caster, olive, palm, coconut, soybean, peanut, cottonseed, linseed, sunflower, safflower, canola, and corn oils are some plant oils with a wide variety of commercial uses. Oils are often divided into three categories according to their qualities, these categories are nondrying, semidrying, and drying. Nondrying oils are slow to oxidize and so remain liquid for a long time. This quality makes them particularly useful as lubricants and as a fuel for lamps. Drying oils, on the other hand, are quite quick to oxidize and become solid, thus, they are often used in paints and varnishes. Linseed oil is a good example of this. Semidrying oils have qualities intermediate between the other two groups. Almost all commercially grown oil seed crops in the temperate zone are of annual plants. The list is quite long of plants that are used in industry, but some of the most common ones are rape (see Figure 21-19), soya, linseed, sunflower, and safflower.

Waxes from plants play a fairly minor role in the world's economy, but several are important. Used in candles, textile sizing, and leather treatments and coatings, vegetable waxes yield to animal, mineral, and synthetic sources in economic importance. The most significant vegetable oil is undoubtedly **carnauba wax**, from a palm tree native to northeastern Brazil (*Copernicia cerifera*). One of the finest automobile waxes, carnauba, is also used in the manufacture of lubricants, chalk, matches, plastics, film, cosmetics, and as an additive to other waxes. It is the hardest natural wax and has the highest melting points of any known wax. **Candelilla** wax comes from the *Euphorbia antisyphilitica*, which grows in the desert areas of northern Mexico. Extracted in small quantities from wild plants, this wax is very similar to carnauba wax in most of its properties. In spite of being a high-quality wax, it will probably never become a cultivated plant producing a large annual yield because of its preferred habitat. It is used primarily for candle production.

Fiber

Cotton (*Gossypium hirsutum*) is probably the most widely used plant fiber. Cotton was first popularized as a replacement for wool in Europe during the Dark Ages. Wool garments harbored disease-carrying lice and fleas, could not be washed expediently, and were scratchy for

Figure 21-19

Rape seed oil has been gaining in popularity over the past decade because it is healthier than other cooking oils.

summer wear. Cotton was soft, washable, reasonably durable, and, once available commercially, inexpensive. Cotton fibers are not like the fiber cells found in stems; rather they are elongated seed epidermal cells (see Figure 21-20).

Figure 21-20

Cotton has been used for making clothes for centuries, and it is still one of the major fibers used in all textiles.

Another fiber producing plant is flax (*Linum*), from which linen, paper money, and cigarette paper are made from. Popular in ancient Egypt, fine linen was even used for wrapping mummies. Sisal is a coarse fiber from the leaves of the *Agave* and is primarily made into binding twine. Hemp fibers from *Cannabis* stems are excellent for canvas and hemp rope, but hemp is not commercially produced in the United States.

Summary

1. The early history of plants was the history of herbalism and medicinal plants. Modern medicine includes little "traditional medicine" in the Western world but a balanced approach in China. Some of the more primitive societies in Africa and South America also depend extensively on their medicine men even today.

2. Medical plants were often surrounded by mysticism and superstition even though many of them contained compounds that did affect the body physiologically. Some of the well-documented medical plants include mandrake, *Duboisia*, and belladonna, all in the Solanaceae. In addition, *Cinchona*, Opium poppy, curare, *Rauwolfia*, *Ephedra*, foxglove, ergot, senna, St. John's Wort, *Gingko biloba*, and cascara have medicinal properties.

3. Many herbal remedies have less dependable curative powers, but many people still depend on these plants for general well-being, minor pain relief, topical cuts and burns, internal organ disorders, and respiratory tract disorders.

4. Psychoactive plants include such stimulants as cocaine, depressants such as the opiates, the true hallucinogens such as marijuana, ergot, fly agaric, sacred mushrooms, peyote, and members of the Solanaceae.

5. Poisonous plants contain active ingredients in one of several classes of compounds. Internal organ poisons include the alkaloids and cardiac and cyanogenic glycosides. Also producing toxicity in the internal organs are plants containing oxalates, resins, and phytotoxins.

6. There are about 75 species of fungus that are poisonous and about 10 that are deadly. Members of the genus *Amanita* are among the most toxic of all mushrooms.

7. The deadly toxic plants include hemlock and water hemlock, monkshood, yew, oleander, jimsonweed, and tobacco. Many others are only slightly toxic.

8. Plants with industrial importance include the latex producers, such as *Hevea*, those that contain oils, and fiber-containing plants, such as cotton, flax, hemp, and sisal.

Something to Think About

1. Where are medicine men still being used?

2. List some of the well-documented medicinal plants.

3. How do people still use herbal medicines?

4. List the psychoactive plants.

5. Name ingredients of poisonous plants.

Suggested Readings

Bresinsky, A. and H. Besl. 1990. *A colour atlas of poisonous fungi.* New York: Wiley.

Ellis, D. 2003. *Medicinal herbs and poisonous plants.* London: Blackie and Son.

Gottlieb, A. 1997. *Peyote and other psychoactive cacti* (2nd ed.). Berkeley, CA: Ronin Publishing.

Gruber, H. 1992. *Growing the hallucinogens : How to cultivate and harvest legal psychoactive plants* (3rd ed.). Berkeley, CA: Ronin Publishing.

Schultes, R. E., et al. 2001. *Plants of the gods: Their sacred healing and hallucinogenic powers.* Rochester, VT: Healing Arts Press.

Internet

Internet sites represent a vast resource of information. The URLs for Web sites can change. Using one of the search engines on the Internet, such as Google, Yahoo!, Ask.com, and MSN Live Search, find more information by searching for these words or phrases: **herbalism, medicinal plants, codeine, heroin, morphine, Opium Wars, hallucinogenic plants, poisonous compounds from plants, and industrial plants.**

CHAPTER 22

Modern Agriculture and World Food: Why Plant Science?

"**Within a few weeks, *Draba*, shown in** Figure 22-1, the smallest flower that blooms, asks, and gets, but scant allowances of warmth and comfort; it subsists on the leaving of unwanted time and space. Reference books give it two or three lines, but never a plate or portrait. Sand too poor, sun too weak for bigger, better blooms are good enough for *Draba* to produce flowers and seeds to survive. After all, it is no spring flower, but only a postscript to hope." (A Sand Country Almanac, with other essays on conservation for Round River by Aldo Leopold. Oxford University Press.)

To the person lacking a background in plant sciences, *Draba* is indeed as Aldo Leopold describes—small, plain, unimportant, and unnoticed. Again, however, Leopold makes these comments tongue-in-cheek, for he knows that *Draba*'s place in the botanical world is as important as larger, more visible, and more "useful" plants. That realization sets apart the plant educated from the average citizen. When more of the world's people view themselves as natural members of the biotic community instead of users of it, then the future balance and stability of the

605

community is secure, as is the future of each of its individual's components. In a sense, then, humans are not more important than the lowly *Draba*; nor are we any less important. Now, it is possible to view the importance of the botanical world to the human world—they are one.

Objectives

After completing this chapter, you should be able to:

- **Discuss how all plants contribute to the natural world**
- **Present a basic knowledge of how plants function**
- **Explain how sociopolitical considerations are world wide in scope**
- **Describe how modern agriculture has changed**
- **Identify which conservation movements have been successful in this global economy**
- **Describe how to help educate people concerning the significance of environmental problems**

Key Terms

deficit irrigation
genetic engineering

integrated pest
management (IPM)

The Mechanization and World Food Supply

Industrialization and the age of mechanization led to quantum leaps in the concentration of peoples; cities evolved rapidly along with new concepts in the production of material goods, communication, and transportation. This shift toward urbanization strained the existing food supplies and encouraged those in agriculture to find more efficient and productive means of supply for the cities. This challenge was met with varying degrees of success, but eventually the size of towns and cities increased.

Large cities developed in parts of the Old World thousands of years ago, and the Mayan civilizations in Central and South America was highly structured in 5000 BC. Beans and squash were being cultivated in the New World 9,000 years ago, and rice was cultivated in Thailand 12,000 years ago. By the time of the Industrial Revolution, agriculture was common throughout the world and established the basis for permanent settlement and civilization. Food plants first domesticated—wheat, barley, rice, corn, and potatoes—remain today as staple crops. Only tomatoes and coffee among major food crops have been domesticated in the past 2,000 years! Although there are as many as half a million species of plants in the world, only a few have been developed as major food crops. Farmers, traditionally reluctant to try new crops, have made up in quantity what they lacked in variety. Although surpluses were never large, farmers were able to produce more food than required. Today in industrialized nations all the food is produced by about 5% of the population.

Food Commodities

Modern human diets depend heavily on the many forms of plant life. The grass seeds have given rise to cultivated cereals, most domestic animals are fed from cultivated crops (except for range animals), and fish continue to feed on aquatic plants of fresh water and the sea. See Table 22-1 for a list of the major world food plants.

Figure 22-1

A small inconspicuous plant, **Draba,** *blooms early in the spring.*
Courtesy of National Forest Service.

Table 22-1	
Major World Food Plants	
Type	**Plant**
Grains	Wheat, rice, corn, sorghum, millet, barley, oats, and rye
Tuber and root crops	Potato, sweet potato, yam, and cassava
Sugar crops	Sugarcane and sugar beets
Protein seeds	Beans, peas, soybeans, and lentils
Oil seeds	Olive, soybean, peanut, coconut, rape, sunflower, and corn
Fruit and berries	Citrus, mango, banana, and apple
Vegetables (fruits eaten as vegetables)	Cabbage, squash, and onion

At least 90% of all human caloric intake is provided by commercially grown plants. Although meat continues to provide a large portion of the diet in countries that can afford it, the tendency is toward a greater consumption of plant products. Since energy losses increase in ascending trophic levels, it is possible to prevent these losses by eating "lower" on the chain, that is, closer to the producer organisms. In developing countries, plant consumption has always been great. In industrialized nations, however, the tendency toward meat consumption has long been associated with a higher standard of living. World food pressures now dictate even to the most developed nations that such eating patterns are no longer acceptable. Americans are rapidly learning that only a portion of what a cow, pig, or chicken consumes is converted into milk, meat, or eggs. A major portion of the feed consumed sustains metabolic function; bone, skin, and other body organs are unimportant to humans. It is necessary to consider, however, that animals do convert a great deal of nonedible and nonharvestable plant material into important productivity.

At the present time, three-grain crops—wheat, rice, and corn—provide about 80% of the total human calorie consumption (along with potatoes, yam, and cassava). Although rich in carbohydrates, these crops are not particularly high in protein or fat. Diets must be supplemented with other foods to achieve some degree of nutritional balance. Vitamins and minerals are usually supplied through vegetables. Even though poor in protein, these six crops do provide considerable protein simply because they are consumed in large quantities. In addition, beans are an excellent source of protein, and peanut, soybean, and sunflower contains both protein and fat.

Sugarcane and sugar beets were, until recently, chiefly a localized food source, but they have become major world crops as a result of better transportation. The growth of sugar beets has continued to replace sugarcane as the major source of sugar for North America.

A few other grain crops are important food sources. Barley, millet, sorghum, oat, and rye are consumed throughout the world. The world food consumption, however, is based on production of only a few species of the myriad possibilities available in the botanical world. It certainly is true that humans have tried and rejected many potential food plants because they were unpalatable, non-nutritious, or toxic. Others were too difficult to cultivate, harvest, transport, or store. On the other hand, human beings are exceedingly conservative when it comes to trying new foods. Most of us reject new foods because we do not like their look, smell, or feel.

Nutritional value is seldom a consideration. Religious and social customs often prohibit certain foods. There are hundreds of nutritious food choices available to the human populations that have not been adopted. "Experimental" new foods or new methods of preparations appear almost every day, but most discoveries fail to obtain general acceptance.

Specialty crops, such as strawberries (see Figure 22-2a), onions, and even mushrooms (see Figure 22-2b), are a recent phenomenon and contribute little to the total human diet; likewise, spices add to the palatability of foods but contribute little to human nutrition.

Figure 22-2a

The consumption of strawberries is increasing; however, they do not contribute to the nutrition of the consumer.

Figure 22-2b

Mushrooms are a specialty crop that does have nutritional value. Though mushrooms are commonly thought to have little nutritional value, many species are high in fiber and provide vitamins such as thiamine, riboflavin and can be used in low calorie diets.

A major portion of the world's population can be divided into those who eat rice and those who eat wheat. Rice is grown in the warmer parts of the world, generally in regions in which predictable monsoons sweep the countryside. In a year without a monsoon, severe famine can affect localized regions, although modern methods of food transportation, distribution, and storage have alleviated much of this problem. In Asia, rice comprises 75% to 85% of the entire diet. Rice has become a major agriculture crop in the Sacramento Valley of California and along the Gulf Coast in Texas and Louisiana. Rice has never become a major part of the American diet, and most rice production is for export. Through genetic engineering, more beta-carotene (vitamin A) has been incorporated into some rice. This will help developing countries feed their people more nutritious rice.

Wheat, shown in Figure 22-3, is the dominant cereal crop in most of the temperate or cooler regions of the world, and its production extended into the subtropics and higher altitudes of the tropics. Before the development of modern plant breeding, wheat had little cold tolerance and was confined to the warmer regions of the globe. Scientists using genetic engineering have also transformed wheat. They have bred cold tolerance into the crop, and now it is grown in the most extreme latitude.

Figure 22-3

Wheat is one of the top three grains consumed in the world.
Courtesy of USDA-NRCS Photo Gallery.

Wheat and rye are considered the bread grains because most of their production goes into flour. Although rye was a major crop a century ago in the United States, today almost all production is centered in Europe and the former Soviet Union. Regional food patterns change over time, and crop production must mirror these changes or fail economically.

Figure 22-4

Corn is also one of the top grains consumed; however, with the push for methanol as an energy source, more corn will be grown for fuel and less for food. Courtesy *of USDA-NRCS Photo Gallery.*

The typical bread wheats are hexaploid (six sets of chromosomes in somatic cells), whereas the macaronis and noodle wheats of Europe and Asia are tetraploid (four sets of chromosomes). Wheat is grown both as a winter annual (planted in the fall) and as a spring annual (planted in the spring). The genetics of the cultivated varieties are different—the winter wheat containing genes for a great deal more cold tolerance. In recent years, livestock grazing on winter wheat has become a major source of income, sometimes exceeding the value of the grain crop. Winter wheat is an excellent source of green forage during the winter, and the grazing forces the plants to tiller (give off more stems) and produce more heads of grain.

In regions that are too warm and dry for the cool season cereals, sorghum and millet are major food crops. Sorghum is particularly important in northern China, India, tropical Africa, and the southern United States. Sorghum flour and millet flour are not considered good for baking, although they are eaten in many countries and yield a fermentation substrate for making beer. Sorghum is currently used as a feed for domestic livestock. The protein content of sorghum is higher than in most cereals, and because it can survive low rainfall, high temperatures, and alkaline soils, its popularity is increasing dramatically worldwide.

Corn, shown in Figure 22-4, was the food staple of the Incas, Mayans, and Aztecs, and today it remains a very important dietary component in the Americas. Other countries are beginning to accept corn as a food crop—slowly. It is well adapted to warm summers. Corn requires more water than many parts of the world can offer, and it is grown in those regions only with supplemental irrigation.

Other grain crops contribute to total world consumption, although rice, wheat, and corn are certainly the most significant. Barley was formerly used extensively as a bread grain, but now about half the entire world production goes into the making of beer. Oats and buckwheat enjoy regional consumption but contribute little to total world caloric intake.

A synthetic grain crop has been introduced during the past years. Plant breeders have hybridized wheat (*Triticum*) and rye (*Secale*) to form Triticale, which has better characteristics of both parents, including higher protein content than bread wheat and excellent baking properties. Triticale has been widely accepted in the regions of the former Soviet Union, but its success in the United States has been limited. The difference in quality and taste of Triticale bread does not seem to justify the development cost still attached to *Triticale* production.

Historical Perspective

The food situation on earth is now and always has been precarious. Approximately three-fourths of the world's population knows poverty and hunger; in contrast, about 10% live in relative affluence. The other 25% live somewhere in between, never quite sure whether the supplies of food, shelter, and fuel will be adequate. The uncertainties are the same as they were for the first *Homo sapiens*, and the population is larger. Modern humans still depend largely on the agricultural success. Insect and disease control is a major problem, and we have surprisingly little appreciation of how organisms interact in nature.

Throughout recorded history, agriculture productivity has increased, but concomitant rises in population have always placed food supply at a dangerously close intersection with demand. Hunger is an ever present and prominent human condition. The stress of famine can induce strange behavior, such as hoarding; stealing; selling children; eating clay, diseased rodents, and bones; murder; suicide; and even cannibalism. Early accounts of the Roman famine of 436 BC and the Indian famine of 1291 AD report that thousands of people drown themselves in rivers. Cannibalism is reported to have occurred in at least 15 famines in England, Scotland, Ireland, Italy, Egypt, and China. As recently as 1921, Russian cemeteries had to be guarded to ensure that hungry thieves did not steal recently buried bodies. Famines were often localized and involved only a few thousand people, but the great Indian famine of 1769 to 1770 cost the lives of approximately 10 million in 1869, 5 million from 1878, and 1 million in 1900. In China, between 9 and 13 million people died in the famine years of 1877 to 1879. In Asia and on the Indian subcontinent, famine has been most often seen because of the unpredictable monsoon rains.

The Irish potato famine of 1846 to 1847 caused the death of 1.5 million people and prompted a massive emigration to the United States. Before the potato famine, Ireland's population was 8 million; the current population is slightly more than 4 million, primarily a result of the long-term effects of the famine.

Through modern transportation and communication efforts, we can anticipate famine and deliver food, supplies, and education materials to relieve its effects. On the other hand, massive aid is not always possible in other than localized situations. War activities often inhibit the flow of food to areas that need it most. Since World War II, for example, supply problems caused the starvation of 2 to 4 million people in the Bengal region of India. From 1969 to 1970, civil war in eastern Nigeria led to several thousand deaths by starvation. There are always droughts, wars, and famine in many developing countries, causing several thousands of deaths by starvation. In spite of a worldwide campaign to send massive food and monetary aid to those regions, there was little relief from the situation until the rains came, and by that time the ecological destruction was excessive.

Various industrialized nations have attempted to buffer the effects of malnutrition and starvation. The United States contributes surplus food for approximately 100 million people per year. Beginning with the enactment of Public Law 480 in 1954, the United States began selling food to poor countries for payment in local currencies rather than in dollars and gold.

During the early 1960s, American grain surplus reached high levels, which were depleted after Public Law 480 came into existence. Land that had been idle was called back into production, and from 1966 to 1967, 20% of the entire U.S. wheat crop was required to reduce famine in India. From the 1980s to the present, American farmers have been asked to reduce production because of temporary surpluses.

One of the social complications of a giveaway food program is the false confidence instilled in the recipients concerning their future. Unless food is given out only in the case of temporary disaster, countries (and individuals) tend to become reliant on the contribution and finally believe that the supply is inexhaustible. Many leaders believe that such false confidence merely results in further population increases and delays the day when food will not be available. Keeping humans alive at a subsistence level so that they may produce yet more offspring raises serious moral concerns about the future of such populations.

All these examples serve to point out the precarious position of world food supplies. Some sources contend that the biosphere has the ability to produce all the food needed for the foreseeable future. Others contended that the level of human population has already surpassed the ability of the world to provide food. The truth lies somewhere in between. New sources of inexpensive energy could improve food production greatly, but such energy sources seem many years away. While we wait, the world's population continues to double at an alarming rate.

Early Agriculture

There is no question that agriculture has progressed steadily ever since the glimmer of domestication first touched the hunters and gathers. Some periods of history have seen rapid progress; others have been slow. The tillage of the rice paddies in Asia, for example, has been highly developed for thousands of years. Control of erosion by terracing has always been a standard practice. In some parts of the world, including Central America and Mexico, elaborate irrigation schemes ensured that water reached the crops at the right time.

In spite of these early advances, it was not until the Industrial Revolution that machinery became available for large-scale agriculture, which relieved a large portion from drudgeries of hand labor (see Figure 22-5). The continued growth of mechanization has increased productivity in almost all crops. Some crops are still labor-intensive, and mechanization has only partially alleviated the need for hand labor.

Figure 22-5

The old agriculture was actually the growing of the crops.

🌀 The New Agriculture

At the turn of the twentieth century, plant breeders and other agriculture and biological scientists began to accomplish impressive gains in crop yields. Productivity has accelerated during the past century, and some yield increases have been truly remarkable. The "North American Breadbasket" was made possible through research and technology.

Basic Research

The elucidation of metabolic processes, including photosynthesis, respiration, and protein synthesis, is only one area of discovery that has led to increased productivity (see Figure 22-6).

An understanding of water and soils has improved conservation practices and irrigation methodology, and genetic research has led to a process of genotype selection to match specific crops with specific environments. Tremendous advances have been made in growth regulation with plant hormones (see the corn and soybean examples in Figure 22-7), and the understanding of how diseases are transmitted has increased yields. Investigation of viruses and their transmission by insects has also been significant.

Figure 22-6

The new agriculture has moved into research laboratories. Agriculture was the first area to use genetic engineering effectively.
Courtesy of Monica Haddix.

Figure 22-7

Corn was genetically engineered to contain a biological control for the corn earworm. Soybeans have been transformed so that Roundup can be sprayed over the field to kill the weeds and not the soybean plant.

Applied Research

Once basic concepts have been elucidated, the processes must be put to the test under field conditions. Not all hypotheses born in a research laboratory survive the rigors of field testing. Some genotype-screening programs may look excellent in laboratory/greenhouse studies, but in the field they sometimes fail because of susceptibility to insects or diseases, low humidity, wind, soil chemistry, weed competition, and other

Figure 22-8

Agriculture has also conducted applied research so that the new advances in agriculture could be used by the growers. Courtesy of USDA Photo Gallery.

factors (see Figure 22-8). Only those gene sources that help the plant withstand the rigors of field competition have a chance of making a contribution to world agriculture.

Once innovative ideas have proven workable, the task of transferring the information to farmers is monumental. Many good ideas of the past have been filed away in a research paper and have never been put to use. For example, the practice of not tilling the fields has proven to be very beneficial because the natural microorganisms of the soil can reestablish when the crop is grown under no-till practices. Even with modern communication and impressive sales campaigns through the media, selling farmers on a new concept, new crop, or even a new variety of the same crop may be difficult. Tradition plays a major role in farming practices and procedures in all societies. Change comes slowly unless technology transfer concepts are applied judiciously.

Perhaps the most impressive example of technology transfer in American agriculture is the land-grant university system, together with the state agricultural experiment stations and their accompanying agriculture extension service. This highly organized sector of the agricultural community managed to "sell" American farmers on innovative technology early in the twentieth century, using the demonstration plots to show that new methods and crops were superior to old ones. Farmers are impressed by success across the fence. Regional experiment stations have brought experimental farms close to every American farmer through the use of open houses, field days, and cooperative research. They have been instrumental in communicating new farming practices. With the continued demise of the family farms in the twenty-first century, farmers need to look very closely to diversifying into other crops and to the new high-technology production and equipment to survive.

At the present time, the Internet can make every corner of the world accessible. New varieties of crops are quickly recognized throughout the country. Fertilizers, seed, pesticide, and irrigation equipment manufacturers spend a large portion of their advertising budget on Web pages and Internet-related activities.

Some times, farmers are eager to accept new ideas, but financial reserves and capital investment inhibit the progress of new technology. Lending institutions are often reluctant to take chances on new innovative ideas, and the wheels of progress may move very slowly indeed. Capital investment is particularly critical in the developing countries, where lack of funds to buy seed or fertilizer may stand between private interests and a high level of productivity.

Perishability is a major factor in food production and delivery. Perhaps some of the reasons for large-scale successes of grain products are their slow perishability and ease of storage and transport. The industry is far better equipped for long-distance transport of a truckload of wheat than for a truckload of lettuce or strawberries.

Basic research has been instrumental in developing techniques to improve the processing and shipping of plant products. Consider, for example, the problem of maintaining fresh corn or pea quality.

Harvesting affects sugar content, and the best produce is obtained if harvesting can be accomplished within a matter of hours after the highest sugar content is reached. The crop must be processed immediately because respiration rates are so high that even minor biochemical changes can affect flavor, texture, and overall quality.

Perhaps the most progressive storage techniques concern apples. Advances in postharvest physiology have been so dramatic that a controlled atmosphere prolongs the storage life of apples up to an entire year with little deterioration of quality. Scientists have spent many years of efforts in plant biochemistry, physiology, genetics, and plant biotechnology. The consequences of such storage and transportation techniques are a more stable market price and therefore a better pricing structure for both growers and consumer.

The Farmer's Bargaining Power

Whether a farmer can afford to maximize yields depends to a large extent on the price of a product. Although food prices have escalated dramatically throughout the world, most of that profit has been realized not by the farmer, but by the processor, packager, distributor, and transporter. Farmers traditionally have been so poorly organized as a political group that they fail to exert a major influence in the political arena or at the marketplace. They usually take the going offer, although modern storage facilities for some farm products allow the farmer to withhold from the market until a better price is offered. Producers of perishable products do not have this luxury, except through frozen or dehydrated foods, and for them the energy costs of withholding from the market are high.

The Limits of Production

The question is often asked whether farmers can continue indefinitely to produce as much food as the world requires. Throughout recorded history this has seldom been possible. Recall that the laws of thermodynamics state that less energy comes out of a system than is put into it, that is no system is perfectly efficient. The conversion of light energy into chemical energy is relatively inefficient.

The total sunlight available at any point on the earth is readily determined. Green plants capture only about 1% of the light energy on an annual basis, and much of the energy falls on land or water without green plants. Even if plants are properly spaced to maximize light trapping, the efficiency of conservation is low. Most of the energy is reradiated or lost to the system as heat. Thus, strictly from a thermodynamic point of view, the light energy available determines the upper limits of productivity. Seldom, however, in commercial agriculture is light the limiting factor in production. On a worldwide basis, the limiting factors are unquestionably water and nutrition. Most of the world's so-called arable land is now in cultivation. The only land left is in the humid tropics, the desert, or mountainous regions too steep to cultivate safely. Although a great deal of land is available in the arid and semiarid parts of the world (somewhere between 25% to 33% of the entire land surface), water is the limiting factor in development, and much of the North American breadbasket land is irrigated. In some cases, irrigation comes from surface water and the other underground water must be pumped to the surface and then transported to the site of application. These pumps are usually located at the individual farms, and pumping may be done at very shallow depths, which requires very little energy, to hundreds of meters, which requires a great amount of energy. Many crops grown in arid regions require 2 to 5 acre-feet of water per year (1 acre-foot of water = 1 acre foot deep, or approximately 325,000 gallons). In some regions, this accounts for 90% of the entire water consumption, with all domestic and industrial uses requiring only 10%. Therefore, irrigated agriculture comes under careful scrutiny in matters of water conservation.

You will recall from Chapter 9 that transpiration accounts for about 99% of a plant's demand, and there are very few realistic suggestions for reducing that water loss. Anything applied to the plant to reduce water loss either seals off the stomata or causes them to close, and if the stomata are closed, no CO_2 can enter, and photosynthesis is greatly reduced. The correct approach appears to be to grow more water-efficient plants, those which produce more biomass while using less water. Sorghum, for example, is far more water efficient than corn if water is limited. If water is not limited, then corn is more efficient.

The Future of Irrigation

The arid and semiarid parts of the western United States typify similar regions of the world. In some states, irrigation has been developed to a higher degree of sophistication with tremendous capital investment, as shown in Figure 22-9. In some states, irrigation water is transported hundreds of miles, as from the Feather River in northern California to the dry southern California desert. A similar scheme has been proposed to transport water from the Arkansas River to the farmlands of western Texas and eastern New Mexico. This particular idea never got started.

Quality is always a concern in the transport of water because irrigation water always contains some salts. In some cases, more than a ton of salt per acre-foot is solubilized in the water. As the water is stored and absorbed by the plant roots, the salt is left behind in the soil or the plant or in some cases leached beyond the root zone. Salt accumulated over a period of years is a major pollutant, and great efforts are made to drain the soil and leach the salts. The Imperial Valley in California is lined with a massive drainage system. This region produces a great percentage of the nation's winter vegetables.

In the central part of the United States, massive underground aquifers hold water stored there for millions of years that is being pumped to the surface for irrigation of the Great Plains. The Ogallala

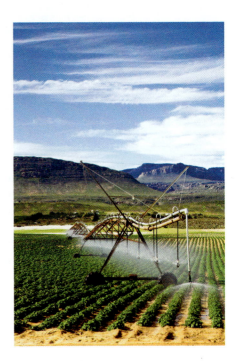

Figure 22-9

Sprinkler irrigation has allowed farming on steeper slops than is often possible with furrow or flood irrigation. Although considerable water is lost to evaporation, new techniques and equipment apply the water closer to the ground.

Aquifer stretches from South Dakota to Texas and has allowed the high plains of Texas to contribute significantly to the nation's supply of food and cotton. Unfortunately, this aquifer is not being recharged, and the "fossil water" is being used up. Some areas have already had wells go dry and others are becoming depleted. Depending on the locale, southern parts of the aquifer will no longer be a source of irrigation in the future.

The suggestion has been made that interbasin transfer of water be accomplished from regions with a surplus of surface water. Many proposals have been made for importing water from Canada, the Missouri River, the Arkansas River, and even from the mouth of the Mississippi River at New Orleans. While technically feasible, the costs of construction and maintenance of such a system are astronomical. The energy costs of pumping alone would be more than the states could afford. Even if such a system were to be approved by voters (it has been voted down once), legal ramifications would probably disallow construction.

The Constitution of the United States relegates the water rights to the individual states, and it seems unlikely that one state would ever be willing to sell its water to another. The projected cost of water delivered to the farmer is absolutely prohibitive for the irrigation of crops.

The future of irrigated agriculture, then, seems dimmed by accurate cost-benefit analysis. In those regions where water delivery is inexpensive, readily available, and dependable, irrigated agriculture is likely to persist. A recently introduced concept, **deficit irrigation**, implies that irrigation is to be used only in emergency situations to save a crop or get over a period of particularly bad weather conditions. Although yields would not be maximized, decreased energy costs of production would offset the loss of yield, and the farmer would make just as much revenue.

In regions that have been irrigated but no longer have the water to do so, the future of agriculture will depend almost entirely on rainfall and the judicious selection of crops. A return to dryland agriculture is forecast for much of the Great Plains, where rainfall is sufficient to support drought-resistant grain crops and cotton. In many cases, revenue to the farmer is just as great as with irrigated agriculture, although the regional and individual farm productivity is decreased. Regions that are forced to switch from high-intensity, dryland agriculture, as shown in Figure 22-10, simply will not be able to provide as large a segment of the world's food supply.

Most of the world's air and semiarid regions fall into desert or grassland biomes. If land has been broken out from a grassland biome, it is likely to support the original or similar vegetation.

Figure 22-10

Marginal land for dryland agriculture, or land intended for irrigated agriculture but without a water supply, fails to support vegetation and often results in severe erosion. Courtesy of USDA Photo Gallery.

If one decides to return the land back to its climax vegetation, reestablishment may be difficult. Under conditions of wind, high temperatures, and unpredictable rainfall, establishment of vegetation under dryland conditions is a difficult assignment. Even so, where dryland agriculture is marginal at best, farmers often consider returning to grassland for livestock or wildlife grazing. If a small amount of irrigation water is still available, the judicious use of that water for grassland establishment may be strongly considered.

The return of native vegetation is even more difficult in the deserts. The competition for water is so great the plants are far apart, and large root systems are essential for survival. When farmers or ranchers remove those plants, it is difficult to achieve reestablishment. Not only is water accumulation a problem, but nutrient cycling in the desert is particularly precarious. Individual plants are often referred to as islands of fertility; the nutrient supply cycle within the plant and in the soil directly under the plant. The desert soil between plants may be sterile, but nutrients abound in the plant. However, once that plant is removed, the nutrients are swept away by the wind or leached out of the root zone by thunderstorms. Once that nutrition has been lost, it is impossible to put it back again. The desert is indeed a delicate resource.

Demography

In 1797, Thomas Malthus published *An Essay on the Principle of Population* in which he put forth the hypothesis that biological species including Homo sapiens, always have the ability to produce more offspring than can be accommodated. Natural systems keep population levels in check, primarily by limiting food supply. No one is likely to be upset when a test tube of bacteria runs out of a nutrient source, or when a plague of grasshoppers runs out of forage plants, but the Malthusian concept strikes home vividly where there is a lack of human food. We have seen that famine and starvation are effective in controlling the size of the human population. The climatic problems involved in food production are compounded by social problems. The message from Malthus, and we have no reason to believe that his hypothesis is not correct, is that if the human species fails to control their own population size, natural forces will intervene and do it for us.

At the beginning of agriculture some 12,000 to 18,000 years ago, the human population was less than 5 million. At the time of Christ, the world had 250 million inhabitants, and that number had grown to 500 million by 1650. In 1850, it had doubled again to 1 billion; and by 1930 it had doubled again in just 80 short years. By 1979, it had doubled again, this time in only 46 years, and at the present rate, doubling will take approximately 40 years. Many factors kept early population levels under control, but the astounding leaps in doubling time are cause for alarm.

Countries that can least afford the population increases (the developing countries) are the ones with the most rapid rises. In Latin America, Africa, and Asia, the annual number of births far exceeds the number of deaths, and the rate of increase is more rapid than 3% per year in some regions. Throughout human history, the birth rate has fluctuated considerably, but changes in the death rate usually have the greatest effect on population growth.

In the developed countries, unlike the underdeveloped countries, population growth is declining because of technological advancements and literacy rather than religious factors. Japan and China have greatly influenced the rate of population through governmental decree and massive birth control campaigns. The availability and understanding of birth control devices and social acceptance of family planning are major factors in control. It appears that a stable food supply and lack of social upheaval are also important in reducing the desire for more children. Anticipation of insecurity in old age is often cited as a reason for large families.

As long as the earth had adequate supplies or was able to bring that land into production to keep up with the population, increases were a relatively simple matter. However, essentially all of the world's arable land is now in cultivation (except for the humid tropics). The problem confronting a move into the arid lands has already been discussed. The only other area with possibilities for expansion is in the humid tropics. The problems there are probably even greater than in desert agriculture. Soil fertility, pest management, storage, and transportation are all major obstacles to agriculture development. In spite of these problems, the tropical rain forests are rapidly being developed, often without rational decision making.

Agriculture is forced to attempt to increase yields on land already in production. In addition to all the problems associated with excessive resource usage and deterioration of the ecosystem, it would appear that the portion of the biosphere capable of supporting agriculture is already being taxed to the limit. What will the future hold?

The Future of Agriculture

Agriculture is realizing that to achieve ecological balance we should abandon the idea of modifying the environment to fit the crop. Instead, we should select or transform crops to fit specific environments, and recent advancements in molecular biology have led to the use of **genetic engineering**, the process of combining portions of the genome of another. Thus, desirable characteristics of totally unrelated organisms have been incorporated into another organism to impart some beneficial effect. For example, corn has been transformed with *bacillus thuringiensis* (a natural parasite of caterpillars). So the corn earworm that devastated corn crops can now be grown without the loss of ears due to the earworm. This is only one of many such transformations; of all the crops grown now, approximately 75% have been transformed for some better traits. Another transformation has produced Roundup-ready soybeans. This allows the farmer to spray for weeds in the field and not kill the soybean plant growing there. This technology is developing so fast that it is hard to keep up. This is the future of agriculture as well as relief for world hunger.

There are hundreds of potential foods and forage plants available but never selected for agriculture production. Introducing a new crop is exceedingly difficult in today's world. Basic research, applied research, and technology transfer must be undertaken before there is any hope of breaking into commercialization. The infrastructure that surrounds each crop is unique to that particular species. Rice, for example, must have specialized planting, cultivating, and harvesting equipment; people familiar with equipment particular to that crop handle the processing, storage, and transportation. Even the marketing and acceptance are unique to certain parts of the world. If the world's rice production should suddenly double, marketing and acceptance would be a major problem throughout the western cultures.

New crops are sometimes suspect because they upset the traditional patterns of production and consumption. On the other hand, economic pressures have caused the world's agricultural producers to take a closer look at innovation, including the new transformed crops. Although new crops are not a panacea for the world's food supply and population ills, they could be highly effective in alleviating hunger and famine in a world caught in its own ecological net.

No single factor will save *Homo sapiens*. We must use our best creative efforts to increase productivity within the ecosystem that surrounds us, while using the same ingenuity to keep our worldwide population at a safe and manageable level. Malthusian theory has worked through recorded history; there is a new race against time—to see whether humans are clever enough to save themselves from extinction.

Gardening: Number One Leisure Activity

Although few people become professional plant scientists, the simple awareness of plants and of their importance makes us all plant scientists in one sense. In addition to a realization of the applied worth of plants to human society, a basic knowledge of plant structure, function, reproduction, distribution, ecology, and cultivation allows us to better appreciate the total role of plants on earth. A plant's role should not be measured exclusively in terms of direct human utility, but in terms of energy flow, oxygen production, erosion control, and overall natural ecological balance (see Figure 22-11).

Figure 22-11

Gardening is the number one leisure activity in the United States. As the baby boomers retire, as families downsize, and as more people develop the empty nest syndrome, gardening will continue to be the number one leisure activity. Courtesy of Alex Wofford.

Learning about the plant kingdom, therefore, is not purely an academic exercise, but rather it contributes to the present and future survival of our planet and the continuation of all the organisms on it, including humans.

The Basic Importance of Plants

The importance of plants to the existence of all other organisms on earth can be summarized by reviewing two aspects of photosynthesis, (1) energy conversion and (2) oxygen production. Plants form the base of the food chain because they combine atmospheric carbon dioxide and water with sunlight energy to produce food energy in the form of carbohydrates. Without green plants, therefore, the animals of the world would have no source of food other than some single-celled aquatic organisms. The availability of plant tissue as a food source has resulted in the development of extensive and involved plant-animal and animal-animal interactions, as well as the evolution of complex forms of land-dwelling organisms.

As a by-product of this energy conversion process, oxygen is released into the atmosphere. With few anaerobic exceptions, all organisms on earth require oxygen to carry on respiration, the process of producing from carbohydrates a form of energy capable of doing biological work. Additionally, the availability of oxygen ultimately resulted in sufficient ozone accumulations to allow life on land to be screened from damaging ultraviolet radiation.

These two functions of plants, combined with their soil stabilization functions, their utility as nesting habitats, nutrient cycling, essential amino acid production, and many other interactions with other organisms, elucidate their central position in the balance of nature. Even lay plant scientists must acknowledge the magnitude of the importance of plants in nature.

Recreation and Aesthetics

Aldo Leopold began the foreword to *A Sand Country Almanac* by stating "There are some who can live without wild things, and some who cannot." More and more people are realizing the worth of wilderness areas and the relaxing beauty of nature. To be able to walk a quiet trail through the woods, to sit on a rock by a river, to notice the Silphium, the *Draba*, the first robin of spring, these are not to be taken for granted but to actively and consciously be preserved. It is in such settings that the brightness of the stars and the contrasting darkness of the voids between them stimulate poetry. It is not without reason that we refer to "escaping" from the big city to pursue recreational activities in the forest, on the mountains, at the seashore—in natural (see Figure 22-12) areas. Even within urban areas, parks can provide a measure of this feeling. Bringing plants into homes and offices is yet another example of the realization that natural surroundings relax and soothe.

Possibly it is instinctive or genetically controlled that humans want to spend time in natural settings. Possibly it is because many

of us exist primarily in the concrete, plastic, and steel world we have built, and natural areas offer a change. Whatever the reason, millions visit the national parks and wilderness areas annually. Too many of these visitors contribute to the decline of such areas by their thoughtless and destructive acts of littering and vandalism. Fortunately, there are many more that follow a good land ethic by leaving these sites as they found them. We are loving our nature to death.

When we enter these areas, we should do so as members of the natural environment, not as controllers or conquerors. This attitude comes from knowing and understanding what altering natural interactions can mean to the ecological balance. Lay plant scientists have such knowledge. They are the stewards of the land.

An often unsuspected bonus of study is a more highly developed aesthetic application. If beauty is in the eye of the beholder, then the better-trained eye has a greater potential for seeing beauty. It is understandable why trained naturalists are often the most emotionally involved in the areas they study. True love can develop on those trails through the woods or on the rock by the river.

Social and Political Considerations

The importance of plants to each individual should be clear. As sources for food, shelter, recreation, and industry, plants obviously serve a purpose for each of us. To all organisms on earth, plants are the most essential. As the basis of the food chain, as a producer of oxygen, and as components in the natural ecosystem balance, plants occupy an irreplaceable position. We should not isolate our new found knowledge, however. The study of plants should improve our ability to function as a component of the biological world. We should now be wiser taxpayers and more thoughtful voters and citizens. Plants are as important in our daily activities as we humans are to each other.

It should always hold true that knowledge—in any discipline—results in an improved awareness of issues, problems, possible solutions, and personal responsibility for participation. Posing questions is a natural extension of the educational experience. Your knowledge of plants should be as much a part of your base of information for living as your actions are an impact on the botanical world around you. Directly and indirectly, both as an individual and as part of a collective society, you can have a positive influence on the world around you and thus on your niche within that world. Even though we question many policies and decisions about ecological issues, land management, increasing population, and the like, there is a large portion of our world that is still beautiful, productive, clean, balanced, and with enormous potential.

Your actions will serve both to solve problems and to ensure the continuation of the many "good" aspects of our world.

World Food Supplies and Political Involvement

One of the consequences of a world of "haves" and "have-nots" is a great deal of resentment toward those people with a relatively high standard of living, sometimes manifested in anger, terrorism, and war. Humans demonstrate responsibility to the individual and family above all else. National pride, religious fervor, and charity pale when a family breadwinner is faced with starvation for his or her family. Psychologists tell us that the bottom line to self-preservation is that the individual will do almost anything to ensure that the family is fed. Too often in history, zealous leaders have taken advantage of this human instinct to rally forces to battle making shallow promises of protection and freedom from hunger.

In the past, ignorance for many people really was bliss. Modern worldwide instant communication has allowed those in the under-developed countries to see and hear about a better life. Their demands are now being heard in the political arena, and local governments are under increasing pressure to provide a fair share of the world's wealth. It has been said that a child born today in India will consume only about one-fortieth as much of the world as will a child born at the same time in the United States. Although that figure may be true for the present, the demand for a better life will no doubt increase as the Indian child's share during his or her lifetime.

One point is clear in the overall picture of global food supplies: The world is overly dependent on North America for its food resources. This is not to say that other countries have totally failed in food production. In recent years, many underdeveloped countries have made major advances in achieving self-sufficiency, at least in certain crops.

Several reasons explain this growing dependence on North America. Increasing population pressures, particularly in the developing countries, have hastened the deterioration of food systems. The increased pressure on traditional farming systems to produce more and more has led to ecological degradation; once fallow land (land cropped only once every 2 or 3 years so that moisture and nutrients can be stored) is now being called into constant production, depleting nutrients at a more rapid rate and often leading to severe erosion. In some parts of the world, newly rich nations, particularly the oil-producing countries, are demanding a greater share of the world's food. The gap between the haves and many of the have-nots continues to widen.

These factors have resulted in North America having significant influence over world food supplies. The United States and Canada are in a position to dictate to the importing nations the terms on which food will be distributed. Serious moral and ethical questions arise from this political power: Does North America have the right to use food as a political force? Can food resources be used to dictate population control, religious expression, and personal philosophy? Should food ever be used as a weapon to beat a recalcitrant country into submission?

There may not be any right or wrong answers to these questions, but they deserve informed thought. There may come a day when such questions will have to be answered. In addition, other questions arise in highly productive agricultural nations.

NEED FOR FOOD

The person without food has but one problem. The person with food has many problems. Today in the United States, about 98% of the people do not have to worry about food, and they have time to create worry about many other real or imagined issues.

Without food, an individual's thoughts focus on food, as Ernest Shackleton found out during the survival of his crew on the failed voyage of the *Endurance*. One of his crew, Frank Worsley, had this to say: "It is scandalous—all we seem to live for and think of now is food. I have never in my life taken half such a keen interest in food as I do now—and we are all alike . . . We are ready to eat anything, especially cooked blubber which none of us would tackle before." (Shackleton, *South: The Story of Sacketon's Last Expedition*; 1931). Next to air and water, food is essential. Food is part of the foundation of Maslow's hierarchy.

For most of human history, civilization struggled to produce sufficient food for survival. For varying reasons, some civilization still struggle for sufficient food for survival. Thanks to technological advances in crops, livestock, storage, management, and preservation, a few people are able to provide sufficient food for the masses in many developed countries.

People rush to blame various issues in countries where people need food. Some blame the population; some blame the economy; some blame politics; and some blame the environment. Providing food to those who don't have it is a complex problem. As the Bishop of Constantinople, John Chrysostom said in 407: "Feeding the hungry is greater work than raising the dead."(www.Chrysostom.org)

If people worldwide are to be fed, our nation and other nations need to make use of the best of technology, genetics, and management available to humankind.

Integrated Pest Management

One of the consequences of high-technology agriculture has been the tremendous money and energy put into an extended use of pesticides. The demands placed on North American agriculture have forced farmers to apply large amounts of insecticides, fungicides, and herbicides so that the monoculture could flourish without competition. Breakdowns in the ecosystems were attributed to an overload from various chemicals; some pesticides were incorporated into fatty tissue of organisms in various complex food webs, affecting reproduction and causing other negative effects. This type of pesticide was eventually banned in the United States. Strict controls and extensive testing have also eliminated many other pesticides from the market.

Renewed research efforts by scientists have led to a philosophical plan called **integrated pest management (IPM)**, which attempts to recognize crop production as an integral part of the ecological system. Relying on ecological principles, this approach seeks to minimize the use of pesticides, while relying on natural systems to assist in production objectives. Included in this system is biological control, or the use of one organism to control another. Although spectacular successes with biological control methods have been demonstrated, often a desirable insect, such as the ladybug shown in Figure 22-13, can be introduced in large numbers to control another insect. Sometimes insects can be used to control a noxious weed; effective control of the prickly pear cactus with an insect that feeds on the pads has led to reclamation of rangelands in Australia and elsewhere. Biological controls can fail to work in certain situations because predators of the introduced species often keep populations too low to be effective. Another precaution concerns the possible introduction of a pest problem even worse than the one being controlled.

Figure 22-13

Integrated Pest Management (IPM) is an environmentally friendly way to rid the gardens from insect pest. Lady bug beetles is one way of controlling aphids and other insect pests. Courtesy of USDA Photo Gallery.

Biologists have known for a long time that some plants produce natural insecticide that discourages herbivores. Chrysanthemum and marigolds contain natural pyrethrums that are extremely effective insecticides. Synthetic pyrethrums have been manufactured and are being used widely as a safer product than insecticides used heretofore.

Another technological advance is the use of *Bacillus thuringiensis,* a natural parasite as shown in Figure 22-14, to get rid of Lepidoptera and other chewing insects. It was first used as an insecticide dust to repel cabbage loopers. Next, large chemical companies started to transform plants through genetic engineering to incorporate this parasite into the DNA of plants that were bothered by Lepidoptera insects. Now, 85% to 95% of all the corn grown is from transformed seeds. This technology has also transformed soybeans with a gene that makes it resistant to herbicide (Roundup) so the weeds can be controlled without harming the crop.

Figure 22-14

Bacillus thuringiensis *is another biological control that helps to control loopers and other pests. This picture shows the parasite the* Bacillus *cause to infect and kill the looper.* Bacillus *is available in stores.* Courtesy of USDA Photo Gallery.

Conservation

The pioneers who settled the United States saw before them an apparently endless vista of forests, grasslands, clear rivers, lakes, seashores—and opportunity. The natural resources seemed boundless and biological diversity vast. The far-reaching extent of the resources probably precluded any consideration they might have had for conservation. Today, however, faced with shortages of fresh water, fossil fuels, clean air, safe food, and recreational sites, we wonder how they could have destroyed so much while settling America. A great deal is written about the need for conservation measures, but there is not enough actually done. Although state and federal agencies have set aside some wilderness areas and nature preserves, much of the public still has a "parks are for the people" attitude. Too many still cannot comprehend the expenditure of public tax dollars to establish tracts of land from which the public is barred. The pioneer spirit lives on: land is to be utilized, cleared, and made "productive" and habitable."

The conservation movement began slowly in the 1940s and 1950s with relatively few champions, who were often regarded as extremists, doomsayers, and against progress. During the 1960s and 1970s, laws were passed regulating air, water, and noise pollution; tracts of land were designated as preserves; and the public was slowly educated about the need for these measures. It was not until the late 1970s and 1980s, however, that this awareness has been used to understand the conservation movement and gain the support of a large segment of the public. With continued increases in public involvement, tax dollars, and legal controls, in the twenty-first century, we should see gains and not ultimately result in a situation unique since hunting and gathering bands wandered the land: Humans in balance and in harmony with the land is part of nature, not apart from it.

Endangered Species

Nothing more vividly symbolizes the impact of humans on the natural ecosystem balance than the number of plant and animal species that have become, or are in danger of becoming, extinct, even considering that the process of natural selection has produced millions of extinctions over the ages.

The habitat destruction shown in Figure 22-15 that has taken place is fairly recent. Between 1800 and 1850, only four plant species were known to have become extinct. Between 1851 and 1900, however, 41 species were lost to extinction and another 45 between 1901 and 1950. Today, there are estimated to be some 25,000 species of plants in the continental United States; 1,200 of these are categorized as threatened, and approximately 1,000 species are in immediate danger of extinction.

Figure 22-15

There has to be growth in our cities; however, it can be done with more thought to the existing environment and save some of what is already growing. It is easier to bulldoze down trees than to build around them. Courtesy of City of Greensboro, North Carolina.

No single habitat type has been altered more significantly than the tropical rainforest. The United States Environment Program estimates that over two-thirds of the world's rain forests have already been destroyed, and as many as 1,000 animals and 25,000 plant species are threatened with extinction. The tropics have an unusual delicate ecological balance. The specialized relationships of simple food webs result in a chain reaction. The removal of a link in the food chain is not an isolated event in such habitats; it can affect the survival of several other species as well.

The unfortunate response of some is "So what? There are plenty of other plant and animal species even if all the endangered ones do become extinct. Of what importance is a snow leopard or silver sword? They do not provide humans with food, medicine, or building supplies." Understanding the answer to this kind of question requires an appreciation of ecological balance. The loss of any given species may not be directly significant to human needs, but that loss may affect the status of some other species in the chain that is important.

Another part of the answer concerns potential. So little is known about most species that humans do not know that one of the lost species might include a potential crop plant, wonder drug source, or fuel reserve. Any unnecessary loss from the natural genetic pool stunts potential. In a world with massive shortages, this ought to be viewed as unacceptable. The propagation and preservation of rare and endangered pliant species in botanical gardens and arboreta is important so that their biology can be studied, and they can then be reintroduced into appropriate protected habitats where available.

Finally, even if there is no applied use for a species, and even if it could be proven that its loss would not have an impact on other plants or animals, the loss of any aspect of the world's variety is reason enough to protect these species. It would be a great loss to future generations to know of certain plants and animals only from pictures. Some species are currently known primarily from zoos and botanical gardens because they are already so scarce in the wild. The wholesale reduction in total habitat size for many species has already resulted in small recluse populations, sometimes protected by government decree, sometimes not.

The critical component in whether zoo protection, wilderness areas, legislation, and any other measures will significantly slow the rate of extinction resulting from human activity destroying available natural habitats is whether people realize the consequences of their actions. If not, then there will be little progress. In underdeveloped and overpopulated countries, the education of the layperson to these problems is especially difficult. The instinct for self-preservation is an immediate response to need; the significance of today's actions on the next generations is far too abstract to have much impact. The degree and the continuation of protection, therefore, are in the hands of the educated public. It is up to those of us who understand and care to speak out to rationally support conservation measures.

Environmental Quality

The most significant issue in human ecology during the past century is overpopulation. The degradation of our environment, as shown in Figure 22-16a and 22-16b, is a direct function of ever-increasing world population size. Resource shortages, pollution, and urban crowding all result from overpopulation.

Figure 22-16a

Water pollution is a very serious condition. Pollution of our water supplies could have very disastrous consequences.

Figure 22-16b

Air pollution is the presence in the atmospheric environment of natural and artificial substances that affect human health or well-being, or the well-being of any other specific organism.

Until humans realize that the laws of natural balance do not apply only to all other species, but also to Homo sapiens, a rational solution to these problems is unlikely. Other solutions are possible but not desirable. Mass starvation, epidemics, or war would effectively reduce the world's human population.

On the other hand, total dependence on continued technological advancement and vastly improved crop strains is not a realistic approach either. The quality of our environment will ultimately be determined by how many demands are made on it, and those reduced demands can happen only when there is a reduced human population.

Fortunately, natural systems have amazing resiliency and buffering capacity. Even ecosystems that seem for all practical purposes to be "dead" have the ability to recover if given enough time for cleaning and restoration. Lake Erie, for example, had become so polluted with industrial wastes during the 1960s that almost all life in that once magnificent lake had died. Once the cleanup began, under strict controls, Lake Erie started to recapture its ecological balance and has been restored to its magnificence.

In an industrial society, the cost of environmental quality is great. Every convenience and time-saving gimmick has its price tag, and we have sometimes cut corners for which we pay dearly at a later time. The "throw-away" attitude has made some material goods obsolete and created mountains of garbage for which there are no landfills. The intentions were valid in the beginning: Paper plates meant less dishwashing (less labor, less water, less detergent, and fewer sanitation problems) and a boon to the fast-food industry. Plastic containers were convenient: readily disposable, lightweight, and required less labor and lower shipping costs, a benefit to the transportation industry. Attention was not paid to the enormous tasks of waste, disposal, and recycling. Degradation of plastic is slow or sometimes impossible, and burning leads to toxic fumes and unsightly air pollution.

Economic and Environmental Trade-Offs

A true democracy maximizes freedom of expression, thoughts, ideals, and the pursuit of personal goals. Citizens as a group decide what restrictions will and will not apply to the society. Modern technology has led industrialized, democratic nations into a trap: On the one hand, freedom of expression dictates to "do your own thing"; to some this includes polluting the environment in its many forms (littering, gases in the air, noise, pesticides). On the other hand, concerned citizens strive to impose restrictions on society as a whole to ensure a more acceptable quality of life. Voters and taxpayers find themselves lining up on one side of an issue at one time and having to reverse themselves at another time. The determining factor is often economics. How much will it cost me and society to achieve these so-called improvements in the quality of life? Unfortunately, one person's quality of life is another person's pollution.

This freedom of expression in the developed countries has led to an environmental degradation unparalleled in history. Cleaning up the environment was simply considered too expensive for most of the paying customers. Factories spewed contaminants in the air and poured toxic wastes into rivers and streams, and increasing pressures for food forced modern agriculture to use more and more pesticides so the weeds, insects, and diseases could be kept under control. There has been environmental awareness legislation passed that has helped a great deal in improving the quality of air, water, and soil resources. Much remains to be done, and changes in political philosophy have caused a roller-coaster effect for many years and many years to come. Ultimately, however, individual responsibility must bring about the changes considered desirable by a majority of the society. Individuals must make these decisions as informed citizens, concerned taxpayers, and most importantly members of the natural community, not exclusively as exploiters of its resources. The future of the botanical world is in human hands, and the future of human society rests in the continuation of that world (see Figure 22-17a and 22-17b).

Figure 22-17a

Proper management of our world by an educated and concerned populous will help maintain ecological balance and plant function.

Figure 22-17b

Our beautiful natural areas will be preserved for generations to come.

🌐 Summary

1. Industrialization led to a rapid increase in population concentration, along with production of material goods, communication, and transportation. Crops were domesticated between 12,000 and 18,000 years ago in many parts of the world, apparently in response to food pressures created by villages and towns.

   The first domesticated food crops—wheat, barley, rice, corn, and potatoes—are still worldwide. Of major food crops, and there are only a few of them, only tomato and coffee have been domesticated in the past 2,000 years. Even though variety is lacking in total number of species, farmers have been very successful in total productivity.

2. Although surpluses were never large, farmers were able to produce more food, than required. Today in industrialized nations all the food is produced by about 5% of the population.

3. At the present time, wheat, rice, and corn provide about 80% of the human caloric intake. Most people receive enough carbohydrates, but protein and fat consumption is severely limited in developing countries.

4. Currently about two-thirds of the world's population suffer from poverty and hunger, 10% live in a relative affluence, and the other 25% live somewhere in between, never being sure of adequate food and shelter. Throughout recorded history, agricultural productivity has increased, but rise in population has always placed food supply at a dangerously low intersection with demand.

5. In years past, U.S. food production has enjoyed major surpluses, which have suppressed prices. Sometimes those surpluses are used for export and foreign aid, but political complications and the recipient's inability to pay for food has caused the producer nations to reduce production.

6. The exceptional productivity of the North American breadbasket reflects progress in basic and applied research, technology transfer, and the ingenuity of farmers. The land-grant college system and the agriculture cooperative extension system have been successful in providing information for farmers; similar plans have met with varying degrees of success in the developing countries.

7. The Green Revolution has brought high-yielding varieties of wheat and rice to a hungry world. These varieties are selected for maximum production under conditions of plentiful water and nutrition; some critics feel that the energy price for that productivity is too high.

8. Lack of bargaining power has forced farmers to settle for less than satisfactory profits, even though food prices continue to rise. Most of those increases are demanded by the infrastructure that harvests, packages, transports, and markets the products.

9. Lack of sufficient arable land has caused concern about agriculture ability to continue to feed a world that doubles in population every 40 years. Although desert land is available, the cost of water transport limits the potential for irrigation with current energy sources. The ecological and environmental problems associated with agriculture in the humid tropics are equally insurmountable.

10. The achievements in high technology, including transformed plants, help to increase the productivity to help feed the world population.

11. New crops never before considered for agriculture might adapt to climates where traditional crops will not grow. Even though these crops developed by genetic engineering have a great potential, the human factor is likewise important in the acceptance process.

12. No single solution will save Homo sapiens. Our best creative efforts must be used to ensure survival within the constraints of the biosphere.

13. All plants, no matter how large or small, are important in the total balance and functioning of the biological world. Obviously critical in photosynthesis, energy conversions, and oxygen production, plants form the base of the food chain.

14. Any person with a basic knowledge of plant structure and function is a lay plant person who can better appreciate the biological, ecological, aesthetic, and applied roles of plants.

15. Knowledge should help form a base of people who act more responsibly toward their natural environment and participate, more wisely, in the decision-making processes that affect their world. Many sociopolitical considerations are worldwide in scope, and yet rational evaluation of resource use can be a decision that affects each individual.

16. Modern agriculture now involves food surpluses in some countries and shortages in others and pest control with dangerous chemicals or biological agents such as IPM and genetic engineering.

17. Conservation movements have resulted in endangered species legislation in the United States and even some international efforts to legally protect the environment in the development of industrial products. The basic problem, however, is too large a world human population for the available resources.

18. Continued and increased efforts, especially by those members of the human population educated in the causes and significance of environmental problems, can result in a return to a biological world in balance and with all its members (including humans) existing as part of the world community, not apart from it.

Something to Think About

1. What organism is responsible for the basic food chain?

2. Why should the knowledge gained from this book help people understand about natural environmental issues?

3. Integrate this knowledge into the workings of global environmental conservation.

4. What makes our modern agriculture so good?

5. How do the conservation movements help us exist in balance with the world community?

Suggested Readings

Gore, A. 2006. *An inconvenient truth.* Emmaus, PA: Rodale.

Lombory, B. 2001. *The skeptical environmentalist: Measuring the real state of the world.* Cambridge, MA: Cambridge University Press.

Lombory, B. 2004. *Global crises, global solutions.* Cambridge, MA: Cambridge University Press.

Suzuki, D. 2001. *Eco-fun: Great projects, experiments, and games for a greener earth.* New Delhi: Sterling Publishers.

Suzuki, D. 2001. *You are the earth: Know the planet so you can make it better.* New Delhi: Sterling Publishers.

Internet

Internet sites represent a vast resource of information. The URLs for Web sites can change. Using one of the search engines on the Internet, such as Google, Yahoo!, Ask.com, and MSN Live Search, find more information by searching for these words or phrases: **sociopolitical conservation, world food supplies, Integrated Pest Management, synthetic pyrethrins, and remote sensing.**

APPENDIX

Due to its location in a book, and because of its name, an appendix is often ignored by the reader. But an appendix contains valuable information that can enhance a reader's understanding and learning. Further, information in an appendix is quick and easy to find.

The information in this appendix includes a variety of useful conversion factors and facts. Armed with this information, the reader can understand more, plan more, and learn more.

TABLE A-1 CONVERSION TABLES FOR COMMON WEIGHTS AND MEASURES	
Common Measures	**Conversion Amounts**
1 pound	454 grams
2.2 pounds	1 kilogram
1 quart	1 liter
1 gram	15.43 grains
1 metric ton	2,205 pounds
1 inch	2.54 centimeters
1 centimeter	10 millimeters or .39 inches
1 meter	39.37 inches
1 acre	.406 hectare

TABLE A-2 WEIGHT CONVERSIONS	
Common Measures	**Conversion Amounts**
8 tablespoons	¼ pound
3 teaspoons	1 tablespoon
1 pint	1 pound
2 pints	1 quart
4 quarts	1 gallon or 8 pounds
2,000 pounds	1 ton
16 ounces	1 pound
27 cubic feet	1 cubic yard
1 peck	8 quarts
1 bushel	4 pecks
Other Conversions	
1 percent	.01
1 percent	10,000 parts per million
1 megacalorie (mcal)	1,000 calories
1 calorie (big calorie)	1,000 calories (small calorie)
1 megacalorie	1 therm

TABLE A-3   STANDARD WEIGHTS OF FARM PRODUCTS PER BUSHEL

Product	Pounds
Alfalfa	60
Apples (average)	42
Barley (common)	48
Beans	60
Bluegrass (Kentucky)	14–28
Bromegrass, orchardgrass	14
Buckwheat	50
Clover	60
Corn (dry ear)	70
Corn & cob meal	45
Corn (shelled)	56
Corn kernel meal	50
Corn (sweet)	50
Cowpeas	60
Flax	56
Millet (grain)	50
Oats	32
Onions	52
Peas	60
Potatoes	60
Ryegrass	24
Rye	56
Soybeans	60
Spelt	30–40
Sorghum	56
Sudangrass	40
Sunflower	24
Timothy	45
Wheat	60
Milk, per gallon	8.6

TABLE A-4   STORAGE AND FEEDING DRY MATTER—LOSSES OF ALFALFA

Storage Method	Storage Loss (percent)	Feeding Loss (percent)
Small bales, stored inside	4	5
Round bales, stored inside	4	14
Hay stacks, stored inside	4	16
Round bales, stored outside	12	14
Hay stacks, stored outside	16	16
Haylage, vertical silo	7	11
Haylage, bunk silo	13	11

TABLE A-5   BUSHEL WEIGHTS AND VOLUMES

Item	Pounds/Cubic Feet	Cubic Feet/Ton
Oats = 32 lb/bu	26	77
Barley = 48 lb/bu	38.4	53
Shelled corn = 56 lb/bu	44.8	45
Wheat = 60 lb/bu	48	42
Corn & cob meal = 70 lb/bu	28	72
Soybeans = 60 lb/bu	48	42
Rye = 56 lb/bu	44.8	45
Soybean oil meal = 54 lb	—	37
Dairy feed = 35 lb	—	57

TABLE A-6   MEASUREMENT STANDARDS, HAY AND STRAW

Item	Average Cubic Feet/Ton	Range Cubic Feet/Ton
Hay, baled	275	250–300
Hay, chopped—field cured	425	400–450
Hay, chopped—mow cured	325	300–350
Hay, long	500	475–525
Straw, baled	450	400–500
Straw, chopped	600	575–625
Hay, loose	480	370–390
Straw, loose	800	750–850

TABLE A-7   FAHRENHEIT TO CENTIGRADE TEMPERATURE CONVERSIONS[1]

°F	°C	°F	°C	°F	°C
100	37.8	77	25.0	54	12.2
99	37.2	76	24.4	53	11.7
98	36.7	75	23.9	52	11.1
97	36.1	74	23.3	51	10.6
96	35.6	73	22.8	50	10.0
95	35.0	72	22.2	49	9.4
94	34.4	71	21.7	48	8.9
93	33.9	70	21.1	47	8.3
92	33.3	69	20.6	46	7.8
91	32.8	68	20.0	45	7.2
90	32.2	67	19.4	44	6.7
89	31.7	66	18.9	43	6.1
88	31.1	65	18.3	42	5.6
87	30.6	64	17.8	41	5.0
86	30.0	63	17.2	40	4.4
85	29.4	62	16.7	39	3.9
84	28.9	61	16.1	38	3.3
83	28.3	60	15.6	37	2.8
82	27.8	59	15.0	36	2.2
81	27.2	58	14.4	35	1.7
80	26.7	57	13.9	34	1.1
79	26.1	56	13.3	33	0.6
78	25.6	55	12.8	32	0.0

[1]Formulas used: $°C = (°F - 32) \times \frac{5}{9}$ or $°F = (°C \times \frac{9}{5}) + 32$

TABLE A-8 CONVERSION FACTORS FOR ENGLISH AND METRIC MEASUREMENTS				
To Convert the English	To the Metric Multiply by	To Convert Metric	Multiply by	To Get English
acres	0.4047	hectares	2.47	acres
acres	4047	square meters	0.000247	acres
BTUs	1055	joules	0.000948	BTUs
BTUs	0.0002928	kilowatt hours	3415.301	BTUs
BTU/hours	0.2931	watts	3.411805	BTU/hours
bushels	0.03524	cubic meters	28.37684	bushels
bushels	35.24	liters	0.028377	bushels
cubic feet	0.02832	cubic meters	35.31073	cubic feet
cubic feet	28.32	liters	0.035311	cubic feet
cubic inches	16.39	cubic centimeters	0.061013	cubic inches
cubic inches	1.639×10^{-5}	cubic meters	61012.81	cubic inches
cubic inches	0.01639	liters	61.01281	cubic inches
cubic yards	0.7646	cubic meters	1.307873	cubic yards
cubic yards	764.6	liters	0.001308	cubic yards
feet	30.48	centimeters	0.032808	feet
feet	0.3048	meters	3.28084	feet
feet/minute	0.508	centimeters/second	1.968504	feet/minute
feet/second	30.48	centimeters/second	0.032808	feet/second
gallons	3785	cubic centimeters	0.000264	gallons
gallons	0.003785	cubic meters	264.2008	gallons
gallons	3.785	liters	0.264201	gallons
gallons/minute	0.06308	liters/second	15.85289	gallons/minute
inches	2.54	centimeters	0.393701	inches
inches	0.0254	meters	39.37008	inches
miles	1.609	kilometers	0.621504	miles
miles per hour	26.82	meters/minute	0.037286	miles per hour
ounces	28.349	grams	0.035275	ounces
fluid ounces	0.02947	liters	33.93281	fluid ounces
liquid pints	0.4732	liters	2.113271	liquid pints
pounds	453.59	grams	0.002205	pounds
quarts	0.9463	liters	1.056747	quarts
square feet	0.0929	square meters	10.76426	square feet
square yards	0.8361	square meters	1.196029	square yards
tons	0.9078	tons	1.101564	tons
yards	0.0009144	kilometers	1093.613	yards
yards	0.9144	meters	1.093613	yards

TABLE A-9   HOW TO COLLECT, PRESS, AND MOUNT PLANTS

Plant mounts make better study material than any manual. A properly dried, pressed, and mounted plant is attractive, easily displayed, and will last a long time. A plant collection makes an interesting conversation piece in the home and can be used as an exhibit at fairs, schools, and other displays.

Equipment

- Digging tool—a shovel, garden digger, or some other digging tool to remove the plant from the soil.
- Trimming tool—a sharp knife or a pair of scissors to cut off woody specimens, to remove excess or old plant material, and to slice thick roots.
- Specimen container—plastic bags are recommended for keeping plants until you can press them.
- Notebook—a field notebook or tablet and a pencil or pen are needed to record all important information about the plant and the location where the plant was found.
- Plant press—a binder-type press, eighteen inches long by twelve inches wide with alternating cardboard, blotter, and folded newspaper is recommended to dry and press the plant. Other items such as magazines will work for pressing if enough weight is placed on top.

Collection Procedures

1. Because some plants bloom in early spring and others bloom in the late fall, you will not be able to collect all the plants at any one time of year. Plan several collection trips throughout the spring, summer, and fall.

2. Choose plant specimens carefully. Select one, or preferably two, of each plant species to be collected.

3. Avoid plants that are off-color, grazed, overmature, diseased, or otherwise not normal.

4. While at the site, record the plant in your field notebook or tablet by giving it a number. Record the plant name (if it is known) and the information that will be needed when completing the plant labels for your mount. Start a numbering system that will work for you. You may want to include the year, such as 99-1, 99-2, and so on. If you use this format, 99 refers to the year, and each different plant species will be numbered consecutively (1,2,3, and so on).

5. When collecting grasses and grass-like plants:
   - Select specimens with seedheads fully emerged from the sheath.
   - Select specimens that are still green including the seedhead.
   - Collect the whole plant, when possible, including a good sample of the roots.
   - Be sure that rhizomes or stolons are attached to the plant if they are typical for that species.

6. When collecting forbs:
   - Select specimens in the flowering stage.
   - Collect the whole plant if possible, including a portion of the root.
   - Some forbs can be collected with both flowers and seeds, or seed pods, on the plant at the same time.
   - Be sure that rhizomes or stolons are attached to the plant if they are typical for that species.
   - Taproots or other thick roots should be sliced away on the underside so that the plant will be fairly flat after pressing.

(Continues)

TABLE A-9 *(Continued)*

Collection Procedures

7. When collecting shrubs and other woody plants:
   - Select a branch about twelve to fourteen inches in length and not over ten inches in width.
   - Collect the plant when it is in bloom.
   - Many shrubs bloom in early spring before they leaf out. In these cases, collect two specimens, one in flower and one after the plant has leafed out.
   - Mount both specimens on the same sheet.
   - It is often useful to include a sample of both the current year and the older bark of woody plants.
   - Roots of large woody plants should not be included on the plant mount.

8. To remove a plant from the soil, dig about six inches straight down around the plant about three inches out from the stem. Carefully lift out the chunk of sod. If the soil is moist, use water to wash away the soil from the roots.

9. Remove all soil particles from the roots. Do not be afraid to wash the roots thoroughly on all the plants collected. In fact, it may take more than one washing. Excess moisture after washing the roots can be removed by firmly pressing the plant between paper towels.

10. Remove the excess plant material from the roots, stems, leaves, and seedheads. For example, by removing several stems from a large bunchgrass or shrub, it is easier to dry and mount a specimen. If plants are very large and bulky, collect a sample of the stem, leaf arrangement, root, and flower or seedhead.

11. Take several plastic bags with you when collecting plants. Put the plants in the bag with a few drops of water (do not overdo it), then seal the bag, and the specimens will stay fresh for several hours. They should be kept out of direct sunlight. If it isn't possible to press all the plants collected, most plants will stay fresh in the plastic bag if kept cool, such as in a refrigerator, for a day or two. However, put only one kind of plant in a bag and number the bag to match your field notebook.

12. Seeds and/or seed pods are very helpful in identifying many plants. A good way to include seeds is to place several seeds in a small, clear plastic, self-sealing envelope attached to the mount sheet. To prevent new infestations, it is also a good idea to carefully remove and burn all other seeds from any undesirable or weedy plant specimens.

Guidelines for Pressing Plants

The object is to quickly dry the plants under firm pressure to retain plant colors and the plant arrangement.

1. Press the plants as soon as possible after collecting. Once a plant wilts, it will not make an attractive mount.

2. Have your press ready to go before you remove a specimen from the plastic bag. Have plenty of newspaper pages folded lengthwise with about a quarter of the upper and lower edges folded toward the center. This will help keep your specimens from sliding out. A supply of corrugated cardboard sheets (cut to fit your press) are also needed. As you fill your press, alternate the cardboard sheets and folded paper (beginning and ending with a sheet of cardboard) to keep the specimens flat and speed the drying process. Although it is not necessary, blotter sheets can be placed between the newspaper and cardboard to speed the drying process.

3. Remove one plant at a time from the plastic bag. Check the plant closely to make sure that all soil is removed from the roots and remove excess moisture with a paper towel.

(Continues)

TABLE A-9 *(Continued)*

Guidelines for Pressing Plants

4. If the plant is less than fourteen inches long, place it between the folded newspaper. Arrange the stems, leaves, roots, and flowers exactly as you want them to appear on the mount. Flowers should be pressed open. Both the upper and lower surfaces of flowers and leaves should be displayed.

5. If the plant is longer than twelve inches, it will be necessary to fold the plant in the shape of a V, N, or W. If the plant is still too large, press a sample of each plant part—stem, leaf, root, and flower or seedhead. For hard-to-handle plants, hold at the stem base firmly and slowly move the plant up and down against the newspaper a few times stopping with an upward stroke. (This will help separate and straighten out the branches and leaves.)

6. Hold the plant firmly in place and fold the upper and lower segments of the newspaper over the plant. While applying pressure to keep the plant in position, write the assigned plant number from your field notebook on the newspaper. Then place the plant into your press (a cardboard sheet should be below and above the folded newspaper).

7. Examine the plant after it has been pressed for twenty-four hours. This is your last opportunity to do some rearranging while the plant is still flexible. Be sure both upper and lower leaf surfaces show. Change the newspaper or blotter paper every day until the plant is thoroughly dry. Remember that succulent (fleshy) plants will take much longer to press.

8. Plants can be removed from the press in seven to ten days. Keep the plants in folded newspaper until you are ready to mount them.

Mounting Plants

After the plant specimens have been pressed and dried, they are ready to be mounted.

1. Herbarium sheets, standard (white) tag board, or poster board are recommended for mounting sheets. Although herbarium sheets usually have to be ordered through biological supply outlets, poster board can be purchased at most stores selling office and school supplies. If you use tag board, four mount sheets can be cut from one board if each sheet is cut at 11″ × 14″, or three sheets can be cut if each sheet is cut at 11½″ × 16½″.

2. Placement of specimen is easy if the plants have been pressed properly. The specimen should be placed upright with roots near the bottom and should provide a pleasing appearance. Leave room in the lower right-hand corner for a 3″ × 5″ mount label.

3. A transparent glue (Elmer's glue is best) is preferred to spot fasten the specimen to the mountsheet. You can also use small strips of gummed cloth. Scotch tape is not recommended. Small weights, such as lead casts, large nails, heavy washers, or large nuts, will hold the plant to the mount sheet while the glue is drying.

4. Each mount requires a label in the lower right-hand corner. The label must be properly filled out.

Guidelines for Storing Plants

1. Mounted plants are usually stored in a cabinet or case to protect them from dust and insects. Although protective material is not required. Some collectors (especially for 4-H projects) use a protective cover to protect the plant material as it becomes brittle. Use a 4–5 mil clear plastic mylar material and do not use Saran Wrap or 1–2 mil clear plastic. Also, your mounts should not be laminated with a clear seal plastic until a botanist has verified the specimen and signed the label.

(Continues)

TABLE A-9   (Continued)

Guidelines for Storing Plants

2. Your plants should be filed in a logical order that makes it easy to find a specific specimen. By filing all specimens by family, then arranging the family members in alphabetical order by genus and species, it is easy to find a specific specimen.

3. It is usually a good idea to store a few mothballs with your plants to protect them from insects.

TABLE A-10   SAMPLE PLANT LABEL

Plant common name _____ Scientific name _____

Collection Site Information

Date collected _____ State _____ Distance _____ (miles)

and _____ (direction) from (nearest town/city) nearest landmark _____

Elevation _____ Slope face _____

Circle one for each item:

Topography: mountains, foothills, breaks, plains, riparian

Slope: nearly level, rolling, moderate, very steep

Abundance: abundant, occasional, very few

Tree overstory: yes, no

Collector _____

Plant number _____ Verified by _____

GLOSSARY

abiotic: Pertaining to nonbiological factors.

abrasion: Mechanical process of gradually breaking down a hard layer, as in a seed coat.

abscisic acid: Plant hormone associated with dormancy, abscission of organs, and water stress.

abscission layer: Detachment of leaves, flowers, or fruits from a plant, usually at a mechanically weak location, termed the *abscission zone.*

absorption: Process of taking in, as uptake by roots.

accessory buds: Buds adjacent to a primary bud and usually smaller in size.

achene: Small, dry, one-seeded indehiscent fruit; the pericarp is easily separated from the seed coat.

acid: H^+ (proton) donor; a substance that associates to release H^+ and thus cause the pH of the solution to be less than 7.0.

acidic: Possessing a relatively large number of hydrogen ions; having a pH of less than 7.0.

aconitine: A highly poisonous alkaloid derived from various aconite species. It is a neurotoxin that opens TTX-sensitive Na+ channels in the heart.

acridine dyes: Organic pigment molecules that are capable of causing permanent genetic changes (mutations).

active transport: Movement of ions or molecules against a concentration gradient using metabolic energy.

adaptation: Conforming to a given set of environmental conditions.

adenine: Nitrogen base found in both DNA and RNA.

adenosine diphosphate (ADP): Building block for ATP; by adding a terminal phosphate group and a large amount of energy, ATP can be formed.

adenosine triphosphate (ATP): Major source of chemical energy for biochemical reactions; metabolic energy is stored primarily in the terminal phosphate ester linkage.

adhesion: Attraction of unlike particles; water particles *adhere* to the surface of clays.

adventitious roots: Structure arising at some location not usually expected, such as on a stem.

aerate: To supply with oxygen.

aerial: Pertaining to being in the air, such as a root projecting from an aboveground stem.

aerobic organisms: Organisms that have to have atmospheric oxygen to live.

aerobic respiration: The process in which glucose is converted into CO_2 and H_2O in the presence of oxygen, releasing large amounts of ATP. This process includes the krebs cycle, electron transport chain, and oxidative phosphorylation.

after-ripening dormancy: Changes that must take place in a seed to overcome dormancy or the dormancy period following seed formation, necessary for embryo changes that insure germination.

agar (ah-ger): Complex polysaccharide made from red algae and used for preparing a semisolid substrate for growing microorganisms.

agent orange: Herbicide used as a defoliant in the Vietnam War; composed of 2,4-D and 2,4,5-T.

aggregate fruit: Fruit developing from numerous simple carpels from a single flower.

aleurone layer (al-u-roan): Group of cells rich in protein granules and located as the outer layer of the endosperm of many grain seeds.

algal bloom: Proliferation of algae due to a nutrient rich medium, usually resulting in a green scum of the water surface.

algin: Polysaccharide derived from brown algae and used for many industrial processes.

alkaline: Denoting substances that release hydroxyl (OH^-) ions into solution; *see* Basic.

alkaloid: Group of nitrogen-containing compounds having diverse structures; many alkaloids have medicinal, hallucinogenic, or toxic properties.

allele: One of the two genes for a given trait at a specific locus on homologous chromosomes.

α (alpha) amylase: Enzyme that converts starch to sugars.

alternate host: Alternate plant required to complete the life cycle of some microorganisms (for example, for *Puccinia graminis tritici*, wheat is the primary host and *Berberis vulgaris* is the alternate host).

alternate leaf arrangement: Leaf arrangement in which there is only one leaf per node.

alternation of generations: Sequence of a diploid sporophyte plant producing haploid spores that develop into gameophyte plants (or stages); gametophytes produce gametes, which fuse to form a sporophyte again.

amber: Fossilized resin of ancient trees.

amino acid (ah-mean-o): Organic molecule including one or more amino ($-NH_2$) and acid ($-COOH$) groups; proteins are made up of these molecules.

amino group ($-NH_2$): Chemical part of molecule that imparts basic properties to an amino acid.

ammonia: Colorless, pungent gas, NH_3, extensively used in a wide variety of nitrogen-containing organic and inorganic chemicals.

anaerobic: Without atmospheric oxygen.

anaerobic respiration: Partial oxidation of pyruvate to lactic acid without atmospheric oxygen.

analgesic: Chemical that reduces the body's sensitivity to pain.

anaphase: Stage of nuclear division in which the chromosomes are pulled to opposite poles while attached to spindle fibers at the centromere.

anchorage: A function of the plant roots.

androsterone: Animal hormone not synthesized by plant.

angiosperm: Group of plants characterized by having flowers as their sexual reproductive structures.

anion: Negatively charged ion in an electrolyte solution, attracted to the anode under the influence of a difference in electrical potential.

annual: Plant that completes its life cycle during one growing season.

annual tree ring: Secondary xylem produced during a single growing season.

anther: Male reproductive organ enclosing and containing the pollen grains.

antheridium: A sperm-producing organ occurring in seedless plants, fungi, and algae.

anthocyanins: Group of water-soluble red to blue flavonoid pigments found in certain plants; especially important pigmentation in flower petals.

anthraquinone glycosides: Derivative of anthracene, a dyestuff. Its derivatives found in aloes, cascara sagrada, senna, and rhubarb act as cathartics.

antibiosis: Inhibition of growth of a microorganism by a substance produced by another microorganism.

antibiotic: Organic molecule naturally produced by one microorganism that retards or prevents the growth of another organism.

anticodon loop: Portion of a tRNA molecule responsible for the anticodon triplet, which pairs with the codon of mRNA.

antiparallel: Opposite in direction, as in the structure of the two strands of DNA.

antipoidal cell: Three haploid cells, at the end of the embryo sac away from the micropyle.

apex: Tip of a structure; a leaf apex, for example, is the tip of the leaf.

apical dominance: Phenomenon leading to controlled growth of lateral shoots; growth occurs primarily at the top of plant.

apoplastic movement: Pertaining to the movement of water in the free space of tissue; free space includes cell walls and intercellular spaces.

archegonium: A multicellular, often flask-shaped, egg-producing organ occurring in mosses, ferns, and most gymnosperms.

asexual: Lacking sexual reproduction; vegetative reproduction.

aspartate: An excitatory amino acid.

atmospheric pressure: Ambient pressure created near the earth's surface by large air cells that circulate around the globe.

atom: The basic unit of matter; the smallest complete unit of the elements, consisting of protons, neutrons, and electrons.

atomic number: Number of protons within the nucleus of an atom, which determines the elemental properties of that atom.

atomic weight: Weight of an atom determined by adding the number of protons and neutrons (the mass of electrons is usually considered to be negligible).

atropine: Drug obtained from belladonna that is administered via injection, eye drops, or in oral form to relax muscles.

autonomic nervous system: Portion of the nervous system that regulates involuntary body functions, including those of the heart and intestine; controls blood flow, digestion, and temperature regulation.

autotrophic (au-to-tro-fik): Organism that produces its own food by photosynthesis; green plant.

auxin: Hormone that has the capacity to induce lengthening of cells.

auxin thresholds: Protein involved in the auxin signaling pathway (e.g., transport and signal).

axil: Lateral; the point of attachment of a bud or shoot other than at the apex of the stem.

axillary bud: Any bud on a stem other than the terminal bud at the shoot apex.

backcross: Crossing a hybrid offspring back to either parent.

bacteria (singular: bacterium): Unicellular microorganisms; typically a few micrometres in length, individual bacteria have a wide-range of shapes.

balsa: Density of dry balsa wood ranges from 100–200 kg/m^3; used to make very light, stiff structures in model building.

baobab tree: African tree having an exceedingly thick trunk and fruit that resembles a gourd and has an edible pulp called monkey bread.

bare root: Not potted; no soil around root.

bark: Portions of a woody plant stem or truck exterior to the vascular cambium.

barley (Hordeum jubatum): Plant grown for forage and grain, used for livestock feed, malt production, and cereal; cultivated since prehistoric times.

basal rosette: Leaves clustered at ground level; can be alternate or whorled, but with extremely short internodes.

basalt: Semiviscous layer of igneous rock underlying the granite continental plates.

basic: Possessing a large number of hydroxyl (OH^-) ions; a pH of more than 7.0.

bedrock: Solid rock that underlies all soil, sand, clay, gravel, and loose material on the earth's surface.

benign: Not dangerous to health; not recurrent or progressive.

benthic (ben-thik): Pertaining to the bottom region of an ocean, lake, or pond.

berry: Fleshy, two or multiple-carpeled ovary, each carpel having many seeds.

β- (beta) carotene: Most abundant of the carotenoids; strong provitamin A activity; unlike vitamin A, it is a strong antioxidant.

biennial: Plant requiring two growing seasons to complete its life cycle.

big bang: Theory is that the entire universe was created at one time.

binomial: Two names, genus and species, comprising the scientific name.

bioassay: Quantitative assay of a particular substance using a portion of or an entire living organism.

bioinformatics: Refers to the use of computer science, mathematics, and information theory to model and analyze biological systems, specifically in systems with genetic material.

biological clock: Internal biological timing system that relates cyclic phenomena within the organism to the diurnal or annual clock.

biological control: Control of pests by disrupting their ecological status, as through the use of organisms that are natural predators, parasites, or pathogens. Also called biocontrol.

bioluminescence: Biological production of light using ATP as an energy source.

biomass: Total amount of organic material produced; usually expressed on a dry weight basis.

biomes: Worldwide groupings of similar ecosystems.

biopharmaceuticals: Medicines produced by biotechnology and genetic engineering.

biosphere: Earth and all of its ecological interactions considered as a single system.

biotechnology: Application of biological and engineering techniques to microorganisms, plants, and animals; sometimes used in the narrower sense of genetic engineering. It is the application of scientific and engineering principles to the processing of materials by biological agents to provide goods and services. The three major techniques are genetic engineering, monoclonal antibody technology, and bioprocessing.

biotic: Pertaining to the living part of the environment.

bisexual: Flower having both stamens and pistil (both sexes); a perfect flower.

bivalent: Two homologous chromosomes associated in a parallel fashion; formed during prophase I of meiosis.

blade: Flattened portion of the leaf.

board feet: Volume of measurement for lumber equal to 144 square inches ($1'' \times 12'' \times 12''$).

bog: Wet, marshy region, such as a peat bog.

bomb calorimeter: Instrument used for measuring the caloric content of any organic materials to CO_2 and H_2O.

boot: One of the growth stages of small grains; during the boot stage, the head can be felt inside the upper leaf sheath and the flag (last) leaf has developed.

bract: Modified leaf or leaflike structure, usually much reduced in size.

brambles: Plants like the raspberry, loganberry, boysenberry, dewberry, and tayberry.

branch roots: Lateral roots arising from the pericycle of a larger root.

broadcast planting: Seeds scattered over a wide area.

bromeliads: Flowering plants native to the tropical and warm temperature regions including both epiphytes and terrestrial.

bronchodilator: Medicine that relaxes the smooth muscles of the airways, allowing the airway to open up (to dilate) since the muscles are not squeezing it shut.

bryophytes: Are all embryophytes ('land plants') that are non-vascular: they have tissues and enclosed reproductive systems, but they lack vascular tissue.

bud: Meristematic shoot located at a terminal or lateral position of a shoot.

bud scales: Modified leaves surrounding and protecting a bud.

budding: Method of asexual propagation characterized by placing a bud of one plant onto the stem of another plant.

buffer: Any substance that absorbs or releases protons (H+) so that the pH of the solution remains stable, even when acid or base is added.

bulb: Underground storage organ characterized by fleshy leaves attached to a stem base.

bulliform cell: Large epidermal cell found on the upper surface of many grass leaves; turgor pressure in these cells controls the lateral rolling of the leaves during water stress.

bunch grass: Grass growing in clumps rather than spreading out.

bundle sheath cells: Cells surrounding the vascular bundle in C4 plants.

buttressing: Stem modification in which the diameter is larger near the ground giving additional mechanical support.

C3 pathway: Calvin cycle of carbon fixation in which CO_2 is incorporated into ribulose 1,5-bisphosphate.

C4 pathway: Also called the Hatch-Slack pathway of photosynthesis; CO_2 is initially fixed in mesophyll cells into malate and aspartate.

callose: Complex carbohydrate found in the sieve areas of sieve tube elements; particularly abundant at the time of injury.

callus: Undifferentiated group of cells formed as a response to wounding (as at the base of a stem) or in tissue culture.

calorie: Unit of heat; one calorie is the amount of heat required to raise the temperature of 1 gm of water 1°C.

Calvin cycle: Series of biochemical, enzyme-mediated reactions during which atmospheric carbon dioxide is reduced and incorporated into organic molecules, eventually some of this forms sugars; in eukaryotes it occurs in the stroma of the chloroplast.

calyx: Referring to all the sepals of a flower collectively.

CAM: *see* Crassulacean acid metabolism.

candelilla wax: Is derived from a shrub native to that country and the southwestern US. Ever wonder why lipstick can stay hard in all but the most sultry of conditions? Thank candelilla wax. It has also been used in both the chewing gum and varnish industries.

cane: Part of the vine left that will produce fruit the next growing season.

canopy: Upper portion of a population of plants; the term is usually associated with forests and agricultural crops.

capillary pores: Small spaces in the soil that become filled with a fluid (such as water) because of the adhesion of particles to the matrix (solid substrate) and the cohesion of the water molecules to themselves.

capsule: Simple fruit that develops from a compound ovary with two or more carpels; capsules dehisce in many ways.

carbohydrate: Organic molecule consisting of a chain of carbon atoms, each having hydrogen (H^+) and (H^-) groups attached in the basic pattern CH_2O; simple sugars and polysaccharides such as starch and cellulose.

carbon dioxide (CO_2): Colorless, odorless incombustible gas; formed during respiration and organic decomposition.

carbon fixation: Process by which CO_2 is incorporated into organic molecules.

carboniferous period: The Carboniferous period was marked by vast, coal-forming swamps (see also bog) and a succession of changes in the earth's surface that, continuing into the Permian Period, ended the Paleozoic era.

carboxyl group: Acid group attached to a molecule—COOH.

cardiac glycosides: Steroid glycosides that have stimulative effects on the heart, for example digitoxin, digoxin, and gitoxin from the foxglove (*Digitalis* spp.); help to support the rate and strength of the heart's contraction when it is failing, and are strongly diuretic (increase urine output).

carnauba wax *(car-now-bah):* High-quality, hard industrial wax extracted from the carnauba palm tree.

carnivore: An animal that feeds on other animals.

carotenoids pigments: Plant cell pigments: red, orange, and yellow lipid soluble pigments found embedded in the membrane of chloroplast and chromoplast.

carpel: Reproduction unit composed of a placental surface and ovules.

carpellate: Unisexual flower having carpels but no stamens.

carrageenan: Polysaccharide extracted from red algae and used for many industrial products.

carrying capacity: Maximum number of organisms that can live in balance within the natural food supply of a given area.

caryopsis: Dry, indehiscent one-seeded fruit in which the pericarp is united to the seed coat.

casparian strip: Suberized layer covering the radial and transverse walls of endodermal cells.

cassava: Edible, starchy root used in making bread or cakes.

cassia: Any of various plants of the genus used medicinally as a cathartic.

catalyst: Any substance that causes a chemical reaction to proceed much faster than it would without the catalyst. In biochemical reactions, enzymes are proteins that act as catalysts.

cations: An ion or group of ions having a positive charge and characteristically moving toward the negative electrode in electrolysis.

cell expansion: Proper development of plant shape is at least partly dependent on regulated cell expansion.

cell membrane: Semipermeable membrane that encloses the cytoplasm of a cell.

cellulase: An enzyme that breaks down cellulose.

cellulose: Primary structural carbohydrate plant cells; composed of many glucose molecules.

cellulose microfibrils: Cellulose molecules oriented parallel to the long axis of the microfibril in a paracrystalline array, which provides great tensile strength; held in place by the wall matrix and their orientation is closely controlled by the protoplast.

central bud: Main bud in a cluster located at one position.

central cell: Also known as the polar cell, the binucleated cell in the center of an embryo sac containing the two haploid polar nuclei.

central nervous system: Portion of the vertebrate nervous system consisting of the brain and spinal cord.

centromere: Most condensed and constricted region of a chromosome, to which the spindle fiber is attached during mitosis.

century plant: Tropical American plants with basal rosettes of fibrous sword-shaped leaves.

chaparral *(shap-a-ral):* Vegetative association typical of the Mediterranean region dominated by smaller, often thorny or roughly branched evergreen trees and shrubs and deciduous trees; included in the temperate deciduous forest biome.

charcoal: Carbonaceous material obtained by heating wood or other organic matter in the absence of air.

chemical energy: Potential energy stored in chemical bonds of molecules.

chemotaxonomy: Using the identification of groups of chemical compounds as genetic indicators in the establishment of taxonomic relationships.

chiasma (pl., chiasmata): Point of physical exchange of equal segments of adjacent nonsister chromatids during the bivalent stage of meiosis.

chinampas: Areas of fertile reclaimed land, constructed by the Aztecs, and made of mud dredged from canals.

chlorophyll: Pigment molecule responsible for trapping light energy in the primary events of photosynthesis.

chloroplasts: Organelles found in cells of the aboveground portions of green plants; specialized for photosynthesis.

cholesterol: Animal steroid found in some plants in low concentrations.

chromatid: One replicate of a chromosome visible in prophase following DNA replication during interphase.

chromatin: Dark-staining nuclear material present during interphase; includes the SNA and nuclear proteins.

chromatophore: Discrete spherical bodies located in the membranes of prokaryotes that carry the enzymes and pigments important in photosynthesis.

chromoplast: Found in colored organs of plants such as fruit and floral petals, to which they give their distinctive colors and or nonphotosynthetic carotenoid-bearing; membrane-bound organelles specialized for carotenoid storage.

chromosome: Microscopic strands within the nucleus of eukaryotic cells that carry DNA, which is responsible for inheritance.

cicutoxin: Active principle of the water hemlock extracted as a poisonous gummy substance.

circulation cells: Air movement in a circular pattern from the earth's surface up into the outer atmosphere and then back down again; produce high and low pressure systems and dictate precipitation patterns at different latitudes.

cis-polyisoprene-: Properties are also close to that of natural rubber.

cladophyll: Stem or branch that resembles a leaf.

climacteric rise: Point during the ripening process of certain fruits in which the respiration rates rise to very high levels.

climatology: Study of climates and the factors influencing them.

climax community: Ultimate vegetative community that any given habitat can support; the final stage in ecological succession.

clone: Genetically identical organisms.

coastal shelf: Shallow region of the ocean surrounding a large land mass.

cocaine: Highly addictive stimulant drug derived from the coca plant that produces profound feelings of pleasure.

coconut palm: Tall palm tree bearing coconuts as fruits; widely planted throughout the tropics.

codeine: Sedative and pain-relieving agent found in opium. Structurally related to morphine but less potent, and constituting approximately 0.5% of the opium extract.

codon: Three-nucleotide sequence of mRNA responsible for coding of a specific amino acid.

coenzymes: Various substances, including certain vitamins and heavy metals, required, in conjunction with a specific enzyme, to bring about a biochemical reaction.

co-evolution: Evolution of two species in concert, such that their survival and reproduction are mutually beneficial.

cofactor: Nonprotein substance required by enzymes for proper function; they may be metallic ions or organic molecules called coenzymes.

cohesive: Attraction of like particles; water molecules *cohere* to each other.

cold damage: Temperature range through which flower bud, leaf, or structure can be expected to occur.

cold frame: Frame house built for growing plants that has no inside heating provided.

coleoptiles: Protective sheath enclosing the shoot tip and embryonic leaves of grasses.

collenchymas: Tissue-type characteristic by primary cell walls with thickened corners.

colonial: Multicellular organism that produces a colony of cells, usually referring to *colonial* algae.

colonization: Pioneer establishment of vegetation on a previously unvegetated area.

colonizer: Organism that initiates the biological "conquest" of soil or rock.

colony: Group of microorganisms growing in a confined area so that the population as a whole has color and texture.

combine: Type of machine that harvests seed crops.

commensalism: Interaction between two species in which one population is benefited but the other is not affected.

common name: Regional name for well-known plants; in the language of the region, rather than in Latin, and not necessarily paralleling any scientific name.

community: All living organisms sharing a given area.

companion cell: Small cell adjacent to a sieve tube element within phloem tissue.

companion crop: Crop grown specifically with another crop because it recovers the land or protects another crop until it can get established; also called a nurse crop.

compartmentalization of enzymes: Division of labor in living cells such that enzymes related to a particular function are packaged and separated from the other cell contents, usually by a membrane.

compensation point: Condition in a living system in which the uptake of

CO_2 equals the release of CO_2; that is, photosynthesis equals respiration.

competition: Demand by two or more organisms for the same resources.

complementary strand: Two polynucleotide chains in which the pairing of adenine is always with thymine (in DNA) or uracil (in RNA), and guanine is always paired with cytosine.

complete flower: Having all four floral parts; sepals, petals, stamens, and pistil.

compost: Partially decayed organic matter used in gardening and farming to enrich the soil and increase water-holding capacity.

compound fruit: Fruit that develops from several ovaries in either a single flower or multiple flowers.

compound leaf: Leaf composed of two or more completely independent blade units called leaflets.

compound pistil: Ovary composed of two or more carpels.

compression: Fossil formed when carbonized plant material is still present in the original shape but is greatly compressed and reduced in size by pressure.

compression wood: Reaction wood produced along the lower side of leaning trees, straightening the trunk by expanding and pushing the tree upright.

concentration gradient: Difference in concentration in two parts of a system.

condensation: To change from a gas to a liquid or solid.

conduction: The flow of thermal energy through a substance from a higher- to a lower-temperature region.

cone: Strobilus; the reproductive structure of a gymnosperm.

conifer: Any of a group of plants that produce a strobilus or cone as a reproductive structure.

container-grown: Plants grown in pots, planters or other containers.

controlled atmosphere: Specified gas or mixture of gases at a predetermined temperature.

controlled burning: Intentionally ignited fire contained within a designated area.

convergent evolution: Groups of unrelated organisms becoming similar in appearance because of evolutionary change in response to similar environments.

coral reef: Hard, rocklike structure in shallow tropical waters; structure is composed of plants and animals encrusted with calcium carbonate.

corepressor: Substance that inhibits production of a particular enzyme.

cork: Secondary tissue produced by cord cambium; the outer part of the periderm.

cork cambium: Secondary cambium giving rise to cork tissue.

corm: Enlarged underground vertical stem.

corolla: Pertaining to all petals of a flower.

cortex: Ground tissue located between the vascular bundles and epidermis of stems and roots.

cortisone: Animal hormone not synthesized by plants.

cosmopolitan: Worldwide in distribution.

cotton: Soft silky fibers from cotton plants in their raw state.

cotyledon: Seed leaf; the first leaf formed in a seed.

covalent bond: Chemical bond between two atoms created as the result of sharing of electrons.

crassulacean acid metabolism (CAM): Type of carbon metabolism in which stomata open only at night, thus conserving water and allowing the leaves to take in CO_2; photosynthesis is completed during the day, allowing a partitioning of reactions in time.

cristae: Folded-membrane inner structure of mitochondria.

cross-pollinated: Pertaining to a flower having pollen deposited on it from a different flower.

cross-protection: Infection of plant tissues by one virus suppresses the disease caused by another closely related strain of the virus; protecting strain must have negligible impact on the host.

crossing over: Physical exchange of equal segments of adjacent nonsister chromatids during the bivalent stage of meiosis; formation of a chiasma.

crown: Topmost portion of a plant.

curare: A variety of poisonous plant extracts from the bark, roots, stems, and tendrils of several woody lianas, including *Chondodendron tomentosum.*

cuticle (*kute-i-kuhl*) thickness: Waxy coating of the epidermis on all above-ground parts of a plant.

cuticular transpiration: Water loss through the cuticles of the epidermis cells, these account for around 5% of water loss.

cutin (*kute-in*): Lipid layer found in the outer walls of epidermal cells.

cuttings: Portions of stems, usually consisting of two or three nodes, used for propagation by the production of adventitious roots.

cyanobacteria: Are mainly responsible for generating the oxygen that is found in the earth's atmosphere today.

cycads: Any of various palmlike gymnospermous cone-bearing evergreen plants, native to warm regions and having large pinnately compound leaves.

cyclic photophosphorylation: Formation of ATP by activation of PS I, but failure to activate PS II. Consequently ATP is formed, but NADP is not reduced.

cyclosis: Spontaneous movement of cytoplasm and some organelles within the cell.

cytochrome: Group of proteins involved in electron transfer in biological systems.

cytokinesis: Cell division; the process accompanying nuclear division in which the cytoplasmic contents are divided between the daughter cells.

cytokinin: Group of hormones that promote growth by stimulating cell division.

cytoplasm: Viscous contents of a cell within the plasma membrane, but generally excluding the nucleus.

cytoplasmic streaming: Flowing of cytoplasm in eukaryotic cells.

cytosine: Nitrogen base found in both DNA and RNA.

day neutral plant: Plant whose flowering is not controlled by the length of day.

daylength: Number of hours that sunlight illuminates a given area on earth; dependent on the angle of the earth relative to the rays of the sun.

deadheading: Removal of old blooms.

decarboxylation: The removal of a single carbon atom as CO_2 from an organic molecule.

deciduous (*de-sid-u-us*): Referring to plants that lose all their leaves during the cool season; as opposed to evergreen plants.

decomposer: Organism that acquires its nutrition by feeding on dead organisms.

deficit irrigation: Concept of low levels of irrigation to achieve a moderate level of productivity; even though yields are not maximized, limited water supplies are conserved.

dehiscent: Mature fruit that splits open to release the seed.

deletion mutant: Mutation in which a base pair is deleted and shifts the sequence out of phase by one pair; also called a frame shift.

denature: To change the configuration of a protein molecule such that it loses specificity and no longer functions as an enzyme.

dendrochronology: Science of studying growth rings of trees to determine past conditions.

dentate: Toothed leaf margin.

deoxyribonucleic acid (DNA): Double-stranded nucleic acid composed of adenine, guanine, cytosine, and thymine, in addition to phosphate and deoxyribose.

deplasmolysis: Dehydration injury.

depressants: Drugs that relieve anxiety and produce sleep; depressants include barbiturates, benzodiazepines, and alcohol.

deoxyribose: Five-carbon sugar contained in DNA.

desert: Arid region that receives annual rainfall of 10 inches (25 cm) or less.

desertification (*de-surt-i-fi-ka-shun*): Process by which fragile, semiarid ecosystems lose productivity because of loss of plant cover, soil erosion, salinization, or waterlogging; usually associated with human misuse.

desiccation (des-i-kay-shun): Process of drying.

destructive distillation: Decomposition of wood by heating out of contact with air, producing primarily charcoal.

determinate growth: Pertaining to a leaf or stem that stops growing after differentiating into a terminal flower.

developmental biology: A large field of investigation that includes the study of all changes associated with an organism as it progresses through life.

Devonian period: A geologic division of the Paleozoic Era following the Silurian Period and preceding the Permian.

diatomaceous earth: Powdery, soil-like material formed by the glass cell walls of dead diatoms deposited on the marine floor.

dichotomous (di-kot-o-mus): Pertaining to the division or forking of a single axis into two branches.

dicotyledon (dicot): Group of angiosperms that produce two cotyledons.

differentiation: Chemical and physical changes associated with the developmental process of an organism or cell.

diffuse-porous wood: Secondary xylem characterized by the same-sized vessels and tracheids through the growing season so that growth rings are difficult or impossible to detect.

diffusion: Random movement of particles from a region of high concentration to a region of low concentration.

digestion: Process by which macromolecules are broken down into smaller molecules.

dihybrid cross: Cross in which the inheritance of two characteristics is studied.

dioecious: Process by which macromolecules are broken down into smaller molecules.

diploid: Nucleus with two sets of identical chromosomes; the sporophyte is diploid, having the 2n number of chromosomes.

dispersal: How seeds distribute themselves before germination is done—by wind, water, animals, and people.

disulfide bond: Linkage between the sulfur atoms of two different amino acids in a protein.

diurnal (dye-urn-al): Pertaining to daily cycles or events.

diurnally: Daily cycle that occurs or is active during the daytime rather than at night.

DNA replication: Use of existing DNA as a template for the synthesis of new DNA strands; in humans and other eukaryotes, replication occurs in the cell nucleus.

doctrine of signatures: Concept that any organic substance carries within itself the likeness of some organ or part of the human economy, as a sign that this particular substance was applicable to disturbances of that organ.

dominant: (1) most prevalent species in a plant community; (2) allele that has its trait expressed; (3) gene that has its phenotypic expression appear in the offspring, regardless of the nature of its allelic partner.

dormant: Having reduced metabolic and respiratory activity.

double fertilization: Reproduction strategy in angiosperms in which two sperm are involved in the fusion with other nuclei.

double helix: Term used to describe the configuration of a DNA molecule. The helix consists of two spiraling strands of nucleotides held together with chemical bonds.

drifts: Color grouping of plants.

drill: Machinery for evenly spacing seeds (grain) in the soil at a uniform depth and giving good soil-seed contact.

drought (drout): Environmental condition in which precipitation is not sufficient to maximize biological productivity.

drupe: Fleshy fruit with a one-carpeled ovary and only one seed; endocarp is hard and stony, tightly enclosing the seed; the mesocarp is fleshy, and the exocarp is soft and thin.

dry fruit: Fruit that has dry ovary tissue unlike fleshy fruits that have a fleshy ovary.

duration: Refers to the number of hours of light a plant receives in a 24-hour period.

dusting: Sprinkling flowers or plants with pesticides to protect them from insects and rodents.

dwarf: Tree that grows to about 4 to 10 feet in height.

early wood: Large, thin-walled xylem cells produced early in the growing season, which appear less dense than latewood.

ecological succession: Sequential replacement of one vegetative community by another through a series of stages; succession ends when the climax community is established.

ecology: Total interrelationships among all the living organisms with each other and with the nonliving components of their environment.

ecosystem: Natural interrelationships among all the living organisms in a given area and with the environmental factors of that area. Ecosystems are self-sustaining, balanced, and self-perpetuating.

edema: The extended swelling in plant organs caused primarily by an excessive accumulation of water. This occurs since the cell walls are composed of flexible cellulose.

effluent: Water from industrial facilities, municipalities, and other such sources.

egg or egg cell: Middle of three haploid cells at the micropylar end of the embryo sac; when fertilized, it will form the zygote.

electrical energy: Carbon and energy enter living organisms through the process of photosynthesis.

electromagnetic spectrum: Entire radiation spectrum from the high-energy levels of cosmic rays to the low-energy level of radio waves; a small portion in the center provides wavelengths visible to humans.

electron: Negatively charged particle that orbits around the atomic nucleus;

number of electrons is always equivalent to the number of protons.

electron transfer: Process of energy transfer in biological systems, usually in small steps with only slight changes in energy levels.

electron transport: Successive passage of electrons from one cytochrome or flavoprotein to another by a series of oxidation-reduction reactions.

element: Substance compound of a single kind of number of protons in the nucleus.

embryo: In plants, that portion of a seed that will form the growing seedling following germination; it has a radicle, apical meristem, and embryonic leaf or leaves.

embryo sac: Mature megagametophyte; contains eight haploid nuclei.

embryonic axis: Main root/shoot body of a seedling.

endangered species: Living species that is in danger of becoming extinct because of small population sizes, poor reproduction, reduced available habitat, or a combination of these factors.

endergonic: Pertaining to a reaction that requires energy input before it will occur; endergonic reactions never occur spontaneously.

endocarp: Interior layer of the fruit wall.

endodermis: Layer of cells directly outside the pericycle and inside the cortex of roots; a portion of the cell layer is suberized by the Casparian strip.

endogenous: Originating or produced within an organism, tissue, or cell.

endophyte: Internal plant fungus.

endoplasmic reticulum (ER): Flattened membrane from a network running throughout the cytoplasm of cells; if ribosomes are attached, it is termed *rough endoplasmic reticulum*; if ribosomes are not present, the membrane is termed *smooth endoplasmic reticulum*.

endosperm: Triploid nutritive tissue resulting form the fusion of a haploid sperm nucleus with the two. haploid polar nuclei in the ovule of angiosperms.

endosymbiosis: Theory that some cellular organelles arose by the incorporation of a prokaryote into the cytoplasm of a eukaryote.

entine: Inner layer of a pollen grain shell.

entire: Smooth leaf margin without teeth, lobes, or undulations.

entropy: Physical concept describing the degree of orderliness in a system.

environment insults: Factor in the physical environment that inhibits the growth and/or development of an organism.

enzyme: Protein that functions as a catalyst in biochemical reactions.

ephedrine: Common ingredient in herbal dietary supplements used for weight loss; ephedrine can slightly suppress your appetite, but no studies have shown it to be effective in weight loss; ephedrine is the main active ingredient of ephedra. Ephedra is also known as Ma Huang, not ephedrine; high doses of ephedra can cause very fast heartbeat, high blood pressure, irregular heart beats, stroke, vomiting, psychoses and even death.

ephemeral *(e-fem-er-uhl)*: Temporary, such as vegetation that completes its life cycle in a short time.

epicotyl: That portion of a seedling above the cotyledonary node.

epidermal cells: Outermost layer of cells covering the entire plant body.

epinasty: Unequal growth of petioles causing the leaf blade to curve downward.

epiphyte *(ep-i-fight)*: Plant that grows on another plant as a support but derives no nutrition from the host.

epistasis: When one gene has a masking effect of the expression of another nonallelic gene, the former is said to be *epistatic*.

equatorial plate: Figure formed by the chromosomes in the center (equatorial plane) of the spindle in mitosis.

ergot: Spore-producing reproductive body of a fungus that infects grain crops; ergot contains lysergic acid and ergonovine.

essential oils: Highly volatile and aromatic oils formed in glands or special cells by some plants; probably involved in pollinator attraction or repulsion of herbivores; used in perfumes, soaps, medicine, and food.

estrone: Animal steroid (hormone) found in some plants in low concentration.

estuary: Marshy region where fresh water from a stream or river merges with salt water from the ocean.

ethylene: Plant hormone that promotes fruit ripening in addition to other physiological responses.

etiolation: Abnormal elongation of stems caused by insufficient light or unbalanced hormonal relationships. *Etiolated* stems often lack chlorophyll.

eukaryotic: Possessing a nucleus.

eutrophication *(you-tro-fi-ka-shun)*: Natural process of dead organisms gradually filling a standing body of water (pond, lake) as eutrophic (nutrient-rich) cycles cause rapid population increases followed by crashes due to shortages of nutrients.

evaporation: Process by which any substance is converted from a liquid state into, and carried off in, vapor.

evapotranspiration: Combined water loss from both leaf surfaces and from the soil surface; the sum of transpiration and evaporation.

evolution: Process by which a lower, simpler form of life is changed through genetic mutations to a higher, more complicated form of life.

exergonic: Pertaining to a reaction that gives off energy and occurs spontaneously.

exine: Hard outer coat of a pollen grain.

exobiology: Study of evidence relative to the possibility of life on other planets.

exocarp: Outer layer of the mature ovary (fruit wall).

exogenous: Developing outside an organism, tissue, or cell.

exon: "Sense" segments of mRNA that contain the actual genetic message for producing a given protein.

extinction: Permanent removal of all individuals of a species from earth.

F1 Generation (first filial): Hybrid offspring produced in the cross-pollination of parent generation plants.

F2 Generation: Progeny of self-pollinated F1 generation plants.

fact: Concept whose truth can be proved; scientific hypotheses are not facts.

facultative anaerobes: Micro-organisms capable of switching pathways of respiration, depending on the presence or absence of oxygen.

family: Taxonomic category composed of one or more related genera.

far red light (Pfr): Light at the extreme red end of the visible spectrum, between red and infrared light; usually regarded as the region between 700 and 800 nanometers.

fascicular cambium: Layer of cambium that develops between the xylem and phloem within a vascular bundle.

fats: Organic molecules containing high levels of carbon and hydrogen, but little oxygen. Oils are merely fats in a liquid state.

fatty acid: Long-chain hydrocarbon with little or no oxygen and terminating in an acid ($-COOH$) group.

fermentation: Process of ethanol formation by partially oxidizing pyruvic acid; no oxygen is involved in this process.

fertile crescent: Crescent-shaped region stretching from Armenia to Arabia, formerly fertile but now mainly desert, considered to be the cradle of civilization.

fertilization: Fusion of two haploid gametes, egg and sperm, producing a diploid zygote.

fiber: Elongated cell type found in many plants and associated as a support tissue of xylem or phloem; highly sclerified and usually dead at maturity.

fiberboard: Building material composed of wood chips or plant fibers bonded together and compressed into rigid sheets.

fibrous root: Root system with many equally sized roots forming a mat, as in grasses; there is no primary taproot.

fiddlehead: Curled fern frond prior to unrolling and elongation; also known as a *crozier.*

field capacity: Soil water storage capacity; the saturated soil profile after gravitational percolation ceases to flow.

filament: Elongated stalk of a stamen.

filial: Offspring generation, for example F1, the first filial generation.

finite resources: Resources that have a limit to their availability; not boundless.

fire ecology: Study of the environmental effects of fire.

first agriculture revolution: Relative suddenness of the beginnings of agriculture.

flaccid: Pertaining to a cell or tissue with less than full turgor pressure.

flavin mononucleotide: Electron acceptor in the electron transport scheme of aerobic respiration.

flavonoids: Group of secondary compounds produced by plants and important in chemical identification of those plants; believed to be contained in petals that reflect ultraviolet patterns and thus as element in pollinator attraction.

fleshy fruit: After fertilization, the ovary develops into fleshy tissue, for example, the apple.

flora: All the plants of a given area.

florigen: Flowering hormone are the terms used for the hypothesized hormone-like molecules that control and/or trigger flowering in plants.

fluorescence: Release of energy at a longer wavelength than the absorption wavelength, but still in the visible spectrum.

fly agaric: (A. muscaria), mushroom having hallucinogenic properties is also a poisonous mushroom.

follicle: Fruit that develops from a single-carpeled ovary and splits down one side when mature.

food chain: Sequence of organisms in which plants are the primary food source for herbivores, which are in turn the food source for carnivores, etc., until the top carnivore level is reached.

food web: Community food chain depicting what species feed on each other and how many interrelationships are involved.

forage: Foodstuffs from the leaves and stocks of plants. These could be grasses, legumes, or other cultivated crops.

forcing: Making bulbs artificially break dormancy so that they will flower when brought into a warm room.

fossil fuels: Organic molecules derived from partially decayed plant and animal matter produced primarily during the Carboniferous period; includes oil, gas, and coal.

fraction I protein: Equivalent to RuBP carboxylase; the primary leaf protein in many green plants.

fragmentation: Method of asexual reproduction by simply breaking into parts.

frame shift: *see* Deletion mutant; Insertion mutant.

fresh water: Water of relatively low salt concentration, as opposed to sea matter or salt water.

frond: Photosynthetic leaf blade of a fern.

fruit: Mature ovary.

funiculus: Stalklike structure connecting an ovule to its placental surface within an ovary.

fused: Flower parts that grow together.

galactose: Six-carbon sugar.

galactosidase: Enzyme responsible for the splitting of lactose into glucose and galactose.

gametangia: Cell or organ in which gametes are formed.

gamete: Sex cell; the mature haploid reproductive cell of either sex.

gametophyte: Haploid, gamete-producing plants in the alternation of generations; undergoes mitosis to produce the haploid gametes, which fuse to form the diploid zygote of the sporophyte.

gamma rays: That portion of the sun's total range of radiation in which rays are shorter than X-rays; below 0.1 nm in length.

gas exchange: Movement of oxygen and carbon dioxide between the cells.

gene: Unit of inheritance; a group of nucleotides on the DNA molecule responsible for the inheritance of a particular character.

gene activators: System that would allow gene expression to be activated by crossing reporter and activator plants.

gene regulation: Process by which genes are turned on and off to regulate growth and development of an organism.

generative nucleus: Produced by the haploid microspore nucleus of a pollen grain, the generative nucleus divides mitotically to form two sperm nuclei.

genetic code: Set of 64 different possible condons (codes) or nucleic acid triplets and their corresponding amino acids; determines which amino acids will be added during protein synthesis.

genetic engineering: Modifying the genetic structure of one organism by splicing in selected genetic information from another organism.

genome: Complete set of genetic information of an organism including DNA and RNA.

genomics: Study of genomes.

genotype: Genetic composition of an organism.

genus: Grouping of closely related species; the first word of a scientific binomial.

geotropism: Bending responses of a plant to the forces of gravity.

germinate: To resume growth and increase metabolic activity, as in a seed.

gibberellin: Group of related compounds that cause single-gene dwarf mutants of corn and peas to elongate normally.

ginkgos: A deciduous, dioecious tree (*Ginkgo biloba*) native to China and having fan-shaped leaves and fleshy yellowish seeds with a disagreeable odor. The male plants are often grown as ornamental street trees. Also called *maidenhair tree*.

girth: Circumference, as of a tree trunk.

glucose: Six-carbon monosaccharide (simple sugar); the primary substrate for respiration.

glycerol: Organic molecule to which fatty acids are attached to form a fat.

glycogen: Primary storage carbohydrate of animal cells.

glycolysis: Series of reactions preceding anaerobic or aerobic respiration in which glucose is oxidized to pyruvic acid.

glycoprotein: Macromolecule composed of a carbohydrate-protein complex.

glycosides: Cyanogenic glycosides, which are based in cyanide, are sedative and relaxant to muscles, but must be used in very small doses, under the care of a qualified herbalist or holistic physician; vary in solubility, most are soluble in water and alcohol.

glyoxysomes: Subcellular microbody present in the cytoplasm of many oil seeds; enzymes packaged in the glyoxysome convert lipids to carbohydrates during the germination process.

golgi apparatus: Organelles consisting of stacks of flattened membranes that function in packaging and synthesis of membranes and cell walls.

gradient: Rate of change with respect to distance of a variable quantity, as temperature or pressure, in the direction of maximum change.

grafting: Method of asexual propagation characterized by placing a shoot (the scion) onto the rootstock (the stock) of another plant.

granite: Igneous rock overlying most of the land masses of the earth.

granum: Stacks of thylakoids that are the site of light reactions in the chloroplast.

grassland: Land where vegetation is primarily grasses.

greenhouse effect: Climatic effect attributed to high levels of CO_2 in the atmosphere that traps incoming infrared radiation and raises the temperature of the earth.

Gregor Mendel: Monk known as the father of genetics.

growth retardants: Chemical applied as a spray or a drench on plants to reduce their size; commonly used by growers to dwarf potted plants such as chrysanthemums.

growth rings: Layers of wood produced around a tree's stem and branches

during each growing season; also called annual rings. The number of annual rings indicates a tree's age, while the thickness of rings provides an estimate of a tree's growth rate. Rings are frequently visible when a tree is cut.

ground meristem: Basic or fundamental tissue of the apical meristem; dermal tissues surround the ground meristem, and the provascular strands are embedded in it.

ground parenchyma: Basic ground tissue consisting of living parenchymal cells.

guanine: Nitrogen base found in both DNA and RNA.

guard cell: One of two epidermal cells associated with the stomatal apparatus.

guttation: Exudation of liquid water from hydathodes fed by vascular xylem traces.

gymnosperms: Plant group that bears naked seeds in cones.

habitat: Biotic and abiotic components making up the home of all the organisms in a given region.

half-life: Time required for half the radiation of a radioisotope to be emitted.

hallucinogen: Compound that produces a mind-altering effect.

halophilic: Pertaining to microorganisms that tolerate high concentrations of saline media.

haploid: Single set of chromosomes (half the full set of genetic material), present in the egg and sperm cells of animals and in the egg and pollen cells of plants.

hardpan: Impervious layer in the soil that restricts root penetration as well as movement of air and water.

hardwood: A term used in reference to all woody dicot, more accurately, wood having a high specific gravity.

hashish: Purified resin prepared from the flowering tops of the female cannabis plant.

haustorial roots: Parasitic plants may form specialized haustorial roots that form an attachment disc to the host during the first stage of colonization.

head: Many individual flowers tightly compressed into the shape of a single large flower; edible or harvestable part of the grain.

heading: One of the growth stages of small grains; when wheat or oat emerges; when the florets are fertilized and the kernels develop.

heartwood: Wood found in the center of a tree trunk; often a darker color due to the accumulation of resins, oils, gums, and other metabolic by-products, which prevent water movement through this tissue.

heavy metals: Metallic chemicals like cadmium, arsenic, copper, and zinc that can be harmful pollutants when they enter soil and water.

heeling in: Temporary planting in a trench.

helix: Spiral; DNA is a long double-stranded molecule wound in a spiral.

hemicellulose: Polysaccharide component of primary cell walls; similar to cellulose, but more easily degraded.

hemoglobin: Blood protein responsible for the transport of oxygen throughout the body.

hemp: *Cannabis sativa*; source for fibers used in making hemp rope.

hemp plant: The tough, coarse fiber of the cannabis plant, used to make cordage. Any of various plants similar to cannabis, especially one yielding a similar fiber.

herbaceous perennials: Without woody tissues; typical of most annuals and biennials; herbaceous perennials die back to the soil level each year.

herbarium: Collection of pressed, dried plant specimens mounted on sturdy (rag) paper and stored for reference and research.

herbicide: Plant growth regulator that inhibits growth or kills a plant when applied at relatively low concentrations.

herbivore: Animals that feed only on plants.

heredity: Transmission of genetically controlled characters from parents to offspring through sexual reproduction.

hereditary factors: If F2 has two, all have to have two; inbred parents only produce one kind of gamete, so they have only one kind of hereditary factor.

heritable: Capable of being inherited, such as physical trait.

heroin: An opiate processed directly from the extracts of the opium.

hesperidium: Berry with a thick leathery peel (exocarp and mesocarp) and a fleshy endocarp arranged in sections.

heterosporous: Producing two distinct types of spores.

heterotrophic: Organism that obtains its food from other organisms.

heterotrophic theory: Theory of the origin of life that proposes that the first organisms obtained nutrition from the spontaneous formation of organic molecules derived from primordial gases.

heterozygous: Genotype for a given phenotypic expression containing a dominant and a recessive allele for that trait.

hexose: Six-carbon sugar.

hill reaction: Splitting of a molecule of water during the light reactions of photosynthesis; the photolysis of water.

hill system: Method of planting strawberries; the hill system is used for everbearers; plants are set 12 to 15 inches apart; because all runners are removed, three rows can grow together to form one large row; three rows are spaced 1 foot apart.

histones: Basic proteins constituting a portion of the nuclear material and functionally associated with DNA.

holdfast: Organ at the base of macroalgae that attaches the stalk to a rocky surface.

homogeneous: Consistent and similar in composition.

homologous pairs: Two chromosomes that are morphologically and structurally identical and that pair during prophase of meiosis.

homologue: One chromosome morphologically and structurally identical and that pair during prophase of meiosis.

homosporous: Producing only one kind of spore.

homozygous: Genotype for a given phenotypic expression containing either two dominant or two recessive alleles for that trait.

hormone: Organic molecule synthesized by a plant that exerts, even in low concentrations, profound regulation of growth and/or development.

humid: Containing or characterized by a great deal of water vapor; "humid air"; "humid weather."

humus: Organic portion of soil derived from partially decayed plant and animal matter.

hybrid: Organism resulting from the fusion of gametes from different parental species.

hydrocooling: Removing heat from freshly harvested fruits and vegetables by immersion in ice water.

hydrolysis: Splitting of a large molecule into smaller molecules by the addition of water.

hydrophilic: Property of a substance that has a tendency to repel water.

hydroponics: System of growing plants in a liquid nutrient solution without soil.

hyoscyamine: Poisonous crystalline alkaloid (isometric with atropine but more potent); used to treat excess motility of the gastrointestinal tract.

hypocotyl: Shoot portion of a seedling below the cotyledonary node.

hypocotyl hook: Hooked portion of a hypocotyl formed at germination that assists a dicot shoot in pulling itself through the soil crust.

hypothesis: Proposed solution of a problem; a theory to explain unproven events or observations.

identical genetic makeup: Cell or collection of cells containing identical genetic material.

igneous: Rock of molten or volcanic origin.

imbibition: Process of taking up water physically.

immunological: Pertaining to the immune response, in which a protein antibody is synthesized by an organism to counteract some pathogenic factor.

imperfect: Has only one set of reproductive parts.

impermeable: Having the property of restricting the passage of substances.

incomplete dominance: Referring to the phenotypic expression for a given trait demonstrating a blending of the genetic messages from the allele partners controlling that trait; no dominant allelic partner.

incomplete flower: Lacking one or more of the four floral parts (sepals, petals, stamens, pistil).

indehiscent: Denoting mature fruit that do not split open to release their seed.

independent assortment: Random alignment of homologous chromosomes in meiosis.

indeterminate: Pertaining to a stem that produces unrestricted vegetative growth; the stem does not terminate in flowering.

indigenous: Native to or originating in a particular area.

indoleacetic acid (IAA): Naturally occurring auxin.

inducer: Compound that causes the induction of a particular enzyme.

inducible enzyme: Enzyme present only when induced by some particular substrate.

indusium: Thin membrane that covers the sori of many ferns; the membrane often breaks at the time of spore maturation.

inferior ovary: Ovary attachment to a modified receptacle in which other floral parts are fused to the ovary.

inflorescence: Flower cluster with a definite arrangement of individual flowers.

infrared: That portion of the sun's total range of radiation having wavelengths immediately longer than the longest of the visible spectrum (red); between approximately 750 nm and 1 m in length.

inhibitors: Agents that block or suppress the activity of cell growth.

inorganic nitrogen: Chemical compound without carbon as its skeleton atom.

insectivorous plant: Plant capable of deriving nutrition by digesting insects.

insertion mutant: Type of mutation in which an extra base pair is inserted and shifts the sequence out of phase by one pair; also called a frame shift.

installation: Refers to putting sod in a permanent location after having been grown elsewhere.

integrated pest management (IPM): Eco-sensitive way to manage pests of all kinds in the garden and landscape; an approach that focuses on the use of various biological controls.

integuments: Two outer layers of an ovule that enclose the nucleus tissue within.

intensity: State or quality of being intense; intenseness; extreme degree; as, intensity of heat, cold, or light.

intercalary meristem: Type of meristem present at the base of the blade and/or sheath of many monocots.

interfascicular cambium: Layer of cambium that develops between vascular bundles and connects with the fascicular cambium to form the vascular cambium of woody tissues.

interkinesis: Activities occurring between meiosis I and meiosis II; similar to interphase but without chromosome replication.

internode: Stem distance between nodes.

interphase: Nuclear condition between one mitosis and the next; chromosomes are not visible, but intense metabolic activity is occurring.

intertidal zone: Region between high and low tide.

intine: Inner wall of a pollen grain that does not contain sporopollenin.

intron: Intragene segments of genetic gibberish that do not code for the production of a given protein; these segments are left out of the mRNA that leaves the nucleus to direct protein synthesis.

involutions: Infolding of membranes to increase surface area.

ion: Atom or molecule that has gained or lost an electron, causing the particle to become electrically charged.

ionize: To split a molecule into two or more parts, each part becoming an electrically charged particle.

irish potato: Edible tuber native to South America; a staple food of Ireland.

irregular: Bilateral symmetry; having only one plane through which the structure (flower) could be cut to result in mirror image halves; zygomomorphy.

irritants: Things that bother the nose, throat, or airways when they are inhaled (not an allergen).

island of fertility: Region in a desert ecosystem directly beneath a tree or shrub; leaf fall and accumulation of litter result in nutrient cycling directly around the plant, even though the bare soil between plants may be depleted of nutrients.

isogamy: Union of gametes or gametangia of equal size.

isoprenes: Five-carbon compounds that are the basic unit of terpenes.

isozyme: Different chemical forms of the same enzyme; thought to be important in adaptation to environmental extremes.

Jarmo: Ancient city located in the foothills of the Zagros Mountains in the Middle East; studied by archaeologists, who have documented the existence of agriculture activities there between 10,000 and 12,000 years ago.

jointing: One of the growth stages of small grains; at jointing stage, the nodes begin to separate and can be felt in the lower part of the stem.

kinetic energy: Energy resulting from the random movement of molecules.

kinetin: Plant-based, nonirritating growth factor extract that has shown to significantly reduce free-radical damage and slows down aging of plant cells.

knots: Defects that weaken the timber and interfere with its ease of working and other properties.

Kranz anatomy: Specialized leaf anatomy found in C4 plants in which the vascular bundle is surrounded by bundle sheath cells.

Krebs cycle: Mitochondrial oxidation of pyruvic acid by considering acetyl coenzyme A with oxaloacetic acid to form a series of 6-, 5-, and 4-carbon organic acids.

lactose: Milk sugar; a disaccharide composed of glucose and galactose.

lateral buds: Produced in the apex of lateral branches.

lateral meristem: Meristem giving rise to secondary plant tissues, such as the vascular and cork cambia; term is sometimes used to refer to an axillary meristem.

laterite: Type of tropical soil in which iron and aluminum oxides cause the soil structure to harden like concrete.

latewood: Small, thick-walled xylem cells produced at the end of the growing season that appear as a dense ring of wood adjacent to the thin-walled early wood.

latitude: Geographical unit used to measure the distance from the equator toward either pole.

lattice: A crystalline-like structure caused by the precise orientation of molecules in a solid or liquid.

laudanum: Narcotic consisting of an alcohol solution of opium or any preparation in which opium is the main ingredient.

layering: Method of asexual propagation in which portions of the stem are wounded and covered with a medium, usually soil, to stimulate the production of adventitious roots.

leaching: Process of removal of ions or molecules by flushing with water.

leaf area index: Numerical index of the ratio of leaf area to ground area in a plant community.

leaf venation: Leaf skeleton as those axes constituting the primary and secondary vines.

leaflet: Individual blade unit of a compound leaf.

leeward: Side away from the direction of a prevailing wind.

legume: Fruit that develops from a single-carpeled ovary and splits along two sides when mature; plants with the characteristic of forming nitrogen-fixing nodules on its roots, in this way making use of atmospheric nitrogen.

lethal: Gene or genotype that is fatal for the individual gene.

leucine: Essential amino acid, C_4H_9 $CH(NH_2)COOH$, obtained by the hydrolysis of protein by pancreatic enzymes during digestion.

leucoplasts: Membrane-bound organelles specialized for starch storage.

levee: Earthen dike used to enclose water.

liana: Large, woody vine common to the tropical forest; it climbs the tall trees and often trails from the canopy.

lichen: Organism composed of a symbiotic association of an ascomycete fungus with algal or cyanobacterial cells.

light: Electromagnetic radiation produced by the sun that heats and illuminates the earth.

light intensity: Strength of light rays; the degree of brightness dependent on the number of photons striking a given area at a point in time; measure of the brightness of light reaching a surface; decreases as the distance from the source of the light increases.

lignified: Impregnated with lignin, such as the secondary cell walls of woody plants.

lignin: Complicated organic molecule found as an important constituent of many secondary cell walls; imparts strength and rigidity to the cellulose microfibrils.

limiting factor: Climatic factor that would first curtail plant growth if unavailable.

limnetic zone: Open-water zone of a lake or pond; extends to the depth of effective light penetration.

limnologist: Scientist who studies freshwater biology.

linkage: Occurrence of alleles for a different trait.

lipid: Fat or oil; composed of fatty acids and glycerol.

litter: Dead organic material such as branches, tree trunks, and dry grass that accumulates on the floor of a forest; litter acts as additional fuel in a forest fire, producing increased destruction.

littoral zone: Shallow-water zone of a lake or pond; light penetrates to the

bottom, and the area is occupied by rooted plants such as water lilies, rushes, and sedges.

lobed: Leaf having deeply indented margins.

locoweed: Any of several leguminous plants of western North America causing locoism in livestock.

locus (pl., loci): Position that a given gene occupies on a chromosome.

long day plant: Plant that flowers when the length of day *exceeds* some critical value.

LSD: Lysergic acid diethylamide; a synthetic hallucinogenic derivative of lysergic acid.

lumen: Central cavity of a cell.

lysosome: Enzymes that break down proteins and other macro molecules.

macromolecule: Very large complex molecule; found only in plants and animals.

macronutrient or macroelements: Actual chemical form or compound, which enters the root system of a plant.

macropores: Large pore spaces caused by invertebrates and larger animals, including reptiles and mammals that permeate the soil; drain water not held by capillarity.

maize: Worldwide name for corn; a tall annual cereal grass bearing kernels on large ears; widely cultivated in America in many varieties; the principal cereal in Mexico and Central and South America since pre-Columbian times.

margin: Edge of a flattened structure; in leaves, the lateral edge of the blade.

marine biome: Of or pertaining to the ocean.

market gardening: Refers to growing a wide variety of vegetables for local or roadside markets.

marsh: Wetland along the shallow margins of the lakes and ponds and in low, poorly drained lands where water stands for several months of the year.

mass: Fundamental unit of measurement equivalent to the weight of a substance when compared with the weight of hydrogen.

matrix potential: Water potential component caused by the attraction of water molecules to a hydrophilic matrix.

meadow: Open clearing in a landscape in which low-growing, herbaceous plants dominate.

mean: Arithmetic average of a set of data points or values; the sum of each data point or value divided by the number of data points.

media: Growing materials in which plants can be started that are loose, well drained, fine textured, low in nutrients, and free of disease.

megagametogenesis: Production of megagametes (large gametes) from megaspore in the ovules of angiosperms.

megagametophyte: Gametophyte stage containing eight haploid nuclei within the embryo sac.

megaspore: Haploid cell produced by meiosis in the ovules of angiosperms; a single megasporocyte produces four megaspores, only one of which remains functional.

megasporocyte: Megaspore mother cell; this diploid cell undergoes meiosis to produce four haploid megaspores in the ovules of angiosperms.

megasporogenesis: Process of megaspores (large spores) being produced via meiosis in the ovules of angiosperms.

meiosis: Process of two sequential nuclear and cellular divisions resulting in four haploid cells from a single diploid cell; occurs in reproductive organs.

membrane: Phospholipid bilayer impregnated with protein and certain other compounds in living organisms; functions in partitioning of cellulose activities.

membrane-bound: Eukaryotic with many membrane-bound organelle . . . but plants have only the large vacuole, the chloroplast, the cell wall, and the plastids.

meristematic cells: Meristem is a tissue in plants consisting of undifferentiated cells. Combine to make meristematic tissues.

meristematic regions: Region of cells capable of division and growth; classified by location as apical, or primary (at root and shoot tips).

mescaline: A norepinephrine-related hallucinogenic drug. Its source is the peyote cactus.

mesic: Referring to a region that receives adequate precipitation to maintain biological productivity.

mesocarp: Middle layer of the fruit wall, located between the exocarp and endocarp.

mesophyll: Parenchymal tissue in the center of a leaf.

messenger RNA (mRNa): Ribonucleic acid transcribed from the DNA template.

metabolic energy: Energy obtained from ATP produced in metabolism.

metabolism: Summation of all the biochemical events within a cell or organism; it includes both synthesis and breakdown of molecules.

metabolities: Chemical substances required in metabolism.

metacentric: Denoting a chromosome with a centromere located at the center.

metamorphic rock: Type of rock, either granite or sedimentary in origin, and structurally changed by high temperature and pressure.

metaphase: Stage of nuclear division in which the chromosomes migrate to the center of the cell.

methane: CH_4, natural gas.

methanogen: Bacterium that obtains energy from CO_2 and H_2 and forms methane.

microelements or micronutrients: Nutrients required in smaller concentrations, but vital to plants.

microfibril: Elongated strand of cellulose molecules.

microgametogenesis: Production of microgametes (small gametes) from microspores in the anthers of flowering plants.

microorganisms: Single-celled living organisms that can be seen only with the aid of a microscope.

micropores: Spaces between soil particles that hold water by means of capillary forces.

micropropagation: Tissue culture.

micropyle: Small opening at one end of an ovale where the integuments come together.

microsporangia: Receptacle in which microspores (male spores) are developed.

microspore: Haploid cells produced by meiosis in the anthers of angiosperms; four microspores are produced from a single microsporocyte.

microspore mother cell: Diploid mother cell that will go through meiosis to produce the (4) haploid microspores.

microsporocyte: Microspore mother cell, this diploid cell undergoes meiosis to produce four haploid microspores in the anthers of angiosperms.

microsporogenesis: Process of microspores (small spores) being produced via meiosis in the anthers of the angiosperms.

microsporophyll: Leaflike structure giving rise to one or more microsporangia.

microtubules: Tiny rodlike structures made of the protein tublin and important in the synthesis of certain membranes.

middle lamella: Cementing layer of pectic substances between primary cell walls.

midrib: Central large vein of a leaf.

millet: Annual grass (*Panicum milaiceum*) cultivated in Eurasia for its grains and in North America for hay.

mimicry: Pertaining to the resemblance of one organism to a totally unrelated organism; the similarity in color, form, or behavior may result in protection or some other advantage to the mimic.

mitochondrion (pl., mitochondria): Membrane-bound organelle associated with the Krebs cycle and electron transport aspects of respiration.

mitosis: Nuclear divisions of somatic cells resulting in two genetically identical daughter nuclei.

modification: Change, as in morphology, usually associated with a functional advantage.

mole: One gram molecular weight of any substance; that is, the molecular weight of any substance in grams.

molecular weight: Sum of all the atomic weights of a molecule.

molecules: Groups of atoms that share electrons and therefore bond together.

monocarpic: Denoting a plant that flowers only once.

monocotyledon (monocot): One of the two primary groups of angiosperms characterized by a single cotyledon, parallel venation of leaves, and floral parts in threes.

monoculture: Agricultural system in which only one crop species is cultivated.

monoecious: Condition of being unisexual with both pistillate and staminate flowers on the same plant.

monohybrid cross: Cross in which the inheritance of only one character at a time is studied.

monosaccharide: Sugar broken down to its simplest functional unit—a single saccharide.

monsoon: Exceptionally heavy precipitation that occurs seasonally in certain parts of the world.

Mormon's tea: Common name of *Ephedra*.

morphine: Pain-relieving and addictive compound derived from the opium poppy (*Papaver somniferum*).

morphology: Study of form and structure of living organisms.

mountain tundra: Portion of tundra vegetation confined to alpine meadows; low-growing grasses, sedges, and forbs with very short growing seasons; permafrost is typical.

mucilaginous: Containing a mucilage, usually composed of mucopolysaccharides.

muck soil: Soil found in old bogs, river deltas, and lake beds; contains high organic matter and has good water-holding capacity; nitrogen is also readily available through mineralization of organic matter.

mulberry tree: Species of deciduous trees native to warm, temperate, and subtropical regions; produces a small multiple fruit, 2–3 centimeters (0.8–1.2 in) long, that is dark purple to black, edible, and sweet.

mulch: Materials such as straw, sawdust, leaves, plastic film, and the like, spread upon the surface of the soil to protect the soil and plant roots from the effects of raindrops, soil crusting, freezing, evaporation, and so on; to apply protective materials to the soil surface.

multigenic (polygenic): Inheritance in which the genetic control for a trait results in the phenotypic expression varying continuously.

multiple fruit: Ovaries of many separate flowers clustered together.

multipurpose plant: Plant used for more than one purpose. For example, the baobab tree has many uses: oil from the seeds, the vitamin C and tartaric acid from the pulp, the leaves for their medicinal value, and the trunk for its fibers to make rope.

muscarine: Toxic quaternary ammonium compound found in species of *Clitocybe* and *Inocybe*; causes perspiration-salivation-lacrymation syndrome.

mushroom: Common name for a group of fungi (basidiomyctes) that produce an aboveground reproductive structure.

mutagen: Substance capable of causing a mutation.

mutation: Permanent heritable change in the genetic code of the DNA.

mutualistic: Having an interrelationship in which both (or all) organisms benefit from their association.

mycelium: Tangled mass of fungal hyphae.

mycorrhiza: Study of form and structure of living organisms.

N-formylmethionine: Effective in the initiation of protein synthesis.

NAD (nicotinamide adenine dinucleotide): Coenzyme that acts as an electron acceptor; particularly important in respiration.

NADP (nicotinamide adenine disnucleotide phosphate): Coenzyme that acts as an electron acceptor; particularly important in photosynthesis.

nastic movements: Movement of plant parts in response either to certain external stimuli or to internal growth stimuli; generally slow.

natural law: Proven event of nature with consistent results, such as the laws of gravity and thermodynamics.

nectar: Sweet syrupy exudate produced by some flowers as an attractant for pollinators.

negative matrix potential: Inside plant cells, the matrix potential is always negative and is caused by attraction of water, although water movement in passive plants can regulate transpirational flow.

negative solute potential: Inside plant cells, water potential is always negative; solutes reduce the free energy of water.

netted venation: Veins reticulated and resembling a fish net.

neutron: Uncharged atomic particle; in most common stable atoms the number of neutrons is equivalent to the number of protons.

niche: Specific and unique set of environmental conditions for a given species.

nicotine: Addictive drug in tobacco; nicotine activates a specific type of acetylcholine receptor.

nitrates: Only form in which nitrogen can be used directly by plants.

nitrites: Intermediate product in the conversion breakdown of ammonium to nitrate; nitrite is very unstable, and is almost immediately converted into nitrate.

nitrogen base: Basic component of nucleic acids; composed of a nitrogen-containing molecule of purine and pyrimidine.

nitrogen fixation: Process by which nitrogen gas (N_2) in the atmosphere is reduced to ammonia by certain microorganisms.

nitrogen gas: Nonmetallic, colorless, odorless, and tasteless inert diatomic gas that makes up 78% of the atmosphere by volume; nitrogen is found in all living tissues.

nitogenase: Enzyme involved in the conversion of atmospheric nitrogen gas into ammonia.

nitrophiles: Plants that absorb any excess of nitrates.

node: Point on a stem at which leaves and buds are attached.

nodule: Tumor-like growth on the roots of certain higher plants that encloses a population of nitrogen-fixing bacteria.

noncyclic photophosphorylation: Formation of ATP using the Z-scheme of photosynthesis.

nonvascular plants: Plants without vascular systems such as algae.

nucellus: Inner somatic tissue of an ovule.

nuclear envelope: Membrane system that surrounds the nucleus of eukaryotic cells; consists of inner and outer membranes, separated by perinuclear space and perforated by nuclear pores.

nuclear pores: Pitted regions in the nuclear envelope through which processed mRNA migrates to the ribosomes.

nucleic acid: Macromolecule composed of a polymer arrangement of nucleotides; DNA is double stranded, and RNA is single stranded.

nucleoli: Bodies in the nucleus that become enlarged during protein synthesis and contain the DNA template for ribosomal RNA.

nucleolus: An area on certain chromosomes associated with the formation of the nucleolus.

nucleotide: Portion of a nucleic acid molecule composed of a nitrogen base (purine or pyrimidine), a pentose sugar (ribose or deoxyribose), and a phosphate.

nucleus: Central "kernel" or membrane-bound organelle of a eukaryotic cell; contains the chromosomes and most of the DNA of the cell.

nut: Dry, indehiscent one-seeded fruit with a hard pericarp (the shell).

nutmeg: Colatile oil contains a substance used for medical purposes; 1 tbsp of powdered nutmeg produces a floating euphoria for between 6 and 24 hours.

nutrients: Food substances necessary for the sustenance of living organisms.

nyctinasty: "Sleep movements" of leaves in response to change in turgor pressure of cells at the base of the petiole.

oat (*Avena sativa*): Species of cereal grain, and the seeds of this plant; used for food for people, and also as fodder for animals, especially poultry and horses; oat straw is used as animal bedding and also sometimes used as animal feed.

obligate anaerobe: Microorganism that can survive only under anaerobic conditions.

ocean desert: Concept of a region in a lack of plants and animals; a region of low biological productivity.

olericulture: Vegetable production.

omnivore: Organism that feeds on other plants and animals.

operator gene: Gene responsible for the activation and deactivation of the structural genes.

operon: Gene regulation of the system consisting of an operator gene and one or more structural genes controlled by the operator.

opposite leaf arrangement: Leaves are attached at a node directly across from one another on the stem.

orbital: Discrete pathways or bands surrounding an atom through which electrons move.

organelle: Subcellular particles that perform some particular function within the cell; compound of carbon and another element or a radical.

organic matter: Living or once living material; compounds containing carbon formed by living organisms.

osmoregulator: Substance, either organic or inorganic, that functions to change the solute potential of a solution and thereby controls the water relations of the solution by osmosis.

osmosis: Diffusion of water molecules across a selectively permeable membrane.

osmotic agents: Induce transcription in eukaryotic organisms.

outcrossing: Flower that must be cross-pollinated to successfully complete the reproductive process.

ovary: Lower, enlarged portion of the female reproductive structure in a flower that gives rise to the fruit.

ovule: Female reproductive structure of a flower enclosed by an ovary; a developing embryo that gives rise to a seed.

oxalates: Group of molecules called organic acids, and are routinely made by plants, animals, and humans.

oxidation: Loss of electrons by an atom.

oxidation phosphorylation: Electron transport system associated with aerobic respiration and mitochondria.

oxidation reduction: Chemical reaction in which an atom or ion loses electrons to another atom or ion.

oxygen (O_2): Essential element in the respiration process to sustain life; colorless, odorless gas makes up about 21% of the air.

oxytocic agent: Having similar actions as those of the natural hormone oxytocin, primarily contraction of smooth muscle such as the uterus.

ozone (O_3): Molecular form of oxygen in equilibrium with O_2; layer of ozone in the stratosphere, which protects the earth form harmful ultraviolet radiation.

palatability: Taste potential for an animal.

paleobotanist: Studies the fossilize record of plants.

Paleozoic era: Lasted from about 540 to 250 million years ago, and is divided into six periods; time of many important events, including the development of most invertebrate groups, the evolution of fish, reptiles, insects, and vascular plants.

palisade: Vertical photosynthetic cells below the upper epidermis in leaf tissue; these cells are a specialized parenchyma.

palmate: Arrangement of leaflets (or lobes on a simple leaf), each originating from a common point, usually the axial end of a petiole.

palmately compound: Arrangement of leaflets in whorled fashion around the top of the petiole, which then attaches to the stem of the plant; resembles the arrangement of fingers attached to the palm of a hand.

paper chromatography: Process of separating small quantities of a substance in a mixture (often a solution) through ultrathin pieces of paper to show the selective absorption.

parallel: Veins that run parallel to each other; a characteristic of the monocot.

parasitic: Denoting an association where one living organism benefits at the expense of another.

parchment: Superior paper resembling sheepskin.

parenchyma: Tissue type characterized by simple, living cells with only primary cell walls.

parthenocarpy: Fruit development without fertilization.

parthenocarpic fruits: Fruits devoid of embryo and endosperm; such as seedless grapes; bred to provide the taste of the fruit without the seed.

pathogen: Any organism that causes a disease of another organism.

peat moss: Relatively sterile, inert medium composed of partially decomposed plants of the genus *Sphagnum*; exceptionally high water-holding capacity.

pectin: Cementing substance of which the middle lamella is composed; primarily calcium pectate and pectic acid.

peduncle: Stalk of a single flower or an inflorescence.

pelagic: Refers to fish and animals that live in the open sea, away from the sea bottom.

pellicle: A protein layer located just inside the plasma membrane in euglenoids.

peltate: Shield-shaped; having a flat circular structure attached to a stalk near the center, rather than at or near the margin.

PEP carboxylase: Enzyme responsible for CO_2 fixation in the primary fixation of C_4 metabolism.

pepo: Berry with the outer wall formed from receptacle tissue fused to the exocarp; the fleshy interior is mesocarp and endocarp.

peptide bond: Chemical bond formed between the amino group of one amino acid and the carboxyl (acidic) group of an adjacent amino acid.

percent purity: 100% minus (percentage inert matter + percentage other crop seed + percentage weed seed).

percolate: Gravity-induced movement of water down through the soil.

perennial: Plant that overwinters and continues to grow for many years; may reproduce every year, or only on rare occasions.

perfect flower: Having both stamens and carpels and therefore bisexual.

perianth: Referring to all the sepals and petals collectively.

pericarp: Fruit wall that develops from the ovary wall.

pericycle: Layer of cells surrounding the xylem and phloem of roots and considered to be a part of the vascular cylinder.

periderm: Protective tissue that replaces epidermis when it is sloughed off during secondary growth; includes cork, cork cambium (phellogen), and phelloderm.

permafrost: Permanently frozen soil in polar regions.

permanent wilting point: Soil moisture content at the point when a given plant's root system can no longer absorb water.

permeability: Property of membranes allowing all substances to pass freely.

peroxisome: Cellular micro-body containing enzymes involved with photorespiration and photosynthesis.

pesticides: Pesticides classified according to the organisms that they are used to control; for example: fungicides, herbicides, insecticides, molluscicides, nematicides, and rodenticides.

petal: Showy flower component attached just inside the sepals; usually colorful to attract pollinators.

petaloid: Modified, flattened filament of a stamen that may resemble a petal.

petiole: Stalklike portion of a leaf connecting the blade to the stem or branch.

petrifaction: Fossil formed when plant parts are infiltrated or replaced by mineral substances such that the structure is preserved but the fossil is actually rock.

peyote: Hallucinogenic drug containing mescaline that is derived from peyote buttons.

pH: (1) Negative logarithm of the hydrogen ion concentration; (2) numerical scale used to measure the acidity or basicity of any substance.

pharmaceuticals: Medicinal drugs.

phelloderm: Tissue laid down to the inside of the phellogen; the inner part of the periderm.

phellogen: Cork cambium.

phenotype: Physical features of an organism; the manifestation of the genotype as influenced by the environment.

phospholipid: Molecule in which glycerol is attached to two fatty acids plus a phosphate group; important part of biological membranes.

photolysis: Splitting of a molecule of water during the light reactions of photosynthesis; the Hill reaction.

photon: Elementary particle of light; a discrete unit of light energy.

photoperiodism: System within organisms that causes certain events, including the onset of reproduction, to be related to the length of day.

photophosphorylation: Formation of ATP utilizing light energy in photosynthesis.

photorespiration: Production of glycolic acid in chloroplasts in the light; glycolic acid may be oxidized by enzymes of peroxisomes.

photosynthesis: Production of carbohydrates by combining carbon dioxide and water in the presence of light energy; occurs in chlorophyll pigments of plants and releases oxygen as a by-product.

photosynthesis bacteria: Bacteria able to carry out photosynthesis; light is absorbed by bacteriochlorophyll and carotenoids.

photosynthetic unit: Group of associated chlorophyll molecules, including antenna molecules and a central chlorophyll *a* collector molecule.

photosystem I: Second part of the Z-scheme in which a chlorophyll *a* molecule absorbs most effectively at 700 nm.

photosystem II: First part of the Z-scheme in which the chlorophyll *a* absorbs most effectively at 680 nm.

phototropism: Bending of a plant toward a unidirectional light source.

phycobilins: Red and blue accessory pigments found in certain photosynthetic organisms.

phytochrome: Protein pigment responsible for the phenomenon of photoperiodism.

phytoplankton: Single-celled plants of aquatic biomes; the food source for zooplankton.

phytotoxins: Substance produced by plants that is similar in its properties to extracellular bacterial toxin.

phytrochrome red (Pr) far red (P$_{fr}$): Is a photoreceptor, a pigment that plants use to detect light. It is sensitive to light in the red and far-red region of the visible spectrum. Many flowering plants use it to regulate the time of flowering based on the length of day and night (photoperiodism) and to set circadian rhythms. It also regulates other responses including the germination of seeds, elongation of seedlings, the size, shape and number of leaves, the synthesis of chlorophyll, and the straightening of the epicotyl or hypocotyl hook of dicot seedlings.

pigment: Molecules that reflect and absorb light at particular wavelengths.

pinnate: Denoting an arrangement of leaflets (or lobes on a simple leaf) along a main central unit.

pinnately compound: Leaflets arranged along a single axis much like a feather.

pistil: Reproductive part of a flower, consisting of female stigma, style, and ovary.

pistillate: Denoting a unisexual flower having a pistil but no stamens (carpellate).

pith: Parenchymal tissue in the center of a stem located interior to the vascular bundles.

placental: Of the tissue to which an ovary is attached.

planed: Leveled or smoothed.

plant growth regulator: Any molecule that exhibits hormone-like effects in a plant, whether synthesized or naturally occurring.

plasma membrane: Cell membrane that holds the cytoplasm and lies against the cell wall (in plant and fungal cells).

plasmids: Circular, double-stranded unit of DNA that replicates within a cell independently of the chromosomal DNA; most often found in bacteria; and used in recombinant DNA research to transfer genes between cells.

plasmodesma (pl., plasmodesmata): Cellular microbody containing enzymes involved with photorespiration and photosynthesis.

plasmodium: Naked body of cytoplasm with many nuclei.

plasmolysis: Osmotic removal of water from the cytoplasm and vacuole, causing the cytoplasm to pull away from the cell wall and clump in the center.

plastids: Any of several pigmented cytoplasmic organelles including chloroplasts, chromoplasts, and leucoplasts.

pleiotropic: Action of single genes that affect the expression of several traits.

plumule: Shoot apex of a seedling, including embryonic leaves.

plywood: Panel products manufactured by gluing together layers of veneer with the grain of alternate layers oriented at right angles to provide strength.

point mutation: Change in the genetic composition of a cell to a change in a single triplet of the DNA template.

polar ice cap: Portions of the globe close to the poles that are permanently covered with ice.

polarity: Establishment of poles or areas of specialization at opposite ends of a cell, tissue, organ, or organism; polarity leads to the differentiation of roots from shoots.

polar nuclei: Two of the eight haploid nuclei of a megagametophyte that migrate, one from each end, to the middle of the embryo sac; two nucleifuse with a sperm nucleus during double fertilization.

pollen: Collective term of pollen grains, the male gametophytes.

pollen grain: Structure into which a haploid microspore develops; contains a haploid tube nucleus and two haploid sperm nuclei at maturity.

pollination: When the pollen grain is dispersed on the stigma.

pollinator: Organism that effects pollination.

polycarpic: Denoting a plant that flowers more than once during its lifetime.

polyembryony: Development of more than one embryo.

polymer: Large molecule formed by the linkage of many smaller, identical molecules.

polymerases chain reaction (PCR): Technique for amplifying DNA sequences in vitro by separating the DNA into two strands and incubating it.

polynucleotide chain: Attachment of one nucleotide to another in a linear fashion.

polypeptide: Any number of amino acids linked by peptide bonds.

polyploid: Term describing a cell with more than two sets (2n) of chromosomes.

polysaccharide: Large molecule formed by joining sugar molecules end to end.

polysome: Group of ribosomes functionally related and held together by a strand, presumable mRNA.

pome: Fleshy fruit derived from a compound inferior ovary; the fleshy edible part is the ripened tissue surrounding the ovary (derived from receptacle and perianth tissue); the ovary matures into the core and contains the seed.

pomologist: Fruit scientist.

population: All individuals of a given species in a given area.

porosity: Degree to which soil, gravel, sediment, or rock is permeated with pores or cavities through which water or air can move.

postharvest physiology: Study of the functioning of crop tissues and cells following harvest; seeks to maintain quality and prevent spoilage of crop.

potential energy: Energy available to do work, but not yet expressed.

prairie: Area of land dominated by grasses with occasional shrubby plants and small trees occurring where the grass cover is broken and with herbaceous perennials during certain seasons.

precipitation: Moisture falling from the atmosphere in the form of rain, sleet, snow, or hail.

precooling: Process of rapidly removing heat from vegetables before shipping.

preplant fumigation: Fumigation of weeds before planting.

pressure potential: Component of water potential caused by the force created by real pressure (turgor pressure) against a membrane.

prickle: Small, spine-like growth that is attached only to the epidermis; can be removed easily; for example roses have prickles.

primary cell wall: Cellulose wall of all plant cells laid down at time of mitosis and cytokinesis.

primary consumer: An organism that consumes a producer organism as a food source; a herbivore.

primary pit fields: Regions within the primary cell wall in which plasmodesmata traverse the cell wall.

primary succession: Plant successional events occurring in a pristine or newly forming habitat.

primary tissue: Any tissue derived from the apical meristem, either shoot or root.

primary xylem: Derived from procambium and divided into the earlier protoxylem and the later metaxylem.

primordia: Organ or tissue in its earliest recognizable stage of development.

principle of independent assortment: Mendel's law states that allele pairs separate independently during the formation of gametes.

principle of segregation: Mendel's law states that allele pairs separate or segregate during gamete formation, and randomly unite.

pristine: Denoting a natural and undisturbed state.

producers: Autotrophic organisms that synthesize organic molecules directly from CO_2 and H_2O.

profundal zone: Deepest portion of a lake.

progymnosperms: Generally looked like gymnosperms, did not produce seeds, reproduced by releasing spores into the environment, but lacked other characteristics of ferns.

prokaryotes: Organisms whose cells have no nucleus, including bacteria and cyanobacteria.

promoter: Segment of DNA that controls a group of structural genes.

prop roots: Adventitious roots arising on a stem above the ground and imparting some mechanical support to plants; angled roots may provide for adsorption.

propagation: Process of increasing in number.

prophase: First stage of nuclear division, characterized by the disappearance of the shorten chromosome.

proplastid: Membrane-bound particles that develop some internal structure; may subsequently develop into chloroplast. chromoplasts, or leucoplasts.

protected ovules: Adaptation such as cones.

protein: Macromolecule composed of a given linear sequence of amino acids.

protein sequencer: Determination of the identity and order of the amino acids present in a protein of interest.

proteomics: Branch of genetics that studies the full set of proteins encoded by a genome.

prothallium cells: Fern that produces both male and female sex cells.

protoderm: Dermal or outer tissue of an apical meristem that gives rise to the epidermis.

proton: Positively charged atomic particles that determine the elemental characteristics of an atom.

protoplast: Living portion of the cell.

protoplast fusion: Technique of enzymatically digesting the cell wall of two distinctly different cells, then treating the plasma membrane so that protoplasts of the two cells fuse. The resulting hybrid may be difficult or impossible to achieve with traditional plant breeding.

protoxylem: First xylem cells formed in the primary xylem.

proviral state: Condition of a host cell after having been transformed.

pruning: Selective removal of parts of a plant, usually woody shrubs or trees.

pseudocopulation: Sexual deception; generally applied to a pollinator attempting to copulate with a flower.

pseudohallucinogens: Produce psychotic and delirious effects without the classic visual disturbances of true hallucinogens.

psilocin: Hallucinogenic drug that is obtained from *Psilocybe mexicana* mushrooms.

psilophyta: First fossil of land plants had both horizontal and vertical shoots.

psychoactive: Having effects on the mind.

pterophyta: Group of non-seed plants with a fossil record dating back to the lower Devonian; consists of about 11,000 living species.

pubescenct: Having hairs or trichomes on the surface.

pulping: Process of partially digesting and breaking up wood fibers to make paper.

pumice: Lightweight white, yellow, or gray stone formed from volcanic glass; used in polishing and cleaning.

punnett square: Visual aid in determining the possible recombination frequencies for the existing gamete types of a cross.

pure opium: A pure form; morphine is derived (refined) from pure opium; heroine is a "synthetic" form of morphine.

pure parents: Deliberate mating of selected parents based on particular genetic traits; inbred line; group of identical purebreeding diploid or polyploidy.

purgative: Tending to cleanse or purge, especially causing evacuation of the bowels.

purine: Double-ring nitrogen base found in the nucleic acids, including adenine and guanine.

pyrimidine: Single-ring nitrogen base found in the nucleic acids including cytosine, thymine, and uracil.

quality of light: Where the particular wavelength of light falls on the light spectrum.

quarter-sawed: To saw into quarters and then into boards, as by cutting alternately from each face of a quarter.

quinine: Bitter-tasting drug obtained from the bark of the cinchona tree; related to coffee and gardenia; used in the treatment of malaria.

radial micellation: Reinforcement of the cell wall of guard cells in specific regions such that the cells curve outward when fully turgid.

radially cut: One of the three standard cuts of wood are transverse (cross sectional), radial, and tangential.

radiation: Energy in the form of waves or particles.

radicle: Primary root of a germinated seed.

radiocarbon dating: Dating method based on the radioactive decay of Carbon-14 (C^{14}) contained in organic materials.

rain-shadow desert: Desert on the leeward side of a mountain range.

ratoon: Shoot growing from the root of a plant, providing a second harvest.

rayon: Manufactured fiber composed of regenerated cellulose, derived from wood pulp, cotton linters, or other vegetable matter.

rays: Parenchymal cells found in secondary xylem and phloem in both woody angiosperms and gymnosperms, which provide for lateral (radial) transport.

reaction wood: Wood produced in response to a tree that has lost it's vertical position, causing the tree to straighten.

receptacle: Modified apex of a stem to which the floral parts are attached.

recessive: Gene masked by its dominant allelic partner, having the recessive phenotype expressed only when both alleles for a given trait are recessive.

recognition surface: Three-dimensional structure of a biological membrane surface that gives specificity due to macromolecules of various sizes and shapes that extend above the lipid layer.

recolonization: Revegetate; reestablishment of a natural community following a natural or unnatural event that removed the existing vegetation.

recombinant DNA (rDNA): Genes produced by genetic engineering manipulations that contain genetic segments from other organisms.

red light (Pr): Since plants use red light for photosynthesis, and reflect and transmit far-red light, the shade of other plants also can make Pfr into Pr.

reduction division: First nuclear division of meiosis during which the paired homologues migrate to opposite poles, resulting in cells with a reduction from a diploid to a haploid number of chromosomes.

regular symmetry: Radial symmetry; having two or more planes through which the structure (flower) could be cut to result in mirror-image halves; actinomorphy.

regulator gene: Gene that inhibits the structural genes of an operon.

relative density: Heterogeneous matrix such as wood; density differences between layers of cell; breakdown in the regular structure of cell walls.

relative humidity (RH): Measure of the amount of water in the atmosphere.

remote sensing: Means of acquiring information using airborne equipment and techniques to determine the characteristics of an area; aerial photographs from aircraft and satellite are the most common form of remote sensing.

renewable natural resources: Natural resources that can be replaced or renewed within a short time span.

repressible enzymes: Enzymes whose production can be a corepressor.

repressor: Compound that binds to and controls the regulator in gene regulation.

reserpine: Active ingredient isolated from the root of the snakeroot plant (*Rauwolfia serpentina*), a small evergreen climbing shrub of the Dogbane family.

resin canal: Long intercellular spaces present in the longitudinal system of vascular cells of pine, spruce, larch, and Douglas fir; adjacent parenchymal cells secrete the resin into the canals.

resins: Carbohydrates synthesized by certain plants in glands, canals, or ducts; insoluble in water; used in various industrial products, including paints and varnishes; aid in resistance of wood to decay.

resistant: Treated so as not to get certain diseases.

respiration: Process of converting food energy in the form of glucose to ATP energy; occurs in the cells of all living organisms and releases carbon dioxide as a by-product; aerobic respiration requires the presence of oxygen, but some organisms can respire anaerobically.

revegetate: Natural or induced replacement of plants into a cleared area; the recurrence of the same plant community that existed prior to clearing.

rhizobium: Soil bacteria that fix nitrogen after becoming established inside root nodules of legumes.

rhizoid: Rootlike absorptive structures on the underside of certain gametophytes.

rhizome: Fleshy, horizontal underground stem.

rhyniophyta: Earliest known division of vascular plants.

ribonucleic acid (RNA): Single-stranded nucleic acid composed of adenine, guanine, cytosine, uracil, phosphate, and ribose.

ribose: Five-carbon sugar important in RNA and many other compounds.

ribosomal RNA (rRNA): Ribonucleic acid involved in the formation of the ribosome.

ribosome: Cellular organelle responsible for the translation part of protein synthesis.

ring-porous wood: Secondary xylem characterized by large vessels and tracheids being produced early in the season (following favorable growing conditions); each term of growth activity is seen as a ring.

RNA polymerase: Enzyme responsible for forming mRNA during transcription.

roll: System of cleaning and preparing ground to receive sod.

rootbound: Condition of a restricted root system.

root cap: Group of cells covering the root meristem that aid the root's penetration through the soil.

root hairs: Root epidermal cells that elongate, increasing the total absorptive surface area.

root pressure: Positive pressure in the xylem due to a negative solute potential and closed stomata.

rootstock: Root onto which a scion or bud is grafted or budded.

rosettes: Consists of a single, compact stem, with long, spiraled leaves distributed in a short portion of the stem.

rough endoplasmic reticulum: Function is to synthesize and export proteins and glycoproteins.

rubber: Elastic material obtained from the latex sap of trees (especially trees of the genera *Hevea* and *Ficus*) that can be vulcanized and finished into a variety of products.

RuBP carboxylase: Enzyme that fixes CO_2 in the Calvin cycle of photosynthesis.

runners: New plants are formed at nodes by runners, which are stems from old plants; the stems grow along the ground.

salicylic acid (sal-ah-sill-ik): Compound with pain-relieving characteristic, found in willow bar and other plants; basic ingredient of aspirin.

samara: One- or two-seeded dry indehiscent fruit in which the ovary wall forms as outgrowth to form a wing.

sandstone: Sedimentary rock formed by the consolidation and compaction of sand, held together by a natural cement such as silica.

sapwood: Functional secondary xylem found between the vascular cambium and the nonfunctional heartwood in the center of the trunk or branch.

savannas: Lands with herbaceous understory, typically graminoids, and with shrub cover between 10% to 30%; tree/shrub cover height exceeds 2 meters.

scarification: Mechanical or chemical degradation of a hard surface, such as seed coats, so that oxygen and water can penetrate the hard layers.

scavengers: Any animal that eats refuse and decaying organic matter.

schizocarp: Dry, indehiscent two- or multiple-carpeled ovary that splits at maturity into separate one-seeded section that falls away.

scientific method: Objective process of approaching a problem. Involves hypothesis establishment, testing, observing the results, revaluation of the hypothesis in light of new knowledge, and retesting to seek repeatability and thus validity of the hypothesis.

scientific name: Latinized binomial (genus an species) unique to each identified organism.

scion: Short length of stem, taken from one plant which is then grafted onto the rootstock of another plant.

sclereids: Stone cells found in tissues varying from pear fruit to the hard shell of nuts.

sclerenchyma: Tissue type characterized by thick, sclerified cell walls; includes both fibers and sclereids (stone cells).

scopolamine: Alkaloid with anticholinergic effects that is used as a sedative and to treat nausea and to dilate the pupils in ophthalmic procedures; transdermal scopolamine is used to treat motion sickness.

seasonal cycles: Periodic, repetitive sequence of events in climate that play out over time; based on the earth's tilt on its axis and rotation around the sun.

secondary cell wall: Cellulose wall, often impregnated with lignin, laid down inside the primary cell wall of many woody species.

secondary compounds: Organic molecules synthesized by certain species of plants and not thought to be directly involved in essential metabolism.

secondary consumer: Organism that consumes a primary consumer; a carnivore.

secondary phloem: Phloem tissue formed by the vascular cambium in woody stems.

secondary succession: Revegetation of cleared land; return to previous community structure.

secondary tissue: Pliable sheet of tissue that covers, lines, or connects the organs or cells of plants.

secondary xylem: Xylem tissue formed by the vascular cambium in woody plants.

sedative: Substance that reduces nervous tension; usually stronger than a calmative.

sedentary: Inactive.

sedimentary rock: Rock formed by cementation and solidification of sedimentary deposits weathered from granite.

seed: Mature ovule within a fruit.

seed coat: Protective layer that develops from the integuments around a maturing ovule.

seedbed: Area of soil cultivated and prepared for planting seeds.

seed ferns: Any plant of the extinct order Pteridospermales; modern ferns reproduce by spores.

seedling: Embryonic product of the germination of a seed; the young shoot and root axis.

seed-scales complex: Spirally arranged scales on a female strobilus in gymnosperms.

sieve cell: Organic solute-conducting cell of the phloem in gymnosperms.

sieve tube members: Sieve areas more specialized than the sieve; sieve-tube members are joined end to end to form a tube that conducts food.

seismonasty: Plant's reaction to being touched, or the movement of a plant in response to a shock, such as shaking, intense heat, wind, or rain.

self-incompatibility-: Conditions of a flower that cannot successfully complete the reproductive process with pollen produced by its own stamens.

self-pollinating: Plant that has its own pollen fall on its own stigma.

semi-dwarf: Tree that grows to about 10 to 15 feet.

senescence: Aging process, usually characterized by the loss of some functional capacity, including reproduction.

sepal: Flower part attached outside the others, enclosing the flower when in bud.

seral: One stage of the communities in an ecological succession.

sessile: Referring to leaf blades without petioles in which the blade is attached directly to the stem.

sex cells: Specialized cells from a female fuse with a specialized cell from a male. Each of these sex cells contains the genetic from their parents.

sexual propagation: Pertaining to the fusion of gametes; sexual reproduction.

sexual reproduction: Gives plant offspring the genetic variability that enables them to adapt to changes in the environment.

shale: Sedimentary rock composed of mud.

shard: Piece of broken clay or ceramic pot placed over a drainage hole of a pot to prevent the loss of soil during watering.

sheath: Base of a leaf in monocots, usually wrapping around the stem.

shoots: Aboveground portions of a plant.

short day plant: Plant that flowers when the length of day is shorter than some critical value.

shrub: Plant that is shrublike in growth habit; usually a short, perennial plant without strong apical dominance.

silent mutation: Permanent genetic change, but one that is never expressed by the phenotype.

silique: Simple fruit that develops from a two-carpeled ovary; at maturity the two halves fall away, leaving the seeds attached to the persistent central wall.

simple leaf: Leaf having a single blade portion; may be highly lobed or dissected.

simple fruit: Fruits in which part or all of the pericarp (fruit wall) is fleshy at maturity; types of fleshy, simple fruits are apples, peaches, and pears.

simple pistil: Composed of one carpel.

sink: Site of collection of metabolites, such as sugar; metabolic sinks may exist anywhere in the plant where organic solutes are being transported by the phloem and stored.

slash and burn: Method of cultivation whereby areas of the forest are burnt and cleared for planting.

smooth endoplasmic reticulum: Cellular organelle; consists of tubules and vesicles that branch forming a network; allows increased surface area for

the action or storage of key enzymes and the products of these enzymes.

sphenophyta: Division of primitive spore-bearing vascular plants; most members of the group are extinct and known only from their fossilized remains.

sociobiologist: Study of the biological basis of human social behavior.

sociopolitical conservation: Policies must be tailored to local contexts, both ecological and sociopolitical.

sod: Grass that has soil and roots attached.

sod-forming: Grasses that grow to produce a densely matted sod from the spread of stolons and rhizomes; covers more uniformly than bunch-type grasses.

softwoods: Functional secondary xylem found between the vascular cambium and the nonfunctional heartwood in the center of the trunk or branch.

soil probe: Tool used to withdraw small cylindrical samples of soil for analysis or for determining moist soil depth.

soil profile: Vertical section of a soil, showing horizons and parent material.

solarization: Vegetation is removed, soil is broken up and then wetted to activate the nematode population. Next, the soil is covered with a sturdy clear plastic film during the warmest

six weeks of summer. High temperatures (above 130°F) must be maintained during this time for best results.

solubility: Relative ability of a solute to be dissolved.

solute: Any substance dissolved in a solvent.

solute potential: Water potential component caused by the presence of solutes in water.

solvent: Liquid matrix in which a solute is dissolved.

somatic cells: All body cells other than sex cells, containing at least the two sets of chromosomes inherited by both parents.

somatic nervous system: Part of the autonomic nervous system.

sorghum: Old World grass (*Sorghum bicolor*), several varieties of which are

widely cultivated as grain and forage or as a source of syrup.

SPAC (soil-plant-air continuum): Concepts and theories that describe the dynamics of water, carbon, and nitrogen in the soil-plant-air continuum.

speciation (*spe-see-s-shun*): Evolutionary processes leading to the development of new species.

species: Group of similar organisms that are capable of interbreeding with each other and can interbreed with members of a different species with only minimal success or not at all; the second word of a scientific binomial, always in Latin.

specific gravity: Ratio of the density of a material to the density of water.

specific heat: Amount of heat required to raise the temperature of 1 gm of any substance 1°C.

sperm nuclei: Each pollen grain produces two sperm nuclei, which effects double fertilization in angiosperms.

spindle fibers: Protein fibers formed during prophase of nuclear division; chromosomes attached to these fibers at the centromere.

spine: Leaf or leaf part modified as a hard, sharp-pointed structure.

spongy parenchyma: Leaf tissue composed of columnar cells containing

numerous chloroplasts in which the long axis of each cell is perpendicular.

sporangiospore: Any spore produced from a sporangium.

sporangium: Spore case; a hollow structure in which spores are produced.

spore: Haploid structures produced by the sporophyte plant via meiosis that develop into the gametophyte.

sporophyte: Diploid, spore-producing plant in the alternation of generations; undergoes meiosis to produce the haploid spores.

staking: Keeping plants in the correct growing position by using wires, wooden posts, or similar supports.

stamen: Male reproductive structure in a flower, consisting of an anther supported on a filament.

staminate: Unisexual flower having stamens but no pistil.

staminodium: Sterile stamen, non-functional anthers and often with petaloid filaments.

starch: Polysaccharide composed of glucose linkages; the primary storage carbohydrate of plants.

stele: Central vascular cylinder of roots and stems of vascular plants.

stem cuttings: Piece of stem is part buried in the soil, including at least one leaf node; produces new roots, usually at the node.

sterols: Complex alcohols (steroid alcohols) that are important in animals as hormones, coenzymes, and precursors for vitamin D.

stigma: Apical portion of the pistil, the surface on which pollen lands and germinates.

stimulants: Drug that increases the activity of the sympathetic nervous system and produces a sense of euphoria or wakefulness.

stimulates: Mode of action of plant hormones.

stolon: Horizontal aboveground stem.

stoma (pl., stomata): Epidermal complex consisting of two guard cells and the pore between them.

stomatal regulation: Complex process depending on microclimate and measured soil water potential.

strata: Layers of sedimentary rock, the oldest rocks occurring at the bottom of the ocean.

stroma: Matrix between the grana in chloroplasts and site of the dark reactions of photosynthesis.

structural genes: Genes that are transcribed into proteins.

style: Central, elongated portions of the pistil between the stigma and ovary.

sulfite process: Process for the digestion of wood chips in a solution of magnesium.

suberin: Lipid material found in the Casparian strip of the endodermis and in the cell walls of cork tissue.

submetacentric: Pertaining to a chromosome with a centromere located between the center and one end of the chromosome.

substrate: Beginning substance from which other molecules are synthesized.

succession: Processes of vegetation change over time, or plant succession, are also the processes involved in plant community restoration.

succulent: Plant characterized by thick, fleshy leaves or stems; an adaptation usually associated with a water or salt stress.

sulfite process: Chemical process for the manufacture of paper pulp; uses an acid bisulfite solution to soften the wood material by removing the lignin from the cellulose.

sunken stomata: Specially adapted structures reduce transpiration.

superior ovary: Attachment of the ovary to the receptacle above and free from the attachment of the other floral parts.

symbiotic: Plant characterized by thick, fleshy leaves or stems; an adaptation usually associated with a water or salt stress.

symplastic movement: Pertaining to the movement of water and solutes through tissues by passing through biological membranes.

synapsis: Coming together of homologous chromosomes to form bivalents during prophase I of meiosis.

synergid: Two haploid cells on either side of the egg cell at the micropylar end of the embryo sac.

synthetic pyrethrums: Type of chrysanthemum, as there are not enough flowers available to meet the demand; synthetic pyrethrums are on the market that are very safe. They are used in IPM rotations.

system toxins: Poisons (toxins) that work on the nervous system.

taiga *(tie-ga)*: Northernmost coniferous forest, characterized by cold winters and heavy snowpack.

tannin: Hemi-cellulose and lignins produce a host of volatile phenols.

tapetum: Nutritive somatic tissue surrounding the microsporocyte.

taproot: Primary root from which secondary roots originate.

taq polymerase: Enzyme frequently used in polymerase chain reaction (PCR), methods for greatly amplifying short segments of DNA.

taro: Early root crop from Asia with swollen underground stems with stem buds that can each produce a new plant—like a potato.

taxine: Drug that inhibits cell growth by stopping cell division; used as treatment for cancer; also called antimitotic or antimicrotubule agents or mitotic inhibitors.

telocentric: Pertaining to a chromosome with a centromere located at one end.

telophase: Stage of nuclear division in which sets of chromosomes finally arrive at the poles of the dividing cell; new nuclear envelope forms at this stage.

temperate deciduous forest: Characterized by a mild or moderate temperature with deciduous trees.

temperature shock treatments: Sudden change in temperature.

template: Genetic message on the sense strand of DNA as determined by the sequence of nucleotides.

tendril: Modified leaf or stem in which only a slender strand of tissue constitutes the entire structure.

tension wood: Reaction wood produced along the upper side of leaning woody trees, straightening the trunk by contracting and "pulling" the tree upright.

terminal bud: Meristematic tissue located at the tip of a stem.

terpenes: Group of secondary compounds composed of two to many isoprene units in a chain or ring; sometimes categorized as hydrocarbons only, sometimes to include terpenoids.

terpenoids: Term referring to all compounds composed of isoprene units.

terrarium: Closed biological system in which plants and animals coexist without external inputs or discharges; H_2O, CO_2, O_2, and nutrients cycle in the closed system.

terrestrial biomes: Pertaining to the land.

testcross: Crossing an organism having a dominant expression for a trait with an organism having a dominant homozygous recessive genotype for that trait to determine the genotype of the organism expressing the dominant phenotype.

thermal stratification: Layering of different temperatures of water or air caused by different densities, less dense floating on more dense layers.

thermonasty: Plant movement caused by temperature: The movement of plant parts in response to a change in temperature, for example, the opening of flowers.

thigmotropism: Turning or bending response of a plant upon direct contact with a solid surface or object.

thorn: Modified stem termination in a sharp point that is attached to the vascular system of the plant.

thorn forest: Type of vegetation in the warm, semiarid zones in which the predominant vegetation is scrubby, has thick leathery leaves, and is characterized by sharp or thorny stems.

thylakoid: Lamellar structure of the grana of chloroplasts.

thymine: Nitrogen base found in DNA.

tiga: Snow forest.

tillers: First side shoots in small grains.

tillering: Production of lateral buds and shoots near the ground to result in a plant with several shoots instead of one; particularly important in grasses and grain crops.

tonoplast: Membrane surrounding the vacuole.

top carnivore: Consumer at the end of a food chain or web; a carnivore that ordinarily has no predator under those ecological conditions.

topography: Surface condition of an area of land; relief features.

topsoil: Uppermost layer of soil, highly weathered and often rich in organic matter nutrients.

total body mass: Measurement of the relative percentages of fat that correlates strongly with the total body fat content.

tracheids: Elongated, spindle-shaped cells that conduct water in the xylem; particularly important in gymnosperms.

tracheophytes: Plants with a well-defined vascular system.

training: Directing plants to grow certain directions by using different methods.

transcription: First stage of protein synthesis in which the template of DNA is transcribed into mRNA.

transfer RNA (tRNA): Small ribonucleic acid molecule involved in the transfer of specific amino acids to a protein being synthesized.

transformation: Incorporation of viral nucleic acid into the nucleic acid of the host cell.

transforming factor: Substance that can be passed from one cell to another and cause a permanent change in heredity.

transgenic: Plants and animals result from genetic engineering experiments in which genetic material is moved from one organism to another, so that the latter will exhibit a desired characteristic.

transitional mutants: Type of mutation in which a single purine-pyrimidine base pair is replaced by another.

translation: Second stage of protein synthesis in which the codon of mRNA pairs with the anticodon of tRNA at the surface of the ribosome.

transpiration: Evaporation of water from the surface of plant leaves and stems.

transpirational pull: Main phenomenon driving the flow of sap in the xylem tissues of large plants.

transplantability: Success of plants being transplanted after starting indoors.

transversional mutant: Type of mutation in which a purine-pyrimidine base pair is replaced by a pyrimidine-purine base pair.

transverse cut: Longitudinal cut of wood.

trellis: Support system of wires or wood for growing certain plants.

trichome: Epidermal protrusion such as a hair or scale.

trihybrid cross: Cross between individuals of the same type that are heterozygous for three pairs of alleles at three different loci.

triplet: Group of three nucleotides on a nucleic acid that codes for a particular amino acid.

trophic: Feeding level; a step in the energy flow through a food chain.

tropical rain forest: Forest that gains more water from precipitation than it loses through evaporation; located in the tropical zone and having an average temperature between 70° and 85°F (21° and 29°C) and average yearly rainfall of more than 80 inches (200 cm).

tropism: Turning or bending movement of an organism or a part toward or away from an external stimulus, such as light, heat, or gravity.

truck cropping: Refers to commercial, large-scale production of selected vegetable crops for wholesale markets and shipping.

tryptophan: Essential amino acid involved in human nutrition; one of the 20 amino acids encoded by the genetic code.

tube cell: Nucleus of a pollen grain believed to influence the growth and development of the pollen tube.

tube nucleus: One division of the microspore nucleus in a pollen grain that is responsible for the formation of pollen tube from the stigma through the style to the ovule.

tuber: Horizontal, underground stem with a very enlarged tip.

tubocurarine: Alkaloid isolated from the bark and stems of *Chondodendron tomentosum* (Menispermaceae); active principle of curare.

tubulin: Protein of which microtubules are composed.

tumor: Spherical mass of cells in which cell division occurs at random and often in an uncontrolled fashion.

tundra: Treeless plain characteristic of the arctic and subarctic regions.

turf: Intertwined fibrous roots of grasses forming a mass with the soil just below ground level.

turgidity: Full, even distended with water taken in by osmosis.

turgor pressure: Real pressure developed in living cells by pressing against a membrane.

turpentine: Solvent that includes two terpenes: camphor and pinene.

ultraviolet: Portion of the sun's total range of radiation having wavelengths immediately shorter than the shortest of the visible spectrum (purple); between approximately 380 and 100 nm.

understory: Vegetation that characterizes the lower level of plants in a forest; the vegetation below the canopy.

unisexual: Condition of having either stamens or pistil, not both; an imperfect flower.

unit membrane: Transmission electron microscopic interpretation of a biological membrane, consisting of two electron-dense lines separated by translucent space.

uracil: Nitrogen base found in RNA.

uredospore (u-reed-o-spore): Binucleate spore produced by rust during the summer.

vacuolar sap: Aqueous substances inside the vacuole.

vacuole: Aqueous cavity within the cytoplasm and surrounded by the tonoplast that stores low molecular weight ions and molecules.

vapor concentration: Gradient equals the water vapor concentration inside the product minus that of the surrounding air.

variability: How plants differ from each other.

vascular bundle: Veins in herbaceous plants and in leaves of all plants, including the specialized cells of both xylem and phloem.

vascular cambium: Lateral meristem characteristic of secondary growth; gives rise to secondary xylem and secondary phloem.

vascular plants: Plants that have lignified tissues for conducting water, minerals, and photosynthetic products through the plant, includes the ferns, club mosses, flowering plants, conifers, and other gymnosperms.

vascular systems: Lignified tissues for conducting water, minerals, and photosynthetic products through the plant.

vector: Agent of transfer, such as insects and other animals in pollinations.

vegetative: Period when the plant grows vigorously and rapidly.

vegetative reproduction: Reproductive process that is asexual and so does not involve a recombination of genetic material. It involves unspecialized plant parts that may become reproductive structures (such as roots, stems, or leaves).

venation: Pattern of vein development in leaves.

veneer: Rotary-cut from a bolt of wood, called a flitch, centered in chucks.

vermiculite: Expanded mica used as a sterile medium for the rooting of cuttings; high water-holding capacity and relatively inert.

vesicles: Membrane-found particles pinched off by constriction of a membrane, as in the Golgi apparatus.

vessel: Primary water-conducting cell-system in the xylem of most angiosperms.

vessel element: Individual cells of xylem that combine end-to-end to form a vessel.

viability: Period of time an organism remains alive; often used to describe the length of time before a seed will fail to germinate.

virus: Crystal-like particle composed of a protein coat and a central core of DNA or RNA, but never both.

viscous: Thick and highly dense.

visible spectrum: That portion of all the sun's radiation that can be perceived as light by humans; between approximately 380 and 100 nm wavelength.

viviparous: Characterized by beginning embryo growth while still attached to the mother plant.

volatile oils: Terpenes composed of two to four isoprene units; also known as essentials oils, such as lemon and peppermint.

volatilization: Process of evaporation; to pass into the atmosphere as gas.

volunteer plants: Plants that may grow following harvest or the next season without being planted.

vulcanization: Process of treating rubber or rubberlike materials with sulphur at great heat to improve elasticity and strength or to harden them.

wadi: Dry riverbed that only contains water during times of heavy rain.

water potential: Relative ability of water molecules to do work by interacting with each other.

water quality: Describes the chemical, physical, and biological characteristics of water, usually in respect to its suitability for a particular purpose.

water stress: Occurs when the demand for water exceeds the available amount during growth.

wax: Lipid material with considerable oxygen inserted in the molecule; high melting point and relatively impermeable to water.

weed: Any plant growing in an area where it is not desired.

wheat: Cereal grass (*Triticum vulgare*) and its grain, which furnishes white flour for bread, and, next to rice, is the grain most largely used by the human race.

white light: Sunlight integrated over the visible portion of the spectrum (4,000–7,000 angstroms) so that all colors are blended to appear white to the eye.

whorled: Three or more leaves attached at a node.

windward: Side toward which the wind is blowing.

winter hardy: Seeds that can be seeded even where temperatures frequently fall to 40°F, below zero, or where mean winter temperatures are about 0°F.

wolfbane: Aconitum known as aconite or monkshood.

world supply: To meet essential food import needs in all countries, considering world price and supply fluctuations and taking especially into account food consumption, each nation can define for itself what is required for survival.

xeric (zee-rick): Characterized by very dry environment.

xerophyte: Plant adapted for growth under arid conditions.

X-ray: Portion of the sun's total range of radiations, which is immediately shorter than ultraviolet; between 0.1 and 100 nm.

X-ray crystallography: Technique for studying molecular and atomic structure of a substance.

Yellow river: River of north China; flows from the west starting at the high plateau of Tibet eastward into the Yellow Sea; complex society of East Asia originated from the valley of this river.

zone of elongation: Area in plant roots where recently produced cells grow and elongate prior to differentiation.

zooplankton (*zo-o-plank-tun*): Single-celled and small multicellular animals that feed on phytoplankton in aquatic ecosystems.

zoospore: Motile asexual spore.

zosterophyllophyta: Division of fossil plants; among the first vascular plants.

Z-scheme: Diagrammatic representation of electron flow through PS II and PS I.

zygospore: Thick-walled, resistant spore resulting from a zygote.

zygote: Diploid cell formed by the fusion of two haploid gametes; the result of fertilization.

INDEX

A

Abiotic, 20, 645
Abrasion, 645
Abscisic acid, 296, 303, 645
Abscission layer, 299, 645
Absence of stomata, 236
Absorption, 81, 85–86, 183, 645
Accessory buds, 645
Accessory meristem, 175
Accessory pigments, 258
Achene, 122, 645
Acid, 645
Acidic, 645
Aconitine, 597, 645
Acoustical properties, of wood, 197
Acridine dyes, 645
Active transport, 230, 645
Adaptations, 78
    definition of, 645
    land plants, 370–371
    vegetative, 80–81
    water loss, 235–237
Adenine, 302, 346, 645
Adenosine diphosphate (ADP), 645
Adenosine triphosphate (ATP), 645
Adhesion, 232–233, 645
Adventitious roots, 83–84, 645
Aerate, 645
Aerial, 645
Aerobic organisms, 254, 645
Aerobic respiration, 146, 271, 645
Aesthetics, plants and, 10–11, 623–624
After-flowering care, of bulbs, 530
After-ripening dormancy, 375, 645
Agar, 297, 645
Agent orange, 645
Aggregate fruit, 122, 645
Agricultural Research Service (ARS), 479
Agriculture, 605–606
    alfalfa, 442–447
    barley, 416–417
    beginnings of, 404–406
    biotechnology and, 349–356
    corn, 410–412
    cotton, 427–430
    crop plants, earliest, 406–410
    dry beans, 430–434
    farmer's bargaining power, 616
    food commodities, 607–611
    future of, 621–622
    gardening and, 622–624
    historical perspective, 611–613
    mechanization and world food supply,
        606–607
    new agriculture, 614–616
    oats, 417–418
    peanuts, 437–442
    peas, 434–437
    potatoes, 420–425
    production limits, 616–621
    research and innovation, 409–410
    rice, 418–420
    social and political considerations, 624–633
    soybeans, 425–427
    summary, 447–448, 634–635
    U.S. agricultural crops, 410–447
    wheat, 412–416

Agroecology, 21–22
Agroecosystems, 21
A horizon, 221, 222
Air. See Soil-plant-air continuum
Air layering, 105
Aleurone layer, 301, 645
Alfalfa, 442–447, 638
Alfalfa stands, 444
Algal bloom, 73, 645
Algin, 645
Alkaline, 645
Alkaloids, 394, 592, 645
Allard, H. A., 292, 293
Alleles, 314–315, 645
Allergens, 593
Alpha-amylase, 301, 645
Alternate host, 645
Alternate leaf arrangement, 89, 645
Alternation of generations, 319–323, 645
Amber, 645
American Association for the Advancement
    of Science, 343
American Pomological Society, 511
American Society for Horticultural Science, 511
Amino acid, 645
Amino group, 645
Ammonia, 35, 645
Anaerobic, 254, 645
Anaerobic respiration, 146, 645
Analgesic, 580, 646
Anaphase, 170, 172, 646
Anaphase I, 317
Anaphase II, 317
Anchorage, 81, 183, 646
Androsterone, 646
Angiosperms, 81, 375, 381–382
    coevolution, 385–393
    definition of, 646
    flower evolution, 382–384
    fruit evolution, 384–385
    life cycle, 319–323
    pollinator specificity, 387–393
Animal dispersal, 123–124
Animal hormones, from plants, 291
Animal-plant co-evolution. See Co-evolution
Anions, 215, 646
Annuals, 131, 132
    definition of, 646
    flowering, 523–529
Annual tree ring, 89, 646
Anther, 646
Antheridium, 370–371, 646
Anthers, 112
Anthocyanins, 258, 646
Anthraquinone glycosides, 583, 646
Antibiosis, 646
Antibiotic, 646
Anticodon loop, 162, 646
Antiparallel, 646
Antipoidal cells, 323, 646
Ants, 391–392
Apex, 96, 98, 646
Aphids, 243
Apical dominance, 299–300, 646
Apical meristems, 173–174, 175
Apidermal cells, 82
Apoplastic movement, 230, 646

Applied research, 614–616
Aquatic biomes, 68
    freshwater biomes, 71–73
    marine biomes, 68–71
    summary, 73–74
Archegonium, 371, 646
Asexual, 101, 646
Aspartate, 265, 646
Assortment, independent, 329
Atmospheric pressure, 646
Atom, 646
Atomic number, 141, 646
Atomic weight, 141–142, 646
Atropine, 581, 646
Attachments, 86–87
Autonomic nervous system, 592, 646
Autotrophic, 37, 646
Auxin, 296, 297–300, 646
Auxin threshold, 299, 646
Avery, Oswald, 156–157
Axil, 87, 646
Axillary bud, 646
Aztecs, 408–409

B

Backcrosses, 326–327, 646
Bacteria, 144, 646
Bacteria chlorophyll, 256
Bacteriophage, 157
Balsa, 197, 646
Baobab tree, 406–407, 646
Bare root, 520, 646
Bargaining power, of farmers, 616
Bark, 188–189, 646
Barley, 416–417, 608, 646
Basal rosette, 89, 646
Basalt, 220, 646
Bases, 96, 98
Basic, 646
Bats, 392–393
Beadle, George, 156
Beans. See Dry beans
Bed renovation, for strawberries, 486
Bedrock, 222, 646
Bees, 388–389
Beetles, 386, 387–388
Belladonna, 581, 590
Benign, 167, 646
Benson, Andrew, 265
Benthic, 646
Bermudagrass, 549
Berries, 646. See also Small fruits; *specific berries*
Beta-carotene, 258, 646
B horizon, 221–222
Biennials, 131, 132, 647
Big bang, 647
Big bluestem, 549
Bilateral symmetry, 116
Binomial, 134, 647
Bioassay, 647
Biochemical evolution, 395
Biocontrol, 353, 355–356
Bioinformatics, 362–363, 647
Biological clocks, 291, 647
Biological control, 207, 647

Biological factories, 350–351
Bioluminescence, 647
Biomass, 39, 647
Biomes, 29, 48–50
    adaptations to, 86
    aquatic, 68–74
    definition of, 647
    summary, 73–74
    terrestrial, 50–68
Biopharmaceuticals, 359, 647
Biosphere, 27–30, 51, 647
Biotechnology, 341–342
    agriculture and, 349–356
    definition of, 343, 647
    future of, 360
    "omics" revolution, 360–364
    policy, public perception, and
        law, 364
    summary, 365
    transgenic plants, 356–360
Biotic, 27, 647
Birds, as pollinators, 390–391
Bisexual, 114, 647
Bivalent, 316–317, 647
Blackberry, 492–493
Blades, 95, 647
Blueberry, 479, 486–488
Blue-black grapes, 478–479
Board feet, 202, 647
Bog, 647
Bomb calorimeter, 647
Boot, 647
Borlaug, Norman, 462, 463
Boron, 218
Botanical classification, 453–457
Botanical names, 133–134
Botany, supermarket, 134–136
Bowers, 417
Boxer Rebellion, 581
Boyer, Herbert, 159
Bract, 647
Bragg, Sir William Henry, 156
Bragg, Sir William Lawrence, 156
Brambles, 488, 647
Branch roots, 81–82, 647
Breeding, selective, 359
Broadcast planting, 647
Bromegrass. See Smooth bromegrass
Bromeliads, 84, 85, 647
Bronchodilator, 381, 647
Bryophytes, 370, 647
Buckwheat, 611
Budding, 100, 102–103, 647
Buds, 174, 647
Bud scales, 88, 647
Buffer, 647
Bulbs, 90, 529–532, 647
Bulliform cell, 180, 647
Bunch grass, 544, 647
Bundle sheath cells, 186, 647
Bushels
    farm products per, 638
    weights and volumes, 639
Butterflies, 389–390
Buttressing, 647

C

C2 pathway, 262–264
C3 pathway, 264–265, 266, 647
C4 anatomy, 266
C4 efficiency, 267
C4 fixation, 266–267
C4 pathway, 264–266, 647
Cacti, 239–240
Calcium, 218

Callose, 647
Callus, 304, 647
Calorie, 268, 647
Calvin cycle, 262–264, 265, 647
Calvin, Melvin, 262, 263, 264, 265
CAM. See Crassulacean acid metabolism
Canada bluegrass, 544
Candelilla wax, 601, 647
Canes, 480, 647
Cannabis, 587–588
Canopy, 52–53, 63, 647
Capillary pores, 648
Capsule, 648
Carbohydrate, 648
Carbon, 218
Carbon cycle, 33–35
Carbon dioxide, 6, 648
Carbon dioxide concentration, 234
Carbon dioxide–oxygen balance, 30–31
Carbon fixation, 266, 268, 648
Carboniferous period, 371, 648
Carboxyl group, 648
Cardiac glycosides, 583, 648
Carnauba wax, 601, 648
Carnivores, 38, 648
Carotene, 258
Carotenoids pigments, 257, 648
Carpel, 112, 648
Carpellate, 114, 648
Carrageenan, 648
Carrying capacity, 29–30, 648
Caryopsis, 648
Cascara, 583
Casparian strip, 183, 648
Cassava, 408, 648
Cassia, 648
Catalyst, 648
Cations, 215, 648
Caucasian bluestem, 549
Cell division (mitosis and cytokinesis), 168–175
Cell elongation, 300
Cell expansion, 175, 648
Cell membranes, 144–145, 648
Cells, 139–141
    organelles and other inclusions, 145–151
    prokaryotic and eukaryotic, 143–145
    summary, 151
Cell structure, 142–143, 160
Cellulase, 648
Cellulose, 195, 648
Cellulose microfibrils, 175, 648
Centigrade-to-Fahrenheit temperature
        conversions, 639
Central bud, 648
Central cell, 648
Central nervous system, 592, 648
Centromere, 170, 648
Century plant, 407, 648
Cereals, 408–409
Chain, 142
Chaparral, 64, 648
Charcoal, 205–206, 648
Chemical composition and properties,
        of wood, 195
Chemical energy, 250, 648
Chemical factors in seed germination, 130
Chemical incompatibility, 118
Chemotaxonomy, 648
Chiasma, 318, 648
Chinampas, 408, 648
Chlorine, 218
Chlorophyll, 256–257, 648
Chloroplasts, 147–148, 258, 648
Cholesterol, 648
C-horizon, 222

Chromatids, 170, 648
Chromatin, 164, 648
Chromatophore, 648
Chromoplasts, 147, 648
Chromosome mapping, 331
Chromosomes, 153–154, 314–315, 649.
        See also Cell division; DNA
Chrysostom, John, 510, 626
Cicutoxin, 596, 649
Cinchona, 579
Circulation cells, 23, 649
Cis-polyisoprene, 599, 649
Cladophylls, 93, 649
Clay, 223–224, 225
Climacteric rise, 649
Climate, fruit and nut trees, 502
Climatology, 22–27, 649
Climax community, 42–43, 649
Clone, 649
Coastal shelf, 649
Cocaine, 587, 649
Coconut palm, 649
Codeine, 580, 649
Code of mutations, accuracy of, 167–168
Coding system, for DNA, 347
Codon, 162, 649
Coenzymes, 649
Co-evolution, 119, 385–393, 649
Co-factor, 217, 649
Cohen, Stanley, 159
Cohesion-adhesion transpiration pull, 232–233
Cohesive, 232–233, 649
Cold damage, 287, 649
Cold frames, 535, 649
Coleoptiles, 297, 649
Collection, of plants, 641–642
Collenchyma, 176, 177, 649
Colloid, 224
Colonial, 649
Colonization, 649
Colonizer, 649
Colony, 649
Color, of wood, 196
Combine, 649
Commensalism, 28–29, 649
Commercial greenhouse environment,
        534–536
Common names, 134, 135–136, 649
Communities, 5, 29, 649
Companion cells, 179, 649
Companion crops, 443, 649
Compartmentalization of enzymes, 145, 649
Compartmentation, 139
Compensation point, 30, 269, 649
Competition, 28, 649
Complementary strand, 649
Complete flower, 114, 649
Complex tissue, 178–180
Compost, 240–241, 649
Compound fruits, 122, 649
Compound leaves, 95, 649
Compound pistil, 112, 649
Compression, 649
Compression wood, 649
Condensation, 22, 649
Conduction, 81, 85–86, 87, 649
Conduction systems, 371
Cone, 377, 649
Coniferous forest, 64–65
Conifers, 375, 376–379, 649
Conservation, 628–629
Container-grown, 649
Control, of plants, 355
Controlled atmosphere, 304, 650
Controlled burning, 207–208, 650
Convergent evolution, 93, 650

Copper, 218
Coral reefs, 70–71, 650
Corepressor, 650
Cork, 206, 650
Cork cambium, 188, 650
Corms, 92, 650
Corn, 410–412, 611
Corn-soybean croprotation, 412
Corolla, 111, 650
Cortex, 181, 650
Cortisone, 650
Cosmopolitan, 650
Costs, sod-related, 570
Cotton, 427–430, 650
Cotyledons, 127, 650
Covalent bond, 650
Crassulacean acid metabolism (CAM), 235,
    266, 267–268, 650
Crick, Francis, 156–157
Cristae, 146, 650
Crop rotation
    corn-soybean, 412
    cotton, 429–430
    peanuts, 439–440
Crops. See Agriculture; U.S. agricultural crops
Crossing over, 318, 650
Cross-pollination, 117, 650
Cross-protection, 353, 650
Crown, 650
Cube, 142
Cultivation
    annuals, 528
    blueberries, 487
    currants and gooseberries, 494
Cultural practices
    fruit and nut production, 508–510
    vegetables, 463–465
Curare, 581–582, 650
Currant, 493–495
Cuticle thickness, 236, 650
Cuticular transpiration, 231, 650
Cutin, 650
Cuts, 203
Cuttings, 104, 650
Cyanobacteria, 35, 144, 650
Cycads, 375, 380, 650
Cycles, effect on plant growth and
    development, 291–295
Cyclic photophosphorylation, 261, 650
Cycling, in ecosystem, 31–37
Cyclosis, 650
Cytochromes, 254, 650
Cytokinesis, 168–169, 172–175, 317, 650
Cytokinins, 296, 302–303, 650
Cytoplasm, 175, 650
Cytoplasmic streaming, 650
Cytosine, 346, 650

D
Dark reactions, 255, 261–267
Darwin, Charles, 297, 298, 354
Darwin, Francis, 297
Darwin Day, 354
Daylength, 24, 27, 650
Day neutral plant, 650
Deadheading, 517, 528, 650
Deadly nightshade. See Belladonna
Deadly plants, 593–601
Decarboxylation, 650
Deciduous, 61, 650
Decomposers, 35, 650
Deficiency symptoms, 218
Deficit irrigation, 618, 650
Dehiscent, 120, 121, 650
Deletion mutants, 167–168, 650

De Materia Medica (Dioscorides), 576
Demography, 620–621
Denature, 650
Dendrochronology, 201–202, 650
Dentate, 96, 650
Deoxyribonucleic acid. See DNA
Deoxyribose, 651
Deplasmolysis, 227, 650
Depressants, 586, 650–651
Dermal tissues, 180–181
Desertification, 59, 61, 651
Deserts, 57–60, 86, 651
Desiccation, 651
Design. See Plant design
Destructive distillation, 205, 651
Determinate growth, 651
Development. See Plant growth and
    development
Developmental biology, 165, 651
Devonian period, 651
Diatomaceous earth, 651
Dichotomous, 651
Dicot roots, 183
Dicots, 132–133, 135–136, 651
Dicot stems, 181–182
Dietary Supplement Health and Education Act
    of 1994 (DSHEA), 586
Dietary supplements, 586
Differentiation, 285, 651
Diffuse-porous wood, 651
Diffusion, 226, 651
Digestion, 127, 651
Dihybrid cross, 328, 651
Dihybrid inheritance, 328–330
Dioecious, 114, 115, 651
Dioscorides, 576
Diploid, 170, 651
Disease resistance, 352–355
Diseases/disease control
    alfalfa, 447
    blueberries, 488
    bulbs, 530–532
    corn, 412
    cotton, 429
    dry beans, 431, 432–433
    fruit and nut production, 509
    grapes, 481
    peanuts, 441
    peas, 436
    perennials, 523
    potatoes, 423
    raspberries, 491–492
    rice, 420
    sod, 566, 567
    soybeans, 426–427
    vegetables, 464
    wheat, 413–415, 416
Dispersal, 122–124, 651
Dissociation, 215–216
District petals, 116
Disulfide bond, 651
Diurnal, 651
Diurnal cycle, 291
Diurnally, 651
Diurnal stomatal closing, 235
Diurnal temperature fluctuations, 24, 286
Division, of perennials, 522
DNA (deoxyribonucleic acid), 144
    coding system, 347
    definition of, 650
    genetic engineering and, 344, 346–347
    genetic gibberish and, 336–337
    genome sequences, 334
    transcription and, 161, 162
    See also Biotechnology; Cell division;
        Genetic engineering; Genomics

DNA jigsaw, 346–347
DNA replication, 156–160, 169, 651
Doctrine of signatures, 578, 651
Dominant, 324–325, 651
Dormant, 129, 651
Double fertilization, 124, 651
Double helix, 159, 651
Drifts, 651
Drill, 651
Drought, 61, 651
Drupe, 500, 501, 651
Dry beans, 430–434
Dry fruits, 120, 121, 651
Drying
    dry beans, 434
    soybeans, 427
Duboisia, 579
Durability, of wood, 196
Duration, 651
Dusting, 530, 651
Dwarf trees, 500, 501, 502, 651

E
Early wood, 199, 651
Ecological succession, 42–44, 651
Ecology, 18–19
    agroecology, 21–22
    angiosperms and, 395
    biosphere, 27–30
    climatology, 22–27
    conifers and, 379
    definition of, 651
    ecological succession, 42–44
    ecosystem cycling, 31–37
    oxygen-carbon dioxide balance, 30–31
    plants and, 20–21
    recolonization, 44
    summary, 44–46
    trophic levels, 37–42
Economic importance
    conifers, 379
    plants, 7–9
    seeds, 125–126
Economic trade-offs, 632–633
Ecosystem, 5, 27–28, 651
Ecosystem cycling, 31–37
Edema, 583, 651
Edible parts, vegetable classification by, 453, 458
Efficiency, of respiration, 276–277
Effluent, 557, 651
Egg cell, 322, 651
Egypt, 405–406
Electrical energy, 250, 651
Electromagnetic spectrum, 259, 651–652
Electron, 141, 652. See also Oxidation
    reduction
Electron transfer, 254, 652
Electron transport, 271, 274–275, 652
Elements, 178, 652
Embryo, 124, 652
Embryonic axis, 130, 652
Embryo sac, 322, 652
Endangered species, 629–630, 652
Endergonic, 253, 652
Endocarp, 120, 652
Endodermis, 230, 652
Endogenous, 295, 652
Endoplasmic reticulum (ER), 148–149, 652
Endosperm, 124, 323, 652
Endosymbiosis, 652
Energy conversions, 250–251
    law of thermodynamics, 252–254
    metabolism, implications of, 277–278
    oxidation reduction, 254–255
    photosynthesis, 255–271

Energy conversions (*Continued*)
  respiration, 271–277
  summary, 278–279
English measurement, 640
Entine, 652
Entire, 96, 652
Entropy, 253, 652
Environmental interactions, seed germination
      and, 129
Environmental quality, 631–632
Environmental trade-offs, 632–633
Environment insults, 286, 652
Enzymes, 160, 652
Ephedra, 582–583
Ephedrine, 381, 582, 583, 652
Ephemeral, 59, 652
Epicotyl, 127, 652
Epidermal cells, 652
Epidermis, 180, 181
Epinasty, 304, 652
Epiphytes, 53, 84, 85, 652
Epistasis, 332–333, 652
Equatorial plate, 172, 652
Ergot, 584, 588–589, 652
Essay on the Principle of Population,
    An (Malthus), 620
Essential oils, 652
Essential plant nutrients, 216–220
Establishment
    alfalfa, 443
    forage grasses, 554–555
    sod, 555–557
Estrogens, 291
Estrone, 652
Estuary, 652
Ethylene, 296, 303–304, 652
Ethylene injury, in
      floriculture, 522–523
Etiolation, 652
Eukaryotic, 143–145, 652
European grapes, 476
Eutrophication, 73, 652
Eutrophic conditions, 73
Evaporation, 23, 652
Evapotranspiration, 288, 652
Evolution, 28, 652
Excitation, 257
Exergonic, 253, 652
Exine, 320, 652
Exocarp, 120, 652
Exogenous, 295, 652
Exons, 337, 653
Experimental biosphere, 51
Explosively dehiscent, 125
Extinction, 653

F

F1 generation, 653
F2 generation, 653
Fact, 12, 653
Facultative anaerobes, 276, 653
Fahrenheit-to-centigrade temperature
      conversions, 639
Family, 653
Family names, 134, 135–136
Farmers, bargaining power of, 616
Farm products per bushel, 638
Far red light, 293–294, 653
Fascicular cambium, 182, 653
Fats, 653
Fatty acid, 653
Feeding loss, of alfalfa, 638
Fermentation, 275–276, 653
Ferns, reproduction of, 372–374
Fertile crescent, 404, 653

Fertilization (reproduction), 113, 653
Fertilizers/fertilization
    alfalfa, 445
    angiosperms, 323
    annuals, 528–529
    applications of, 238–239
    blackberries, 493
    blueberries, 488
    bulbs, 530
    corn, 412
    currants and gooseberries, 494
    dry beans, 431–432
    forage grasses, 550–551, 555
    fruits and nuts, 503, 508
    grapes, 481
    peanuts, 440
    peas, 435
    perennials, 521
    potatoes, 422
    raspberries, 490
    rice, 419
    sod, 560–562
    strawberries, 484–485
    wheat, 416
Fiber, 601–602, 653
Fiberboard, 206, 207, 653
Fibers, 178
Fiberwood, 205
Fibrous roots, 81–82, 653
Fiddlehead, 372, 653
Field capacity, 224–225, 653
Filaments, 112, 653
Filial, 653
Finite resources, 653
Fire ecology, 653
Fire management, 56–57
First Agricultural Revolution, 409, 653
Flaccid, 233, 238, 653
Flavin mononucleotide, 275, 653
Flavonoids, 291, 653
Fleshy fruit, 653
Flies, 391
Flooding, for rice, 419–420
Flora, 653
Floral parts, 111–115
Floriculture
    ethylene injury in, 522–523
    trends in, 517
    value of, 516, 517
Florigen, 653
Flower bud formation, on strawberries, 482
Flower business, 515
Flower evolution, 382–384
Flowering and foliage plants, 513–514
    bulbs, 529–532
    commercial greenhouse environment,
        534–536
    floriculture, value of, 516, 517
    flower business, 515
    flowering annuals, 523–529
    flowering herbaceous perennials, 516–523
    flowering house plants, 532–534
    summary, 536
Flowering annuals, 523–529
Flowering herbaceous perennials, 516–523
Flowering hormone, 305
Flowering house plants, 532–534
Flower modification, 116
Flower morphology, 111–120
Flower shape, 116
Flower stalks (strawberries), removal of, 485
Fluid mosaic, 145
Fluorescence, 257, 653
Fly agaric, 589, 653
Foliage plants. *See* Flowering and foliage plants
Follicle, 653

Food, 605–606
    agriculture's future, 621–622
    farmer's bargaining power, 616
    gardening and, 622–624
    historical perspective, 611–613
    mechanization and world food supply,
        606–607
    new agriculture and, 614–616
    production limits, 616–621
    social and political considerations, 625–626
    summary, 634–635
    value of, 510
Food chain, 30, 38–39, 653
Food commodities, 607–611
Food pyramid, 41–42
Food storage, 81, 85, 87
Food supply, global, 606–607, 670
Food web, 39–41, 653
Forage, 653
Forage grasses, 538–540
    Canada bluegrass, 544
    Garrison grass, 549
    Kentucky bluegrass, 541–543
    management of, 550–553
    no-till and minimum-till seeding, 553–555
    orchard grass, 544
    perennial ryegrass, 544–545
    perennial warm-season grasses, 549–550
    prairie grass, 548–549
    reed canary grass, 545, 546
    smooth bromegrass, 545–546, 547
    summary, 570–571
    tall fescue, 546–548
    timothy, 548
Forcing, 532
Ford Foundation, 462
Forest, 207–208. *See also* Coniferous forest;
        Temperate deciduous forest; Tropical
        rain forest
Forest management policies, 207
Forestry, 207–208
Fossil fuels, 653
Four-arm Kniffin, 480
Foxglove, 583, 584
Fraction I protein, 653
Fragmentation, 653
Frame shift, 653
Franklin, Rosalind, 157
Fresh water, 653
Freshwater biomes, 71–73
Fronds, 372, 653
Frost heaving, 521
Frost protection, for strawberries, 485–486
Frost susceptibility, of fruit and nut trees, 502
Fruit
    definition of, 653
    evolution of, 384–385
    types of, 500–502
    See also Small fruits
Fruit and nut production, 498–499
    business, 500
    cultural practices, 508–510
    harvesting and storage, 510–511
    national grower/horticultural
        organizations, 511
    planting, 507–508
    site and soil preparation, 507
    summary, 511
    tree selection, 502–507
    types and varieties, 500–502
Fruiting dates, of fruit and nut trees, 503–507
Fruit Laboratory (Beltsville, Maryland), 479
Fruit morphology, 120–124
Fruit trees
    cultural practices, 508–510
    selection of, 502–507

subtropical and tropical, 501
temperate, 501
Fuel, wood as, 204–205
Function
leaf, 95–100
nutrients, 218
root, 81–85
stem, 86–94
Funiculus, 653
Fused, 116, 653

G

Galactose, 653
Galactosidase, 653
Gamagrass, 549
Gametangia, 653
Gametes, 112, 653
Gametophyte, 319, 371, 653
Gamma rays, 653–654
Gardening, 453, 622–624
Garner, W. W., 292, 293
Garrison grass, 549
Gas exchange, 95, 181, 654
Gene, 160, 164–168, 654
Gene activators, 295, 654
Gene interaction, 331–332
General care, of house plants, 533
Generations. See Alternation of generations
Generative nucleus, 654
Gene regulation, 165–166, 654
Gene splicing, 345
Gene synthesizers, 334–336
Genetic code, 346, 654
Genetic engineering, 333–334,
        341–342, 344, 621
agriculture and, 349–356
definition of, 343, 654
DNA coding system, 347
DNA jigsaw, 346–347
future of, 360
genetic code, 346
history of, 344
necessary knowledge, 344–345
"omics" revolution, 360–364
policy, public perception, and law, 364
recombinant DNA technology, 348–349
summary, 365
transgenic plants, 356–360
Genetic gibberish, 336–337
Genetics, molecular, 333–337
Genome, 165, 361, 654
Genome sequences, publishing, 334
Genomics, 361–362, 654
Genotype, 284, 654
Gentophytes, 375
Genus, 134, 654
Geotropism, 306–307, 654
Germinate, 654. See also Seed germination
Germplasm Resources Information Network
        (GRIN), 373
Gibberellins, 296, 301–302, 654
Gingkos, 375, 379–380, 654
Girth, 654
Glucose, 165, 654
Glucose molecules, 143
Glycerol, 654
Glycogen, 654
Glycolic acid metabolism, 150
Glycolysis, 271, 272–273, 654
Glycoproteins, 142, 654
Glycosides, 592, 654
Glyoxysomes, 145, 150, 654
Gnetophytes, 381
Golgi apparatus, 145, 149–150, 654
Goodyear, Charles, 600

Gooseberry, 493–495
Gradients, 227, 654
Grafting, 102–103, 654
Grains, 203
Granite, 220, 654
Granum, 147, 654
Grape hybrids, 476
Grapes, 476–481
Grasses. See Forage grasses; Sod
Grasslands, 60–64, 654. See also Savannas
Gravel, 223
Great Barrier Reef, 70–71
Green grapes, 478
Greenhouse effect, 34–35, 654
Greenhouse environment, commercial,
        534–536
Green Revolution, 462–463
Ground meristem, 654
Ground parenchyma, 654
Ground tissue, 181
Grower organizations. See National
        grower/horticultural organizations
Growing area size, for vegetable
        production, 458
Growing season
dry beans, 431
perennials, 517
potatoes, 431
vegetable classification by, 453, 458
Growth. See Plant growth and development
Growth patterns, of fruit and nut trees,
        503–507
Growth retardants, 304, 654
Growth rings, 187, 199–200, 654
Guanine, 346, 654
Guard cells, 181, 654
Guttation, 230, 654
Gymnosperms, 80, 371, 375–376
conifers, 376–379
cycads, 380
definition of, 654
gingko, 379–380
gnetophytes, 381

H

Habitat, 28, 654
Half-life, 654
Hallucinogens, 586, 587–591, 654
Halophilic, 654
Handling, of peas, 436–437
Haploid, 319, 654
Hardpan, 556, 654
Hardwoods, 197, 654
Harvesting
alfalfa, 445–446
blueberries, 488
cotton, 430
currants and gooseberries, 495
dry beans, 433
fruits and nuts, 510–511
grapes, 481
peas, 436–437
potatoes, 424
rice, 420
sod, 568–570
soybeans, 427
strawberries, 486
vegetables, 465–467
Hashish, 588, 655
Haustorial roots, 85, 655
Hay measurement standards, 639
Head, 655
Heading, 655
Heartwood, 196, 200, 655
Heat conduction, through wood, 197

Heavy metals, 168, 655
Heeling in, 483–484, 655
Helix, 346, 655
Hemlock, 596
Hemicellulose, 655
Hemoglobin, 156, 655
Hemp, 205, 655
Hemp plant, 407, 655
Henbane, 590
Herbaceous perennials, 131, 516–523, 655
Herbalism, history of, 576–578
Herbal remedies, 584–586
Herbarium, 655
Herbicides, 655
Herbivores, 38, 655
Herbs, 468–471
Hereditary factors, 324–325, 655
Heredity, 655
Heritable, 655
Heroin, 580, 655
Hesperidium, 655
Heterosporous, 655
Heterotrophic, 38, 655
Heterotrophic theory, 655
Heterozygous, 326, 655
Hexose, 655
Hill reaction, 259–260, 655. See also Photolysis
Hill system, 484, 655
Histones, 166, 655
Holdfast, 53, 655
Home gardening, 453
Homestead Act (1862), 409
Homogeneous, 655
Homologous pairs, 170, 655
Homologues, 170, 655
Homosporous, 655
Homozygous, 326, 655
Hooke, Robert, 139
Hormone interactions, 304–305
Hormones, 166, 295–297
abscisic acid, 303
animal, 291
auxin, 297–300
cytokinins, 302–303
definition of, 655
ethylene, 303–304
flowering, 305
gibberellins, 301–302
growth retardants, 304
Horticultural organizations. See National
        grower/horticultural organizations
Houseplants, 239, 532–534
Human Genome Project, 336
Human intervention, on plant growth and
        development, 290
Humidity, 22, 234–235, 533, 655
Humus, 655
Hybrid, 655
Hydration, 126
Hydrocooling, 467, 655
Hydrogen, 218, 219–220
Hydrolysis, 655
Hydrophilic, 655
Hydroponics, 246–247, 655
Hyoscyamine, 581, 655
Hypocotyl, 127, 655
Hypocotyl hook, 655
Hypothesis, 14, 655
Hypothesis testing, 14

I

Ice, in freshwater biomes, 72–73
Identical genetic makeup, 172, 655
Igneous, 220, 655
Imbibition, 126–128, 656

Immunological, 656
Imperfect flowers, 114, 656
Impermeable, 656
Incomplete dominance, 327–328, 656
Incomplete flowers, 114, 656
Indehiscent fruits, 120, 121, 656
Independent assortment, 329, 656
Indeterminate, 656
Indian grass, 549
Indigenous, 656
Indoleacetic acid (IAA), 298, 656
Indoor planting, of annuals, 525–526
Indoor plants, flowering, 532–534
Inducer, 656
Inducible enzyme, 656
Indusium, 656
Industrial plants, 599–602
Industrial Revolution, 613
Inferior ovary, 113, 386, 656
Inflorescence, 113, 656
Infrared, 656
Inheritance, 324–333
Inhibitors, 295, 656
Innovation, agricultural, 409–410
Inoculation, against Rhizobium, 427
Inorganic nitrogen, 35, 656
Insect-flower coevolution, 385–387
Insectivorous plants, 99–100, 656
Insect pollination, 119–120
Insects/insect control
    alfalfa, 446
    blueberries, 488
    bulbs, 530–532
    corn, 412
    cotton, 429
    dry beans, 433
    forage grasses, 555
    fruit and nut production, 509
    grapes, 481
    peas, 436
    perennials, 523
    potatoes, 423
    raspberries, 491
    rice, 420
    sod, 566
    vegetables, 464–465
Insertion mutants, 167, 656
Installation, 563–564, 656
Integrated pest management (IPM), 627–628,
    656. See also Pests/pest management
Integuments, 322, 375, 656
Intensity, 656. See also Light intensity
Intercalary meristem, 656
Interfascicular cambium, 656
Interkinesis, 317, 656
Internal organ poisons, 592
International Apple Institute, 511
International Dwarf Fruit Tree Association, 511
Internode, 87, 181, 656
Interphase, 164, 169, 170, 656
Intertidal zone, 656
Intine, 320, 656
Intron, 656
Involutions, 656
Ion, 656
Ionization, 215–216, 656
Irish potato, 408, 656
Irish potato famine (1846–1847), 612
Iron, 218
Irregular, 116, 656
Irrigation
    fruit and nut production, 508
    future of, 617–620
    sod, 557
    See also Watering
Irritants, 592, 656

Island of fertility, 656
Isoflavones, 291
Isogamy, 656
Isoprenes, 656
Isozymes, 395, 656

J

Jarmo, 404, 656
Jimsonweed, 598
Jointing, 656

K

Kentucky bluegrass, 541–543
Kinetic energy, 226, 656
Kinetin, 302, 656
Klages, K. H. W., 21
Knots, 203, 656
Kranz anatomy, 186, 657
Krebs cycle, 271, 273–274, 657

L

Lactic acid, 275
Lactose, 165, 657
Lakes, 72. See also Freshwater biomes
Land plants, adaptations of, 370–371
Land preparation. See Site preparation
Lateral buds, 88, 174, 657
Lateral meristems, 175, 657
Laterite, 657
Latewood, 199, 657
Latex, 599–601
Latitude, 657
Lattice, 657
Laudanum, 579, 657
Law, biotechnology and, 364
Law of thermodynamics, 252–254
Layering, 104, 105, 657
Leaching, 657
Leaf anatomy, 184–189
Leaf area index (LAI), 269–270, 657
Leaflet, 657
Leaf margins. See Margins
Leaf morphology and function, 95–100
Leaf shape, 96, 97–98
Leaf structure, 95–96, 97–98
Leaf venation, 95, 657. See Venation
Leeuwenhoek, Anton van, 139
Leeward, 24, 657
Legumes, 657
Leopold, Aldo, 4–5, 6, 12, 623
Lethal, 167, 657
Leucine, 166, 657
Leucoplasts, 147, 657
Levee, 657
Lianas, 53–54, 657
Lichen, 657
Light, 24–27
    applications of, 239
    definition of, 657
    flowering house plants, 532
    plant growth and development, 288–290
    seed germination and, 128–129
Light duration, 289
Light intensity, 24–25, 289–290, 657
Light quality, 289, 290, 664
Light reactions, 255, 259
Lignan, 291
Lignified, 129, 657
Lignin, 195, 657
Liming, forage grasses, 550–551
Limiting factor, 657
Limnetic zone, 72, 657
Limnologist, 657

Linkage, 329, 330, 657
Lipid, 657
Litter, 657
Littoral zone, 72, 657
Loam, 224
Lobed leaf margins, 96, 657
Locoweed, 657
Locus, 657
Lodging control, for wheat, 413
Long day plant, 657
LSD (lysergic acid diethylamide), 168, 588,
    589, 657
Lumber, 202–204
Lumen, 144, 196, 657
Luster, of wood, 196
Lysosomes, 145, 150, 657

M

MacLeod, Colin, 156–157
Macroelements (macronutrients), 217, 657
Macromolecule, 657
Macropores, 222, 657
Magnesium, 218
Maidenhair tree, 379
Maintenance, of perennials, 521
Maize, 408, 657
Malthus, Thomas, 620
Mandrake, 578–579, 590
Manganese, 218
Margins, 96, 98, 657
Marijuana, 587–588
Marine biomes, 68–71, 657
Market gardening, 453, 657
Marketing, of sod 570
Marsh, 657–658
Mass, 658
Matrix potential, 228, 658
Maturity, of peanuts, 441
McCarty, Maclyn, 156–157
McLintock, Barbara, 318
Meadow, 658
Mealy bugs, 242
Mean, 658
Measurement
    alfalfa storage and feeding losses, 638
    bushel weights and volumes, 639
    English and metric, 640
    farm products per bushel, 638
    hay and straw measurement standards, 639
    temperature conversions, 639
    weights and measures, 637–638
Measures, conversion tables for, 637
Mechanization, world food supply
    and, 606–607
Media, 525, 658
Medicinal plants, 578–584
Megagamete, 319
Megagametogenesis, 321–323, 658
Megagametophyte, 658
Megaspore, 658
Megasporocyte, 658
Megasporogenesis, 321–323, 658
Meiosis, 314–319, 658
Meiosis I, 316–317
Meiosis II, 317
Membrane, 658
Membrane-bound, 145, 658
Membrane selectivity, 145
Mendel, Gregor, 324–330, 654
Meristematic cells, 658
Meristematic regions, 173, 658
Meristematic tissue, 87
Meristems, 173–175
Mescal beans, 591
Mescaline, 590, 658

Mesic, 99, 658
Mesic grasslands, 60
Mesocarp, 120, 658
Mesophyll, 658
Messenger RNA (mRNA), 161, 162, 164, 336–337, 347, 658
Metabolic activity, 169
Metabolic energy, 658
Metabolism
    C2 photosynthetic pathway, 262–264
    C3 photosynthetic pathway, 264–265, 266
    C4 photosynthetic pathway, 264–266
    definition of, 658
    implications of, 277–278
    water in, 214–215
Metabolites, 658
Metacentric, 658
Metamorphic rock, 220, 658
Metaphase, 170, 171–172, 658
Metaphase I, 316–317
Metaphase II, 316, 317
Metaphase plate, 172
Methane, 35, 658
Methanogen, 658
Methionine, 166
Metric measurement, 640
Michigan Agricultural Experiment Station, 417
Microbes, manipulation of, 353
Microelements (micronutrients), 217, 658
Microfibrils, 143, 658
Microgametogenesis, 319–321, 658
Microorganisms, 658
Micropores, 223, 658
Micropropagation (tissue culture), 359–360, 658
Micropyle, 322, 658
Microsporangia, 377, 658
Microspore, 377, 658
Microspore mother cells, 377, 658
Microsporocyte, 658
Microsporogenesis, 319–321, 658
Microsporophyll, 658
Microtubules, 145, 150–151, 658
Middle lamella, 143, 658
Midrib, 95, 658
Miescher, Johann Friedrich, 156
Millet, 408, 608, 611, 658
Mimicry, 658
Mineral poisons, 593
Minerals, 221
Minimum-till seeding, for forage grasses, 553–555
Mitochondria, 145, 146, 272, 658
Mitosis, 168–172, 659
Modification, 659
Modified leaves, 96, 99–100
Modified roots, 84, 85
Modified stems, 90–94
Moisture
    seed germination and, 129
    wood and, 196–197
Mole, 659
Molecular forces, 250
Molecular genetics, 333–337
Molecular weight, 659
Molecules, 139–142, 151, 659
Molybdenum, 218
Monkshood, 597
Monocarpic plants, 132, 659
Monocot, 132–133, 135, 659
Monocot roots, 183
Monocot stems, 182–183
Monoculture, 208, 270, 659
Monoecious, 114, 115, 659
Monohybrid crosses, 328, 659
Monosaccharide, 659

Monsoon, 659
Mormon's tea, 381, 659
Morphine, 580, 659
Morphological adaptations, to reduced water loss, 236
Morphology
    definition of, 659
    flower, 111–120
    fruit, 120–124
    leaf, 95–100
    root, 81–85
    seed, 124–130
    stem, 86–94
    vegetative, 80–81
Morrill Land Grant College Act, 409
Mosquitoes, 392
Moths, 389–390
Motion, 250
Mountain tundra, 659
Mounting, 643
Movements. See Plant movements
Mowing, 562–563
Mucilaginous, 659
Muck soils, 556, 659
Mulberry tree, 406, 659
Mulching, 463
    annuals, 528
    blueberries, 487–488
    currants and gooseberries, 494
    definition of, 659
    strawberries, 485
Multicellular green alga, 369–370
Multigenic (polygenic), 333, 659
Multiple fruits, 122, 659
Multipurpose plant, 406, 659
Muscarine, 589, 659
Mushrooms, 589, 659
Mutagen, 168, 659
Mutations, 167–168, 318–319, 659
Mutualistic relationships, 29
Mycelium, 659
Mycoestrogens, 291
Mycorrhiza, 86, 87, 659

N
NAD, 254, 659
NADP, 659
Nastic movements, 307–308, 659
National Genetic Resources Program (NGRP), 373
National grower/horticultural organizations
    fruit and nut production, 511
    small fruits, 497
National Institutes of Health, 364
National Plant Germplasm System (NPGS), 373
National Turfgrass Evaluation Program (NTEP), 558
Natural law, 659
Nature, plants and, 5–6
Nectar, 116, 659
Negative matrix potential, 659
Negative solute potential, 659
Nematode damage, in sod, 567–568
Nerve poisons, 592
Net primary productivity, 270–271
Netted venation, 95, 659
Neutrons, 141–142, 659
New agriculture, 614–616
N-formylmethionine, 166, 659
Niche, 28, 659
Nicotine, 599, 659
Nile River, 405
Nitrate, 35, 659
Nitrite, 37, 659

Nitrogen, 217, 218, 418, 422
Nitrogenase, 660
Nitrogen base, 659
Nitrogen cycle, 35–37
Nitrogen fixation, 35–36, 352, 659
Nitrogen gas, 35, 660
Nitrophiles, 407, 660
Nodes, 87, 660
Nodulation, 85–86
Nodule, 660
Noncyclic photophosphorylation, 261, 660
Nonvascular plants, 80, 660
North American Bramble Growers Association, 497
North American grapes, 476
North American Strawberry Growers Association, 497
No-till cropping systems
    corn, 412
    forage grasses, 553–555
    oats, 418
    wheat, 416
Nucellus, 375, 660
Nuclear envelope, 145, 660
Nuclear pores, 145–146, 162, 660
Nucleic acid, 660
Nucleoli, 171, 660
Nucleolus, 146, 660
Nucleotides, 158, 346, 660
Nucleus, 145–146, 660
Nutmeg, 586, 660
Nutrient absorption and conduction, 85–86
Nutrient requirements, for vegetables, 459
Nutrients, 216–220, 660
Nutrition, water and, 214–220
Nutritional quality, of plants, 355
Nuts, 500–502, 660. See also Fruit and nut production
Nut trees, 502
    cultural practices, 508–510
    selection of, 502–507
Nyctinasty, 307, 660

O
Oats, 409, 417–418, 608, 611, 660
Obligate anaerobe, 660
Ocean currents, 69. See also Marine biome
Ocean desert, 660
Oils, 601
Old World grapes, 476
Oleander, 598
Olericulture, 453, 660
"Omics" revolution, 360–364
Omnivores, 38, 660
On the Origin of Species (Darwin), 354
Operator gene, 660
Operon, 165, 660
Opium poppy, 579–581
Opium Wars, 580–581
Opposite leaf arrangement, 89, 660
Orbital, 660
Orchard grass, 544
Organelles, 145–151, 660
Organic compounds, 183
Organic matter, 33–34, 660
Osmoregulator, 217, 660
Osmosis, 226–227, 660
Osmotic agents, 166, 660
Osmotic potential, 228
Outcrossing, 117, 660
Outdoor planting, of annuals, 526
Ovary, 112, 660
Overgrazing, 61
Ovule, 112, 375, 660
Oxalates, 592, 660

Oxidation, 254, 660
Oxidation phosphorylation, 660
Oxidation reduction, 254–255, 660
Oxygen, 6
    definition of, 660
    functions and deficiency symptoms, 218
    seed germination and, 128
    water and, 219–220
Oxygen–carbon dioxide balance, 30–31
Oxytocic agent, 584, 660
Ozone, 369, 660

P

Palatability, 394–395, 544, 660
Paleobotanists, 381, 660
Paleozoic era, 371, 660
Palisade, 185, 660
Palmate leaf venation, 95, 660
Palmately compound, 95, 661
Paneling, 206, 207
Paper, 205
Paper chromatography, 262, 661
Parallel leaf venation, 95, 661
Parasitic relationships, 28, 661
Parchment, 205, 661
Parenchyma, 176, 177, 178, 661
Parthenocarpic fruits, 299, 661
Parthenocarpy, 119, 661
Particleboard, 206, 207
Pathogen, 661
Pauling, Linus, 156
Peanuts, 437–442
Peas, 434–437
Peat moss, 661
Pectin, 661
Peduncle, 661
Pelagic, 661
Pellicle, 661
Peltate, 96, 661
PEP carboxylase, 661
Pepo, 661
Peptide bond, 661
Percent purity, 552, 661
Percolation, 33, 661
Perennials, 131, 132
    definition of, 661
    flowering, 516–523
    ryegrass, 544–545
    warm-season grasses, 549–550
Perfect flowers, 114, 661
Perianth, 111, 661
Pericarp, 120, 661
Pericycle, 183, 230, 661
Periderm, 188, 661
Permafrost, 65, 661
Permanent wilting point, 225, 661
Permeability, 661
Peroxisomes, 145, 150, 661
Pest control, biological, 355–356
Pesticides, 168, 661
Pests/pest management
    alfalfa, 446
    blueberries, 488
    bulbs, 530–532
    cotton, 429
    dry beans, 433
    grapes, 481
    IPM, 627–628
    nut production, 509
    potatoes, 423–424
    raspberries, 491–492
    rice, 420
    sod, 564–565
    types of, 241–246
    vegetables, 464–465
    See also Insects/insect control

Petal, 661
Petaloid, 661
Petal modifications, 116–117
Petiole, 95, 661
Petrifaction, 661
Peyote, 590, 591, 661
Pfr (phytochrome far-red), 293–294
pH, 215–216, 418, 459, 661
Pharmaceuticals, 661
Phelloderm, 661
Phellogen, 661
Phenotype, 284, 661
Phloem, 178, 179–180
Phospholipid, 661
Phosphorus, 218
Photolysis, 259–260, 661
Photons, 256, 661
Photo oxidation, 257–258
Photoperiod, for potatoes, 422
Photoperiodism, 292–295, 661
Photophosphorylation, 259, 260–261, 262, 661
Photorespiration, 267, 661
Photosynthesis, 5, 255–256
    CAM photosynthesis pathway, 267–268
    carbon dioxide and, 34
    carbon fixation, ecological aspects of, 268
    chloroplast, 258
    dark reactions, 261–267
    definition of, 661
    leaf area index (LAI), 269–270
    light reactions, 259
    net primary productivity, 270–271
    photolysis, 259–260
    photophosphorylation, 260–261
    pigments, 256–258
Photosynthetic bacteria, 256, 661
Photosynthetic unit, 260, 661–662
Photosystem I (PSI), 261, 262, 662
Photosystem II (PSII), 261, 262, 662
Phototropism, 662
Phycobilins, 257, 662
Physiological adaptations, to reduced water loss, 236–237
Physiological factors, in seed germination, 130
Phytochrome, 289, 293–294, 295
Phytochrome red, 662
Phytoestrogens, 291
Phytoplankton, 41
Phytotoxins, 592, 662
Pigments, 256–258, 662
Pinnate, 662
Pinnately compound, 95, 662
Pistil, 111, 662
Pistillate, 114, 115, 662
Pith, 181, 662
Placental, 112, 662
Planed, 556, 662
Plant body, water movement throughout, 231
Plant design (nonreproductive structures), 78–80
    leaf morphology/function, 95–100
    root morphology/function, 81–85
    stem morphology/function, 86–94
    summary, 106
    vegetative morphology and adaptations, 80–81
    vegetative reproduction, 100–105
    water and nutrient absorption and conduction, 85–86
Plant design (reproductive structures), 109–111
    botanical names, 133–134
    development (seedling to adult), 130–132
    flower morphology, 111–120
    fruit morphology, 120–124
    monocots and dicots, 132–133

seed morphology, 124–130
    summary, 136–137
    supermarket botany, 134–136
Plant disease resistance. *See* Disease resistance
Plant growth and development, 153–155, 282–284
    biotechnology and, 351–352
    cell division (mitosis and cytokinesis), 168–175
    DNA replication, 156–160
    genes, 164–168
    leaf anatomy, 184–189
    limitations of (stresses), 285–295
    plant hormones, 295–305
    plant movements, 305–308
    principles of, 284–285
    protein synthesis, 160–164
    seedling to adult, 130–132
    summary, 189–191, 308–309
    tissues, 175–184
Plant growth regulators, 296, 662. *See also* Hormones
Plant hormones. *See* Hormones
Planting
    annuals, 525–526
    blackberries, 492–493
    blueberries, 487
    bulbs, 529
    corn, 412
    cotton, 430
    currants and gooseberries, 494
    dry beans, 431, 432
    fruits and nuts, 507–508
    grapes, 479
    peanuts, 441
    peas, 435
    perennials, 520
    raspberries, 490
    sod, 558–563
    strawberries, 483–484
    vegetables, 460–462
    See also No-till cropping systems; Soil; Tillage
Plant label (sample), 644
Plant movements, 305–308
Plant nutrients, essential, 216–220
Plant palatability, 394–395
Plants
    aesthetic and recreational significance of, 10–11
    collection procedures, 641–642
    control of, 355
    deadly, 593–601
    ecology and, 20–21
    economic importance of, 7–9
    herbalism and, 576–578
    herbal remedies, 584–586
    importance of, 623
    industrial, 599–602
    medicinal, 578–584
    mounting, 643
    nature and, 5–6
    nutritional quality of, 355
    poisonous, 591–599
    pressing, 642–643
    psychoactive, 586–591
    society and, 15–16
    storage of, 643–644
    transgenic, 356–360
    See also Flowering and foliage plants; Forage grasses; Fruit and Nut production; Small fruits; Sod; Vascular plants
Plant science 2–3
    Leopold, Aldo, 4–5
    plants, aesthetic and recreational significance of, 10–11
    plants, economic importance of, 7–9

plants and nature, 5–6
plants and society, 15–16
scientific method, 12–14
summary, 16
technology and science, 11–12
Plant-soil-water relationships, 212–214
  applications, 237–246
  hydroponics, 246–247
  soil, 220–225
  soil-plant-air continuum (SPAC), 228–237
  summary, 247–248
  water and nutrition, 214–220
  water movement, 226–228
Plant-water relationships, 351
Plasma membrane, 175, 662
Plasmids, 348, 662
Plasmodesma, 143, 662
Plasmodium, 662
Plasmolysis, 227, 662
Plastids, 145, 146–148, 662
Pleiotropy, 333, 662
Plumule, 127, 662
Plywood, 206, 662
Point mutations, 167–168, 662
Poisonous plants, 591–599
Polar ice cap, 662
Polarity, 306, 662
Polar nuclei, 662
Policy, biotechnology, 364
Polish, of wood, 196
Political and social considerations, 624–625
  conservation, 628–629
  economic and environmental trade-offs,
    632–633
  endangered species, 629–630
  environmental quality, 631–632
  food, 625–626
  integrated pest management, 627–628
Pollen, 112, 662
Pollen grain, 662
Pollinating systems, 117–120
Pollination, 503, 662
Pollinators, 119, 662
Pollinator specificity, 387–393
Polycarpic, 132, 662
Polyembryony, 378, 662
Polygenic inheritance. See Multigenic
    inheritance
Polymer, 662
Polymerase chain reaction (PCR), 335–336, 662
Polynucleotide chain, 662
Polypeptide, 662
Polyploid, 662
Polysaccharide, 662
Polysome, 662
Pome, 500, 501, 662
Pomologist, 500, 662
Population, 29, 662
Pore, 181
Porosity, 222–223, 662
Postharvest physiology, 303–304, 663
Potassium, 218
Potatoes, 420–425
Potential energy, 663
Potting, 241
Potting soil, 239
Pr (phytochrome red), 293–294
Prairie, 663
Prairie grass, 548–549
Precipitation, 22–24, 663
Precooling, 467, 663
Prediction, 14
Preplant fumigation, 556, 663
Pressing, 642–643
Pressure potential, 228, 663
Prickles, 93–94, 663
Priestly, Joseph, 30, 255

Primary cell wall, 143, 144, 663
Primary consumers, 39, 663
Primary phloem, 178
Primary pit fields, 663
Primary succession, 43, 663
Primary tissue, 663
Primary xylem, 178, 663
Primordia, 181, 663
Principle of independent assortment, 329, 663
Principle of segregation, 325, 663
Pristine, 663
Producers, 663
Production limits, in agriculture, 616–621
Profundal zone, 72, 663
Progymnosperms, 376, 663
Prokaryotes, 663
Prokaryotic cells, 143–145
Promoter, 165, 663
Propagation
  definition of, 663
  grapes, 479
  perennials, 523
Prophase, 170, 171, 663
Prophase I, 316
Prophase II, 317
Proplastid, 663
Prop roots, 84, 663
Protected ovules, 386, 663
Protein, 663
Protein sequencer, 334–335, 663
Protein synthesis, 160–164, 166–167
Proteomics, 363–364, 663
Prothallium cells, 377, 663
Protoderm, 663
Protons, 141–142, 663
Protoplast, 663
Protoplast fusion, 663
Protoxylem, 663
Proviral state, 663
Pruning
  blackberries, 493
  blueberries, 488
  currants and gooseberries, 494–495
  definition of, 663
  fruit and nut production, 508–509
  grapes, 480–481
  raspberries, 491
Pseudocopulation, 389, 663
Pseudohallucinogens, 586–587, 663
Psilocin, 589, 663
Psilocybin, 589
Psilophyta, 663
Psychoactive, 586–591, 663
Pterophyta, 663
Pubescence, 96, 236, 663
Public Law 480, 612
Public perception, of biotechnology, 364
Pulping, 663
Pumice, 663
Punnett square, 326, 663
Pure opium, 663
Pure parents, 324, 664
Purgative, 583, 664
Purine, 664
Pyrimidine, 664

Q

Quality of light, 289, 290, 664
Quarter-sawed cuts, 203, 664
Quinine, 579, 664

R

Radially cut, 664
Radial micellation, 234, 664
Radial symmetry, 116

Radiation, 664
Radical cut, 203
Radicle, 664
Radiocarbon dating, 202, 664
Rain forest. See Tropical rain forest
Rain-shadow desert, 24, 664
Raspberry, 488–492
Ratoon, 664
Rauwolfia, 582
Rayon, 206, 664
Rays, 664
Reaction wood, 200–201, 664
Receptacle, 111, 664
Recessive, 324–325, 664
Recognition surface, 664
Recolonization, 44, 664
Recombinant DNA (rDNA), 334, 348–349, 664
Recreation, plants and, 623–624
Recreational significance, of plants, 10–11
Red grapes, 478
Red light, 293–294, 664
Reduction division, 664
Reed canary grass, 545, 546
Regular density, 664
Regular flower shapes, 116
Regular symmetry, 664
Regulator gene, 664
Relative density, 197
Relative humidity (RH), 467–468, 664
Remote sensing, 664
Renewable natural resources, 664
Replication. See DNA replication
Repressible enzymes, 664
Repressor, 664
Reproduction
  conifers, 377–379
  ferns (nonseed plants), 372–374
  vegetative, 100–105
  See also Sexual reproduction
Reproductive growth, of peanuts, 438–439
Reproductive structures. See Plant design
    (reproductive structures)
Research, agricultural, 409–410, 614–616
Reserpine, 582, 664
Resettlement plans (1930s), 409
Resin canals, 199, 664
Resins, 592, 664
Resistance, 352–355, 664
Respiration, 6, 271–272
  definition of, 664
  efficiency of, 276–277
  electron transport, 274–275
  fermentation, 275–276
  glycolysis, 272–273
  Krebs cycle, 273–274
  mitochondrion, 272
  substrate for, 277
Returns, sod-related, 570
Revegetate, 664
Rhizobium, 86, 664
Rhizobium inoculation, 427
Rhizoid, 664
Rhizomes, 90–91, 664
Rhyniophyta, 664
Ribonucleic acid (RNA), 664
Ribose, 664
Ribosomal RNA, 664
Ribosomes, 145, 148–149, 347, 665
Rice, 408, 418–420, 610, 611
Ring-porous wood, 665
Rivers. See Freshwater biomes
RNA polymerase, 162, 665
Rockefeller Foundation, 462
Rocks, 220–221
Roll, 564, 665
Rootbound, 241, 533, 665
Root cap, 665

Root crops, 407–408
Root growth, of peanuts, 438
Root hairs, 82–83, 183, 665
Root meristems, 175, 183
Root pressure, 230, 665
Roots
    dicot and monocot, 183
    meristematic activity in, 174
    morphology and function, 81–85
    secondary growth in, 187–188
    water and nutrient absorption and
        conduction, 85–86
Rootstock, 102, 500, 502, 665
Root structure, 81–82
Rosettes, 131, 665
Rough endoplasmic reticulum, 665
Rubber, 599, 600–601, 665
RuBP carboxylase, 665
Runners, 481, 665
Rye, 409, 608, 610, 611
Ryegrass, perennial, 544–545

S

Sacred mushrooms, 589
Sahara Desert, 57, 59
Salicylic acid, 665
Samara, 665
Sand, 223, 224, 225
Sand County Almanac, A (Leopold),
        4–5, 6, 623
Sandstone, 665
Sapwood, 200, 665
Saussure, Théodore de, 255
Savannas, 54–57, 665
Scales, 245–246
Scarification, 123, 665
Scavengers, 39, 665
Schizocarp, 665
Scientific method, 12–14, 665
Scientific names, 134, 135–136, 665
Scion, 102, 665
Sclereids, 177, 178, 665
Sclerenchyma, 176, 177–178, 665
Scopolamine, 581, 590, 665
Seasonal cycles, 291, 665
Secondary cell wall, 143, 144, 665
Secondary compounds, 665
Secondary consumer, 665
Secondary growth, 186
    roots, 187–188
    wood, 195–202
Secondary phloem, 182, 187, 665
Secondary succession, 43, 665
Secondary tissue, 175, 665
Secondary xylem, 182, 187, 665
Second Opium War, 581
Sedatives, 587, 665
Sedentary, 665
Sedimentary rock, 220, 665
Seed, 371
    definition of, 665
    dry beans, 431
    economic importance of, 125–126
    potatoes, 423
Seedbeds, 459, 665
Seed coat, 375, 665
Seed ferns, 371, 665
Seed germination, 124, 126–130, 525–526
Seeding equipment and methods, for forage
        grasses, 554
Seeding rates and mixtures,
        for alfalfa, 444
Seeding year management, of forage
        grasses, 550
Seedling, 665

Seedling-to-adult plant development, 130–132
Seed morphology, 124–130
Seed plants, 375
    angiosperms, 381–393
    gymnosperms, 375–381
    plant palatability, 394–395
Seed quality, of forage grasses, 552
Seed-scales complexes, 377, 665
Seed selection, annuals, 525
Segregation, principle of, 325
Seismonasty, 307, 665
Selective breeding, 359
Self-incompatibility, 117–118, 666
Self-pollination, 117, 666
Semi-dwarf trees, 500, 501, 502, 666
Senescence, 131, 666
Senna, 583
Sepals, 111, 666
Seral, 42, 666
Sessile, 666
Setting, 527
Setting out, 527
Sex cells, 112, 666
Sexual propagation, 666
Sexual reproduction, 100–101, 312–314
    angiosperm life cycle, 319–323
    definition of, 666
    inheritance, 324–333
    meiosis, 314–319
    molecular genetics, 333–337
    summary, 337–338
Shackleton, Ernest, 510, 626
Shale, 666
Shape. See Leaf shape
Shard, 666
Sheath, 96, 666
Shoot meristems, 175
Shoots, 666
Short day plant, 666
Shrinkage, of wood, 196–197
Shrubs, 666
Sieve cell, 665
Sieve tube members, 179, 665
Silent mutations, 167, 666
Silique, 666
Silt, 223, 224
Simple fruits, 120–121, 666
Simple layering, 104, 105
Simple leaf, 95, 666
Simple pistil, 112, 666
Simple tissues, 176–180
Sink, 179, 666
Site preparation
    annuals, 525
    fruit and nut production, 507
    peanuts, 440–441
    perennials, 519
    sod, 555–557
    vegetables, 459–460
Site selection
    alfalfa, 442–443
    annuals, 525
    blackberries, 492
    blueberries, 486–487
    bulbs, 529
    currants and gooseberries, 494
    grapes, 477
    perennials, 517, 519
    raspberries, 489
    sod, 555
    strawberries, 481
    vegetable production, 458
Slash and burn, 405, 666
Small fruits, 474–475
    blackberry, 492–493
    blueberry, 486–488

currant and gooseberry, 493–495
grapes, 476–481
national grower/horticultural organizations, 497
raspberry, 488–492
strawberry, 481–486
summary, 497
Smith, Hamilton, 159
Smooth bromegrass, 545–546, 547
Smooth endoplasmic reticulum, 666
Snow forests, 65
Social considerations. See Political and social
        considerations
Society, plants and, 15–16
Sociobiologists, 10, 666
Sociopolitical conservation, 666
Sod, 538–540, 666
    costs and returns, 570
    definition of, 666
    disease control, 567
    harvesting, 568–570
    insect control, 566
    installation, 563–564
    irrigation, 557
    marketing, 570
    nematode damage, 567–568
    pest management, 564–565
    production, 555
    site preparation and establishment, 555–557
    site selection, 555
    summary, 570–571
    turf selection and planting, 558–563
    weed control, 565
Sod-forming, 545, 666
Softwoods, 197, 666
Soil, 220–225
    applications of, 239–240
    bulbs, 529
    cotton, 430
    dry beans, 431–432
    oats, 418
    peanuts, 439
    peas, 435
    plant growth and development, 288
    potatoes, 422
    soybeans, 427
    See also Plant-soil-water relationships
Soil development, 221
Soil fertility, for alfalfa, 442–443
Soil-plant-air continuum (SPAC),
        228–237, 666
Soil preparation
    annuals, 525
    fruit and nut production, 507
    perennials, 519–520
Soil probe, 557, 666
Soil profile, 221–222, 666
Soil temperature
    peanuts, 437
    peas, 435
Soil texture, 222–225
Solarization, 464, 666
Solubility, 666
Solute, 666
Solute potential, 227–228, 666
Solvent, 666
Somatic cells, 164, 666
Somatic nervous system, 592, 666
Sonoran Desert, 58
Sorghum, 409, 608, 611, 666
Sound, 250
Soybean, 425–427
Soybean-corn crop rotation, 412
Specialty crops, 609
Speciation, 666
Species, 134, 666
Specific gravity, 197–198, 666

Specific heat, 666
Sperm nuclei, 666
Sphenophyta, 666
Spider mites, 244
Spindle fibers, 151, 169, 666
Spines, 58–59, 94, 666
Spongy parenchyma, 185, 666–667
Sporangiospore, 667
Sporangium, 667
Spore, 667
Sporogenous cells, 319–320
Sporophyte, 319, 371, 667
Spring barley, 417
Spring oats, 418
Staking, 667
    annuals, 528
    perennials, 521
Stamens, 111, 112, 667
Staminate, 114, 667
Staminodium, 667
Starch, 147, 667
Stele, 183, 667
Stem
    dicot, 181–182
    monocot, 182–183
    primary growth of, 181–184
Stem crops, 407–408
Stem cutting, 104, 667. See Cuttings
Stem morphology and function,
        86–94
Stem structure, 87–90
Sterols, 667
Stigma, 112, 667
Stimulants, 586, 587, 667
Stimulates, 295, 667
Stobilus, 377
Stolons, 91–92, 667
Stoma, 181, 667
Stomatal regulation, 232, 233–234, 667
Storage, 81, 85, 87
    dry beans, 434
    fruits and nuts, 510–511
    peas, 436–437
    plants, 643–644
    potatoes, 424–425
    vegetables, 467–468
Storage loss, of alfalfa, 638
Storage taproot, 81
Strata, 667
Straw measurement standards, 639
Strawberry, 479, 481–486
Streams. See Freshwater biomes
Stresses, on plant growth and development,
        285–295
Stroma, 147, 667
Structural genes, 165, 667
Style, 112, 667
Suberin, 667
Submetacentric, 667
Substrate, 277, 667
Succession, 43, 667
Succulent, 58, 93, 667
Succulent leaves, 237
Sugar beets, 608
Sugarcane, 608
Sulfite process, 205, 667
Sulfur, 218
Sunken stomata, 236, 667
Superior ovary, 113, 667
Supermarket botany, 134–136
Super trees, 208
Switchgrass, 549
Symbiotic relationships, 29, 86, 87, 667
Symplastic movement, 230, 667
Synapsis, 667
Synergids, 323, 667

Synthesis, 183
Synthetic pyrethrums, 667
Synthetics, wood in, 206
System toxins, 667

T
Table grapes, 477–479
Taiga (snow forests), 65, 667
Tall fescue, 546–548
Tannin, 196, 667
Tapetum, 320, 667
Taproots, 81–82, 667
Taq polymerase, 335, 336, 667
Taro, 407–408, 667
Tatum, Edward, 156
Taxine, 597, 667
Technology
    agriculture and, 410
    recombinant DNA, 348–349
    science and, 11–12
    transgenic, 358
Tehuacán Valley, 405
Telocentric, 667
Telophase, 170, 667
Telophase I, 317
Telophase II, 317
Temperate deciduous forest, 61, 667
Temperate shock treatments, 667
Temperature, 24
    corn, 411
    dry beans, 431, 432
    flowers and foliage, 515
    house plants, 533
    peanuts, 437
    peas, 435
    plant growth and development, 286–287
    potatoes, 421–422
    seed germination and, 128
    transpiration and, 234
Temperature conversions, 639
Temperature shock treatments, 166
Template, 667
Tendril, 667
Tension wood, 201, 668
Terminal buds, 88, 668
Terpenes, 668
Terpenoids, 668
Terrarium, 668
Terrestrial biomes, 50
    coniferous forest, 64–65
    definition of, 668
    deserts, 57–60
    experimental biosphere, 51
    grasslands, 60–64
    savannas, 54–57
    summary, 73–74
    tropical rain forest, 51–54
    tundra, 65–68
Testcrosses, 327, 668
Thermal stratification, 72, 668
Thermodynamics, 252–254
Thermonasty, 307, 668
Thigmotropism, 307, 668
Thinning, 527
Thorn apple, 590
Thorn forest, 56, 668
Thorns, 93–94, 668
Thylakoids, 147, 668
Thymine, 346, 668
Tiga, 668
Tillage
    cotton, 430
    dry beans, 431
    soybeans, 427
    See also No-till cropping systems; Soil

Tiller, 408, 668
Tillering, 668
Timber losses, reducing, 207–208
Timothy, 548
Tip layering, 104, 105
Tissue culture. See Micropropagation
Tissues, 175–176
    dermal, 180–181
    development and, 181–184
    simple, 176–180
    summary, 190
Tobacco, 599
Tobacco mosaic virus (TMV), 353
Tonoplast, 148, 668
Top carnivore, 668
Topography, 668
Topsoil, 668
Total biomass, 269
Total body mass, 668
Tracheids, 178–179, 668
Tracheophytes, 668
Training
    blackberries, 493
    definition of, 668
    grapes, 480
    raspberries, 490
Transcription, 161–162, 668
Transfer RNA (tRNA), 162, 163, 164,
        347, 668
Transformation, 668
Transforming factor, 668
Transgene, 357–358
Transgenic, 356–360, 668
Transitional mutants, 167, 668
Translation, 161, 162–164, 668
Transpiration, 23, 231–235, 668
Transpirational pull, 668
Transplantability, 460, 461, 668
Transplantation
    annuals, 526
    perennials, 522
Transverse cut, 203, 668
Transversional mutants, 167, 668
Tree ring. See Annual tree ring
Tree selection, 502–507
Trellis, 480, 668
Trichomes, 180, 668
Trihybrid cross, 328, 668
Trimerophytes, 376
Triplets, 160, 668
Triticale, 611
Trophic, 29, 37–42, 668
Tropical rain forest, 51–54, 668
Tropics, 405
Tropism, 668
Truck cropping, 453, 668
True potato seed, 423
Tryptophan, 166, 298, 668
Tube cell, 377, 668
Tube nucleus, 668
Tubers, 92, 408, 668
Tubocurarine, 581, 669
Tubulin, 150, 669
Tumors, 142, 669
Tundra, 65–68, 669
Turf, 669
Turfgrass Federation, Inc., 558
Turfgrass selection, 558–563
Turgidity, 33, 238, 669
Turgor pressure, 142–143, 669
Turpentine, 669

U
Ultraviolet, 168, 669
Understory, 669

Unisexual, 114, 669
Unit membrane, 669
U.S. agricultural crops, 410–447
    alfalfa, 442–447
    barley, 416–417
    corn, 410–412
    cotton, 427–430
    dry beans, 430–434
    oats, 417–418
    peanuts, 437–442
    peas, 434–437
    potatoes, 420–425
    rice, 418–420
    soybeans, 425–427
    wheat, 412–416
U.S. Department of Agriculture (USDA),
    364, 558
Uracil, 669
Uredospore, 669

V
Vacuolar sap, 148–149, 669
Vacuole, 145, 148–149, 669
Van Leeuwenhoek, Anton, 139
Vapor concentration, 468, 669
Variability, 669
Varieties
    blackberries, 492
    currants, 493
    dry beans, 431
    flowering indoor plants, 534
    gooseberries, 494
    grapes, 477
    nuts and fruits, 500–502
    peas, 434
    raspberries, 489
    rice, 419
    soybeans, 425
    strawberries, 482–483
    vegetable, 454–457, 458
Variety selection
    alfalfa, 443
    peanuts, 439
    wheat, 412–413
Vascular bundles, 181, 185, 669
Vascular cambium, 102, 182,
    186–187, 669
Vascular plants, 80–81, 367–369, 370
    angiosperms, 381–393
    definition of, 669
    evolution and distribution of, 371–374
    gymnosperms, 375–381
    land, migration to, 369–371
    plant palatability, 394–395
    summary, 395–397
Vascular systems, 80, 669
Vectors, 116, 669
Vegetable production
    scope of, 453
    steps in, 458–459
Vegetables, 451–452
    categorization of, 453–458
    cultural practices, 463–465
    Green Revolution, 462–463
    harvesting, 465–467

    planting, 460–462
    site preparation, 459–460
    storage, 467–468
    summary, 471
Vegetative growth
    definition of, 669
    fruit trees, 506–507
    peanuts, 438
Vegetative morphology and adaptations,
    80–81
Vegetative reproduction, 100–105, 669
Veins, 185
Venation, 95, 669
Veneer, 206–207, 669
Vermiculite, 525, 669
Vesicles, 149–150, 669
Vessel, 178, 669
Viability, 126, 669
Virus, 669
Viscous, 669
Visible spectrum, 24, 669
Vitamins, 166
Viviparous, 125, 669
Volatile oils, 669
Volatilization, 669
Volumes, bushel, 639
Volunteer plants, 669
Vulcanization, 600, 669

W
Wadis, 405, 669
Warm-season grasses, perennial, 549–550
Water
    applications of, 237–238
    biotechnology and, 351
    nutrition and, 214–220
    seed germination and, 129
    See also Plant-soil-water relationships
Water absorption and conduction, 85–86
Water cycle, 32–33
Water dispersal, 122–123
Water hemlock, 596–597
Watering
    annuals, 527–528
    blueberries, 488
    house plants, 533
    perennials, 520–521
    raspberries, 491
    strawberries, 485
    See also Irrigation
Water loss, adaptations to, 235–237
Water movement, 226–228, 231
Water-plant relationships, 351
Water potential, 227–228, 669
Water quality, 288, 669
Water rights, 618
Water stress, 287–288, 669
Watson, James, 156–157
Waxes, 601, 669
Weathering, 223
Weed, 669
Weed control
    alfalfa, 447
    blackberries, 493
    cotton, 429

    dry beans, 432
    forage grasses, 552–553
    grapes, 481
    peas, 435
    potatoes, 423–424
    raspberries, 491
    rice, 419
    sod, 565
    strawberries, 485
    vegetables, 463–464
Weight conversions, 637
Weights
    bushels, 639
    farm products per bushel, 638
Went, Fritz W., 297
Wheat, 408, 412–416, 610–611, 669
White flies, 245
White light, 259, 669
Whorled leaf arrangement, 89, 669
Wind, transpiration and, 234
Wind dispersal, 122
Wind pollination, 118–119
Windward, 24, 669
Winter barley, 417
Winter hardy, 670
Winter protection, for perennials, 521
Winter wheat, 611
Wolfbane, 597, 670
Wood, 187, 193–194
    forests and forestry, 207–208
    structure and secondary growth,
    195–202
    summary, 208–209
    uses of, 202–207
World food supply, 606–607, 670
Worsley, Frank, 510, 626
Wright, Frank Lloyd, 195

X
Xeric, 60, 670
Xerophytes, 236, 670
X-ray, 168, 670
X-ray crystallography, 159, 670
Xylem, 178–179

Y
Yellow River, 405, 670
Yew, 597, 598

Z
Zagros Mountains, 404
Zinc, 218
Zone of differentiation, 176
Zone of elongation, 176, 670
Zooplankton, 41, 670
Zoospore, 670
Zosterophyllophyta, 670
Z-scheme, 261, 670
Zygospore, 670
Zygote, 124, 670